BREN

JERALD D. PARKER
Oklahoma State University

JAMES H. BOGGS
Oklahoma State University

EDWARD F. BLICK
University of Oklahoma

INTRODUCTION TO FLUID MECHANICS AND HEAT TRANSFER

ADDISON-WESLEY PUBLISHING COMPANY
Reading, Massachusetts · Menlo Park, California
London · Amsterdam · Don Mills, Ontario · Sydney

Third printing, February 1974

Copyright © 1969 by Addison-Wesley Publishing Company, Inc. Philippines copyright 1969 by Addison-Wesley Publishing Company, Inc.

All rights reserved. No part of this publication may be reproduced, stored in a retrieval system, or transmitted, in any form or by any means, electronic, mechanical, photocopying, recording, or otherwise, without the prior written permission of the publisher. Printed in the United States of America. Published simultaneously in Canada. Library of Congress Catalog Card No. 79-85-385.

ISBN 0-201-05710-7
FGHIJKLMNO-AL-8987654321

PREFACE

This text grew out of the experience at Oklahoma State University and the University of Oklahoma of teaching an undergraduate course in which the subject matter of fluid mechanics and heat transfer were combined. The engineering faculty at both universities saw the need to integrate the two subject areas to keep pace with the increasing need to compact more material into the basic undergraduate engineering courses. The final report of the recent ASEE study, *Goals of Engineering Education,* lists "transfer and rate processes including heat and mass transfer and some phases of fluid mechanics" as subject matter that is strongly suggested for study by the engineering student. The text is designed to meet that need and should be useful to students in any branch of engineering.

With a few exceptions, the material in this text is not original with the authors. The arrangement of the material, however, and the emphasis given to certain topics make this text different from most other textbooks. The material has been tested in the classroom at both universities and has been taught by teachers with previous experience in teaching courses in fluid mechanics, heat transfer, aerodynamics and compressible fluid flow. The text is suitable for third or fourth year engineering students. Although a knowledge of elementary thermodynamics is desirable, the material can be presented to students who have had no formal training in that subject. It is assumed that the student has had some exposure to differential equations.

The integrated approach to heat transfer and fluid mechanics has several advantages. The most important single advantage is that it permits a learning approach which proceeds from the general to the specific case. This permits the student to better understand and to be aware of significant assumptions made in the solution of a specific problem. The conservation principles (conservation of mass, momentum and energy) constitute the common basis from which all material is derived. In almost every case the introduction of a section in the text starts out with a consideration of these conservation principles. The common mathematical forms of some equations which describe different physical phenomenon are pointed out to the student but not overly emphasized.

With the integrated approach subject material such as the Reynolds analogy is more easily understood by the student. It is an easy step for the student to go from the concept of similarity in regard to velocity profiles to similarity in regard to temperature profiles. Likewise there is efficiency in going immediately from the use of integral methods in boundary layer solutions to integral methods in heat transfer solutions.

A number of fringe benefits result from the integrated approach, including a common nomenclature for all the subject matter and a common definition of terms. The nomenclature and definitions are generally those found in the common undergraduate thermodynamics texts. Also, most heat transfer texts find it necessary to repeat certain fluid mechanics material because of the assumption that this material may not have been adequately covered in a previous fluid mechanics course. In the case of an integrated course such repetition is not necessary.

The early chapters of the text cover the conservation principles from both the overall and differential standpoint. The later chapters are concerned with specific applications of these conservation laws, coupled with certain phenomenological relationships. Vector notation is introduced but used sparingly, and extensive vector manipulation is not required of the student. There are a number of examples throughout the text to aid the student in applying the concept to practical situations. The large number of problems which appear at the end of each chapter can be used to give the student a more thorough understanding of the subject matter covered in that particular chapter.

The material in the text can be covered in a variety of ways. The text can be used to teach a standard three-hour semester course by omission of certain material. In using the text for a three-hour course the instructor would probably wish to use the first ten chapters plus Chapters 14 and 15. In a four-hour course, such as the one taught at the University of Oklahoma, additional chapters may be covered, or the material in these twelve chapters may be covered in a little more detail. The text has been used for a five-hour course at Oklahoma State University, with material in the first 15 chapters being covered rather thoroughly. Also there is sufficient material in the text for two semester courses of three credit hours each.

Material in the text can be covered in a variety of orders. For example, the instructor may wish to introduce the material on overall heat transfer coefficients, conduction, and radiation heat transfer (Chapters 13, 14 and 15) early in the course. The text material is written so that these latter chapters may be taught immediately after Chapter 4 without confusing the student. The authors prefer going straight through the book, of course. The material on high speed flow and heat transfer, multi-phase behavior, heat exchangers, mass transfer, and the special topics of Chapter 17 may be omitted or selected at random to meet the needs of the particular curriculum.

The authors are indebted to many people who have contributed in various ways to this text and we express our thanks for this help. As in the case of many textbooks, we have borrowed heavily from other writers in the field of fluid mechanics and heat transfer. In some cases it is impossible to determine where the material originated. Some problems and examples in this text may resemble those in other textbooks. In such cases we ask forgiveness, for much of the material has come out of old examinations and classroom notes, and in many cases it was difficult to determine the origin of the material.

The authors have been influenced greatly by the teachers under whom they studied at the University of Oklahoma, at Oklahoma State University, and at Purdue University. In particular the O.S.U. authors would like to express their appreciation to Professors W. L. Sibbitt, M. Jakob, and R. J. Grosh at Purdue University. A number of outstanding persons in the field of heat transfer have participated in the annual heat transfer conferences held at Oklahoma State University. These conferences and the principal participants have had a strong influence on the authors. These persons have included Byron Short, Warren Rohsenow, D. Q. Kern, Myron Tribus, E. R. G. Eckert, George Dussinberre, Frank Kreith, A. L. London, Stuart Churchill, R. L. Pigford, John Clark, E. B. Christiansen, A. C. Mueller, and Peter Griffith. In addition Dr. Kenneth J. Bell of the School of Chemical Engineering has been most helpful. The authors are also thankful to Professors Bart Turkington and Jim Harp of the University of Oklahoma for teaching with the material and for reviewing early versions of the manuscript. A particular thanks is due Mrs. Mildred Avery and Mrs. Betty Stewart for typing some of the early editions of the manuscript and to Mrs. Alice Norton for the final typing of the several versions of the manuscript. A note of appreciation also goes to Professors John Wiebelt, Tom Love, Glen Zumwalt, Ronald Panton, Bill Tiederman, and Don Haworth for reading and commenting on portions of the manuscript. A number of graduate and undergraduate students have made valuable comments on the manuscript and their help is also acknowledged.

January, 1969   J.D.P.
                                                                                                             J.H.B.
                                                                                                             E.F.B.

CONTENTS

Chapter 1 Introduction

1-1 The role of fluid mechanics and heat transfer in engineering . 1
1-2 Units and dimensions 3
1-3 Concepts from mechanics 5
1-4 Concepts from thermodynamics 9
1-5 The Fourier law 13

Chapter 2 Fluid Statics

2-1 The concept of pressure 21
2-2 Compressibility and thermal expansion coefficient 26
2-3 The pressure field in a static fluid 31
2-4 Buoyancy and stability 36

Chapter 3 Basic Concepts of Fluid Mechanics and Heat Transfer

3-1 Coordinate systems 46
3-2 Velocity and acceleration 48
3-3 The substantial derivative 50
3-4 Fluid stress and energy conversion 53
3-5 Laminar and turbulent flow 61
3-6 General approach to problems 64

Chapter 4 The Conservation Equations

4-1 Conservation of mass 68
4-2 Streamlines 77
4-3 Conservation of momentum 81
4-4 Conservation of energy 92

Chapter 5 Dimensional Analysis

5-1 Dimensionless groups 110
5-2 Similitude 133
5-3 Interpretation of dimensionless groups 136

Chapter 6 Ideal Fluid Flow and Heat Transfer

6-1 Incompressible fluid in frictionless flow 142

Chapter 7 Fundamentals of Fluid Flow and Heat Transfer in Viscous Fluids

- 7-1 The boundary layer concept. 166
- 7-2 The entrance region in conduits 174
- 7-3 Determination of pressure drop in conduit flow 179

Chapter 8 Laminar Flow and Heat Transfer

- 8-1 Laminar flow and heat transfer in tubes 184
- 8-2 The entrance region in laminar flow 191
- 8-3 Similarity methods in laminar boundary flow 197
- 8-4 Integral methods for boundary layer problems 208
- 8-5 Reynolds analogy. 212

Chapter 9 Turbulent Flow and Heat Transfer

- 9-1 Time-smoothing 221
- 9-2 Flow along a flat plate 227
- 9-3 Turbulent flow in conduits 230
- 9-4 Flow around bodies 248
- 9-5 Flow across tube banks 260

Chapter 10 Free Convection

- 10-1 Free convection on vertical surfaces 272
- 10-2 Free convection from horizontal cylinders 285
- 10-3 Free convection in enclosed spaces 288

Chapter 11 High-speed Flow and Heat Transfer

- 11-1 The speed of sound 299
- 11-2 The Mach number and flow regimes 302
- 11-3 Isentropic flow of a perfect gas 305
- 11-4 Normal and oblique shock waves 313
- 11-5 Diabatic flows and effects of friction 320
- 11-6 Constant area adiabatic flow with friction 323
- 11-7 Flow along a flat plate 326

Chapter 12 Multiphase Behavior

- 12-1 Bubble and droplet behavior 337
- 12-2 Pool boiling 345
- 12-3 Liquid-gas flows 355
- 12-4 Cavitation 366
- 12-5 Condensation 368

Chapter 13 Analysis of Heat Exchangers

- 13-1 The overall heat transfer coefficient 382
- 13-2 The heat exchanger and mean temperature difference . . . 391

13–3	The NTU approach to the thermal design of heat exchangers .	401
13–4	Heat transfer rates and pressure drop in heat exchangers . .	410

Chapter 14 Conduction Heat Transfer

14–1	Steady conduction in solids	425
14–2	Numerical solutions in steady conduction	432
14–3	Conduction with internal energy conversion	440
14–4	Extended surfaces with convective boundaries	444
14–5	Transient conduction.	451
14–6	Numerical solutions in transient conduction	466

Chapter 15 Thermal radiation

15–1	Black and nonblack bodies	485
15–2	Configuration factors and radiant exchange	492
15–3	Radiant exchange between nonblack surfaces	499
15–4	Thermal control of spacecraft	503
15–5	Combined conduction, convection, and radiation	506
15–6	Gas radiation	509

Chapter 16 Mass Transfer

16–1	Fick's law	519
16–2	Diffusivities	522
16–3	Bulk motion and diffusion	523
16–4	Mass transfer coefficient	526
16–5	Species conservation equation	526
16–6	Mass transfer cooling	528

Chapter 17 Special Topics

17–1	Liquid metals	534
17–2	Non-Newtonian fluids	539
17–3	Lubrication	541
17–4	Fluid-solid flows	544
17–5	Flow through porous media.	549
17–6	Heat transfer at cryogenic temperatures	552
17–7	Magnetohydrodynamics	557
17–8	Flow in open channels	559
17–9	Rarefied gas dynamics	562

Appendix 1	List of Symbols and Abbreviations	571
Appendix 2	Useful Conversion Factors	578
Appendix 3	Thermophysical Properties	580
	Index	601

CHAPTER 1

INTRODUCTION

Fluids are constantly encountered in everyday life. A fluid blanket of air surrounds each of us. Water is vital for sustaining our life. Weather is controlled by the movement of the air and water over the earth's surface. Vehicles of transportation move through either air or water.

Engineers, in utilizing the materials and forces of nature, deal constantly with fluid behavior. Many industrial systems involve the transporting and processing of fluids. Fluids are used in cooling and heating processes, in lubrication, and in the transmission of power. They are important as electrical insulators and conductors and play an important role in energy conversion processes, such as those used in hydroelectric power plants. Many of the fuels that man uses are in fluid form. The study of both stationary and moving fluids is called *fluid mechanics*. This text will be devoted to this subject and to the closely related subject of heat transfer.

Heat is a term used to describe one of the forms of energy that crosses the boundaries of a thermodynamic system. *Heat* is defined as thermal energy in transition; temperature is the driving force for a flow of heat. *Heat transfer* is the study of the rate of exchange of energy due to temperature difference. An understanding of heat transfer is important to engineers as they attempt to change the materials of nature into more useful forms, as they work to maintain the comfort of man, and as they utilize various processes for converting energy from one form to another.

1-1 THE ROLE OF FLUID MECHANICS AND HEAT TRANSFER IN ENGINEERING

The term *fluid* refers to a substance which continues to deform in the presence of a shearing stress. In contrast, when a shearing stress is applied to a solid, the solid may deform slightly and then come to rest, in equilibrium with the forces acting on it. A fluid may be classified as either a liquid or a gas.

The differences between liquids and gases are usually apparent, but concise definitions for the two terms are difficult to formulate. A fixed mass of liquid will have a volume which will vary only slightly with temperature and pressure and will freely adjust to the interior shape of any vessel into which it is poured. A liquid has a free surface as one boundary if its volume

2 INTRODUCTION

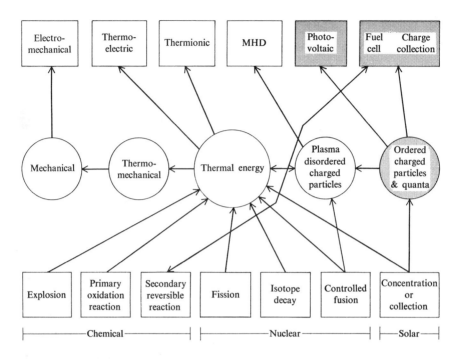

Fig. 1–1 Energy conversion terminating in electricity. (Adapted from "Unconventional Electric Energy Conversion Methods," General Electric Company.)

is less than that of the container. On the other hand, a gas has no definite free surface and will usually expand to fill the entire vessel in which it is contained. The volume of a fixed mass of gas will vary over a wide range, depending upon the temperature and pressure.

The molecules in gases are relatively far apart and have translational energies much greater than their attraction energies. In a liquid the reverse is true. Liquid molecules are still mobile but they are relatively close together, and their translational energies are less than their attraction energies.

While the physical properties of gases can be explained on a molecular basis by kinetic theory and statistical thermodynamics, as yet there is no adequate kinetic theory of liquids. However, for most engineering purposes it is permissible and convenient to treat fluids not from the *molecular* viewpoint but from the *continuum* viewpoint, since interest usually is not in the motions of the molecules but in the gross behavior of the fluid. The continuum viewpoint implies that any property (such as pressure, temperature, or density) is the result of the action of a very large number of molecules in the neighborhood of a point. Fluids may be treated as continual whenever the smallest volume of fluid of interest contains enough molecules to make statistical averages meaningful.

The importance of heat transfer to energy conversion processes can be seen in Fig. 1–1. This figure shows the steps which are involved in energy conversion terminating in electricity, the most useful form of energy in our modern society. Electricity, like all forms of energy, is derived from chemical, nuclear, or solar energy, the three primary energy sources. The dominant role which thermal energy plays in these conversions is obvious from the figure. Conversions to and from thermal energy are usually accompanied by the transfer of heat; thus, heat transfer plays a dominant role in the production of electricity.

Perhaps not so apparent, but very important, are the interrelationships between many fluid and heat transfer processes. Almost all flowing fluids are subjected to some heat transfer, and, likewise, many heat transfer processes involve fluid directly or indirectly. Fortunately, the fluid flow and heat transfer problems can often be studied independently by making a few simplifying assumptions. In many situations, however, the problems of fluid behavior and heat transfer must be considered simultaneously.

The laws of heat transfer and fluid mechanics can be developed from the concepts of solid mechanics and thermodynamics with the introduction of only a very few additional phenomenological relationships.

1–2 UNITS AND DIMENSIONS

Because of the importance of the proper use of the concept of units and dimensions in this book, a brief review of these subjects will be given. The concept of *unit* came naturally as a consequence of man's learning to count. Undoubtedly counting evolved because of a need to obtain a relative idea of the number of eggs, stones, or sets of similar objects or to count the paces between two geographical features. The development of a standard of length more precise than the pace followed, given priority perhaps by the desire to measure shorter lengths, such as a fraction of a pace. Gradually a standard evolved for the measurement of length (the meter). Mass and time standards also evolved.

Various units are in use for measurement in specific systems. Each unit is related to the standard of measurement for the specified characteristic in a precise and reproducible way. Units used to measure length, for example, include the inch, centimeter, foot, mile, and yard.

The designation *dimension* is applied to a term or symbolic representation of the class of units which might be used for measuring a specific physical characteristic. For example, the symbol L is used for all units used to measure length. The dimension L (or length) is used for descriptive purposes, while the unit is a specific quantitative measure in reference to a standard.

Inasmuch as the concept of length is used to measure both area and volume (an arbitrary choice, since one might, for example, conceive of area as the fundamental dimension), the dimensions of area are $L \cdot L$ or L^2, and the dimensions of volume are $L \cdot L \cdot L$ or L^3. Certain dimensional quantities may thus be thought of as *primary*, i.e., length L (where the symbol is raised only to the power one) or *secondary* (where the symbol is raised to the power two, three, etc., as in area, L^2, and volume, L^3). Also the concept of *derived* dimensions arises where the quantity has dimensions which include more than one primary dimension, for example, velocity, L/t, where t stands for the dimension time.

In this text the symbol ≙ will be used to designate a dimensional equation. Using this symbol, distance ≙ length, or d ≙ L, can be read as "distance has the dimension length." Also

$$\text{acceleration} \triangleq \frac{\text{length}}{(\text{time})^2} \quad \text{or} \quad a \triangleq \frac{L}{t^2}$$

can be read as "acceleration has the dimensions of length divided by time squared."

The problem of conversion from one unit or system of units to another often occurs. The recommended scheme for performing such conversions involves arranging the equation which relates the two units or two systems of units so that all factors and dimensions appear on one side of the equality, and unity appears on the other. One may either multiply or divide a term of an equation or value of a property by unity without changing its value. The use of the conversion factor is a matter of checking for the appropriate cancellation of units and multiplying or dividing to obtain the desired result.

Example 1–1. Convert 72 inches to feet.

Solution. It is known that the following holds:

$$1.0 \text{ ft} = 12 \text{ in.}$$

When both sides are divided by feet,

$$1.0 = 12 \frac{\text{in.}}{\text{ft}}.$$

Thus 12 in./ft is a conversion factor, and multiplication or division by it will not change the value of the given data.

Therefore, if 72 in. is divided by the factor,

$$\frac{72 \text{ in.}}{12 \frac{\text{in.}}{\text{ft}}} = 6.0 \text{ ft}.$$

In the solution of problems the student will avoid many pitfalls if he forms the habit of carrying the units of any factor or quantity which appears in the equation. A constant checking of units is done by using the technique of considering the name of the unit to be subjected to the usual algebraic manipulation, as in the above example.

1-3 CONCEPTS FROM MECHANICS

The concepts of mass and force are used extensively in mechanics and will be reviewed here. The term *mass*, for which the symbol is m, is used to describe the amount of material under consideration. Ignoring relativistic effects, the mass of a particular system depends only upon the number and kind of molecules making up the system.

All mass is subject to Newton's laws of motion. Newton's second law may be used to define force, for which the symbol is **f**. Printing a symbol in heavy black type (like **f** in the preceding sentence) indicates that the term is a *vector quantity*, that is, a quantity having both magnitude and direction. Thus, force is a vector quantity. Newton's second law, in its simplest form, states that force is proportional to the product of mass and acceleration:

$$\mathbf{f} \sim m\mathbf{a}, \tag{1-1}$$

where **a** is the acceleration of the system. Therefore, a force accelerates a mass to which it is applied. Force, velocity, and acceleration are vector quantities and, thus, have directional characteristics. The acceleration of a system is always in the same direction as the resultant of the forces which are applied to the system.

The system of units used to describe any physical system is arbitrary. It is possible to select a system of units for use in Newton's second law, Eq. (1–1), in which one unit of force would give one unit of mass an acceleration of one unit. The constant of proportionality in relation (1–1) would be unity, and an equation could then be written

$$\mathbf{f} = m\mathbf{a}. \tag{1-2}$$

The *principle of dimensional homogeneity* requires that the dimensions of each term in any equation be the same. Therefore, the dimensions of force must be equivalent to the dimensions of the product of mass and acceleration. The acceleration can be described in terms of the dimensions of length divided by the square of time.

Therefore, from Eq. (1–2),

$$\text{force} \triangleq \frac{(\text{mass})(\text{length})}{(\text{time})^2} \quad \text{or} \quad \text{mass} \triangleq \frac{(\text{force})(\text{time})^2}{\text{length}}. \tag{1-3}$$

By use of Eq. (1–3), any quantity having dimensions involving only force, mass, length, and time may be described with dimensions involving only force, length, and time (FLt) or only mass, length, and time (MLt). A system which uses mass, length, and time as the fundamental dimensions and force as a derived dimension ($F = ML/t^2$) is called an *absolute system*. Two such systems involving metric units are in common use. In the centimeter-gram-second (cgs) system, the derived unit of force, the gram-centimeter per second squared, is called the *dyne*. A dyne of force will accelerate one gram of mass one centimeter per second squared.

The Eleventh General Conference on Weights and Measures (1960) has recommended the *International System of Units*, abbreviated SI, for all scientific, technical, practical, and teaching purposes. The SI system uses the units of meter, kilogram, and second for length, mass, and time. In the meter-kilogram-second (mks) system, the derived unit of force, the kilogram-meter per second squared, is called the *newton*. A newton of force would accelerate one kilogram of mass one meter per second squared. The *standard kilogram mass* is the mass of a particular cylinder of platinum-iridium alloy kept at Sèvres, France. The length of the meter is exactly 1,650,763.73 wavelengths of the radiation in vacuum corresponding to the unperturbed transition between the levels $2p_{10}$ and $5d_5$ of the atom of krypton 86, the orange-red line. The second is exactly 1/31,556,925.9747 of the tropical year that began on January 1, 1900, at 12 hours, ephemeris time.

Table 1–1

DIMENSIONAL AND UNIT SYSTEMS

Dimensions	Absolute English System MLt	Absolute Metric System (cgs) MLt	International System (SI) MLt	Technical English System FLt	Engineering English System FMLt
Force	poundal	dyne	newton	pound force	pound force
Mass	pound mass	gram	kilogram	slug	pound mass
Length	foot	centimeter	meter	foot	foot
Time	second	second	second	second	second

Note: M = the dimension mass; L = the dimension length; F = the dimension force; t = the dimension time. Additional dimensions and units of temperature (T), electrical charge (Q), and luminous intensity (I) may be used in each system.

A less commonly used absolute system involving the pound mass, foot, and second defines a unit of force called the *poundal*. The three absolute systems just described are listed in Table 1–1, along with two other important systems. Temperature, electrical charge, and luminous intensity may be used as additional dimensions when needed.

A system which uses force, length, and time as the fundamental dimensions and mass as a derived dimension ($M = Ft^2/L$) is in fairly common use in English-speaking countries, particularly in the field of aerodynamics. The unit of force, the pound force, is assumed to be an established quantity. The derived unit of mass, the pound force-second squared per foot, is called a *slug*. Such a system is referred to as the *Technical English System*, or alternately, the *British Gravitational System*.

In most engineering work in English-speaking countries, it is common to define and to use both the dimensions of mass and force, in addition to length and time (an MFLt system). Such a system, involving the units of pound mass, pound force, foot, and second, is called the *Engineering English System*. In such a system, where both mass and force are arbitrarily chosen, the acceleration produced by a unit force on a unit mass would not necessarily be unity. The *pound mass* is an arbitrarily established quantity of matter. A *standard pound mass* is kept by the Board of Trade in London, England. The *pound force* is the pull of earth's gravity on a pound mass at sea level at 45°N latitude. It is known that a pound force will cause a pound mass to accelerate at 32.174 ft/sec², which is the acceleration of any free falling mass at the standard location described above. This acceleration is called the *standard acceleration of gravity* and is designated by the symbol g_0:

$$g_0 = 32.174 \text{ ft/sec}^2. \tag{1-4}$$

Newton's second law, Eq. (1-2), can be written in terms of this standard acceleration for a pound force and pound mass:

$$f = mg_0$$

$$1 \text{ lb}_f = 1 \text{ lb}_m \, 32.174 \text{ ft/sec}^2,$$

or

$$1 = 32.174 \frac{\text{lb}_m \text{ ft}}{\text{lb}_f \text{ sec}^2}. \tag{1-5}$$

Equation (1-5) is a unity conversion factor, as was described in Section 1-2. It can be used to eliminate either lb_m or lb_f from a quantity. In Eq. (1-5), the number 32.174 with the grouping of dimensions on the right hand side of the equation has been called the *dimensional constant*, g_c:

$$g_c = 32.174 \frac{\text{lb}_m \text{ ft}}{\text{lb}_f \text{ sec}^2}. \tag{1-6}$$

It is always permissible to multiply or divide any or all terms in an equation by g_c, or any power of g_c, in order to attain dimensional homogeneity. The dimensional constant will be written into the equations appearing in this text only when it is desirable to use both the units of lb_f and lb_m in a single equation. It will be the responsibility of the student to maintain consistent units in any equation which is used, whether g_c appears or not.

Example 1-2. Convert $62.0 \text{ lb}_m/\text{ft}^3$ to a quantity with units of lb_f instead of lb_m.

Solution.

$$62 \frac{\text{lb}_m}{\text{ft}^3} \cdot \frac{1}{32.174} \frac{\text{lb}_f \text{ sec}^2}{\text{lb}_m \text{ ft}} = 1.93 \frac{\text{lb}_f \text{ sec}^2}{\text{ft}^4}.$$

Since the lb$_f$ second squared per foot is the unit of the slug,

$$62 \frac{\text{lb}_m}{\text{ft}^3} = 1.93 \frac{\text{slugs}}{\text{ft}^3},$$

or

$$1 \text{ slug} = 32.174 \text{ lb}_m.$$

One slug, therefore, is obviously equivalent to 32.174 pounds mass.

The mass per unit volume, or *density*, in the example above is of interest in the fields of heat transfer and fluid mechanics. The symbol used for density is ρ. Whereas the mass of a fixed number of molecules does not change, the density may change as the molecules are confined in a greater or lesser volume. Density is the reciprocal of the specific volume,

$$\text{density} = \frac{1}{\text{specific volume}}, \tag{1-7}$$

and is a thermodynamic property.

Like all properties, the density of a substance may vary from one point in the substance to another. In order to define the density at a point, one must make the assumption that the substance behaves as a continuum. The assumption of continuum behavior is very important in the study of heat transfer and fluid flow and will be utilized in all cases except where the size of the system under study is of the same order of magnitude as the average distance molecules travel between collisions. Such is the case in high-vacuum systems and in the atmosphere at high altitudes; these cases will be given special consideration in a separate section.

The concept of weight is frequently of interest. The *weight* of a mass is the force exerted by the gravitational field of the earth on that mass. At sea level and 45°N latitude the earth exerts a force of one pound on one pound of mass. The acceleration caused by the force of gravity is called the gravitational acceleration. The *standard acceleration of gravity*, g_0, which exists at the standard conditions described above, has already been discussed.

The *local acceleration of gravity*, g, is a variable quantity depending upon location (latitude and altitude). Values of g for various locations are shown in Table 1–2.

A formula for determining the local acceleration of gravity for variations in altitude on the earth's surface is given by

$$g = g_0 \left(\frac{3963}{3963 + \text{altitude in miles}} \right)^2. \tag{1-8}$$

The weight of an object at any location depends upon the local acceleration of gravity. If the weight is given in force units and the mass is given in mass units, then

$$\text{weight} = \text{mass} \frac{g}{g_c}. \tag{1-9}$$

Table 1-2
EFFECT OF LOCATION ON LOCAL ACCELERATION OF GRAVITY

Location	g (ft/sec^2)
Sea level at the equator	32.088
Sea level at 45°N latitude	32.174
Sea level at theNorth Pole	32.258
40,000 ft altitude at 45°N latitude	32.050
500 miles altitude	25.369
2000 miles altitude	14.211
Surface of moon	5.47

The weight per unit volume, or *specific weight*, is related to the density as follows:

$$\text{specific weight} = \rho \frac{g}{g_c}. \tag{1-10}$$

The g_c in Eq. (1-10) is necessary only if the specific weight is desired in terms of force per unit volume and the density is in mass per unit volume. If ρ is expressed in force units, such as lb$_f$-sec^2/ft^4, then the g_c is unnecessary.

The *specific gravity* of a substance is the specific weight of the object divided by the specific weight of water at some specified condition and at the same location. The terms weight, specific weight, and specific gravity will not be used extensively in this text.

Many of the laws of fluid mechanics can be developed from Newton's second law—Eq. (1-2)—with no reference to thermodynamics or heat transfer. However, in those cases where significant variation in properties occurs, additional laws governing thermal characteristics must be utilized. The next section will review those concepts of thermodynamics necessary to solve such problems.

1-4 CONCEPTS FROM THERMODYNAMICS

Classical thermodynamics is restricted to the behavior of systems in equilibrium and to the changes of a system from one equilibrium state to another. The laws of thermodynamics do not enable us to predict the rates at which such changes will occur. For example, the laws of thermodynamics might be used to predict the amount of heat required to change a given system from one specified state to another, but these laws cannot predict the time required for the change of state to occur. The subject of heat transfer deals with heat flow rates and with systems that are not in equilibrium. Therefore, concepts not included in thermodynamics must be introduced.

Nevertheless, some of the concepts of thermodynamics are quite useful in the study of fluid mechanics and heat transfer. One of the basic definitions in thermodynamics is that of a property. Properties define the state or condition of a system that is in thermodynamic equilibrium. One important property, density, has already been discussed.

The concept of a *local property* is used extensively in heat transfer and fluid mechanics. This implies that properties may vary from point to point within a substance. The local value of density for example would be the value at a specified point of interest. Whenever properties are specified at a point, there is an assumption made that local thermodynamic equilibrium exists in the vicinity of the point.

Temperature, a property used extensively in both thermodynamics and in heat transfer, is difficult to define precisely. The temperature of a substance is related to the molecular motions of the substance, with higher temperature corresponding to higher levels of molecular activity. Rather than define temperature, the concept of temperature equality as expressed by the zero*th law of thermodynamics* is often used. The zero*th* law states that *when two systems are in thermal equilibrium with a third system, they are in thermal equilibrium with each other.* Systems in thermal equilibrium are then said to have the same temperature. By the choice of an arbitrary set of rules defining a temperature scale and the selection of an arbitrary temperature-measuring device (or thermometer), quantitative meaning can be given to the term temperature. The concepts embodied in the *second law of thermodynamics* permit us to define an *absolute temperature scale*, a scale independent of any thermometric substance or device. On an absolute scale the lowest obtainable equilibrium temperature is designated zero, or *absolute zero*.

The units of temperature will be briefly reviewed. A temperature scale commonly used in scientific work and in engineering outside of the United States and Great Britain is the *Celsius* or *centigrade* scale, with degrees designated °C. By international agreement the triple point of water is assigned a value of 0.01°C, and the magnitude of the degree is defined in terms of an ideal gas thermometer. On this scale, the steam point (temperature of water and steam in equilibrium at a pressure of one standard atmosphere) is 100°C; the ice point (temperature of ice and water in equilibrium with saturated air at one standard atmosphere) is 0°C. An absolute scale having the same size degree as the centigrade scale is called the *Kelvin scale*. Since the Kelvin scale is an absolute scale, zero degrees Kelvin (0°K) is absolute zero. Absolute zero is 273.15° below zero on the centigrade scale. The relationship between the Kelvin and centigrade scales is

$$°K = °C + 273.15. \qquad (1\text{--}11)$$

It is important to emphasize that one degree change on the centigrade scale is equivalent to one degree change on the Kelvin scale. The Kelvin degree is the unit used in the International System of Units.

For practical temperature measurement, it is important to have a scale based on easily reproducible fixed-temperature points that agree as closely as possible to those on the absolute temperature scale. Such a scale, the *International Practical Temperature Scale* (IPTS), is defined by a set of fixed points and by rules governing interpolation and extrapolation of these values. It is widely used and accepted in most scientific work throughout the world. The absolute temperature scale and the IPTS are in exact agreement at one fixed point, the triple point of water. The fixed points of IPTS are given in Table 1–3 below.

Table 1–3

FIXED POINTS OF THE INTERNATIONAL PRACTICAL TEMPERATURE SCALE, 1968

Fixed point	°K	°C
Triple point of hydrogen	13.810	
Boiling point of hydrogen	20.280	
Triple point of oxygen	54.361	
Boiling point of oxygen	90.188	
Triple point of water	273.16	0.01
Boiling point of water		100.000
Freezing point of zinc		419.58
Freezing point of silver		916.93
Freezing point of gold		1064.43

The most common temperature scale in the United States is the *Fahrenheit scale*. On this scale the temperature of the ice point is 32°F and the temperature of the steam point is 212°F. Therefore, a change of one degree Fahrenheit is $\frac{1}{180}$ of the temperature change between the ice point and the steam point. An absolute temperature scale having the same size degree as the Fahrenheit scale is called the *Rankine scale*. Absolute zero is 459.67° below zero on the Fahrenheit scale. The relationship between the two scales is

$$°R = °F + 459.67. \qquad (1\text{--}12)$$

It should be remembered that a temperature change of one degree Fahrenheit is equivalent to a change of one degree Rankine.

Since both the Rankine and Kelvin scales are absolute scales, absolute zero is identical on each scale. The relationship between the Rankine and Kelvin scales is

$$°K = \tfrac{5}{9}°R. \qquad (1\text{--}13)$$

Between the centigrade and Fahrenheit scales, the relationship is

$$°F = \tfrac{9}{5}°C + 32. \tag{1-14}$$

The relationship between the four scales is shown in Fig. 1–2.

Temperature often appears in the *equation of state*, which is used to relate the thermodynamic properties of pure substances. When used in an equation of state, the temperature must always be expressed as absolute temperature (°K or °R).

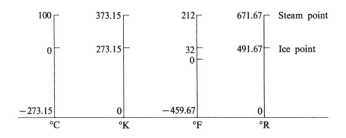

Fig. 1–2 Relation between temperature scales.

Both *mechanical* and *thermal* forms of energy are discussed in thermodynamics, and both are important in the study of heat transfer and fluid flow. The mechanical energy which crosses the boundary of a system is referred to as *work*. The thermal energy which crosses the boundary is referred to as *heat*. Both thermal and mechanical energy can be stored in a system in either of two forms, kinetic or potential energy. The *first law of thermodynamics* states that these forms of energy are equivalent and implies the concept of *conservation of energy*.

The units in which energy is expressed will vary. For example, thermal energy is often expressed in *British thermal units*, or Btu's, whereas mechanical energy is often expressed in foot-pounds force or ft-lb_f. By international agreement, a value has been selected for the British thermal unit so that the relationship between it and the foot-pound of force is

$$1 \text{ Btu} = 778.28 \text{ ft lb}_f. \tag{1-15}$$

This definition of the Btu will be used throughout the text. By this definition the Btu is approximately the amount of energy required to raise the temperature of one pound mass of water from 59.5°F to 60.5°F. The factor 778.28 ft lb_f/Btu is given the symbol *J* and is commonly called the *mechanical equivalent of heat*.

Other units of energy, such as the calorie, the watt-hour, and the horsepower-hour, are in common use and will be introduced in certain problems and examples. Other conversion factors between the various units for energy may be found in the appendix.

1-5 THE FOURIER LAW

Heat is that energy which crosses the boundary of a system due to a temperature difference. Heat is a form of energy in transition and is never considered to be stored or contained within a substance. In the heat transfer literature the term heat is used to describe energy passing through a given substance; for example, reference may be made to heat flowing through a wall. In such a case, the definition of heat given above is still acceptable if the substance is considered to be made up of many infinitesimally small systems, each receiving and giving up heat. Heat may be transferred between a system and its surroundings by two basic mechanisms:

1. *Conduction*—Conducted heat is thermal energy transferred through a material due to the existence of a temperature gradient within the material. In order for conducted heat to cross the boundaries of a thermodynamic system a temperature gradient must exist in the material at the boundary. Heat conduction is the transfer of thermal energy between adjacent molecules of a substance, supplemented in some cases by the flow of free electrons.

2. *Radiation*—Radiation is thermal energy which crosses the boundary of a system in electromagnetic waves. The electromagnetic waves have their origin in the thermal energy of the radiating substance and give up this energy in thermal form to the substance absorbing the radiation. Thus, by this process, thermal energy is transferred from one substance to another. Radiant thermal energy can pass through certain types of substances and can also pass through a perfect vacuum. A temperature difference must exist between the system and its surroundings in order for a net flow of radiated heat to occur across the boundary of the system.

In many practical engineering problems the energy transported across the boundary of a system due to thermal radiation can be neglected. In this book these types of problems will be considered first. Methods for treating problems where thermal radiation is important will be presented in a later chapter.

If radiation is neglected, the only heat crossing the boundaries of a system will be conducted heat.* This phenomenon is described for substances

---

* The term *convection* is frequently used to describe a third mechanism of heat transfer. Strictly speaking, convection is transport of stored thermal energy (internal energy) due to fluid motion. It is not considered a basic heat transfer mechanism in this text.

having continuum properties by the *Fourier-Biot conduction equation*, named after J. B. Biot, who first formulated it, and J. B. J. Fourier, who used it extensively to develop the field of conduction heat transfer. It is most commonly referred to today as the *Fourier equation:*

$$\frac{\dot{q}_n}{A} = -k\frac{\partial T}{\partial n}. \tag{1-16}$$

The Fourier equation states that the rate of heat flow $\dot{q}_n$ in a given direction n per unit area normal to that direction is proportional to the temperature gradient $\partial T/\partial n$ in that direction. The proportionality factor, k, defined by the equation, is called the *thermal conductivity* of the material. Heat flow in the positive n direction is considered positive; therefore, the minus sign in Eq. (1–16) is necessary in order to meet the second law requirement that heat must flow from a higher to a lower temperature. (The temperature gradient is negative for a positive flow.)

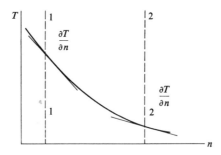

Fig. 1–3 A temperature gradient.

Figure 1–3 illustrates a temperature profile that might exist at some instant of time within a substance in a nonsteady flow of heat in a material for which the thermal conductivity is not a function of temperature or position and is constant. Two parallel reference planes within the substance are shown as dotted lines 1 and 2. The temperature gradient in the n direction at plane 1 is the slope of the straight line drawn tangent to the temperature curve at plane 1. The temperature gradient at plane 2 is also shown. Notice that the slope, or gradient, at both planes 1 and 2 is negative. From Eq. (1–16), the heat transfer by conduction would be positive, or in the direction of increasing n.

Since the gradient at plane 1 is greater than that at plane 2, a larger amount of energy is entering the material at plane 1 than is leaving at plane 2. In the absence of other means of removing energy, this would result in a net accumulation of energy and an accompanying change in thermodynamic properties of this material between planes 1 and 2.

In many materials the value of k depends on the direction of heat flow. Materials in which k is independent of direction are said to be isotropic, which is the most usual case. Wood is an example of anisotropic material.

The thermal conductivity of a pure substance is a property, and its value for a particular substance depends only upon the value of the other properties. The two properties usually chosen to fix the value of a specified property are temperature and pressure since they are easily measured. In most cases, k is much less dependent on pressure than on temperature, so that the thermal conductivity may be tabulated as a function of temperature for a given substance, and the dependence on pressure may be neglected. The dependence on pressure cannot be ignored at extremely high pressure. For normal engineering considerations, the thermal conductivity is independent of the temperature gradient.

It is quite common to tabulate values of thermal conductivities for non-homogeneous substances such as wood, powders, insulating materials of various types, and soils. Such values depend upon the history or prior treatment of the material and may vary considerably from one sample to another. In this case, k cannot be considered to be a property in the true sense, and is referred to as the *apparent thermal conductivity*.

In tabulating values of thermal conductivity, the proper choice of units must be made. Equation (1–16) can be rewritten as

$$k = -\frac{\dot{q}_n/A}{\partial T/\partial n}. \tag{1-16a}$$

Expressed in this way, the thermal conductivity is seen to be the ratio of heat flux to temperature gradient. The term *heat flux* is used to describe the rate of heat flow per unit area normal to the flow. In engineering work in English-speaking countries, the heat flux $\dot{q}_n/A$ is usually expressed in Btu/hr ft². The temperature gradient is often given in °F/ft. Thus, the desired units of thermal conductivity would be

$$\frac{\text{Btu}/(\text{hr ft}^2)}{°\text{F/ft}} \quad \text{or} \quad \frac{\text{Btu}}{\text{hr ft °F}}.$$

In nearly all scientific work and in engineering work in most non-English-speaking countries, the metric system is used. In this system, the heat flux is usually expressed in calories/sec cm² and the temperature gradient in °C/cm. Thus the proper units of thermal conductivity would be

$$\frac{\text{cal}/(\text{sec cm}^2)}{°\text{C/cm}} \quad \text{or} \quad \frac{\text{cal}}{\text{sec cm °C}}.$$

It can be easily shown that

$$1 \frac{\text{cal}}{\text{sec cm °C}} = 241.9 \frac{\text{Btu}}{\text{hr ft °F}}.$$

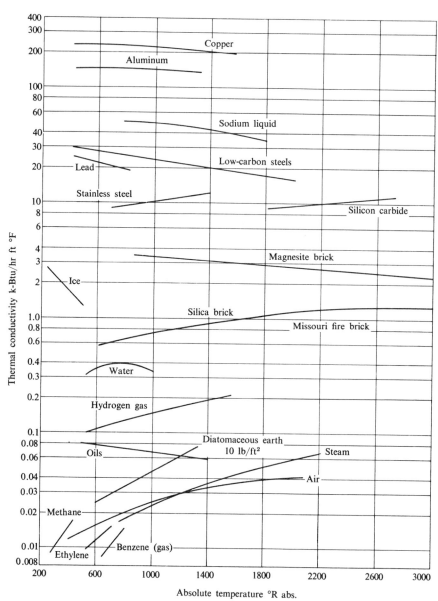

Fig. 1-4 Values of thermal conductivity of common substances at various temperatures. (From J. H. Boggs, "Physical Properties of Materials," *Oil and Gas Equipment*, Feb., 1960. Reproduced by permission of the publisher.)

Values of thermal conductivity for various engineering materials are given in Fig. 1–4 as functions of temperature at a pressure of one atmosphere. The dependency of k on pressure can be ignored except for gases at very high pressure and at high-vacuum conditions. Note that gases have the lowest value of thermal conductivity and pure metals the highest. Liquids have intermediate values, water having the highest value of any liquid except the liquid metals.

The Fourier equation is valid whether the material is stationary or in motion since the heat flux does not include energy transport due to flow or due to mass diffusion. Such forms of energy transfer are not heat but represent transfer of thermal energy stored in the moving material.

The significance of the Fourier equation to heat transfer problems cannot be overemphasized. It implies that heat can flow through a given substance by conduction only if a temperature gradient exists. Conversely, in the absence of thermal radiation, surfaces on which the temperature gradients in the normal direction are zero ($\partial T/\partial n = 0$) are surfaces across which no heat transfer due to conduction occurs. Such surfaces are referred to as *adiabatic surfaces*.

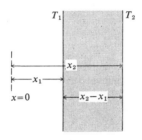

Fig. 1–5 Nomenclature for one-dimensional heat transfer.

If, as in Fig. 1–5, the temperature distribution in a substance depends only upon one coordinate direction, such as x, then any plane normal to x would have a uniform temperature. If the temperature at two of these parallel planes is known and if the thermal conductivity of the substance is constant, then the heat flux between the planes can be determined for any given spacing between the planes. In such cases, since $\dot{q}/A$ and k are constant, and the normal direction is assumed to be x, then Eq. (1–16) can be integrated to give

$$\frac{\dot{q}}{A} = -k\left(\frac{T_2 - T_1}{x_2 - x_1}\right) = k\left(\frac{T_1 - T_2}{x_2 - x_1}\right), \tag{1–17}$$

where T_1 and T_2 are the respective temperatures at distances x_1 and x_2 measured from a reference point in the direction of the heat flow. Note that the minus sign can be eliminated by rearrangement of Eq. (1–17). It can be seen also that for this steady state case, with constant k, the variation of temperature is linear with distance.

The assumption that k is constant, made in obtaining Eq. (1–17), will work for many cases where $(T_1 - T_2)$ is not great or where k is a weak function of the temperature. However, the variation of k may be compensated for by evaluating k at some intermediate temperature. It is quite easy to show that if the variation of k is assumed to be linear with temperature, then the proper value of k for use in Eq. (1–17) is that corresponding to the average temperature $(T_1 + T_2)/2$.

Example 1–3. A large slab of material 5 cm thick has a thermal conductivity of 0.05 cal/sec cm °C. One surface of the slab is at 80°C and the opposite surface is at 30°C. Estimate the rate at which heat flows across the slab per square foot of slab surface.

Solution. It will be assumed that the heat flow is one-dimensional across the slab; that is, the plate is very large or is insulated so that there are no heat losses at the edges. Assuming also that k is constant and using Eq. (1–17),

$$\frac{\dot{q}}{A} = k\left(\frac{T_1 - T_2}{x_2 - x_1}\right) = 0.05 \frac{\text{cal}}{\text{sec cm °C}} \frac{(80 - 30) \text{ °C}}{5 \text{ cm}}$$

$$= 0.5 \frac{\text{cal}}{\text{sec cm}^2}.$$

Since

$$1 \frac{\text{cal}}{\text{sec cm}^2} = 13{,}272 \frac{\text{Btu}}{\text{hr ft}^2},$$

$$\frac{\dot{q}}{A} = 6636 \frac{\text{Btu}}{\text{hr ft}^2}.$$

Example 1–4. The temperature gradients at an instant of time in a material are measured at two parallel planes 1.5 in. apart. The gradients are found to be 1500°F per foot and 900°F per foot. Assuming a constant thermal conductivity of 10 Btu/hr ft °F in the material, estimate the rate of energy accumulation between the two planes due to these gradients and at that particular instant of time. Express the answer in Btu per hour per cubic foot of material. (See Fig. 1–6.)

Solution. The rate of heat flow at face A at that instant is

$$\dot{q}/A_A = (-1500)(10) = -15{,}000 \text{ Btu/hr ft}^2.$$

The rate of heat flow at face B at that instant is

$$\dot{q}/A_B = (-900)(10) = -9000 \text{ Btu/hr ft}^2.$$

The net rate of heat flow into the region between the planes at that instant is

$$\dot{q}/A_A - \dot{q}/A_B = -15{,}000 - (-9000) = -6000 \text{ Btu/hr ft}^2.$$

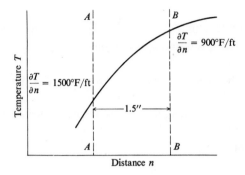

Fig. 1-6 Temperature variation in a material at an instant of time.

The net rate of accumulation of energy per cubic foot at that instant is

$$\frac{\text{energy accumulation rate}}{\text{ft}^3} = \frac{-6000 \text{ Btu/hr ft}^2}{(1.5/12) \text{ ft}} = -48{,}000 \text{ Btu/hr ft}^3.$$

For the specified conditions, the region between the planes is losing energy as is seen by the negative answer, and therefore the internal energy of the material in that region will drop if it is assumed that no other forms of energy transfer or conversion will take place.

REFERENCES

1. E. A. Mechtly, "The International System of Units, Physical Constants and Conversion Factors," NASA SP-7012 (1964).
2. L. G. Rubin, "Temperature-Concepts, Scales and Measurement Techniques," Leeds and Northrup Tech Memo T-538.
3. Joseph Fourier, *The Analytical Theory of Heat*, Dover, New York (1955).
4. J. H. Boggs, "Physical Properties of Materials," *Oil and Gas Equipment* (Feb. 1960).
5. N. V. Tsederberg, *Thermal Conductivity of Gases and Liquids*, ed. by R. D. Cess, M.I.T. Press, Cambridge, Mass. (1965).

PROBLEMS

1. Determine the force necessary to accelerate 10 lb_m at the rate of 10 ft/sec².
2. One gallon of a liquid is found to weigh 6.5 lb_f at standard conditions. Calculate (a) the density and (b) the weight of the same liquid on the moon.
3. It is desired to accelerate 10.8 kg at a constant acceleration of 20 cm/sec². What force, in dynes, will be required?

4. What force, in pounds, would be necessary to overcome the pull of earth's gravity at 500 miles altitude on a satellite having a mass of 1800 lb_m?
5. A mass of 30 slugs accelerates at a constant acceleration of 10 ft/sec². What net force is acting on the mass?
6. An object is to be designed so that it will have a weight of 30 lb_f on the surface of the moon. What mass must it have?
7. What force must be exerted on a mass of 100 lb_m at 2000 miles altitude to give it an acceleration of 8.0 ft/sec² away from the earth?
8. Convert the following temperatures and fill in the blanks:

°C	°K	°F	°R
300	—	—	—
—	300	—	—
—	—	300	—
—	—	—	300

9. At what point on the centigrade scale is the temperature numerically equal to that on the Fahrenheit scale?
10. Calculate the number of joules in 1.0 ft lb_f of energy.
11. Ten cubic inches of material weighs 3 lb_f in a place where $g = 31.0$ ft/sec². What is the density of the material?
12. A 3000 lb_m car is traveling at 50 mph. Calculate the number of Btu's of energy that must be dissipated to bring the car to a complete stop.
13. A large plane slab of material 2 in. thick has a thermal conductivity of 0.005 cal/sec cm °C. Determine the temperature difference across the slab, in °F, necessary to have a heat flux of 200 Btu/hr ft² pass through the slab.
14. Determine the average thermal conductivity of a material, $\frac{1}{4}$ in. thick, which conducts 5 kw of energy per square foot of area when the temperature difference across the material is 80°F.
15. Show that the following relationship exists between thermal conductivity values in the two indicated systems:

$$1.0 \frac{\text{cal}}{\text{sec cm °C}} = 241.9 \frac{\text{Btu}}{\text{hr ft °F}}.$$

16. Determine the heat flow through each square foot of insulation on an oxygen tank if the insulation is 2 in. thick, has a thermal conductivity of 3.5 microwatt/cm °K and has temperatures of -320°F and 80°F on its faces. Express the answer in Btu/hr ft².
17. A 10-lb mass is located on the moon's surface. Calculate the power necessary to raise the mass a vertical distance of 10 feet in 5 seconds against the moon's gravitational pull. Express your answer in watts.

CHAPTER 2

FLUID STATICS

Before we undertake a presentation of the laws governing the flow of fluids, it will be worthwhile to consider the characteristics of a fluid at rest, or in a static condition. An understanding of fluid statics is useful since there are many situations in which engineers are concerned with fluids at rest. The containment of a fluid in a vessel or behind a dam, the design of a ship or submarine, the measurement of pressure and flow, all require some knowledge of the subject. It will also be seen that an understanding of fluid statics is necessary to properly develop an understanding of fluid flow.

2-1 THE CONCEPT OF PRESSURE

In a thermodynamic system consisting of a large number of molecules in the fluid state, the molecules are free to move about within the boundaries of the system. The boundaries may be real, such as the wall of a container, or imaginary, such as would be the case if the system were an element of fluid surrounded by similar fluid. The boundary is referred to as a *control surface*. The concept of a control surface is developed further in Chapter 4.

For the case of the real, or solid, boundary the moving molecules will strike the walls of the container and change direction. The acceleration resulting from the direction change causes a force to be exerted on the wall of the container as required by Newton's laws of motion. If the fluid is assumed to be a continuum, a large number of molecules bombard the boundary and give rise to a continuous force against any reference area. Taking the smallest area over which a continuum condition may be assumed to exist, the *fluid pressure* is defined as the normal force per unit area which exists on that small area of the boundary. In the case of an imaginary boundary separating the fluid from a similar fluid, the pressure force is normal to the imaginary surface area and exists even though no solid surface is present. In liquids, intermolecular forces also influence the pressure.

For an infinitesimally small fluid element in static equilibrium with surrounding fluid, it can be shown that the normal force per unit area of surface (the pressure) is the same in all directions. Figure 2-1 shows an arbitrary triangular fluid element of unit depth perpendicular to the plane of the figure. For this static element the only forces present are those at the surface of the element (*surface forces*) and that force due to the weight of the element (*body force*). A body force is a force acting on each small mass within an element or object due to presence of a force field, such as a gravitational or magnetic field. According to the definition of a fluid given in Section 1–1, shear forces cannot exist in a static fluid. Therefore, the forces acting on the surfaces of the element must be normal to each surface.

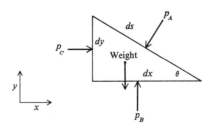

Fig. 2–1 Forces acting on a static fluid element.

The average surface forces per unit area on the faces of the element are designated p_A, p_B, and p_C, as shown. Summing the forces in the x and y directions gives, for static equilibrium:

$$\sum f_x = p_C \, dy - p_A \, ds \sin \theta = 0, \tag{2-1}$$

$$\sum f_y = -\rho \frac{g}{g_c} \frac{dy \, dx}{2} - p_A \, ds \cos \theta + p_B \, dx = 0. \tag{2-2}$$

The dimensions of the fluid element may be made as small as desired. Eventually the weight term will be of higher order than the other terms and in the limit may be neglected. Since $ds \sin \theta = dy$, Eq. (2–1) can be written

$$p_C \, dy - p_A \, dy = 0 \quad \text{or} \quad p_C = p_A.$$

Also since $ds \cos \theta = dx$,

$$-p_A \, dx + p_B \, dx = 0 \quad \text{or} \quad p_A = p_B.$$

Therefore, $p_A = p_B = p_C$. The mean of the normal surface forces per unit area is defined as the pressure. Thus, the pressure is equal to each of the normal surface stresses in a static fluid. Because the angle θ was arbitrary, it may be argued that the pressure p becomes the same in all directions as the element is made very small. The shape of the element is also arbitrary. The conclusion is that the pressure at a point in a static fluid is independent of direction. This is usually referred to as *Pascal's law*. Only the *magnitude* of the pressure must be specified at each point. Because of this characteristic, pressure is referred to as a scalar quantity. The local pressure in a static fluid may vary with both time and position, or

$$p = p(x,y,z,t). \qquad (2\text{–}3)$$

In many engineering situations a liquid and its vapor may exist in thermodynamic equilibrium. The same pressure exists in both phases, and the temperature is uniform throughout.* Under such conditions there is no net change in the amount of either phase. For this condition of equilibrium, a definite and predictable pressure exists for each temperature of the fluid. This pressure is referred to as the *vapor pressure*. The vapor pressure is tabulated at various temperatures for most substances of interest. For example, Keenan and Keyes [1] give the vapor pressure of water in addition to the density and other thermodynamic properties.

In many cases a liquid may exist in equilibrium with its own vapor when another gas and/or vapor may be present. In such cases, the pressure in the liquid may be greater than the vapor pressure for that temperature, since the total pressure results from the pressure exerted by both gases. An increase in total pressure, resulting from the other gas, increases the vapor pressure of a substance, but such increase is usually small [2].

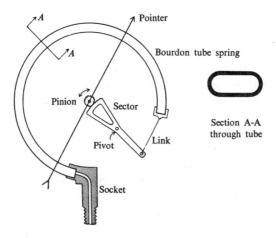

Fig. 2–2 Schematic of Bourdon-tube gage elements.

The term pressure, as it has been used so far, refers to *absolute pressure*. The *gage pressure* is the difference between the absolute pressure and the surrounding ambient pressure, or *atmospheric pressure*. This pressure difference is measured by pressure gages such as those of the Bourdon-tube type shown schematically in Fig. 2–2. A pressure difference between the

---

* For small elevation changes, any nonuniformity of pressure due to the weight of the fluid above may be neglected.

inside and outside of the hollow Bourdon tube causes the tube to tend to straighten out, moving the pointer. The pointer movement is proportional to the pressure difference, which, in the case of atmospheric pressure on the outside of the tube, is the gage pressure. *Vacuum* is defined as the difference between atmospheric pressure and an absolute pressure which is less than the atmospheric pressure. All these terms are graphically presented in Fig. 2–3.

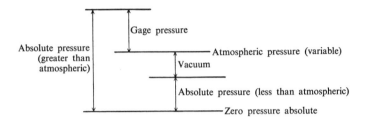

Fig. 2–3 Pressure terminology.

Pressure may be measured by a wide variety of instruments, including devices such as dead-weight testers, manometers, capsule and Bourdon-tube gages, potentiometric gages, reluctive gages, capacitive gages, piezoelectric gages, and pressure-sensitive wires and rare earths [6]. Figure 2–4 shows some typical pressure transducers. The particular device to use in a given situation may be determined by factors such as the accuracy and response required, the pressure range to be covered, the type of output reading desired, ruggedness, and economy. Manometers will be discussed in Section 2–3.

Since pressure is a force per unit area, the dimensions of pressure are force divided by length squared. In the United States the unit of pressure most commonly used is pounds force per square inch (psi). The designation of absolute pressure is *psia* and of gage pressure, *psig*.

The accepted standard atmospheric pressure is equal to 14.6959 psia. The variation of pressure with altitude is shown in the appendix.

Example 2–1. A pressure gage on a fuel tank reads 45 psig. What is the absolute pressure in the tank in pounds force per square foot? Assume that atmospheric pressure is 14.7 psia.

Solution.

$$\text{absolute pressure} = \text{gage pressure} + \text{atmospheric pressure}$$
$$= 45 + 14.7 = 59.7 \text{ psia}.$$

$$59.7 \frac{\text{lb}_f}{\text{in}^2} \times 144 \frac{\text{in}^2}{\text{ft}^2} = 8600 \frac{\text{lb}_f}{\text{ft}^2} \text{ absolute}.$$

Fig. 2-4 Typical pressure transducers. [Photographs courtesy of H. E. Sostman and Company (a); Statham Instruments, Inc. (b); Kistler Instrument Corporation (c); and the Instrument Systems Division, Whittaker Corporation (d).]

Example 2-2. A vacuum gage on an altitude simulator reads 13.8 psi vacuum. The ambient pressure is 14.6 psia. What is the absolute pressure in the simulator? What altitude does that represent?

Solution.
$$\text{absolute pressure} = \text{atmospheric pressure} - \text{vacuum}$$
$$= 14.6 - 13.8 = 0.8 \text{ psia}.$$

According to the data in the appendix, this would correspond to an altitude of about 65,000 feet.

Equation of State

Pressure is a thermodynamic property which may be related to other properties existing at that same point in a fluid at the same time. An equation which defines the state or condition of the fluid in terms of thermodynamic properties is called an *equation of state*. For a pure substance, it is convenient to define the state in terms of pressure, density, and temperature. The equation of state in such a case would be the equation that relates these three properties, or

$$f(p,\rho,T) = 0. \tag{2-4}$$

The most common example of an equation of state is that for an *ideal gas*, applicable to real gases only under certain limited conditions. It may be expressed as

$$p - \rho R_g T = 0, \tag{2-5}$$

where R_g is the gas constant. Other useful equations of state are found in the literature, but such equations are usually more complex than Eq. (2-5).

2-2 COMPRESSIBILITY AND THERMAL EXPANSION COEFFICIENT

All fluids normally change in density as the pressure and/or temperature changes. If it is assumed that a fluid can be described by the equation of state that relates density, pressure, and temperature:

$$f(\rho,p,T) = 0, \tag{2-6}$$

then the change in density can be related to the change in pressure and temperature by the total derivative

$$d\rho = \left(\frac{\partial \rho}{\partial p}\right)_T dp + \left(\frac{\partial \rho}{\partial T}\right)_p dT. \tag{2-7}$$

The fractional change in density is

$$\frac{d\rho}{\rho} = \frac{1}{\rho}\left(\frac{\partial \rho}{\partial p}\right)_T dp + \frac{1}{\rho}\left(\frac{\partial \rho}{\partial T}\right)_p dT. \tag{2-8}$$

If the temperature is held constant, then the fractional change in density that will occur as the pressure changes by the small amount dp is given by the first term on the right in Eq. (2–8). The coefficient of dp is the fractional rate of change of density with pressure at constant temperature, and is called the *isothermal compressibility*. Letting the symbol K_T stand for this quantity,

$$K_T = \frac{1}{\rho}\left(\frac{\partial \rho}{\partial p}\right)_T. \tag{2-9}$$

If the equation of state is expressed in terms of density, pressure, and entropy, where s stands for entropy,

$$f(\rho, p, s) = 0, \tag{2-10}$$

then the fractional change of density might be expressed as

$$\frac{d\rho}{\rho} = \frac{1}{\rho}\left(\frac{\partial \rho}{\partial p}\right)_s dp + \frac{1}{\rho}\left(\frac{\partial \rho}{\partial s}\right)_p ds. \tag{2-11}$$

The coefficient of dp in this case is called the *isentropic compressibility* and is given the symbol K_s:

$$K_s = \frac{1}{\rho}\left(\frac{\partial \rho}{\partial p}\right)_s. \tag{2-12}$$

The compressibilities may be approximated in finite terms for small changes of density:

$$\bar{K}_T = \frac{1}{\rho_0}\left(\frac{\Delta \rho}{\Delta p}\right)_T \qquad \bar{K}_s = \frac{1}{\rho_0}\left(\frac{\Delta \rho}{\Delta p}\right)_s, \tag{2-12a}$$

where ρ_0 is the initial density.

Both the isothermal and the isentropic compressibility are properties which, for a pure substance, depend only on the value of two other independent properties. For example, values of K_T or K_s may be tabulated for particular values of temperature and pressure. From Eq. (2–12a) it can be seen that the dimensions of K are length squared over force. A common unit is ft^2/lb_f.

The reciprocal of compressibility is called the *bulk modulus of elasticity* and is given the symbol B. It is also defined for either the isothermal or isentropic case:

$$B_T = \rho\left(\frac{\partial p}{\partial \rho}\right)_T \qquad B_s = \rho\left(\frac{\partial p}{\partial \rho}\right)_s. \tag{2-13}$$

The bulk modulus may also be approximated in finite terms:

$$\bar{B}_T = \rho_0\left(\frac{\Delta p}{\Delta \rho}\right)_T \qquad \bar{B}_s = \rho_0\left(\frac{\Delta p}{\Delta \rho}\right)_s. \tag{2-14}$$

The term *tangent bulk modulus* is frequently used for the quantities described by Eq. (2–13), and the term *secant bulk modulus* is used for the quantities described by Eq. (2–14).

Table 2-1

ISOTHERMAL BULK MODULUS OF WATER, Psi*

Pressure, (psi)	Temperature, °F				
	32	68	120	200	300
15	292,000	320,000	332,000	308,000	
1,500	300,000	330,000	312,000	319,000	218,000
4,500	317,000	348,000	362,000	338,000	271,000
15,000	380,000	410,000	420,000	405,000	350,000

* From *Handbook of Fluid Dynamics* by Streeter. Copyright 1961 by McGraw-Hill Book Company. Used with permission of McGraw-Hill Book Company.

Values of isothermal bulk modulus for water are given in Table 2-1 [3]. Values of adiabatic (isentropic) bulk modulus for a common hydraulic fluid, MIL-5606, are given in Fig. 2-5 [4]. Both the table and the figure give *point*, or tangent, values of the bulk modulus.

The concept of bulk modulus is not only useful in the study of fluids but is also used extensively in the theory of elasticity [5]. For isotropic elastic bodies, the bulk modulus is related to the modulus of elasticity E (Young's modulus) by Poisson's ratio m:

$$B = \frac{E}{3(1 - 2m)}. \qquad (2\text{–}15)$$

Young's modulus and Poisson's ratio are not significant in fluid problems. The units of both E and B in common use are $lb_f/in.^2$, or psi.

Water is often said to be incompressible, and for most engineering applications this is a good assumption. However, at 68°F and 14.7 psia, the density of water is 62.4 lb_m/ft^3, and, at 68°F and 80,000 psia, the density is 72.5 lb_m/ft^3. Serious error could result at high pressures when this effect is ignored. The isothermal bulk modulus of water at 68°F and 14.7 psia is seen in Table 2-1 to be approximately 320,000 psi.

The isothermal bulk modulus of an ideal gas is determined easily by the use of Eq. (2-14) and the equation of state. The result, for an ideal gas, is

$$B_T = p \quad \text{(ideal gas)}. \qquad (2\text{–}16)$$

The isothermal bulk modulus is equal to the absolute pressure for the ideal gas, and the isothermal compressibility is the reciprocal of the absolute pressure.

Another property useful in the study of both static and moving fluids is the *thermal expansion coefficient* β, the fractional change in density with

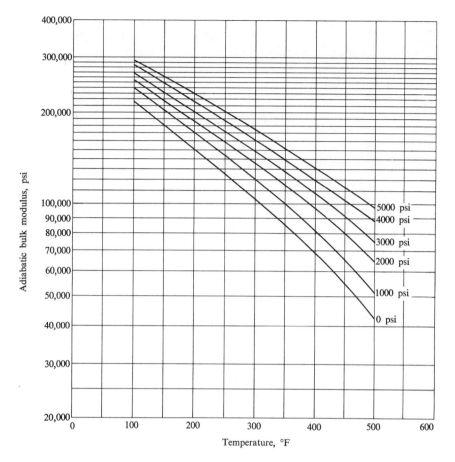

Fig. 2–5 Adiabatic bulk modulus of Mil-5606 hydraulic fluid (Courtesy of American Society for Testing and Materials. Used with permission).

temperature at constant pressure. This is the negative of the coefficient of dT in Eq. (2–8):

$$\beta = -\frac{1}{\rho}\left(\frac{\partial \rho}{\partial T}\right)_p. \qquad (2\text{–}17)$$

The thermal expansion coefficient is sometimes referred to as the *coefficient of volume expansion* or the *coefficient of cubical expansion*. For isotropic bodies, it is related to the *coefficient of linear expansion* α by the relation

$$\beta = 3\alpha. \qquad (2\text{–}18)$$

Most liquids have a positive thermal expansion coefficient. Water is unique in that it has a negative value of β between 0° and 4°C, or, in other words, at atmospheric pressure water has a maximum density at 4°C. It will expand if either heated or cooled from that temperature. A plot of β for saturated water at various temperatures is given in the appendix.

The dimension of β is obviously reciprocal temperature, or $\beta \triangleq (1/T)$. For an ideal gas, it follows from Eq. (2–5) that

$$\beta = \frac{1}{T} \quad \text{(ideal gas)}, \tag{2-19}$$

or, in other words, the thermal expansion coefficient of an ideal gas is equal to the reciprocal of the absolute temperature.

The change of volume corresponding to a change in temperature is an important effect in many fluid processes. The variations of density that result may cause fluid motion due to forces created in the fluid. This is referred to as *free convection* or *natural convection* and is discussed in more detail in Chapter 10.

Example 2–3. Calculate the change in density that occurs when water at 68°F is compressed isothermally from 14.7 psia to 500 psia. Assume that the bulk modulus of water is constant over this range and is equal to 320,000 psi.

Solution. If it is assumed as a first approximation that the change in density is small compared to the original density, Eq. (2–14) may be employed:

$$\Delta \rho = \bar{K}_T \rho_0 \Delta p = \frac{\rho_0 \Delta p}{\bar{B}_T} = \frac{62.4 \frac{\text{lb}_m}{\text{ft}^3} (500 - 14.7) \frac{\text{lb}_f}{\text{in.}^2}}{320,000 \frac{\text{lb}_f}{\text{in.}^2}}$$

$$= 9.46 \times 10^{-2} \frac{\text{lb}_m}{\text{ft}^3}.$$

This example might also be worked using the tables of Keenan and Keyes [1], but interpolation would usually be required.

If the change in density is a significant fraction of the original density, then Eq. (2–9) should be utilized to obtain the solution. This requires integration, which leads to an exponential expression:

$$\left(\frac{\partial \rho}{\partial p}\right)_T = \rho K_T.$$

This partial differential equation has the solution

$$\ln \rho = K_T p + f(T).$$

The function of temperature $f(T)$ is arbitrary. For the initial condition and the final condition

$$\ln \rho_1 = K_T p_1 + f(T_1),$$
$$\ln \rho_2 = K_T p_2 + f(T_2).$$

Subtracting the second equation from the first and noting that, since T is constant,

$$f(T_1) = f(T_2),$$

$$\ln \rho_2 - \ln \rho_1 = \ln \frac{\rho_2}{\rho_1} = K_T(p_2 - p_1),$$

$$\frac{\rho_2}{\rho_1} = e^{K_T(p_2-p_1)},$$

$$\rho_2 - \rho_1 = \rho_1(e^{K_T(p_2-p_1)} - 1).$$

Substitution of the values $K_T = (320{,}000 \text{ psi})^{-1}$, $\rho_1 = 62.4 \text{ lb}_m/\text{ft}^3$, $p_2 = 500$ psia, and $p_1 = 14.7$ psia leads to an expression that is difficult to evaluate because of the small exponent of e. Expanding e in a power series and approximating its value by the use of the first two terms leads to an answer:

$$\rho_2 - \rho_1 = \rho_1 \left\{ \left[1 + K_T(p_2 - p_1) + \frac{(K_T(p_2 - p_1))^2}{2} + \cdots \right] - 1 \right\}$$

$$\approx \rho_1 [K_T(p_2 - p_1)]$$

$$= \frac{62.4 \text{ lb}_m/\text{ft}^3 \, (500 - 14.7) \text{ lb}_f/\text{in.}^2}{320{,}000 \text{ lb}_f/\text{in.}^2}.$$

$$\Delta\rho = 9.46 \times 10^{-2} \text{ lb}_m/\text{ft}^3.$$

This is the same answer that was obtained above, using Eq. (2–14). Thus, the use of Eq. (2–14) is identical to neglecting higher-order terms in the series expansion obtained from Eq. (2–9).

2–3 THE PRESSURE FIELD IN A STATIC FLUID

It has been stated in the previous section that the only forces acting in a static fluid are the normal surface forces (pressure) and the body forces, such as the force caused by the gravitational effect on the mass of the liquid. A force balance may be used to obtain the equation that describes the variation of pressure with altitude in the fluid.

Assume that a coordinate system is defined so that distance z is measured from an arbitrary datum plane which is normal to the direction of the pull of gravity. Assume further that a fluid element of dimension dA by dz is located at some arbitrary distance z with dA normal to the z direction (Fig. 2–6). The body force or force due to the weight of the element is equal to the product of the volume of the element and the specific weight of the fluid. If the density ρ is expressed in units of mass, the force is

$$\text{body force} = -dA \, dz \, \rho \, \frac{g}{g_c}. \tag{2–20}$$

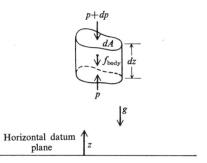

Fig. 2-6 Forces on a static fluid element.

The net surface force, or pressure force, in the z direction is that due to the pressure difference dp between the upper and lower surfaces, where an increase in p with increasing z is taken to be positive.

Note that there are no shear forces in the stationary fluid. The only surface force acting on the element is that due to pressure.

Summing forces in the vertical (z) direction,

$$p\, dA - (p + dp)\, dA - \rho \frac{g}{g_c} dA\, dz = 0. \qquad (2\text{-}21)$$

After simplifying and dividing by dA,

$$dp = -\rho \frac{g}{g_c} dz. \qquad (2\text{-}22)$$

Equation (2-22) is the differential equation of a static fluid. It is easily integrated (for $\rho = $ constant) to give

$$p = -\frac{g}{g_c} \rho z + C_1, \qquad (2\text{-}23)$$

where C_1 is the constant of integration which depends upon the pressure existing at some specified value of z. Equation (2-23) shows that the pressure increases in a fluid as one moves in the direction of the gravitational pull. For a static liquid this means that the pressure increases as the depth increases. Anyone who has dived in water or has descended in an unpressurized airplane has experienced such a pressure change, sometimes with a sharp ear pain. The pressure is uniform at all horizontal locations where $z = $ constant.

If Eq. (2-23) is solved for C_1,

$$p + \frac{g}{g_c} \rho z = C_1. \qquad (2\text{-}24)$$

The term p in Eq. (2-24) is simply the static pressure. The term $(g/g_c)\rho z$ is the *elevation pressure*. The sum of these two pressures, the left side of Eq.

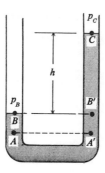

Fig. 2–7 A simple U-tube manometer.

(2–24), is referred to as the *piezometric pressure*. The piezometric pressure is constant throughout a static fluid and it is numerically equal to the static pressure at the datum ($z = 0$) line, assuming that the datum is within the fluid.

For two points, z_1 and z_2 in a static fluid where the pressures are p_1 and p_2 and where the density ρ may be assumed constant, Eq. (2–23) becomes

$$(p_2 - p_1) = \frac{g}{g_c}\rho(z_1 - z_2),$$

or

$$\frac{(p_2 - p_1)}{\rho}\frac{g_c}{g} = (z_1 - z_2). \tag{2–25}$$

The dimension of each term in Eq. (2–25) is length. Pressure expressed in such terms is referred to as the *head*. A head of 20 feet of water, for example, would be equivalent to the pressure difference created by a static column of water 20 ft high. Equation (2–23) can be written

$$\frac{p}{\rho}\frac{g_c}{g} + z = C_2. \tag{2–26}$$

The first term on the left is the *pressure head*, the second term is the *potential head*, or *elevation head*, and the sum of both terms is the *piezometric head*. The piezometric head is constant throughout a static fluid.

Manometers

A *manometer* is a device used to measure a pressure differential by measuring the height of a column of fluid supported by that pressure differential. A *U-tube manometer* is shown in Fig. 2–7. The two pressures of interest are sensed at the two tube ends and cause a difference in the heights of the two columns of liquid. This difference in height, h, is proportional to the pressure difference ($p_B - p_C$). The manometer is based on the simple concept that the pressures at two points of the same elevation in a continuous static column of arbitrary shape are equal. In Fig. 2–7, for example,

$$p_A = p_{A'} \quad \text{and} \quad p_B = p_{B'} \quad \therefore \quad p_B - p_C = p_{B'} - p_C,$$

but

$$p_{B'} - p_C = h\rho \frac{g}{g_c} = p_B - p_C. \tag{2-27}$$

In the case where a liquid is present above the surface of the manometer fluid, as in Fig. 2–8, Eq. (2–27) must be modified to account for the effect of the elevation head of the second fluid. In such a case, referring to Fig. 2–8 and making use of Eq. (2–24) with the datum line along the line B–B',

$$p_1 = p_B - \rho_A \frac{g}{g_c} h_3,$$

and

$$p_2 = p_{B'} - \rho_B \frac{g}{g_c} h_1 - \rho_A \frac{g}{g_c} h_2.$$

Here we see that p_1 and p_2 are the static pressures on the left and right side of the U-tube respectively. Since

$$h_1 = h_3 - h_2 \quad \text{and} \quad p_B = p_{B'},$$

then

$$p_1 - p_2 = \frac{g}{g_c} h_1 (\rho_B - \rho_A). \tag{2-28}$$

The two fluids in this case must be such that $\rho_B > \rho_A$, and they do not tend to react or go into solution with each other.

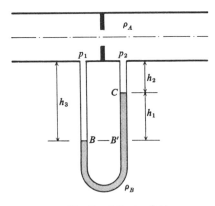

ρ_A = Density of flowing fluid

ρ_B = Density of manometer fluid

Fig. 2–8 Use of manometer to measure pressure drop across an orifice.

Example 2–4. A mercury manometer is used to measure the pressure drop across an orifice through which water flows (Fig. 2–8). The manometer indicates a reading (h_1) of 20 in. Hg. What is the pressure drop ($p_1 - p_2$) in psi? Water is assumed to exist in the manometer lines above the mercury, where B and C define the water-mercury interfaces.

Solution. Assume $\rho_{H_2O} = 62.4 \ \text{lb}_m/\text{ft}^3$, $\rho_{Hg} = 848 \ \text{lb}_m/\text{ft}^3$, and $g = 32.174 \ \text{ft}/\text{sec}^2$. From Eq. (2–28), then

$$p_1 - p_2 = \frac{g}{g_c} h_1 (\rho_{Hg} - \rho_{H_2O})$$

$$= \frac{32.174 \ \frac{\text{ft}}{\text{sec}^2}}{32.174 \ \frac{\text{lb}_m \ \text{ft}}{\text{lb}_f \ \text{sec}^2}} (20) \ \text{in.} \ (848 - 62.4) \frac{\text{lb}_m}{\text{ft}^3} \ \frac{1 \ \text{ft}^3}{1728 \ \text{in}^3}$$

$$= 9.1 \ \text{psi.}$$

A device which is quite useful for determining the absolute pressure of the atmosphere is the *barometer*, shown in simplified form in Fig. 2–9. A tube, initially full of some liquid, has its open end placed in a dish of the same liquid. The fluid in the column will fall until the pressure at the bottom of the tube is equal to the pressure of the liquid in the dish at that same level. From Fig. 2–9, the following equation can be derived:

$$p_v + h\rho \frac{g}{g_c} = p_{atm}, \qquad P = h\rho \qquad (2\text{–}29)$$

where h is measured from the free surface of the liquid in the open dish to the free surface of the liquid in the column, p_v is the vapor pressure of the substance used in the barometer, and p_{atm} is the atmospheric pressure

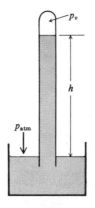

Fig. 2–9 A simple barometer.

exerted on the free surface of the liquid in the dish. The vapor pressure can, in many instances, be neglected relative to the atmospheric pressure, and the height of the column of liquid can be related directly to the atmospheric pressure.

Example 2–5. A barometer similar to that in Fig. 2–9 has mercury as the fluid ($\rho = 848$ lb_m/ft^3). When the column of liquid in the barometer is 29.5 inches, what is the atmospheric pressure? Assume $g = 32.174$ ft/sec². The vapor pressure of mercury at 68°F is 0.0000232 psia and may be neglected.

Solution.

$$p_{atm} = (29.5 \text{ in.})(848)\frac{lb_m}{ft^3}\left(\frac{32.174}{32.174}\right)\frac{\frac{ft}{sec^2}}{\frac{lb_m \text{ ft}}{lb_f \text{ sec}^2}}\frac{1 \text{ ft}}{12 \text{ in.}}$$

$$= 2080\frac{lb_f}{ft^2} = 14.5 \text{ psia.}$$

2–4 BUOYANCY AND STABILITY

Buoyancy

The principles discussed in the previous section regarding pressure in a static fluid lead to an important law of buoyancy, sometimes referred to as the *principle of Archimedes:*

A body immersed or floating in a fluid is acted upon by an upward buoyant force which is equal to the weight of the fluid displaced, and which acts through the center of gravity of the displaced volume, the "center of buoyancy."

The term *upward* implies a direction opposite to the direction of pull of the gravitational force.

The principle is easily proved by considering a body completely immersed in a fluid of constant density ρ. The net upward vertical pressure force df_z on a vertical prism of cross-sectional area dA_z is:

$$df_z = p_2 \, dA_z - p_1 \, dA_z,$$

or, since

$$p_2 - p_1 = -\rho\frac{g}{g_c}(z_2 - z_1),$$

$$df_z = -\rho\frac{g}{g_c}(z_2 - z_1)\, dA_z,$$

where $(z_2 - z_1)\, dA_z$ is the volume of the vertical prism, so that by integration over the entire body surface, the total upward (buoyant) force is equal to $\rho(g/g_c)$ Vol. Vol is the volume of the submerged body, or in other words, the volume of the liquid displaced, and $\rho(g/g_c)$ Vol is the weight of liquid displaced.

2-4 BUOYANCY AND STABILITY

Considering the horizontal prism in Fig. 2–10, it is obvious that there is no net component of force in the horizontal plane.

The principle of Archimedes is useful in many situations, such as determining the density of an irregularly shaped solid object. When the object is immersed in a fluid of known density, the loss in weight can be easily measured and related to the volume of the object.

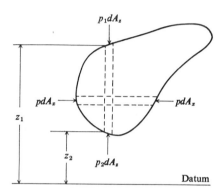

Fig. 2–10 Buoyancy of a submerged volume.

Example 2–6. A stone is found to weigh 18.0 lb_f when weighed in air and 12.6 lb_f when weighed completely immersed in water (density of water is 62.4 lb_m/ft^3). Assuming that the local acceleration of gravity is 32.1 ft/sec², determine the density of the stone. Neglect the buoyancy effect in air.

Solution. The loss in weight of the stone is $18.0 - 12.6 = 5.4\ lb_f$. This loss in weight is equal to the weight of water displaced:

weight of displaced water = (specific weight of water)(volume of stone)

$$= \rho_w \left(\frac{g}{g_c}\right) \text{(volume of stone)}.$$

By definition,

$$\text{density of stone} = \frac{\text{mass of stone}}{\text{volume of stone}}$$

$$= \frac{18.0\ lb_f \left(\frac{g_c}{g}\right)}{\frac{5.4\ lb_f}{62.4\ lb_m/ft^3}\left(\frac{g_c}{g}\right)} = 208\ \frac{lb_m}{ft^3}.$$

The buoyant force on an object immersed in air may not always be neglected, particularly in precision determinations of the weight of objects of low density. Many objects, such as a gas-filled balloon, are actually lighter than air and will exert an upward force on any restraints.

When an object immersed in a liquid with a free surface has a mass less than the liquid which it displaces, the buoyant force will lift the object until it is only partially submerged, or in other words, it will float on the surface. It is also easy to show by a force balance that a floating object displaces an amount of liquid equal to the weight of the object.

Example 2–7. Determine the water line on a rectangular barge 3 ft high by 10 ft wide by 20 ft long if the barge has a mass of 20,000 lb_m. Assume that the density of the water in which the barge floats is 62.4 lb_m/ft^3.

Solution. 20,000 lb_m of water would have a volume of

$$\frac{20,000\ lb_m}{62.4\ \frac{lb_m}{ft^3}} = 321\ ft^3.$$

The depth of the barge in the water equals

$$\frac{321\ ft^3}{(10 \times 20)\ ft^2} = 1.605\ ft.$$

At a floating depth (or draft) of 1.605 ft, the weight of water displaced by the barge is equal to the weight of the barge.

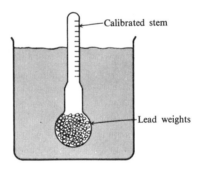

Fig. 2–11 Hydrometer.

Another example of the use of the principle of Archimedes is in the measurement of the specific weight of a liquid by means of a *hydrometer*, like that shown in Fig. 2–11.

The height at which the hydrometer of fixed weight floats in the fluid depends only on the specific weight of the fluid. The stem of the hydrometer may, therefore, be marked off directly in specific weight or specific gravity. A common use of hydrometers is for checking the fluid in an automobile cooling system to determine the ratio of water to antifreeze.

Stability

In some situations it is important to determine whether an object submerged or floating in a fluid is stable. An object has *linear stability* if a small linear displacement in any direction sets up restoring forces which tend to return the object to its original position. For example, in Fig. 2–12 is shown a weight on a plane horizontal surface with springs loading it from both positive and negative x directions. Displacements in either x direction cause restoring forces, and thus the weight has *horizontal stability* in the x direction. A floating object has *vertical stability* since it will return to its original position if displaced vertically. It has *neutral stability* with respect to horizontal displacements since no forces are created by small horizontal displacements.

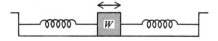

Fig. 2–12 Example of horizontal stability.

An object has *rotational stability* if a small angular displacement is accompanied by a restoring effect which tends to return the object to its original position. A sphere of uniform material resting on a flat horizontal surface or floating in a liquid has neutral rotational stability since no forces are created by a small angular displacement. Some objects may be stable with respect to rotation however. This can be seen in Fig. 2–13, which shows the cross section of a floating object (a ship) and a submerged object (a submarine).

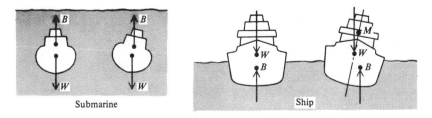

Fig. 2–13 Stability of submerged and floating objects.

In both cases the objects are acted upon by two forces: (1) the weight of the object, acting downward through the center of gravity of the object, and (2) the buoyant force, acting upward through the center of gravity of the displaced fluid (center of buoyancy).

In the case of the submarine, the center of buoyancy is above the center of gravity. A small angular displacement creates a restoring couple and the submarine rights itself. Thus, it is rotationally stable. In all submerged objects the center of buoyancy must be above the center of gravity for rotational stability to exist.

In the case of the floating ship, the center of gravity is shown to be above the center of buoyancy. In order for rotational stability to exist the center of buoyancy must shift properly with angular displacement. The buoyant force must intersect the center line of the ship at a point *above* the center of gravity for stability. The distance between the intersection, or the metacenter (shown as point M in Fig. 2–13), and the center of gravity is a measure of the stability and is referred to as the *metacentric height*. An incorrect distribution of the cannons aboard the *Vasa*, pride of the Swedish navy in 1628, created an unstable situation and caused sinking of the ship in the harbor as it was embarking on its maiden voyage.

Buoyancy effects can also exist in fluids due to nonuniform density distributions. Lighter or less dense fluids tend to rise above more dense fluids. In the case of a fluid layer heated from above and cooled below, a stable situation exists. In the opposite case, however, fluid motions are created by the less dense fluid as it rises above the more dense fluid.

Static fluids in which horizontal density variations exist are also unstable since pressure variations at a given depth are created. These pressure variations lead to motion of the fluid. The heating of a vertical surface submerged in a fluid creates such horizontal density gradients and leads to a fluid motion referred to as free convection. This will be discussed in detail in Chapter 10.

REFERENCES

1. J. H. Keenan and F. G. Keyes, *Thermodynamic Properties of Steam*, Wiley, New York (1936).
2. J. F. Lee and F. W. Sears, *Thermodynamics*, 2d ed., Addison-Wesley, Reading, Mass. (1965).
3. R. L. Daugherty, "Fluid Properties," *Handbook of Fluid Dynamics*, McGraw-Hill, New York (1961).
4. R. L. Peeler and J. Green, "Measurement of Bulk Modulus of Hydraulic Fluids," ASTM Bulletin, No. 235 (Jan. 1959), p. 51.
5. S. Timoshenko and J. N. Goodier, *Theory of Elasticity*, 2d ed., McGraw-Hill, New York (1951).
6. M. H. Aronson, *Pressure Handbook*, Instruments Publishing Co., Pittsburgh (1963).

PROBLEMS

1. As the pressure on a liquid is changed isothermally from 1000 to 5000 psia, it is noted that the density changes from 56.0 to 56.6 lb_m/ft^3. Estimate the mean isothermal compressibility K_T.

2. Derive expressions for K_T, K_S, and β in terms of specific volume v instead of density ρ.

3. Van der Waals' equation of state for a gas is

$$(p + a\rho^2)\left(\frac{1}{\rho} - b\right) = RT,$$

where a, b, and R are constants for any particular gas. Derive expressions for K_T and β using the Van der Waals equation of state and show that they reduce to the proper expressions for an ideal gas when a and b equal zero.

4. A hydraulic control system contains 2.78 cubic feet of hydraulic fluid in the tubing and components of the high-pressure section. The pressure in this section is changed from 14.7 to 5000 psia. Estimate the amount of additional fluid (in gallons) that must be added to this section to account for compressibility of the hydraulic fluid. Neglect expansion of the tubing and components and assume a density of 56 lb_m/ft^3 and a mean bulk modulus of 180,000 psi.

5. A thermometer uses alcohol as the working fluid. The bulb contains 0.002 in.^{3} and is attached to a stem having a bore cross-sectional area of 0.00002 in.^{2} At 100°F the top of the liquid is found to be 1.88 in. above the top of the bulb. Where will the top of the liquid be at 200°F? Assume that β is constant and equal to 4.5 × 10^{-4} per °F, that the glass bore does not change with temperature, and that the pressure change within the liquid is small. Assume that the density of alcohol is 50 lb_m/ft^3.

6. For the thermometer in Problem 5, where should the mark for 100°C be located?

7. An alcohol thermometer is examined and found to have a stem-bore cross section of 0.000025 in.^{2} and markings of 20°F per inch along the stem. Estimate the amount of alcohol in the bulb. Neglect tube expansion and assume that the pressure change is small, that β is constant and equal to 4.5 × 10^{-4} per °F, that the mean density of alcohol is 50 lb_m/ft^3, and that the stem volume is much smaller than the bulb volume.

8. The wreckage of the submarine *Thresher* was eventually found at a depth of approximately 8000 ft in the Atlantic Ocean. A special search vessel was used to survey the ocean floor in this area. What force was exerted on a 6-in. diameter viewing port of the search vessel?

9. For the fluid system shown in the figure, assume that the fluid is light oil at 100°F, the pistons are weightless, and

$$g = 32.174 \text{ ft/sec}^2.$$

The force

$$F_1 = 10 \text{ lb}_f.$$

Find the force F_2 in lb_f.

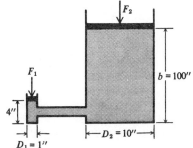

10. The pressure in the diving bell shown in the figure will change with depth of submersion L. Find an expression for the height of liquid y as a function of h and L.

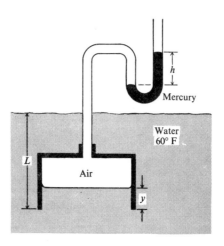

11. The temperature of the atmosphere decreases almost linearly from sea level to 35,000 ft at a rate of 3.5°F for each 1000 ft. Assuming standard conditions of 14.7 psia and 59°F at sea level, calculate the expected pressure at 35,000 ft and compare your result with the value of 498 lb_f/ft^2 as given in the standard atmosphere tables of the International Civil Aviation Organization.

12. A Bourdon-tube type pressure gage is located 70 ft below the top of a water tank. If the top of the tank is vented to the atmosphere, which of the following gages should be used: 0–20 psi, 0–50 psi, or 0–100 psi? What is the absolute pressure if the gage reading is 40 psig and the local barometric pressure is 29.50 in. Hg?

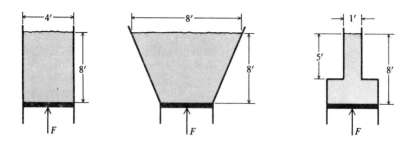

13. Each vessel in the accompanying figure contains water ($\rho = 62.4$ lb_m/ft^3). The bottom of each vessel is a 4-ft square. Compute the force exerted by the fluid pressure on the horizontal bottom of each vessel.

14. Determine the resultant force due to water pressure acting on one side of the 5 × 8 ft rectangular area shown in the figure.

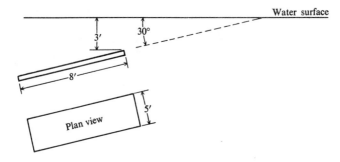

15. Find the location of the resultant force on the rectangular plate in Problem 14.

16. Determine the magnitude of the horizontal and vertical forces due to fluid pressure on one side of the curved plate in the figure. The plate is a quadrant of a circle, with width W.

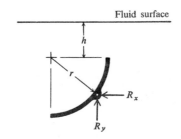

17. A ship has a cross section of 4000 ft² at the water line when the draft is 10 ft. How many pounds of cargo will increase the draft 2 inches? Assume that salt water has a density of 64.0 lb_m/ft^3.

18. Estimate the buoyant force acting on an average-size man due to the fact that he is immersed in an ocean of air. (Assume that air density is 0.0765 lb_m/ft^3.

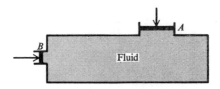

19. Determine the diameter of piston B in the fluid press in the figure. The force on piston B is 20 lb_f and the weight sustained by piston A is 1500 lb_f. Piston A has a diameter of 1 ft.

20. Calculate the gage pressure at A in the figure. The oil density is 52 lb_m/ft^3, and the density of mercury is 848 lb_m/ft^3.

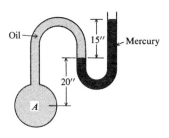

21. Calculate the density of an ideal gas (R_g = 48.3 ft lb_f/lb_m °R) at a pressure of 150 mm of mercury absolute and 80°F. Assume that the weight of the mercury is 848 lb_f/ft^3.

22. Assume that a barge 20 ft long has a constant cross section as shown in the figure. If the barge weighs 25,000 lbs, how deep will it sit in the water?

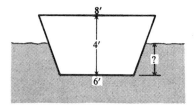

23. A rectangular gate 7 ft long and 3 ft high lies in a vertical plane with its center 10 ft below a water surface. Calculate the magnitude and location of the force on the plate.

24. A thin triangular plate of 4 ft base and 3 ft altitude has its base horizontal and lies in a 45° plane from the vertical. The apex lies below the base and is 7 ft below a water surface. Calculate the magnitude, location, and direction of the force on the plate. Does the same force exist on both sides of the plate?

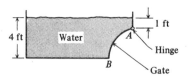

25. Find the magnitude and direction of the force due to the water acting on the gate AB, which is a quarter circle. The width of the gate is 6 ft.

26. A balloon has a solid weight (including crew but not gas) of 600 lbs$_f$ and contains 14,000 cu ft of hydrogen. How many pounds of ballast must be added to keep the balloon from ascending? Assume the pressure of air and hydrogen to be 14.7 psia; temperature of air and hydrogen 50°F.

CHAPTER 3

BASIC CONCEPTS OF FLUID MECHANICS AND HEAT TRANSFER

In beginning a study of fluid dynamics and heat transfer, it is necessary to define terms and to specify the usage of symbols. It is important to be able to locate points in a fluid and the motion of the fluid with exactness and brevity. The stresses existing in the fluid and their relation to the fluid motion must be described conveniently and with no misunderstanding. The recognition of the differences between various flow patterns must be understood before the choice of a general approach to a problem can be made.

These topics will now be covered to furnish the background needed for subsequent chapters.

3-1 COORDINATE SYSTEMS

In the solution of problems in heat transfer and fluid flow, it is desirable to be able to provide geometric identification of various points in the system under study. This is done by the use of a coordinate system. Each point in space is identified by specifying its location relative to the origin of the system. If the coordinate system is nonaccelerating and nonrotating, it is called an *inertial system*. It is necessary to specify both the location of the origin and a system for locating points relative to that origin.

The most common coordinate system is the *rectangular* or *Cartesian coordinate system*. Three mutually perpendicular planes are assumed to pass through the origin. A point is identified by specifying the distances x, y, and z from each of these planes respectively. This is illustrated in Fig. 3-1(a). The point $P(x,y,z)$ is located a distance x from the z-y plane, a distance y from the x-z plane, and a distance z from the x-y plane. The figure illustrates a *right-hand coordinate system*. If the fingers on the right hand are assumed to curve around the z axis while moving from the positive x axis to the positive y axis, then the thumb will point in the positive z direction. The right-hand coordinate system is most common and will be used throughout this text.

Another coordinate system in wide use is the *cylindrical coordinate system*, illustrated in Fig. 3-1(b). Each point in space is located relative to the origin by specifying a distance z from the origin along a line or axis passing through the origin, a distance r normal to that line, and an angle θ

relative to a fixed reference plane passing through the line. Such systems are quite useful in dealing with fluid flow and heat transfer problems in cylinders, pipes, and tubes. In a right-hand system a positive angle θ is measured from the reference plane in the direction of the fingers as the right hand is wrapped around the z axis with the thumb in the positive z direction.

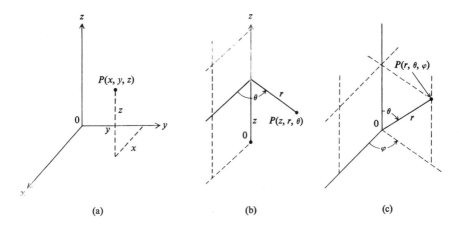

Fig. 3-1 (a) Rectangular coordinate system. (b) Cylindrical coordinate system. (c) Spherical coordinate system.

A third coordinate system of interest is the *spherical coordinate system* where points are located by giving a distance from the origin r and specifying direction in terms of two angles θ and φ. The two angles are measured in mutually perpendicular planes, both passing through the origin, one with an arbitrary orientation, and the other passing through the line connecting the origin and the point. This is illustrated in Fig. 3-1(c).

Once a point has been located by specifying its coordinates in one of the above systems, it may be identified in terms of either of the other coordinate systems. If the first two systems are assumed to have the same origin and if the reference plane of the cylindrical system is assumed to be in the x–z plane of the rectangular system (see Fig. 3-2), the transformation equations for the two systems are

$$x = r \cos \theta, \tag{3-1}$$

$$y = r \sin \theta, \tag{3-2}$$

$$z = z. \tag{3-3}$$

The spherical coordinates may be transformed into rectangular coordinates. If it is assumed that the origins coincide and the two planes of the

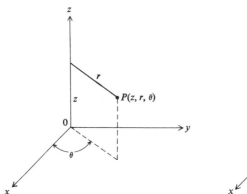 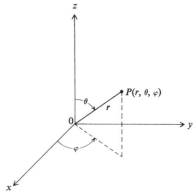

Fig. 3-2 Cylindrical coordinate system superimposed on a rectangular coordinate system.

Fig. 3-3 Spherical coordinate system superimposed on a rectangular coordinate system.

spherical system coincide with the x–z plane and the plane formed by r and the z axis (Fig. 3–3), then the equations are

$$x = r \sin \theta \cos \varphi, \qquad (3\text{–}4)$$

$$y = r \sin \theta \sin \varphi, \qquad (3\text{–}5)$$

$$z = r \cos \theta. \qquad (3\text{–}6)$$

3-2 VELOCITY AND ACCELERATION

Specification of the location of a particular point in space was discussed in the previous section. It is convenient to be able to describe the time rate of change of position of a particular element of fluid. This time rate of change of position, **ds/dt**, is called *velocity* and will be given the symbol **V**. Velocity is a *vector* quantity, which must be described by both a magnitude and a direction. Both the magnitude and direction may vary with position and time. In a rectangular coordinate system where position is described in terms of x, y, and z and time is designated by the symbol t, the velocity may be written as

$$\mathbf{V} = \mathbf{V}(x,y,z,t).$$

As with any vector, it is convenient to define the velocity vector in terms of components or projections along the axes of a space coordinate system. For every point in a fluid, the velocity vector may be represented diagrammatically by an arrow, with the direction of the arrow corresponding to the direction of the velocity at that point and the length of the arrow corresponding to the magnitude, using any convenient scale.

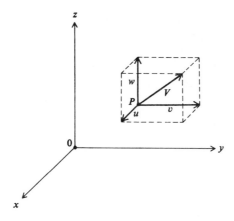

Fig. 3–4 Representation of velocity components in a rectangular coordinate system.

The velocity components of the vector **V** in a rectangular coordinate system consist of the projections of **V** on the x, y, and z axes as indicated in Fig. 3–4.

The velocity vector at point P is shown as an arrow with the tail located at the point in space under consideration. The length of the arrow is determined by the magnitude of the velocity in the fluid at that point. The magnitude of the components of **V** in the x, y, and z directions are designated as u, v, and w. If **i, j,** and **k** are considered to be unit vectors (magnitude equal to one) in the x, y, and z directions respectively, then **V** may be expressed as

$$\mathbf{V} = \mathbf{i}u + \mathbf{j}v + \mathbf{k}w. \tag{3-7}$$

When x, y, and z describe the instantaneous position of a particular fluid element in space, then $dx/dt = u$, $dy/dt = v$, and $dz/dt = w$.

The magnitude of each velocity component, a scalar quantity, can be expressed as a function of position and time, $u(x,y,z,t)$, $v(x,y,z,t)$, and $w(x,y,z,t)$. The magnitude of the velocity at each point at any time is obtained by use of analytic geometry and Fig. 3–4:

$$V = \sqrt{u^2 + v^2 + w^2}. \tag{3-8}$$

Notice that specification of u, v, and w at every point in a fluid completely describes the velocity throughout the fluid at that time. Thus the velocity vector field can be described in terms of the scalar component fields. In many cases the coordinate system can be oriented relative to the fluid motion so that one or two of the components of velocity are always zero. This simplifies the problem considerably.

It is, therefore, important to select the proper type of coordinate system for a particular problem so that simplifications may be made where possible. If the cylindrical coordinate system described in Fig. 3–2 is used, then velocity vector components defined by z, r, and θ are appropriate. These will be designated as u_z, u_r, and u_θ respectively. For example, at point P in Fig. 3–5, the local velocity vector $\mathbf{V}$ has a component u_z parallel to the z axis, a component u_r parallel to r, and a component u_θ perpendicular to both u_z and u_r. In the spherical coordinate system, the velocity components u_r, u_θ, and u_φ are those corresponding to the r, θ, and φ directions respectively in Fig. 3–6.

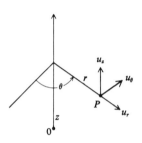
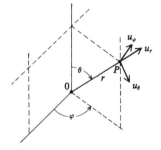

Fig. 3–5 Velocity components in a cylindrical coordinate system.

Fig. 3–6 Velocity components in a spherical coordinate system.

The acceleration or time rate of change of velocity is important in many fluid problems as was pointed out in Section 1–3. Since acceleration is the time derivative of a vector quantity, it is also a vector and is given the symbol **a**:

$$\mathbf{a} = \frac{d\mathbf{V}}{dt}. \tag{3-9}$$

Using Eq. (3–7), Eq. (3–9) may be expressed as

$$\mathbf{a} = \mathbf{i}\frac{du}{dt} + \mathbf{j}\frac{dv}{dt} + \mathbf{k}\frac{dw}{dt}. \tag{3-10}$$

The components of acceleration can thus be stated in terms of the time derivatives of the velocity components.

3–3 THE SUBSTANTIAL DERIVATIVE

Often the magnitude of some property of a fluid, such as pressure or velocity, depends upon both position and time. That is, at every position in the fluid at any instant of time, the magnitude of the pressure p, for example, has some

definite value. This may be written in a rectangular coordinate system as
$$p = p(x,y,z,t). \tag{3-11}$$
The spatial or position coordinates x, y, and z are independent of time t. However, a pressure-sensing device, such as a pressure transducer of the resistance type, may be moving in space. Any such sensing device will be called an *observer*. Thus, an observer may be considered to be moving about in space, and the position of the observer $(x, y,$ and $z)$ is time-dependent. This is because it takes a definite finite time to change from one position to another.

For such a case the total derivative of p with respect to time can be written as
$$\frac{dp}{dt} = \frac{\partial p}{\partial t}\frac{dt}{dt} + \frac{\partial p}{\partial x}\frac{dx}{dt} + \frac{\partial p}{\partial y}\frac{dy}{dt} + \frac{\partial p}{\partial z}\frac{dz}{dt}. \tag{3-12}$$

Equation (3-12) expresses the total rate of change of pressure with time as the point of observation of p is moved an infinitesimal distance in the fluid. The derivatives dx/dt, dy/dt, and dz/dt describe the motion of the observer. If the observer moves along with a selected particle of the fluid, that is, if the observer's motion is the same as the fluid particle being observed, then $dx/dt = u$, $dy/dt = v$, $dz/dt = w$, and Eq. (3-12) can be written
$$\frac{Dp}{Dt} = \frac{\partial p}{\partial t} + u\frac{\partial p}{\partial x} + v\frac{\partial p}{\partial y} + w\frac{\partial p}{\partial z}. \tag{3-13}$$

Here the symbol D/Dt replaces d/dt because the derivative is, in a sense, a special total derivative where the observer moves with the substance observed. This special type of derivative is called the *substantial derivative*, and is quite useful in the study of fluid mechanics. In addition to being called the *substantial derivative*, D/Dt is sometimes called the *material derivative*, or the *particle derivative*, and sometimes *the derivative following the motion of the fluid*. The concept of an observer moving with the substance observed is called the Lagrange viewpoint, as contrasted to the Euler viewpoint, where the observer is fixed in space. Both viewpoints will be used in various applications in this text.

The general operator form of the substantial derivative is
$$\frac{D}{Dt} = \frac{\partial}{\partial t} + u\frac{\partial}{\partial x} + v\frac{\partial}{\partial y} + w\frac{\partial}{\partial z}. \tag{3-14}$$

The substantial derivative may be expressed for any quantity which is a function of time and/or position. The substantial derivative of the velocity vector, for example, will be used quite often.

The substantial derivative may be thought of as expressing the observed time rate of change of a quantity caused by two factors: (1) the change in that quantity with time at each point, and (2) the change in that quantity with time due to motion. The latter change is due to the fact that an elapse of time is required to move from one position to another position where the quantity has a different value.

This might be illustrated by imagining a person swimming in a large pipe filled with a flowing fluid (Fig. 3–7). We will assume that flow is in the positive x direction.

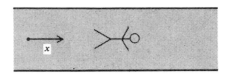

Fig. 3–7 Illustration of substantial derivative of pressure.

Normally there is a pressure drop in the direction of flow, or in other words, dp/dx is negative. The swimmer's velocity is given as dx/dt and, therefore, the rate of change of pressure with time for the swimmer would be

$$\frac{dp}{dt} = \frac{dp}{dx}\frac{dx}{dt}.$$

If the swimmer were swimming downstream, the rate of change of pressure with time would be negative. If he were swimming upstream, the rate of change of pressure with time would be positive. If the swimmer moved with the fluid so that $dx/dt = u$, then the pressure change for the swimmer would be

$$\frac{Dp}{Dt} = u\frac{dp}{dx}.$$

If, in addition, the pressures at all points were changing with time and this change at the location of the swimmer were given by $\partial p/\partial t$, then the rate of pressure change observed by the swimmer moving with the current would be

$$\frac{Dp}{Dt} = \frac{\partial p}{\partial t} + u\frac{dp}{dx}.$$

When velocity is the observed quantity, the term *local acceleration* is sometimes applied to the rate of change of velocity with time at a point, $\partial \mathbf{V}/\partial t$. The term *convective acceleration* refers to the change in velocity of a bit of fluid as it moves from one point to another, assuming that the velocity at each point is steady with time ($\partial \mathbf{V}/\partial t = 0$). The total acceleration of a fluid element is the sum of the local and convective accelerations, or

$$\mathbf{a} = \frac{D\mathbf{V}}{Dt} = \frac{\partial \mathbf{V}}{\partial t} + u\frac{\partial \mathbf{V}}{\partial x} + v\frac{\partial \mathbf{V}}{\partial y} + w\frac{\partial \mathbf{V}}{\partial z}. \tag{3-15}$$

3-4 FLUID STRESS AND ENERGY CONVERSION

In mechanics it is often necessary to consider the forces acting upon the fluid. These forces affect the motion of the fluid and therefore the transport of both momentum and energy. The forces are also important in processes where work is converted into stored energy of the fluid.

The net force acting on a fluid element may consist of *surface forces* (such as pressure), which act over the surface of the element, and *body forces* (such as gravitational force) which act on the entire mass. Both surface and body forces are vector quantities which may be conveniently described in terms of their components in three coordinate directions.

In the most general case, the components of both the body force and the surface force will vary with time and with position in the fluid. Usually the local body force components throughout the fluid can be described with little difficulty. Gravity, for example, normally exerts the same force on each element of mass, and the body force per unit volume due to gravity would be constant.

The description of the local surface force components is more complex. The description is most easily applied to a control volume which is a cube of dimension $(dx)(dy)(dz)$ described in rectangular coordinates. It is convenient to express the surface force components in terms of force per unit area of surface, or stress, acting on each face of the control volume. On each face the stress is assumed to be an averaged value across that face, having one normal and two tangential components. Thus, there are three stress components per face or a total of eighteen stress components on the six faces of the cubical element.

The Greek letter tau τ with a double subscript will be used in designating the components of stress. The first subscript designates the axis to which a particular face is perpendicular. The second subscript designates the axis to which a particular stress component is parallel. Using this convention, the normal stress components on the faces of the element are designated with repeated subscripts, τ_{xx}, τ_{yy}, and τ_{zz}. The two shear or tangential stress components on each face are designated by unlike subscripts. For example, τ_{xy} refers to a stress component acting in the y direction on a face normal to the x axis. Figure 3–8 shows the use of this convention to designate the stresses acting on three faces of the cubical element. The three faces on which the stress components are shown are considered to be *positive faces* since they are on the side of the cube corresponding to a positive coordinate direction.

A standard convention for stress sign is followed. On the positive face of the element the stress is positive in a positive coordinate direction. On the sides of the cube facing the negative coordinate direction a positive stress is in a negative coordinate direction. Figure 3–8 shows the stress components on the positive faces drawn in the positive direction.

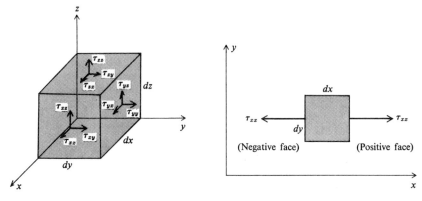

Fig. 3–8 Convention for stress components.

Fig. 3–9 Normal stress components on the x faces.

Since each stress component may vary with position in the fluid, the value of a particular stress component may be different on opposite parallel faces. For example, the normal component on the x faces of the cube, τ_{xx}, may be different on the positive and negative face (Fig. 3–9).

If the value of a stress component is specified for one face of the infinitesimal fluid element, then the value of that same type of stress component at the opposite face of the element can be derived in terms of the first stress component. This is done by using a Taylor series expansion about the point of interest and dropping terms of the order of $(dx)^2$ and higher. For a function of x, y, and z, $f(x,y,z)$, the Taylor Series is

$$f(x + dx, y, z) = f(x,y,z) + \frac{\partial f}{\partial x} dx + \frac{\partial^2 f}{\partial x^2} \frac{(dx)^2}{2!} + \cdots$$

The function f in this case is τ_{xx}. Thus, the stress components shown in Fig. 3–9 can be expressed as τ_{xx} on the left face and $[\tau_{xx} + (\partial \tau_{xx}/\partial x) dx]$ on the right face (Fig. 3–10). As the dimensions of the element, and thus dx, are

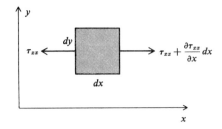

Fig. 3–10 Normal stress components expressed in terms of value at the left face.

made infinitesimally small, the left and right faces of the element approach each other. In the limit as the left and right faces approach each other, τ_{xx} becomes the normal x-directed stress component at the point of interest.

It can be seen that the surface stress at a point in a fluid can be described in terms of nine quantities, made up of three components acting on each of the three pairs of faces of an infinitesimally small fluid element assumed to be located at that point. These nine quantities are sometimes referred to as the *stress tensor*. The stress tensor is a quantity that describes the surface stress at each point in a fluid at each instant of time. In rectangular coordinates the stress tensor is described by the following *stress matrix:*

$$\begin{pmatrix} \tau_{xx} & \tau_{xy} & \tau_{xz} \\ \tau_{yx} & \tau_{yy} & \tau_{yz} \\ \tau_{zx} & \tau_{zy} & \tau_{zz} \end{pmatrix} \qquad (3\text{-}16)$$

The six shear stress components are not entirely independent. By considering the cubical element and taking moments about each of the three coordinate axes, it can be shown that

$$\tau_{xy} = \tau_{yx}, \qquad \tau_{xz} = \tau_{zx}, \quad \text{and} \quad \tau_{yz} = \tau_{zy}. \qquad (3\text{-}17)$$

Thus, there are only three independent shear stress components at any point in the general case.

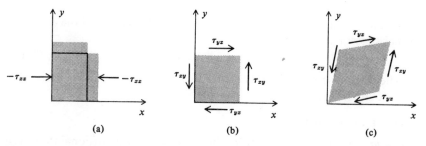

Fig. 3-11 (a) Distortion of element due to normal stress. (b) Element before distortion by shear. (c) Element after distortion by shear.

Normal forces tend to change the *size* of a fluid element of fixed mass. If the local normal force components are unequal, for example if $\tau_{xx} \neq \tau_{yy}$, they will also tend to change the *shape* of the element from a cube to a rectangular prism, as in Fig. 3-11(a).

Shear forces always tend to distort the shape of the element. This shearing distortion, shown in Figs. 3-11(b) and 3-11(c), changes the shape of the cubical element of fixed mass to one having sides parallel but no longer perpendicular. Stresses which change the shape of an element are referred to as *deviatoric stresses*.

The mean of the normal stresses is defined as the fluid pressure p for the general case, where a minus is necessary as a sign convention:

$$-p = \frac{\tau_{xx} + \tau_{yy} + \tau_{zz}}{3}. \tag{3-18}$$

This definition of pressure is exactly the same as in the case of static fluids. In the case of fluids in motion, the definition is an arbitrary but convenient one. The quantity p at some point in a flow at a particular time does not depend upon the type of coordinate system used to define the point. For example, if the pressure as defined above is determined at some point in a flow field described by rectangular coordinates, the value of p would not differ if the point were to be described by cylindrical coordinates. A quantity which exhibits this kind of behavior is referred to as an *invariant* in the notation of the fluid stresses.

The normal stresses τ_{xx}, τ_{yy}, and τ_{zz} are usually divided into two parts, the mean normal stress and the deviatoric normal stress (the portion of the normal stress which tends to change the shape of the element). The deviatoric normal stress is simply the deviation of that particular normal stress from the mean value:

$$\tau_{xx} = -p + \tau'_{xx}, \quad \tau_{yy} = -p + \tau'_{yy}, \quad \text{and} \quad \tau_{zz} = -p + \tau'_{zz}. \tag{3-19}$$

It can be seen from Eqs. (3–18) and (3–19) that the sum of the deviatoric normal stresses is zero:

$$\tau'_{xx} + \tau'_{yy} + \tau'_{zz} = 0 \tag{3-19a}$$

The deviatoric normal stresses are a useful convenience in the derivations which follow.

Stokes' Law of Friction

Stokes' law of friction states that the fluid stresses are proportional to the rate of change of strain. Fluids having this behavior are said to be *Stokesian* in nature, and many practical situations approximate this case. By applying Stokes' law to an infinitesimally small three-dimensional element, it can be shown that the relationships between the local deviatoric stresses and the local velocity gradients in the fluid are, for rectangular coordinates [2]:

$$\begin{aligned}
\tau'_{xx} &= \mu\left(2\frac{\partial u}{\partial x} - \frac{2}{3}\,\text{div }\mathbf{V}\right), & \tau_{xy} = \tau_{yx} &= \mu\left(\frac{\partial u}{\partial y} + \frac{\partial v}{\partial x}\right), \\
\tau'_{yy} &= \mu\left(2\frac{\partial v}{\partial y} - \frac{2}{3}\,\text{div }\mathbf{V}\right), & \tau_{xz} = \tau_{zx} &= \mu\left(\frac{\partial u}{\partial z} + \frac{\partial w}{\partial x}\right), \\
\tau'_{zz} &= \mu\left(2\frac{\partial w}{\partial z} - \frac{2}{3}\,\text{div }\mathbf{V}\right), & \tau_{yz} = \tau_{zy} &= \mu\left(\frac{\partial v}{\partial z} + \frac{\partial w}{\partial y}\right).
\end{aligned} \tag{3-20}$$

3-4 FLUID STRESS AND ENERGY CONVERSION

The term div **V** is called the *divergence* of **V** and is defined in rectangular coordinates by

$$\text{div } \mathbf{V} = \frac{\partial u}{\partial x} + \frac{\partial v}{\partial y} + \frac{\partial w}{\partial z}. \tag{3-20a}$$

Stokes' law of friction may be written in cylindrical coordinates:

$$\tau'_{rr} = \mu\left(2\frac{\partial u_r}{\partial r} - \frac{2}{3}\text{div } \mathbf{V}\right),$$

$$\tau'_{\theta\theta} = \mu\left[2\left(\frac{1}{r}\frac{\partial u_\theta}{\partial \theta} + \frac{u_r}{r}\right) - \frac{2}{3}\text{div } \mathbf{V}\right],$$

$$\tau'_{zz} = \mu\left(2\frac{\partial u_z}{\partial z} - \frac{2}{3}\text{div } \mathbf{V}\right),$$

$$\tau_{r\theta} = \tau_{\theta r} = \mu\left[r\frac{\partial}{\partial r}\left(\frac{u_\theta}{r}\right) + \frac{1}{r}\frac{\partial u_r}{\partial \theta}\right],$$

$$\tau_{\theta z} = \tau_{z\theta} = \mu\left(\frac{\partial u_\theta}{\partial z} + \frac{1}{r}\frac{\partial u_z}{\partial \theta}\right),$$

$$\tau_{zr} = \tau_{rz} = \mu\left(\frac{\partial u_z}{\partial r} + \frac{\partial u_r}{\partial z}\right),$$

where

$$\text{div } \mathbf{V} = \frac{1}{r}\frac{\partial}{\partial r}(ru_r) + \frac{1}{r}\frac{\partial u_\theta}{\partial \theta} + \frac{\partial u_z}{\partial z}.$$

In spherical coordinates the relations for Stokes' law of friction become:

$$\tau'_{rr} = \mu\left(2\frac{\partial u_r}{\partial r} - \frac{2}{3}\text{div } \mathbf{V}\right),$$

$$\tau'_{\theta\theta} = \mu\left[2\left(\frac{1}{r}\frac{\partial u_\theta}{\partial \theta} + \frac{u_r}{r}\right) - \frac{2}{3}\text{div } \mathbf{V}\right],$$

$$\tau'_{\varphi\varphi} = \mu\left[2\left(\frac{1}{r\sin\theta}\frac{\partial u_\varphi}{\partial \varphi} + \frac{u_r}{r} + \frac{u_\theta}{r}\cot\theta\right) - \frac{2}{3}\text{div } \mathbf{V}\right],$$

$$\tau_{r\theta} = \tau_{\theta r} = \mu\left[r\frac{\partial}{\partial r}\left(\frac{u_\theta}{r}\right) + \frac{1}{r}\frac{\partial u_r}{\partial \theta}\right],$$

$$\tau_{\theta\varphi} = \tau_{\varphi\theta} = \mu\left[\frac{\sin\theta}{r}\frac{\partial}{\partial \theta}\left(\frac{u_\varphi}{\sin\theta}\right) + \frac{1}{r\sin\theta}\frac{\partial u_\theta}{\partial \varphi}\right],$$

$$\tau_{\varphi r} = \tau_{r\varphi} = \mu\left[\frac{1}{r\sin\theta}\frac{\partial u_r}{\partial \varphi} + r\frac{\partial}{\partial r}\left(\frac{u_\varphi}{r}\right)\right],$$

where

$$\text{div } \mathbf{V} = \frac{1}{r^2}\frac{\partial}{\partial r}(r^2 u_r) + \frac{1}{r\sin\theta}\frac{\partial}{\partial \theta}(u_\theta \sin\theta) + \frac{1}{r\sin\theta}\frac{\partial u_\varphi}{\partial \varphi}.$$

The factor μ in the above equations is the proportionality factor between the local shearing stress and the local rate of strain. Note that in Eq. (3–20) the velocity gradients such as $\partial u/\partial z$ represent rates of strain. The term μ is defined by these equations and is commonly called by several names such as *coefficient of viscosity*, *dynamic viscosity*, *absolute viscosity*, or, more briefly, the *viscosity*. Viscosity is related to the ability of a fluid to diffuse momentum on a molecular scale.

Viscosity is a fluid property which varies with temperature and pressure. The dependence on pressure in usually small. Values of viscosity for various fluids are given in the appendix. All real fluids have a coefficient of viscosity which is greater than zero and are, therefore, referred to as viscous fluids. In *ideal fluids*, which are defined as those having a viscosity coefficient equal to zero, the only stress present is the fluid pressure. Viscosity is necessary in order for a fluid to sustain a shearing stress.

In Eq. (3–20), a quantity called the second viscosity coefficient or the *bulk coefficient of viscosity* has been neglected. This quantity is identically zero for monatomic gases and is negligibly small for most other fluids of engineering interest. In studies of acoustical vibrations in liquids and shock-wave phenomena, the bulk viscosity is important; where incompressible flow is assumed, the terms containing this quantity may be neglected.

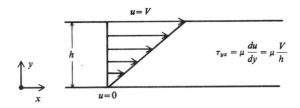

Fig. 3–12 Parallel flow between infinite plates, showing constant shear stress.

Newton's Law of Viscosity

In the special case of parallel flow where only one velocity component (such as u) exists and no velocity gradient exists in the flow direction ($\partial u/\partial x = 0$), then the Stokes' relations, Eq. (3–20), reduce to

$$\tau_{yx} = \mu \frac{du}{dy}. \qquad (3\text{–}21)$$

Note the similarity in mathematical form between Eqs. (3–21) and the Fourier equation, Eq. (1–16). This similarity has important consequences as will be seen later in the study of analogies.

A flow situation like that described in Eq. (3–21) is shown in Fig. 3–12, where two infinite parallel plates are separated by a fluid. The upper plate is assumed to move at a velocity V relative to the lower plate. If the viscosity is constant, the velocity profile is linear according to Eq. (3–21), since a constant shear stress τ_{xy} exists in the fluid.

Equation (3–21) is often referred to as *Newton's law of viscosity* after Sir Isaac Newton. This relation is often used to define viscosity and is the basis for one method of experimental determination of viscosity. Fluids which obey Eq. (3–21) are often called *Newtonian fluids*. For a brief discussion of non-Newtonian fluids, see Section 17–2.

When the shearing stress is given in units of pounds force per square foot and the strain rate in units of feet per second per foot, then the coefficient of viscosity would have the units of pounds force-second per square foot as indicated below:

$$\tau \equiv \mu \frac{du}{dy},$$

$$\mu \equiv \frac{\tau}{du/dy} \equiv \frac{lb_f/ft^2}{(ft/sec)/ft} \equiv \frac{lb_f \, sec}{ft^2}.$$

From the definition of g_c, Eq. (1–6),

$$1 \, lb_f \, sec/ft^2 = 32.2 \, lb_m/ft \, sec.$$

This relation can be used as a conversion factor for viscosity between the FLt and the MLt systems.

Example 3–1. Water at 100°F has a viscosity of $0.458 \times 10^{-3} \, lb_m/ft \, sec$. What is the value of the viscosity in the FLt system?

Solution

$$0.458 \times 10^{-3} \frac{lb_m}{ft \, sec} \times \frac{1}{32.2} \frac{lb_f \, sec^2}{lb_m \, ft} = 1.42 \times 10^{-5} \frac{lb_f \, sec}{ft^2}.$$

In the metric system a common unit of viscosity, the dyne sec/cm², is called a *poise*. Since the viscosity of water at 68.4°F is *one* centipoise (= 0.01 poise) the value of viscosity expressed in centipoise is an indication of the viscosity of a fluid relative to that of water at 68.4°F.

Energy Conversion in Fluid Flow

Whenever a fluid element is distorted or its size changed, energy conversions will occur. For example, work is always involved whenever a force acts over some distance. This work may be converted into some other form of energy; for example, it may result in an increase in the kinetic energy, potential energy, or in the internal energy (stored thermal energy) of the fluid element. A development of the equations for such energy conversion is needed for a good understanding of the process.

For this purpose, a fluid element is selected (Fig. 3–13) having dimensions dx by dy by dz. The element is assumed to be a fixed mass of fluid which is followed by the observer (Lagrangian viewpoint). The boundaries of the element in this case define a *closed thermodynamic system*, that is, no mass crosses the boundaries. This is in contrast to the control volume concept where mass is allowed to flow through a fixed volume in space. It is further assumed that no heat transfer occurs and, therefore, only work crosses the boundaries.

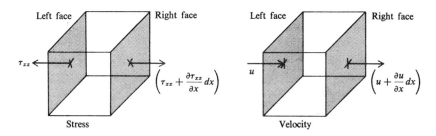

Fig. 3–13 Normal stress and x-velocity component on x faces.

The work done per unit time on a face of the element is the force on that face multiplied by the velocity component in the direction of that force. The normal stress in the x direction is assumed to be τ_{xx} at the left face and is, therefore, $[\tau_{xx} + (\partial \tau_{xx}/\partial x)\, dx]$ at the right face. The velocity is u at the left face and $[u + (\partial u/\partial x)\, dx]$ at the right face. Thus,

net rate of work on element on x faces due to normal stress

$$= \text{(force on left face)(velocity of left face)}$$
$$+ \text{(force on right face)(velocity on right face)} \quad (3\text{--}22)$$

$$= \left[-\tau_{xx} u + \left(\tau_{xx} + \frac{\partial \tau_{xx}}{\partial x}\, dx \right)\left(u + \frac{\partial u}{\partial x}\, dx \right) \right] dy\, dz.$$

By carrying out the multiplication indicated in Eq. (3–22) and by neglecting the higher order terms, the following equation results:

net rate of work on element on x faces due to normal stress

$$= \frac{\partial}{\partial x}(\tau_{xx} u)\, dx\, dy\, dz. \qquad (3\text{--}23\text{a})$$

In a similar manner the following can be derived:

net rate of work on element on x faces due to shear stress

$$= \frac{\partial}{\partial x}(\tau_{xy} v)\, dx\, dy\, dz + \frac{\partial}{\partial x}(\tau_{xz} w)\, dx\, dy\, dz. \qquad (3\text{--}23\text{b})$$

net rate of work on element on y faces due to normal stress

$$= \frac{\partial}{\partial y}(\tau_{yy}v)\,dx\,dy\,dz. \tag{3-23c}$$

net rate of work on element on y faces due to shear stress

$$= \frac{\partial}{\partial y}(\tau_{yx}u)\,dx\,dy\,dz + \frac{\partial}{\partial y}(\tau_{yz}w)\,dx\,dy\,dz. \tag{3-23d}$$

net rate of work on element on z faces due to normal stress

$$= \frac{\partial}{\partial z}(\tau_{zz}w)\,dx\,dy\,dz. \tag{3-23e}$$

net rate of work on element on z faces due to shear stress

$$= \frac{\partial}{\partial z}(\tau_{zx}u)\,dx\,dy\,dz + \frac{\partial}{\partial z}(\tau_{zy}v)\,dx\,dy\,dz. \tag{3-23f}$$

The net rate at which work is done on *all* six faces of the fluid element per unit volume and time is found by summing Eqs. (3-23a-f) and dividing by the volume of the element $(dx\,dy\,dz)$:

net work rate per unit volume acting on the fluid element due to surface force

$$= \frac{\partial}{\partial x}(u\tau_{xx} + v\tau_{xy} + w\tau_{xz}) + \frac{\partial}{\partial y}(u\tau_{yx} + v\tau_{yy} + w\tau_{yz}) \\ + \frac{\partial}{\partial z}(u\tau_{zx} + v\tau_{zy} + w\tau_{zz}). \tag{3-24}$$

In this general case, all of the nine possible fluid stress components are included. Equation (3-24) will be of use in developing the material in the next chapter.

3-5 LAMINAR AND TURBULENT FLOW

Basic to an understanding of the flow of fluids is the observation, first recorded by Osborne Reynolds [1], of two distinct modes or types of flow. These modes are called *laminar flow* and *turbulent flow*.

Laminar flow is also called *viscous flow* or *streamline flow*. In laminar flow there is no mixing of minute particles of fluid on a macroscopic scale. The only mixing which occurs is on a molecular level. In laminar flow the velocity at any point is either constant or varies with time in some nonrandom manner. If a small stream of dye is inserted into a laminar flow, as was first done in 1883 by Reynolds, the dye flows from the source in a continuous, unbroken line with very little mixing of the dye with the surrounding fluid (Fig. 3-14a).

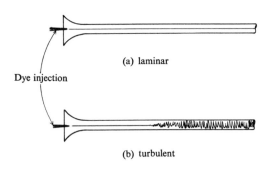

Fig. 3-14

In turbulent flow, however, there is a definite mixing in the fluid on a macroscopic basis, and the local velocity varies with time in a random fashion. Small fluid particles are mixed by the velocity fluctuations, which occur both parallel and transverse to the mean flow. If a small stream of dye is inserted into a turbulent flow, the dye disperses rapidly from the source in a disorderly manner and soon mixes with the surrounding fluid (Fig. 3-14b).

Smoke rising from a cigarette in still air offers an opportunity for observation of both laminar and turbulent flow. The smooth, orderly flow of smoke near the cigarette is laminar. This orderly flow breaks up after a short distance into disorderly, fluctuating turbulent flow.

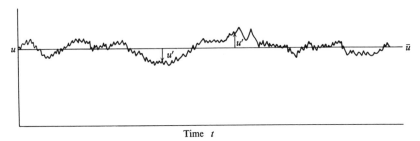

Fig. 3-15 Variation of a velocity component with time.

A typical plot of the variation of the velocity component u with time in a turbulent fluid at a given location is shown in Fig. 3-15. The *instantaneous velocity component* u is shown with a random variation with time about a *mean velocity component* $\bar{u}$. At any instant of time the difference between the instantaneous velocity component and the mean velocity component is referred to as the *fluctuating velocity component*, which is designated by a prime. Thus the three components of velocity in turbulent flow may be written

$$u = \bar{u} + u' \quad v = \bar{v} + v' \quad w = \bar{w} + w'. \quad (3\text{-}25)$$

The fluctuating component can be seen from Fig. 3-15 to have both positive and negative values. From the definition of the fluctuating component, the time average of these values is zero. Thus,

$$\overline{u'} = \frac{1}{t_2 - t_1} \int_{t_1}^{t_2} u' \, dt = 0, \quad \text{also} \quad \overline{v'} = 0, \quad \text{and} \quad \overline{w'} = 0. \quad (3\text{--}26)$$

The time interval $(t_2 - t_1)$ used to define an average must be sufficiently large to obtain a true average.* Since the time average of the fluctuating components is zero in each case, it is common practice to describe the magnitude of the fluctuating component by a root mean square value.

$$u'(\text{RMS}) = \sqrt{\overline{u'^2}}. \quad (3\text{--}27)$$

The root mean square value of *all three* fluctuating components will be *nonzero* in turbulent flow, even in those cases where one or two of the mean velocity components are zero (such as one- and two-dimensional turbulent flow).

The *intensity of turbulence* is defined in terms of the fluctuating components,

$$\text{intensity of turbulence} = \frac{\sqrt{\tfrac{1}{3}(\overline{u'^2} + \overline{v'^2} + \overline{w'^2})}}{|\mathbf{V}|}, \quad (3\text{--}28)$$

where $\mathbf{V}$ is some fixed reference velocity. The intensity of turbulence is sometimes called the *relative intensity, degree of turbulence,* or *turbulence level*. It indicates the relative magnitudes of the fluctuating velocity components compared to a mean reference velocity, usually the "undisturbed" stream velocity.

Many fluid flow and heat transfer problems are studied experimentally in wind tunnels. The intensity of turbulence varies widely in different wind tunnels. An average value of intensity in a wind tunnel is probably around 0.001. It is difficult to obtain intensities below 0.0002 in wind tunnels.

It will be shown that the intensity of turbulence is an important consideration in the study of turbulent flows and turbulent heat transfer, sometimes having a significant effect on the fluid behavior.

The pressure and temperature in a turbulent fluid can be considered in the same manner as the velocity components. Thus,

$$T = \bar{T} + T' \quad \text{and} \quad p = \bar{p} + p'. \quad (3\text{--}29)$$

The methods for predicting whether a flow is laminar or turbulent will be described later. It will be found that turbulent flow is more common than laminar flow in most practical engineering situations. However, laminar flow is simpler to analyze mathematically and occurs in some very important cases.

---

* In *unsteady* turbulent flow, $\bar{u}$ may vary with time.

3-6 GENERAL APPROACH TO PROBLEMS

In attempting to predict the behavior of a proposed design in engineering, several approaches are possible.

One approach is to use *empirical equations* based upon previous experience in similar situations. This approach to a problem is quite satisfactory for situations which do not differ greatly from those used to develop the empirical information. The empirical approach must be used in some cases where no other method is available. The shortcoming of this method is that it may not be used with any degree of confidence in situations that are entirely new or different.

In design situations where no previous experience in the proposed area is available, the engineer may choose to perform laboratory experiments to develop the needed information. In such cases the use of *dimensional analysis* is most helpful. It permits the use of experimental equipment of a scale which differs from actual equipment and also permits a reduction in the number of measurements required. An example of this method is the use of models in a wind tunnel to predict the behavior of full-scale aircraft. The method has limitations in that not all desired conditions can be simulated in the laboratory and not all physical phenomenon can be easily scaled. Nevertheless, the method is powerful and very commonly used.

Another method for predicting behavior of a proposed engineering system is by the use of *analogies.* An analogy is a similarity in some respects between things otherwise unlike. Some engineering systems behave in some ways similar to an entirely different type of system. For example, the flow of electricity and the flow of heat in a solid conductor are mathematically similar. The engineer may study the behavior of a certain selected electrical system and predict the behavior of a similar thermal system by analogy. This method is particularly useful since electrical circuits are easily assembled and electrical measurements may be easily made.

The *analytical approach* often yields valuable results. The physical laws governing the engineering system may be applied and predictions made from these laws. The reliability of this method is limited only by the number of simplifying assumptions made and by the mathematical techniques used to obtain desired information. This method is particularly useful in predicting behavior of physical systems where no experience is available, and where the analogies cannot be applied. In fluid mechanics and heat transfer, the physical laws which are important in predicting behavior may be classified as follows:

1. Conservation of mass
2. Conservation of momentum
3. Conservation of energy (first law of thermodynamics)

4. Second law of thermodynamics
5. Phenomenological equations

A partial list of some phenomenological laws includes:

1. Equation of state (for gases)
2. Newton's viscosity law
3. Hooke's law (for elastic solids)
4. Fourier's heat conduction law
5. Stefan-Boltzmann's radiation law
6. Newton's law of cooling
7. Fick's law of diffusion
8. Ohm's law (for electrical conductors)

A solution to a problem may be obtained by formulating the above laws as they apply to control volumes or systems, and deriving suitable equations. In some cases these laws are expressed in terms of differential equations, valid at each point in the substance under study. By appropriate simplification of the equations and by application of suitable boundary and initial conditions, the equations may be solved for the desired information. The solution in some cases may be approximate in nature; for example, certain differential equations yield only to numerical solutions.

The next chapter is devoted to the development of the conservation equations in forms useful for the study of fluid mechanics and heat transfer. These conservation equations and appropriate phenomenological equations will be used extensively throughout the remainder of the text. In situations where the analytical approach cannot be applied conveniently, the conservation equations will be found useful in the development of empirical equations, dimensional analysis, and analogies.

REFERENCES

1. O. Reynolds, "An Experimental Investigation of the Circumstances which Determine whether the Motion of Water Will Be Direct or Sinuous, and of the Law of Resistance in Parallel Channels," *Phil. Trans. Roy. Soc.*, London, **174A**, 935 (1883).

2. J. O. Hinze, *Turbulence*, McGraw-Hill, New York (1959).

3. S. Corrisin, S. J. Kline, and A. H. Shapiro, "The Occurrence of Turbulence," National Committee for Fluid Mechanics Films, Encyclopaedia Britannica Educ. Corp., Chicago.

4. J. Lumley, "Deformation of Continuous Media," National Committee for Fluid Mechanics Films, Encyclopaedia Britannica Educ. Corp., Chicago.

PROBLEMS

1. The following points are located in a rectangular coordinate system. Determine their coordinates in a cylindrical and in a spherical coordinate system oriented as shown in Figs. 3–2 and 3–3.

	x	y	z
a)	3	5	4
b)	4	4	6
c)	−3	2	8
d)	8	−12	10

2. Determine the velocity magnitude for the following velocity components (given in ft/sec):
   a) $u = 15, v = 0, w = 10$.
   b) $u = 20, v = 25, w = 50$.
   c) $u_z = 10, u_r = 5, u_\theta = 0$.
   d) $u_r = 25, u_\theta = 0, u_\varphi = 0$.

3. A pressure transducer is lowered vertically into a tank of oil ($\rho = 56.0 \text{ lb}_m/\text{ft}^3$) at the rate of 100 ft/min. What is the rate of change of pressure with time sensed by the transducer in psi per second?

4. Fluid flows with an average velocity of 20 ft/sec down a tube. The pressure *drop* in the direction of flow is uniform and equal to 0.16 psi per foot of length. What is the rate of change of pressure with respect to time for an observer moving with the fluid? What is the rate of change of pressure with time for an observer moving upstream at a velocity of 10 ft/sec?

5. Derive the expression for the substantial derivative in cylindrical coordinates.

6. Derive the expression for the substantial derivative in spherical coordinates.

7. A two-dimensional velocity field is described by

   $$V = 2x^2 y \mathbf{i} + 2xy^2 \mathbf{j} \text{ ft/sec.}$$

   Find the total acceleration at $x = 2$ and $y = 3$.

8. A shaft of 1 in. outside diameter is turning at 1000 rpm inside a bearing. A thin film of lubricant (0.005 in. thick) is between the shaft and the bearing. Assume

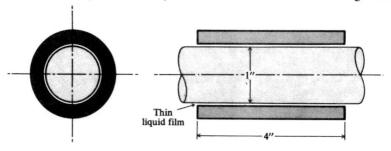

that the film is thin compared with the radius of the shaft so that Newton's law of friction is valid in the lubricant film, and that the flow pattern is similar to that shown in Fig. 3–12. Calculate the torque necessary to overcome the fluid friction. Assume that the viscosity of the lubricant is 0.002 lb_m/ft sec.

9. Change the following units of viscosity to equivalent values having units of lb_m/ft sec.
    a) 0.63 centipoise
    b) 8.45 lb_m/ft hr
    c) 5×10^{-5} lb_f sec/ft^2
    d) 7.6 dynes sec/cm^2

CHAPTER 4

THE CONSERVATION EQUATIONS

Everyone in his daily life uses principles of conservation to account for various related events that take place during a given time interval. These multifarious events range from the drudgery of balancing one's checking account (conservation of monetary value) to the painful shrinking of one's waistline by counting calories (conservation of mass and energy). The purpose of this chapter is to develop the basic equations for conservation of mass, momentum, and energy.

4-1 CONSERVATION OF MASS

Integral Conservation Equation

Before deriving a specific conservation relation, it is desirable to first derive a general overall balance equation. This equation will then be specialized for mass, momentum, and energy. The conservation principles may be applied to either a control volume or a system. The *control volume* is any volume in space set aside for study. It can be rigid (i.e., a jet engine) or nonrigid (i.e., a balloon being filled), and it may or may not be stationary. The boundary of a control volume is called the *control surface*. A *system* is any specified amount of mass that is set aside for study. A system may change energy content, shape, and position, but it always has the same amount of mass. Both the control volume approach and the system approach are useful and each has its advantages. The choice of which approach to use usually depends upon the problem under consideration. A useful relationship between a control volume and the system contained within the control volume at some instant of time will now be developed.

Let B represent some *extensive property* (a property proportional to mass) of a system. Examples of an extensive property include energy, momentum, volume, and weight. Let b represent B per unit mass (hence b is a specific property such as volume per unit mass or energy per unit mass). Thus for a system occupying a certain volume,

$$B = \iiint b\rho \, d(\text{Vol}),$$

where the triple integral stands for integration over the entire volume.

4-1 CONSERVATION OF MASS

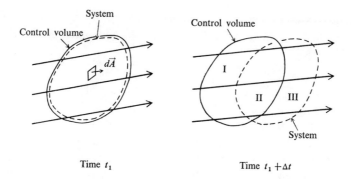

Fig. 4–1 Illustration of system and control volume.

Suppose a system and control volume are coincident at time t_1, as in Fig. 4–1. The rate of change of B in the control volume is given by:

$$\frac{\partial B_{c.v.}}{\partial t} = \frac{\partial}{\partial t}\iiint_{c.v.} b\rho \, d(\text{Vol}) = \lim_{\Delta t \to 0} \frac{B(t_1 + \Delta t)_{c.v.} - B(t_1)_{c.v.}}{\Delta t}. \quad (4\text{–}1)$$

Since the system is a fixed amount of mass being observed, the rate of change of B with time for the system is described by the substantial derivative (Section 3–3). Thus, the rate of change of B of the system is

$$\left(\frac{DB}{Dt}\right)_{sys} = \lim_{\Delta t \to 0} \frac{B(t_1 + \Delta t)_{sys} - B(t_1)_{sys}}{\Delta t}. \quad (4\text{–}2)$$

The system is moving with the fluid velocity $\mathbf{V}$. The control volume is free to deform to any arbitrary shape and it can move with any arbitrary velocity. From Fig. 4–1 it can be seen that $B_{III}(t_1 + \Delta t)_{sys}$ represents the efflux of B out of the control volume in time Δt and that $B_I(t_1 + \Delta t)_{c.v.}$ represents the influx of B into the control volume in time Δt. Consider a small element of area dA of the control surface. Its orientation and size is described by a vector $d\mathbf{A}$ as shown in Fig. 4–1. Using the convention of a positive $d\mathbf{A}$ pointing outward from the control volume, then $(b\rho\mathbf{V}_r \cdot d\mathbf{A})$ is the outward efflux of B through $d\mathbf{A}$ per unit time, where $\mathbf{V}_r$ is the relative velocity of the fluid as seen by a point on the surface of control volume:

$$\mathbf{V}_r = \mathbf{V} - \mathbf{V}_c.$$

$\mathbf{V}_c$ is the local velocity of the control surface at a given point and $\mathbf{V}$ is the local fluid velocity. The net rate of efflux through the control volume is obtained by integrating the local efflux of B over the entire control surface:

$$\iint_{c.s.} b(\rho\mathbf{V}_r \cdot d\mathbf{A}) = \lim_{\Delta t \to 0} \frac{B_{III}(t_1 + \Delta t)_{sys} - B_I(t_1 + \Delta t)_{c.v.}}{\Delta t}. \quad (4\text{–}3)$$

Equations (4–1), (4–2), and (4–3) can now be combined, recognizing that $B(t_1)_{\text{c.v.}} = B(t_1)_{\text{sys}}$ and that the limit of a sum is equal to the sum of the limits. The result is

$$\left(\frac{DB}{Dt}\right)_{\text{sys}} = \frac{\partial}{\partial t} \iiint_{\text{c.v.}} b\rho \, d(\text{Vol}) + \iint_{\text{c.s.}} b(\rho \mathbf{V}_r \cdot d\mathbf{A}). \tag{4–4}$$

Equation (4–4) is the integral conservation equation for finite size control volumes. It relates the rate of change of an extensive property, such as B, of a system at some time t_1 to the changes for a control volume which is coincident at that same time. Equation (4–4) will be used to derive important equations describing conservation of mass, momentum, and energy. It is important to remember that the velocity $\mathbf{V}_r$ in Eq. (4–4) is the velocity of the fluid relative to the boundary of the control volume.

Conservation of Mass for a Finite System

One of the basic physical laws useful in predicting fluid behavior is the *law of conservation of mass*, which states that mass can be neither created nor destroyed. An exception occurs when mass is converted into energy or energy into mass. Since the change in mass in these situations is usually very small, such energy-mass conversions will not be considered in this text.

When chemical reactions or changes of phase occur in a fluid, the principle of conservation of mass may be applied to each chemical species or to each phase. The equations which result must include terms to account for the change between chemical species or between phases. Equations describing conservation of mass will now be derived for a single-phase substance with no chemical change. This will be done by first developing the overall mass balance equations and then considering the equations in differential form.

The overall mass balance equation is developed from Eq. (4–4). Letting $B = m =$ mass of a fluid, which is an extensive variable, the corresponding intensive variable b is then equal to unity ($b = 1$), since b is defined to be B per unit mass. Applying the principle of conservation of mass, the mass of an arbitrary system will not change by definition. Substituting $b = 1$ into Eq. (4–4), the overall mass balance equation is

$$\frac{Dm}{Dt} = 0 = \iint_{\text{c.s.}} \rho \mathbf{V}_r \cdot d\mathbf{A} + \frac{\partial}{\partial t} \iiint_{\text{c.v.}} \rho \, d(\text{Vol}),$$

or

$$\iint_{\text{c.s.}} \rho \mathbf{V}_r \cdot d\mathbf{A} = -\frac{\partial}{\partial t} \iiint_{\text{c.v.}} \rho \, d(\text{Vol}). \tag{4–5}$$

In effect this equation says that the net flow rate of mass out of a control surface (the left-hand term) is equal to the rate of decrease of mass in the control volume (the right-hand term).

4-1 CONSERVATION OF MASS

For steady flow there is no net flow rate of mass across the control surface, and Eq. (4–5) reduces to

$$\iint_{c.s.} \rho \mathbf{V}_r \cdot d\mathbf{A} = 0. \tag{4-6}$$

For incompressible flow, $\rho = $ constant, and

$$\iint_{c.s.} \mathbf{V}_r \cdot d\mathbf{A} = 0. \tag{4-7}$$

The application of the overall mass balance equation can be illustrated by several examples.

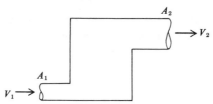

Fig. 4–2 Example 4–1.

Example 4–1. Figure 4–2 shows fluid entering a device at the left and leaving at the right. Assume that the inlet flow and the outlet flow are steady relative to the control volume and assume that ρ and V are constant over the inlet and outlet area.

Solution. Applying Eq. (4–6) for this case,

$$\iint_{c.s.} \rho \mathbf{V}_r \cdot d\mathbf{A} = \iint_{A_1} \rho \mathbf{V}_r \cdot d\mathbf{A} + \iint_{A_2} \rho \mathbf{V}_r \cdot d\mathbf{A} = 0.$$

Since ρ and V are constant at both inlet and outlet,

$$\iint_{c.s.} \rho V_r \cdot d\mathbf{A} = \rho_1 \mathbf{V}_1 \cdot \iint_{A_1} d\mathbf{A} + \rho_2 \mathbf{V}_2 \cdot \iint_{A_2} d\mathbf{A} = 0,$$

or

$$\rho_1 \mathbf{V}_1 \cdot \mathbf{A}_1 + \rho_2 \mathbf{V}_2 \cdot \mathbf{A}_2 = 0.$$

$\mathbf{V}_1$ and $\mathbf{A}_1$ are parallel but have opposite directions and $\mathbf{V}_2$ and $\mathbf{A}_2$ are parallel and have the same directions. Then using the rules for vector dot product,

$$-\rho_1 V_1 A_1 + \rho_2 V_2 A_2 = 0,$$

or

$$\rho_1 V_1 A_1 = \rho_2 V_2 A_2.$$

Using Eq. (4–6) to solve this example problem is like using an elephant gun to hunt fleas. A greatly simplified approach could have been used. The method of reasoning, however, will be helpful in the solution of more complicated problems.

The quantity ρVA is the mass rate of flow $\dot{m}$ crossing the cross-sectional area A normal to the velocity. In some cases of interest, the velocity V might vary across the cross section.

In the flow through a pipe, for example, the velocity is not uniform across the pipe cross section but varies from zero at the wall to some maximum value at the centerline. If any small area dA normal to the flow is arbitrarily chosen, then the mass flow rate through that area is

$$d\dot{m} = \rho u \, dA, \tag{4–8}$$

where ρ and u are the local density and local velocity normal to the area. The total mass flow rate across the entire pipe cross section is obtained by integration of Eq. (4–8) over the total area of the cross section:

$$\dot{m} = \int_A \rho u \, dA. \tag{4–9}$$

The average velocity is that velocity which is related to the total mass rate of flow through a cross-section area A by the expression

$$\dot{m} = \rho u_{\text{avg}} A. \tag{4–10}$$

Combining Eqs. (4–9) and (4–10), assuming the density is constant,

$$u_{\text{avg}} = \frac{1}{A} \int_A u \, dA. \tag{4–11}$$

Example 4–2. Fluid with constant density flows through a round tube at a steady rate. The velocity distribution across the tube at a particular cross section, $u(r)$, is

$$\frac{u(r)}{u_{\text{max}}} = 1 - \left(\frac{r}{R}\right)^2,$$

where R is the radius of the tube, r is the distance from the tube axis, and $u_{\text{max}} = 60$ ft/sec. Calculate the average velocity of the fluid at that cross section.

Solution.

$$u_{\text{avg}} = \frac{1}{A} \int_A u \, dA = \frac{1}{\pi R^2} \int_0^R u_{\text{max}} \left(1 - \left(\frac{r}{R}\right)^2\right) 2\pi r \, dr$$

$$= \frac{(2)(60)}{R^2} \int_0^R \left(r - \frac{r^3}{R^2}\right) dr = \frac{120}{R^2} \left(\frac{R^2}{2} - \frac{R^2}{4}\right) = 30 \text{ ft/sec}.$$

It can be seen that

$$u_{\text{avg}} = \tfrac{1}{2} u_{\text{max}}.$$

For other velocity profiles a different relation between u_{avg} and u_{max} would exist.

It is the average velocity of flow in a conduit that is measured by flow measuring devices such as venturi and orifice flow meters (see Chapter 6, Problems 15, 16, and 17). Another type of flow measuring device, the turbine meter, also measures the average velocity; its speed of rotation being proportional to the total volume flow. A turbine meter is shown below in Fig. 4–3.

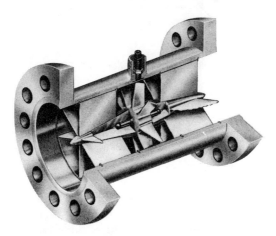

Fig. 4–3 A cut-away view of a turbine flowmeter (Courtesy of Halliburton Company).

Conservation of Mass in Differential Form

For solutions requiring knowledge of *local* conditions, in other words, knowledge of the value of properties at specified points in the fluid itself, conservation equations are more useful when expressed in differential form. It is not difficult to show that Eq. (4–4) reduces to an identity for an infinitesimal volume. Thus the same differential equations will result whether a system approach or a control volume approach is used in these derivations.

The conservation of mass equations will now be derived for an infinitesimal control volume by use of a step-by-step accounting for all mass flowing into the volume (Fig. 4–4). Equation (4–5) could be used to obtain a conservation of mass equation for an infinitesimal control volume directly by use of Gauss' theorem, but the following detailed derivation may give the reader more insight into the resulting differential equation.

Assume a cubical volume of dimensions $(dx)(dy)(dz)$ with one corner located at some arbitrary point (x,y,z), described by a rectangular coordinate system in a fluid, as shown in Fig. 4–4.

74   THE CONSERVATION EQUATIONS

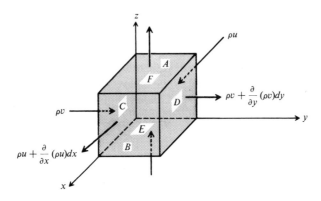

Fig. 4-4 Three-dimensional rectangular element, or control volume.

The control volume, which is fixed in size but infinitesimally small, will be considered fixed in space at a location where mass crosses the boundary. The flow of fluid into and out of the control volume is described by the velocity vector **V** which is variable with position x, y, z and with time t. The velocity components of **V** in the x, y, and z directions are described by the symbols u, v, and w respectively.

Expressions for the net rate at which mass flows across each of the six faces of the control volume will now be derived. With the orientation of the cubical element shown in Fig. 4-4, only one of the velocity components of **V** will be perpendicular to each face; the other two will lie in the plane of that face. The rate $\dot{m}_A$ at which mass flows across face A, which is perpendicular to the x axis and which passes through the origin, is

$$\dot{m}_A = (\text{density})(\text{velocity})(\text{area}) = \rho u \, dz \, dy.$$

The density and velocity are assumed to be appropriate average values across that face. The rate at which mass crosses face B (the parallel face a distance dx away from face A) can be expressed by using the Taylor's series expansion for a function about a point (x,y,z) as was done in Section 3-4. The mass flow rate across face B at $(x + dx)$ is (neglecting higher-order differentials):

$$\dot{m}_B = -\left\{ \rho u \, dy \, dz + \frac{\partial}{\partial x}[(\rho u)(dy \, dz)] \, dx \right\}.$$

The minus sign is used to indicate fluid leaving the element. Similarly the fluid flow rates across the faces perpendicular to the y axis are

$$\dot{m}_C = \rho v \, dx \, dz$$

and

$$\dot{m}_D = -\left\{ \rho v \, dx \, dz + \frac{\partial}{\partial y}[(\rho v)(dx \, dz)] \, dy \right\}.$$

The rates of flow across the two faces perpendicular to the z axis are

$$\dot{m}_E = \rho w \, dx \, dy$$

and

$$\dot{m}_F = -\left\{ \rho w \, dx \, dy + \frac{\partial}{\partial z}[(\rho w)(dx \, dy)] \, dz \right\}.$$

The net rate at which mass can be stored in the element due to density changes is

$$\dot{m}_S = \frac{\partial \rho}{\partial t} dx \, dy \, dz, \qquad (4\text{--}12)$$

where ρ is the average density of the fluid in the element (which becomes the density at a point as the element is made very small).

Equating the rate of mass storage to the net rate of inflow of mass,

$$\dot{m}_S = \dot{m}_A + \dot{m}_B + \dot{m}_C + \dot{m}_D + \dot{m}_E + \dot{m}_F. \qquad (4\text{--}13)$$

Substituting for the quantities in Eq. (4–13) and dividing each term by $(dx \, dy \, dz)$,

$$\frac{\partial \rho}{\partial t} = -\left[\frac{\partial(\rho u)}{\partial x} + \frac{\partial(\rho v)}{\partial y} + \frac{\partial(\rho w)}{\partial z} \right]. \qquad (4\text{--}14)$$

Equation (4–14), which satisfies the requirements of conservation of mass at any arbitrary position in a fluid, is referred to as the differential equation for conservation of mass, or the *continuity equation*, and is expressed here in rectangular coordinates. It must be satisfied at all points (x, y, z) within a fluid at all times. The equation can be simplified for the case of steady flow in which the density does not vary with time at any point in the fluid:

$$\frac{\partial(\rho u)}{\partial x} + \frac{\partial(\rho v)}{\partial y} + \frac{\partial(\rho w)}{\partial z} = 0. \qquad (4\text{--}15)$$

Where the density is constant,

$$\frac{\partial u}{\partial x} + \frac{\partial v}{\partial y} + \frac{\partial w}{\partial z} = 0. \qquad (4\text{--}16)$$

Equation (4–16) is sometimes written in terms of the divergence,

$$\text{div } \mathbf{V} = 0. \qquad (4\text{--}17)$$

The divergence was defined in Eq. (3–20a):

$$\text{div } \mathbf{V} = \frac{\partial u}{\partial x} + \frac{\partial v}{\partial y} + \frac{\partial w}{\partial z}. \qquad (4\text{--}18)$$

It is sometimes useful to rearrange the conservation of mass equation into a form describing the system viewpoint. Equation (4–14) can be expanded by taking the derivatives to give

$$\frac{\partial \rho}{\partial t} + u\frac{\partial \rho}{\partial x} + \rho\frac{\partial u}{\partial x} + v\frac{\partial \rho}{\partial y} + \rho\frac{\partial v}{\partial y} + w\frac{\partial \rho}{\partial z} + \rho\frac{\partial w}{\partial z} = 0. \quad (4\text{-}19)$$

Rearranging and using the definition of the substantial derivative and divergence,

$$\frac{D\rho}{Dt} = -\rho \text{ div } \mathbf{V}. \quad (4\text{-}20)$$

Equation (4–20) is a form of the conservation of mass equation which describes the variation of density with time as seen by an observer moving along with a particular element of fluid. This is descriptive of a system viewpoint as contrasted to the control volume viewpoint described by Eq. (4–14).

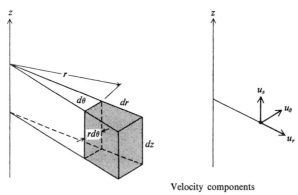

Fig. 4-5 Volume element in cylindrical coordinates.

Cylindrical Coordinates

A typical volume element for cylindrical coordinates is shown in Fig. 4–5 with dimensions $rd\theta$, dz, and dr. The continuity equation in cylindrical coordinates can be derived either by a mass balance on the element or by using a transformation of coordinates applied to Eq. (4–14). The resulting form of the continuity equation is

$$\frac{\partial \rho}{\partial t} + \frac{1}{r}\frac{\partial}{\partial r}(\rho r u_r) + \frac{1}{r}\frac{\partial}{\partial \theta}(\rho u_\theta) + \frac{\partial}{\partial z}(\rho u_z) = 0, \quad (4\text{-}21)$$

where u_z, u_r, and u_θ are the components of velocity in the z, r, and θ directions respectively, as shown in Fig. 4–5.

For the case of constant density the continuity equation in cylindrical coordinates is

$$\frac{\partial u_r}{\partial r} + \frac{u_r}{r} + \frac{1}{r}\frac{\partial u_\theta}{\partial \theta} + \frac{\partial u_z}{\partial z} = 0. \quad (4\text{-}22)$$

Spherical Coordinates

In spherical coordinates the continuity equation is

$$\frac{\partial \rho}{\partial t} + \frac{1}{r^2}\frac{\partial}{\partial r}(\rho r^2 u_r) + \frac{1}{r \sin \theta}\frac{\partial}{\partial \theta}(\rho u_\theta \sin \theta) + \frac{1}{r \sin \theta}\frac{\partial}{\partial \varphi}(\rho u_\varphi) = 0. \quad (4\text{-}23)$$

The conservation of mass equation in differential form involves three unknowns for the case of constant density and four unknowns for the case of variable density. Generally, solutions to the equation cannot be obtained independently of momentum and energy considerations, except for extremely simple cases. The value of the conservation of mass equation in differential form is in its use in conjunction with the differential equations representing conservation of momentum and energy. This usefulness will be illustrated in later chapters.

4–2 STREAMLINES

In describing fluid flow it is convenient to use the concept of streamlines. A *streamline* is a line in a fluid flow field which, at every point along the line, has the direction of the local velocity of the fluid at some instant of time. The streamlines form the *flow pattern* at any instant of time. If the flow is steady, then the streamlines describe the motion of individual fluid particles. For steady flow, a streamline by definition is a line across which no fluid flows.

The motion of an individual particle over a finite time interval is called its *pathline*. In steady flow the streamlines and pathlines are identical.

The concept of a streakline is also useful. A *streakline* is the current location of fluid particles, all of which passed through a fixed point in the fluid at some previous time. Injection of smoke, dye, or particles in a fluid stream reveal these streaklines. They are identical to streamlines for steady flow.

Streamlines are useful for at least two reasons: (1) They permit a description of a flow field in pictorial form, and (2) they permit simplification of the analytical description of two-dimensional flows. By using streamlines a vector field (velocity) may be described by a scalar function. Figure 4–6 shows the streamlines for flow in several simple, steady-flow situations. The path of the fluid particles is easily seen.

The use of the streamline to simplify the analytical treatment of two- and three-dimensional flows is accomplished by defining a coordinate system keyed to the streamlines. One coordinate of the system is always tangent to the streamline and the other two coordinates are normal. Since the fluid flows along (or tangent to) the streamline and never normal to it, the fluid motion is always one-dimensional in such a system. The velocity magnitude is given by $V = ds/dt$, where s is distance measured along the streamline.

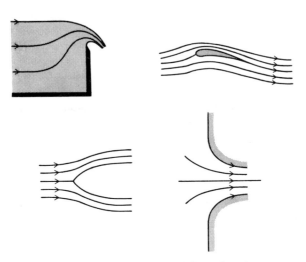

Fig. 4-6 Streamlines in typical flow situations.

It is important to note that, although no fluid flows normal to a streamline, there may exist a component of acceleration normal to the streamline due to change of direction of the velocity vector along the streamline. The magnitude of this normal acceleration is given by $a_n = V^2/r$, where r is the local radius of curvature of the streamline. The local acceleration of the fluid is the vectorial sum of the normal acceleration and the tangential acceleration, dV_s/dt.

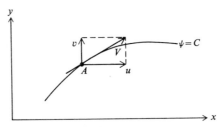

Fig. 4-7 Description of fluid velocity at a point on a streamline.

The streamline can be described in terms of the local fluid velocity components. Consider, for example, in rectangular coordinates the two-dimensional flow situation described by Fig. 4-7. In a coordinate system keyed to the streamline, the flow is one-dimensional. In a rectangular coordinate system, however, the flow is two-dimensional, having both x and y components.

At a point A in the figure, the local velocity **V** along the streamline can be described in terms of the two components of velocity u and v in the x and y

directions:

$$\mathbf{V} = \frac{d\mathbf{s}}{dt} = \mathbf{i}\frac{dx}{dt} + \mathbf{j}\frac{dy}{dt}, \qquad (4\text{-}24)$$

where

$$u = \frac{dx}{dt} \quad \text{and} \quad v = \frac{dy}{dt}. \qquad (4\text{-}25)$$

Both these equations can be combined since they describe the same fluid element and, therefore, the same time period, dt:

$$dt = \frac{dx}{u} = \frac{dy}{v} \quad \text{or} \quad -u\,dy + v\,dx = 0. \qquad (4\text{-}26)$$

Equation (4-26) is the differential equation of the two-dimensional streamline. From concepts of mathematics alone, it is known that Eq. (4-26) is an exact differential equation if

$$\frac{\partial u}{\partial x} = -\frac{\partial v}{\partial y}. \qquad (4\text{-}27)$$

However, Eq. (4-27) is known to be an equality since it is the equation of continuity for two-dimensional incompressible flow. Therefore, a solution is of the form $\psi = \psi(x,y)$, where the function ψ is called the *stream function*. The differential of the stream function ($d\psi$) is the exact differential of Eq. (4-26):

$$d\psi = u\,dy - v\,dx. \qquad (4\text{-}28)$$

For a given streamline, $d\psi = 0$, and each streamline is, therefore, a line representing a constant value of ψ.

From calculus, the expression for the total differential of the function ψ, which is a function of x and y, is

$$d\psi = \frac{\partial \psi}{\partial x}dx + \frac{\partial \psi}{\partial y}dy. \qquad (4\text{-}29)$$

Equating the coefficients of dx and dy in Eqs. (4-28) and (4-29) gives

$$\frac{\partial \psi}{\partial x} = -v \quad \text{and} \quad \frac{\partial \psi}{\partial y} = u. \qquad (4\text{-}30)$$

Equation (4-30) is often used to define the *stream function*. It is easily shown that the Eq. (4-30) satisfies the two-dimensional incompressible continuity equation.

A significant physical meaning of ψ and its relationship to the velocity components can be illustrated by an examination of Fig. 4-8 and Eq. (4-30).

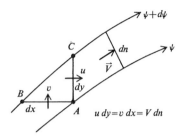

Fig. 4-8 Physical significance of stream function.

In Fig. 4-8, two streamlines an infinitesimal distance apart are shown. The values of the stream functions for the two streamlines are assumed to be ψ and $\psi + d\psi$. As an observer moves from point A to point B (negative dx) the change of ψ or $d\psi$ would be simply $-(\partial \psi/\partial x)\, dx$ since $dy = 0$. From Eq. (4-30) then,

$$d\psi = -\frac{\partial \psi}{\partial x} dx = v\, dx.$$

But $(v\, dx)$ is the volume rate of flow per unit depth across AB. Since there is no x component of flow across AB, $d\psi$ represents the total volume rate of flow per unit depth between the two streamlines. The flow direction is always from the observer's left to right as he moves from one streamline to another having a larger value of ψ. In finite terms, the volume rate of flow per unit depth between two streamlines is equal to the difference in the value of the stream functions of the two streamlines (see Fig. 4-8). Since ψ is a constant along any streamline, the volume rate of flow between any two streamlines is constant. From this it can be seen that the closer together two given streamlines become, the higher is the velocity of the fluid between them. This is illustrated in Fig. 4-9 where fluid is shown flowing through a nozzle.

The streamlines, represented by constant values of ψ, are shown spaced relatively far apart at the entrance to the flow nozzle. As the area for flow

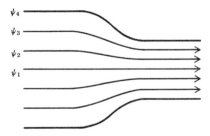

Fig. 4-9 Streamlines in a converging nozzle.

decreases in the nozzle, the streamlines are forced closer together. According to Eq. (4-30), this results in an increased velocity in the region of decreased area. The streamlines, therefore, show the regions of relatively high and low velocity as well as indicate the directions of the fluid motion.

Fig. 4-10 A streamtube.

The concept of a *streamtube* is also useful in describing steady flow situations. Consider, for example, a family of streamlines which bound the small area dA_1 in Fig. 4-10. The same streamlines bounding the small dA_1 also bound the small area dA_2. Therefore, all of the fluid flowing through dA_1 also flows through dA_2 since no fluid can cross the streamlines bounding the areas. These streamlines make up a streamtube. As the areas bounded by the streamtube become infinitesimally small, the velocities across any given area can be considered to be uniform. The velocity in the tube will be considered to vary only along the streamtube. Therefore, even though the fluid velocity vectors may have finite values for all three components, the streamtube fluid can be described in one-dimensional terms. This represents a useful simplification in many cases. For example, the equation for one-dimensional conservation of mass,

$$d\dot{m} = \rho(V)(dA) = \text{constant}, \qquad (4\text{-}31)$$

can be used to describe the flow through the streamtube where V is the velocity along the tube at any dA. It is again apparent from this equation, that, for constant density, as the streamlines spread apart (increasing dA in the direction of flow), the velocity of the fluid will decrease.

4-3 CONSERVATION OF MOMENTUM

Newton's second law of motion—Eq. (1-1)—states that the time rate of change of momentum of a given mass is proportional to the net force acting on the mass. This is sometimes referred to as the *law of conservation of momentum*. This law must be considered in any physical problem where there is motion. Since many engineering problems involve moving fluids, Newton's second law of motion will be developed into forms that are useful for the analysis of such problems. Both overall and differential forms of the momentum equation will be derived.

Overall Momentum Balance

The overall conservation of momentum for fluid flow can be obtained by application of the general conservation equation, Eq. (4–4), and Newton's second law applied to a fluid system moving with nonaccelerating coordinates:

$$\sum \mathbf{f} = \frac{D(m\mathbf{V})}{Dt}, \qquad (4\text{--}32)$$

where $\sum \mathbf{f}$ is the resultant of the forces acting on the system. $m\mathbf{V}$ is the linear momentum of the system and, since it is a vector, may be broken up into scalar components:

$$\sum f_x = \frac{D(mu)}{Dt}, \quad \sum f_y = \frac{D(mv)}{Dt}, \quad \sum f_z = \frac{D(mw)}{Dt}. \qquad (4\text{--}33)$$

Referring back to the general overall conservation equation, Eq. (4–4), let

$$\mathbf{B} = m\mathbf{V} \quad \text{and} \quad \mathbf{b} = \frac{\mathbf{B}}{m} = \mathbf{V}. \qquad (4\text{--}34)$$

$\mathbf{B}$ and $\mathbf{b}$ are now vector quantities, but this in no way changes the validity of Eq. (4–4). The time rate of change of momentum of the system as it follows the fluid can now be written as

$$\frac{D(m\mathbf{V})}{Dt} = \frac{\partial}{\partial t} \iiint_{\text{c.v.}} \mathbf{V} \rho \, d(\text{Vol}) + \iint_{\text{c.s.}} \mathbf{V}(\rho \mathbf{V}_r \cdot d\mathbf{A}). \qquad (4\text{--}35)$$

Using Eq. (4–32), the conservation of momentum equation can be written

$$\sum \mathbf{f} = \frac{\partial}{\partial t} \iiint_{\text{c.v.}} \mathbf{V} \rho \, d(\text{Vol}) + \iint_{\text{c.s.}} \mathbf{V}(\rho \mathbf{V}_r \cdot d\mathbf{A}), \qquad (4\text{--}36)$$

or in the component form:

$$\sum f_x = \frac{\partial}{\partial t} \iiint_{\text{c.v.}} u \rho \, d(\text{Vol}) + \iint_{\text{c.s.}} u(\rho \mathbf{V}_r \cdot d\mathbf{A}), \qquad (4\text{--}37\text{a})$$

$$\sum f_y = \frac{\partial}{\partial t} \iiint_{\text{c.v.}} v \rho \, d(\text{Vol}) + \iint_{\text{c.s.}} v(\rho \mathbf{V}_r \cdot d\mathbf{A}), \qquad (4\text{--}37\text{b})$$

$$\sum f_z = \frac{\partial}{\partial t} \iiint_{\text{c.v.}} w \rho \, d(\text{Vol}) + \iint_{\text{c.s.}} w(\rho \mathbf{V}_r \cdot d\mathbf{A}). \qquad (4\text{--}37\text{c})$$

The external forces ($\sum \mathbf{f}$) will in general be composed of four types of forces:

1. Body forces (forces due to fields, such as gravity, electric, or magnetic) applied to the material inside the control volume.
2. Externally applied forces (to solid portions of control surfaces).
3. Shear or friction forces (on fluid portions of control surfaces).
4. Pressure forces (on fluid portions of control surfaces).

The several examples that follow should be helpful in understanding the usefulness of the overall momentum equation.

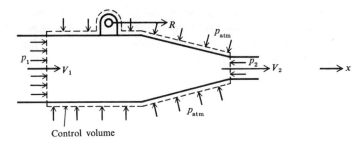

Fig. 4-11 Example 4-3.

Example 4-3. Water is flowing in the stationary nozzle shown in Fig. 4-11. Assuming the velocity profiles are uniform over each cross section, find the force R required to hold the pipe in equilibrium. Assume also that the flow is steady and the force R is located so as to prevent rotation of the nozzle.

Solution. First draw an appropriate control volume around the section (see dotted lines). Next arbitrarily assign a positive direction for reference purposes. Now any velocities or forces in this direction will be considered positive. Summing the forces in the x direction:

$$\sum f_x = p_1 A_1 - p_2 A_2 - p_{\text{atm}}(A_1 - A_2) + R$$

Notice that the atmospheric pressure force component acts on a projected area $(A_1 - A_2)$ of the control surface. One could rearrange the above equation to the following form:

$$\sum f_x = (p_1 - p_{\text{atm}})A_1 - (p_2 - p_{\text{atm}})A_2 + R,$$

or

$$\sum f_x = p_1' A_1 - p_2' A_2 + R,$$

where $p' = p - p_{\text{atm}}$ = gage pressure. The rate of change of momentum of the control volume is given by

$$\iint_{c.s.} u(\rho \mathbf{V}_r \cdot d\mathbf{A}) + \frac{\partial}{\partial t} \iiint_{c.v.} \mathbf{V} \rho d(\text{Vol}) = \rho A_2 V_2^2 - \rho A_1 V_1^2 + 0.$$

Applying the conservation of momentum equation, Eq. (4–37a),

$$p_1' A_1 - p_2' A_2 + R = \rho A_2 V_2^2 - \rho A_1 V_1^2 = \rho A_1 V_1 (V_2 - V_1),$$

where $\rho A_1 V_1 = \rho A_2 V_2$. Therefore,

$$R = \rho A_1 V_1 (V_2 - V_1) + p_2' A_2 - p_1' A_1.$$

If a control volume is exposed to the atmosphere it is usually advantageous to work momentum problems in terms of gage pressure rather than absolute pressure. This saves one the trouble of accounting for the atmospheric pressure force term separately.

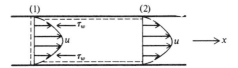

Fig. 4–12 Example 4–4.

Example 4–4. Determine the value of average shear stress τ_w on the wall of the pipe of length L in Fig. 4–12. Assume steady incompressible flow.

Solution. First draw a control volume inside the pipe and assign a positive x direction:

$$\sum f_x = p_1 A - p_2 A - \tau_w \pi D L.$$

$$\iint_{c.s.} u(\rho \mathbf{V}_r \cdot d\mathbf{A}) = \left(\rho \int u^2 \, dA \right)_2 - \left(\rho \int u^2 \, dA \right)_1.$$

Let

$$\beta = \int_A \frac{u^2 \, dA}{V_{\text{avg}}^2 A},$$

or

$$\iint_{c.s.} u(\rho \mathbf{V}_r \cdot d\mathbf{A}) = \beta_2 \rho A V_2^2 - \beta_1 \rho A V_1^2,$$

since $\rho A V_1 = \rho A V_2$, then $V_1 = V_2 = V_{\text{avg}}$. The momentum equation can now be written as

$$p_1 A - p_2 A - \tau_w \pi D L = \rho A V_{\text{avg}}^2 (\beta_2 - \beta_1).$$

If the velocity profiles at section (1) and section (2) are similar, $\beta_1 = \beta_2$, and

$$\tau_w = \frac{\pi D^2}{4\pi DL}(p_1 - p_2) = \frac{D}{4L}(p_1 - p_2).$$

An important concept brought out in this example is the *momentum correction factor* β. For laminar flow in a circular pipe with a parabolic velocity profile, $\beta = \frac{4}{3}$, for one-dimensional ideal flow (constant velocity profile) $\beta = 1$, and for turbulent flow in a circular pipe $\beta \cong 1.05$.

Example 4–5. In Fig. 4–13, determine the forces R_x and R_y required to hold the plate in equilibrium. Assume that the resultant of the forces is such that no rotation occurs.

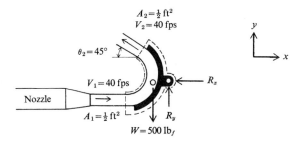

Fig. 4–13 Example 4–5.

Solution. Again a control volume is drawn and positive directions chosen. The selection of a control volume determines the weight of fluid contained. In this case it is assumed that the size of the control volume is such that it contains 500 lb$_f$ of fluid. Pressures in a free jet are equal to the local atmospheric pressure (unless the flow is supersonic), hence there is no net pressure force on the control volume, and

$$\sum f_x = -R_x \quad \text{and} \quad \sum f_y = -W + R_y.$$

For the x direction,

$$\iint_{\text{c.s.}} u(\rho \mathbf{V}_r \cdot d\mathbf{A}) = -\beta_2 \rho A_2 V_2^2 \cos 45° - \beta_1 \rho A_1 V_1^2,$$

and for the y direction,

$$\iint_{\text{c.s.}} v(\rho \mathbf{V}_r \cdot d\mathbf{A}) = \beta_2 \rho A_2 V_2^2 \sin 45° - 0.$$

Assuming uniform velocity profiles, then $\beta_1 = \beta_2 = 1$. For the x direction,

$$-R_x = -\rho A_1 V_1 (V_2 \cos 45° + V_1) = -\frac{62.4}{32.2}(0.5)(40)(28.2 + 40)$$
$$= -2660 \text{ lb}_f,$$

or

$$R_x = 2660 \text{ lb}_f.$$

For the y direction,

$$R_y = W + \rho A_2 V_2^2 \sin 45° = 500 + \frac{62.4}{32.2}(0.5)(40)^2(0.707),$$

or

$$R_y = 1600 \text{ lb}_f.$$

Note that the same technique could be used to solve the problem where the plate might have a uniform velocity. In such a case the velocity V_r is not equal to the jet velocity but equal to the relative velocity between the jet and the plate.

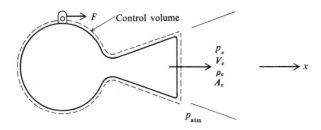

Fig. 4-14 Example 4-6.

Example 4-6. Determine the equation of the thrust force T that is produced by the rocket engine in Fig. 4-14. (Note: In the general case, the exit pressure p_e may be different from the local atmospheric pressure p_{atm} if the exit velocity is greater than the speed of sound.)

Solution. The control volume is drawn around the rocket engine, and a positive x direction is chosen.

$$\sum f_x = F - p'_e A_e,$$

where p'_e is the exit gage pressure $= (p_e - p_{\text{atm}})$.

$$\iint u(\rho V_r \cdot d\mathbf{A}) = \rho_e A_e V_e^2.$$

Applying the momentum equation,

$$F = \rho_e A_e V_e^2 + p'_e A_e.$$

The thrust force T is equal to F but acts in the opposite direction.

The Differential Momentum Equations

The need for a differential form of the momentum equations develops in a variety of cases. If the value of properties at each point in a fluid varies or if the properties are known to vary with time, the differential form of the momentum equations is particularly useful. The differential equation could again be obtained directly from Eqs. (4–36) or (4–37) by use of Gauss' theorem. To give a better physical understanding, however, a separate derivation of this form of the momentum equation will now be presented.

Instead of assuming a volume of rigid dimensions at a fixed point in space (a control volume) as was done in the derivation of the continuity equation, a particular mass of fluid (a system) will be observed as it moves through space. This is the Lagrangian viewpoint, or the viewpoint of an observer moving with the fluid. The same final differential equations would result by assuming that the observer is stationary and the fluid is flowing past. As has been stated in Section 4–1, either viewpoint gives the same result for infinitesimally small control volumes or systems.

Consider the mass in a system defined by a control surface of dimension $(dx)(dy)(dz)$, where each dimension is considered to be infinitesimally small. The mass of the system is constant, so Newton's second law can be written

$$\text{(mass of system)} \frac{D\mathbf{V}}{Dt} = \text{net force on system,} \qquad (4\text{–}38)$$

where the time derivative of the velocity vector $D\mathbf{V}/Dt$ is the acceleration of the system.

Dividing Eq. (4–38) by the volume of the system and classifying the forces as either body or surface forces,

$$\rho \frac{D\mathbf{V}}{Dt} = \frac{\text{net force}}{\text{volume}} = \frac{\text{net body force}}{\text{volume}} + \frac{\text{net surface force}}{\text{volume}}. \qquad (4\text{–}39)$$

This equation can be written in terms of the x, y, and z components of velocity and force to give three scalar equations:

$$\rho \frac{Du}{Dt} = \left(\frac{\text{net body force}}{\text{volume}}\right)_x + \left(\frac{\text{net surface force}}{\text{volume}}\right)_x, \qquad (4\text{–}40\text{a})$$

$$\rho \frac{Dv}{Dt} = \left(\frac{\text{net body force}}{\text{volume}}\right)_y + \left(\frac{\text{net surface force}}{\text{volume}}\right)_y, \qquad (4\text{–}40\text{b})$$

$$\rho \frac{Dw}{Dt} = \left(\frac{\text{net body force}}{\text{volume}}\right)_z + \left(\frac{\text{net surface force}}{\text{volume}}\right)_z. \qquad (4\text{–}40\text{c})$$

The net surface forces per unit volume can be expressed in terms of the stress components by reference to Fig. 4–15. The normal x-directed stress components are shown acting on the two x faces, the faces perpendicular to

88 THE CONSERVATION EQUATIONS

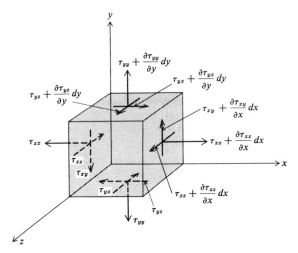

Fig. 4-15 Stress components in rectangular coordinates.

the x axis. The force on each face is the product of stress and area. The net positive x component of force due to those two normal forces is $(\partial \tau_{xx}/\partial x)\,(dx\,dy\,dz)$, and the net positive x component of normal force per unit volume is $\partial \tau_{xx}/\partial x$.

The only other surface forces acting on the element in the x direction are due to the shear forces acting on the y and z faces. Assuming that the shear force on the negative y face is $-\tau_{yx}\,dx\,dz$ and that the positive face is $[\tau_{yx} + (\partial \tau_{yx}/\partial y\,dy)]\,dx\,dz$, then the net force in the x direction due to shear on the y faces is $\partial \tau_{yx}/\partial y\,(dx\,dy\,dz)$, and likewise the net x force per unit volume on the z face due to the shear is $\partial \tau_{zx}/\partial z$. Therefore, the net surface force per unit volume acting on the element in the x direction is

$$\left(\frac{\text{net surface force}}{\text{volume}}\right)_x = \frac{\partial \tau_{xx}}{\partial x} + \frac{\partial \tau_{yx}}{\partial y} + \frac{\partial \tau_{zx}}{\partial z}. \qquad (4\text{-}41\text{a})$$

In a similar manner,

$$\left(\frac{\text{net surface force}}{\text{volume}}\right)_y = \frac{\partial \tau_{xy}}{\partial x} + \frac{\partial \tau_{yy}}{\partial y} + \frac{\partial \tau_{zy}}{\partial z}, \qquad (4\text{-}41\text{b})$$

$$\left(\frac{\text{net surface force}}{\text{volume}}\right)_z = \frac{\partial \tau_{xz}}{\partial x} + \frac{\partial \tau_{yz}}{\partial y} + \frac{\partial \tau_{zz}}{\partial z}. \qquad (4\text{-}41\text{c})$$

X, Y, and Z are components of the body force per unit volume in the x, y, and z directions and they are usually caused by gravitational, magnetic, or electric force fields. That area of fluid mechanics where the effects of body forces due to magnetic and/or electric fields are of primary importance and may not be neglected is called *magnetohydrodynamics* (MHD) and is discussed in Section 17–7.

By substituting Eqs. (4–41a-c) into Eqs. (4–40a-c), the following are obtained:

$$\rho \frac{Du}{Dt} = X + \frac{\partial \tau_{xx}}{\partial x} + \frac{\partial \tau_{yx}}{\partial y} + \frac{\partial \tau_{zx}}{\partial z} \quad x \text{ direction,} \quad (4\text{-}42a)$$

$$\rho \frac{Dv}{Dt} = Y + \frac{\partial \tau_{xy}}{\partial x} + \frac{\partial \tau_{yy}}{\partial y} + \frac{\partial \tau_{zy}}{\partial z} \quad y \text{ direction,} \quad (4\text{-}42b)$$

$$\rho \frac{Dw}{Dt} = Z + \frac{\partial \tau_{xz}}{\partial x} + \frac{\partial \tau_{yz}}{\partial y} + \frac{\partial \tau_{zz}}{\partial z} \quad z \text{ direction.} \quad (4\text{-}42c)$$

Equations (4–42a-c) relate the acceleration components of a fluid element to the net body and surface forces. It is desirable at this point to express the surface force terms in more useful forms, i.e., in terms of velocity and pressure.

If the fluid is Stokesian in nature (many fluids such as air, water, and oil are), then substituting Eqs. (3–19) and (3–20) into Eqs. (4–42a-c), the momentum equation may be expressed in terms of local fluid velocities:

x direction:

$$\rho \frac{Du}{Dt} = X - \frac{\partial p}{\partial x} + \frac{\partial}{\partial x}\left[\mu\left(2\frac{\partial u}{\partial x} - \frac{2}{3}\text{div }\mathbf{V}\right)\right] + \frac{\partial}{\partial y}\left[\mu\left(\frac{\partial u}{\partial y} + \frac{\partial v}{\partial x}\right)\right]$$

$$+ \frac{\partial}{\partial z}\left[\mu\left(\frac{\partial w}{\partial x} + \frac{\partial u}{\partial z}\right)\right], \quad (4\text{-}43a)$$

y direction:

$$\rho \frac{Dv}{Dt} = Y - \frac{\partial p}{\partial y} + \frac{\partial}{\partial y}\left[\mu\left(2\frac{\partial v}{\partial y} - \frac{2}{3}\text{div }\mathbf{V}\right)\right] + \frac{\partial}{\partial z}\left[\mu\left(\frac{\partial v}{\partial z} + \frac{\partial w}{\partial y}\right)\right]$$

$$+ \frac{\partial}{\partial x}\left[\mu\left(\frac{\partial u}{\partial y} + \frac{\partial v}{\partial x}\right)\right], \quad (4\text{-}43b)$$

z direction:

$$\rho \frac{Dw}{Dt} = Z - \frac{\partial p}{\partial z} + \frac{\partial}{\partial z}\left[\mu\left(2\frac{\partial w}{\partial z} - \frac{2}{3}\text{div }\mathbf{V}\right)\right] + \frac{\partial}{\partial x}\left[\mu\left(\frac{\partial w}{\partial x} + \frac{\partial u}{\partial z}\right)\right]$$

$$+ \frac{\partial}{\partial y}\left[\mu\left(\frac{\partial v}{\partial z} + \frac{\partial w}{\partial y}\right)\right]. \quad (4\text{-}43c)$$

Equations (4–43a-c) are called the *Navier-Stokes equations* and represent the conservation of momentum for the three-dimensional flow of a Stokesian fluid.

The assumption of constant values of viscosity and density permits these equations to be reduced to:

$$\rho \frac{Du}{Dt} = X - \frac{\partial p}{\partial x} + \mu \left(\frac{\partial^2 u}{\partial x^2} + \frac{\partial^2 u}{\partial y^2} + \frac{\partial^2 u}{\partial z^2} \right) \quad x \text{ direction,} \quad (4\text{-}44a)$$

$$\rho \frac{Dv}{Dt} = Y - \frac{\partial p}{\partial y} + \mu \left(\frac{\partial^2 v}{\partial x^2} + \frac{\partial^2 v}{\partial y^2} + \frac{\partial^2 v}{\partial z^2} \right) \quad y \text{ direction,} \quad (4\text{-}44b)$$

$$\rho \frac{Dw}{Dt} = Z - \frac{\partial p}{\partial z} + \mu \left(\frac{\partial^2 w}{\partial x^2} + \frac{\partial^2 w}{\partial y^2} + \frac{\partial^2 w}{\partial z^2} \right) \quad z \text{ direction.} \quad (4\text{-}44c)$$

Each term in Eqs. (4-44a-c) represents a component of force per unit volume. The left-hand term in each equation is the *inertia* term, or force due to acceleration. The first terms on the right are *body force* terms, the second are *pressure gradient* terms, and the last are *viscous friction* terms.

The momentum equations, Eqs. (4-44a-c), which assume constant density and viscosity, relate four dependent variables (the three velocity components and the pressure) to the independent variables describing position and time. With suitable boundary conditions and use of the equation of mass conservation, solutions may be obtained in some cases. It should be noted that for this case (constant μ and ρ) an energy equation is not needed. The complexity of Eqs. (4-44a-c) requires the use of numerical techniques for all but the most simple problems.

When μ and/or ρ are considered to be variable, the appropriate form of the momentum equation, such as Eqs. (4-43a-c) must be used with other equations, such as an equation for conservation of energy and phenomenological equations.

Cylindrical Coordinates

In cylindrical coordinates, Eqs. (4-44a-c) are:

z direction:

$$\rho \left(\frac{\partial u_z}{\partial t} + u_r \frac{\partial u_z}{\partial r} + \frac{u_\theta}{r} \frac{\partial u_z}{\partial \theta} + u_z \frac{\partial u_z}{\partial z} \right)$$

$$= Z - \frac{\partial p}{\partial z} + \mu \left(\frac{\partial^2 u_z}{\partial r^2} + \frac{1}{r} \frac{\partial u_z}{\partial r} + \frac{1}{r^2} \frac{\partial^2 u_z}{\partial \theta^2} + \frac{\partial^2 u_z}{\partial z^2} \right),$$

r direction: (4-45a)

$$\rho \left(\frac{\partial u_r}{\partial t} + u_r \frac{\partial u_r}{\partial r} + \frac{u_\theta}{r} \frac{\partial u_r}{\partial \theta} - \frac{u_\theta^2}{r} + u_z \frac{\partial u_r}{\partial z} \right)$$

$$= R - \frac{\partial p}{\partial r} + \mu \left(\frac{\partial^2 u_r}{\partial r^2} + \frac{1}{r} \frac{\partial u_r}{\partial r} - \frac{u_r}{r^2} + \frac{1}{r^2} \frac{\partial^2 u_r}{\partial \theta^2} - \frac{2}{r^2} \frac{\partial u_\theta}{\partial \theta} + \frac{\partial^2 u_r}{\partial z^2} \right),$$

(4-45b)

θ direction:

$$\rho\left(\frac{\partial u_\theta}{\partial t} + u_r\frac{\partial u_\theta}{\partial r} + \frac{u_\theta}{r}\frac{\partial u_\theta}{\partial \theta} + \frac{u_r u_\theta}{r} + u_z\frac{\partial u_\theta}{\partial z}\right)$$
$$= \vartheta - \frac{1}{r}\frac{\partial p}{\partial \theta} + \mu\left(\frac{\partial^2 u_\theta}{\partial r^2} + \frac{1}{r}\frac{\partial u_\theta}{\partial r} - \frac{u_\theta}{r^2} + \frac{1}{r^2}\frac{\partial^2 u_\theta}{\partial \theta^2} + \frac{2}{r^2}\frac{\partial u_r}{\partial \theta} + \frac{\partial^2 u_\theta}{\partial z^2}\right),$$
(4-45c)

where u_z, u_r, and u_θ are the velocity components in the z, r, and θ directions and Z, R, and ϑ are the corresponding body forces per unit volume.

Spherical Coordinates

In spherical coordinates, Eqs. (4-44a-c) are:

r direction:

$$\rho\left(\frac{\partial u_r}{\partial t} + u_r\frac{\partial u_r}{\partial r} + \frac{u_\theta}{r}\frac{\partial u_r}{\partial \theta} + \frac{u_\varphi}{r\sin\theta}\frac{\partial u_r}{\partial \varphi} - \frac{u_\theta^2 + u_\varphi^2}{r}\right)$$
$$= -\frac{\partial p}{\partial r} + \mu\left(\nabla^2 u_r - \frac{2}{r^2}u_r - \frac{2}{r^2}\frac{\partial u_\theta}{\partial \theta} - \frac{2}{r^2}u_\theta\cot\theta - \frac{2}{r^2\sin\theta}\frac{\partial u_\varphi}{\partial \varphi}\right) + R,$$

θ direction: (4-46a)

$$\rho\left(\frac{\partial u_\theta}{\partial t} + u_r\frac{\partial u_\theta}{\partial r} + \frac{u_\theta}{r}\frac{\partial u_\theta}{\partial \theta} + \frac{u_\varphi}{r\sin\theta}\frac{\partial u_\theta}{\partial \varphi} + \frac{u_r u_\theta}{r} - \frac{u_\varphi^2\cot\theta}{r}\right)$$
$$= -\frac{1}{r}\frac{\partial p}{\partial \theta} + \mu\left(\nabla^2 u_\theta + \frac{2}{r^2}\frac{\partial u_r}{\partial \theta} - \frac{u_\theta}{r^2\sin^2\theta} - \frac{2\cos\theta}{r^2\sin^2\theta}\frac{\partial u_\varphi}{\partial \varphi}\right) + \vartheta, \quad (4\text{-}46\text{b})$$

φ direction:

$$\rho\left(\frac{\partial u_\varphi}{\partial t} + u_r\frac{\partial u_\varphi}{\partial r} + \frac{u_\theta}{r}\frac{\partial u_\varphi}{\partial \theta} + \frac{u_\varphi}{r\sin\theta}\frac{\partial u_\varphi}{\partial \varphi} + \frac{u_\varphi u_r}{r} + \frac{u_\theta u_\varphi}{r}\cot\theta\right)$$
$$= -\frac{1}{r\sin\theta}\frac{\partial p}{\partial \varphi} + \mu\left(\nabla^2 u_\varphi - \frac{u_\varphi}{r^2\sin^2\theta} + \frac{2}{r^2\sin\theta}\frac{\partial u_r}{\partial \varphi}\right. \quad (4\text{-}46\text{c})$$
$$\left. + \frac{2\cos\theta}{r^2\sin^2\theta}\frac{\partial u_\theta}{\partial \varphi}\right) + \varphi,$$

where

$$\nabla^2 = \frac{1}{r^2}\frac{\partial}{\partial r}\left(r^2\frac{\partial}{\partial r}\right) + \frac{1}{r^2\sin\theta}\frac{\partial}{\partial \theta}\left(\sin\theta\frac{\partial}{\partial \theta}\right) + \frac{1}{r^2\sin^2\theta}\frac{\partial^2}{\partial \varphi^2},$$

and where u_r, u_θ, and u_φ are the velocity components in the r, θ, and φ directions and R, ϑ, and φ are the corresponding body forces per unit volume.

4-4 CONSERVATION OF ENERGY

The concept of the conservation of energy is used extensively in thermodynamics and is the basis of the first law of thermodynamics. All engineering systems must adhere to this law and, as a consequence, these concepts must be considered in any analysis.

In situations where there is no exchange of thermal energy (no heat flow) and where there is negligible change in stored thermal energy of the fluid, conservation is automatically considered with application of momentum principles. No separate equation describing conservation of energy is necessary. With heat transfer, however, or with significant changes of stored thermal energy in the fluid, an equation relating these thermal energy exchanges with mechanical energy exchanges is needed. Such an equation is usually referred to as the *energy equation*. The equation will be derived in both the overall and the differential form.

Overall Energy Balance

The conservation of energy equation for finite-sized systems can be obtained by application of the general conservation equation, Eq. (4-4), and the first law of thermodynamics. The first law of thermodynamics for a system of mass m, following the motion of the fluid, is

$$m \frac{D}{Dt}(e + KE + PE) = \dot{q} - \dot{W}. \tag{4-47}$$

The three terms in the parentheses of Eq. (4-47) are the internal energy, kinetic energy, and potential energy per unit mass respectively; the quantities e and PE are measured relative to some arbitrary datum; $\dot{q}$ is the net rate of heat entering the system; and $\dot{W}$ is the net rate of work done by the system. In general, $\dot{q}$ can be composed of conduction and radiation terms. The work term represents the work done by the system as it expands and flows, exerting normal and shear forces on the surrounding solid or fluid surfaces. The work may be conveniently divided into three categories—shaft work, shear work, and flow work.

Shaft work is that work done by the system on moving mechanical parts such as pistons, propellers, and turbines. Shaft work may cross the system boundaries in one of several means, such as a rotating shaft, a moving lever or cable, or an electric current.

Shear work is the work done by a fluid as it exerts a shear force on a moving system boundary (Section 3-4). Shear work is defined only where a moving fluid is part of the boundary of the system. By properly selecting the system boundaries, the accounting for shear work may usually be avoided in a finite size system. For example, if flow is only normal to the system boundaries, the shear work is zero. In an infinitesimal system, shear work is considered, as in Eq. (3-24).

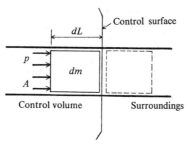

Fig. 4-16 Flow work.

Flow work is that work necessary to move a fluid across the control surface describing the system boundaries at some instant. For example, pressure is usually considered to be the only normal force acting on the system boundaries. For the surface shown in Fig. 4-16, the work done on the surroundings by mass dm as it moves across the control surface to the dotted position is

$$d(\text{flow work}) = pAdL. \qquad (4\text{-}48)$$

However, AdL is the volume $d(\text{Vol})$ of the mass dm, and the flow work is, therefore,

$$d(\text{flow work}) = pd(\text{Vol}), \qquad (4\text{-}49)$$

and, for unit mass,

$$\left(\frac{\text{flow work}}{\text{mass}}\right) = p/\rho. \qquad (4\text{-}50)$$

To obtain the flow work per unit time, Eq. (4-49) should be divided by time, and the volume rate expressed in terms of velocity.

$$d(\text{flow work per unit time}) = p(\mathbf{V} \cdot d\mathbf{A}). \qquad (4\text{-}51)$$

The total flow work rate can be obtained by integrating Eq. (4-51) over the system control surface:

$$\dot{W}_{\text{flow}} = \iint_{\text{c.s.}} p(\mathbf{V} \cdot d\mathbf{A}). \qquad (4\text{-}52a)$$

Since $\mathbf{V}_r = \mathbf{V} - \mathbf{V}_c$ (Section 4-1), Eq. (4-52a) can be rewritten as

$$\dot{W}_f = \iint_{\text{c.s.}} p(\mathbf{V}_r \cdot d\mathbf{A}) + \iint_{\text{c.s.}} p(\mathbf{V}_c \cdot d\mathbf{A}). \qquad (4\text{-}52b)$$

If the control surface is not moving, the last term of Eq. (4-52b) becomes zero.

Some texts use the term *flow energy*, or *energy in transition*, in place of flow work.

The general conservation law, Eq. (4-4), with b replaced by $(e + KE + PE)$, can be substituted into the left side of Eq. (4-47) to obtain the law of conservation of energy:

$$\iint_{c.s.} \rho(e + KE + PE)(\mathbf{V}_r \cdot d\mathbf{A}) + \frac{\partial}{\partial t} \iiint_{vol} \rho(e + KE + PE)\, d(Vol) = \dot{q} - \dot{W}, \quad (4\text{-}53a)$$

where $\dot{q} = \dot{q}_{conduction} + \dot{q}_{radiation}$.

By using the definition of enthalpy,

$$j = e + \frac{p}{\rho}, \quad (4\text{-}54)$$

Equation (4-53) may be written with the flow work term conveniently combined with the energy flow term. For nonmoving boundaries, the result is

$$\iint_{c.s.} \rho(j + KE + PE)(\mathbf{V}_r \cdot d\mathbf{A})$$
$$+ \frac{\partial}{\partial t} \iiint_{vol} \rho(e + KE + PE)\, d(Vol) = \dot{q} - \dot{W}_{shaft} - \dot{W}_{shear}. \quad (4\text{-}53b)$$

The steady flow energy equation (steady flow implies that all partial derivatives with respect to time are zero) for the system shown in Fig. 4-17 may be written as:

$$\left[(\rho V j)_{avg} + \left(\frac{\rho V^3}{2}\right)_{avg} + g(\rho z V)_{avg}\right]_2 A_2$$
$$- \left[(\rho V j)_{avg} + \left(\frac{\rho V^3}{2}\right)_{avg} + g(\rho z V)_{avg}\right]_1 A_1 = \dot{q} - \dot{W}_{shaft}, \quad (4\text{-}55)$$

where properly averaged values over both cross sections are defined by

$$(\rho V j)_{avg} = \frac{1}{A} \iint \rho V j\, dA,$$

$$(\rho V^3)_{avg} = \frac{1}{A} \iint \rho V^3\, dA,$$

$$(z \rho V)_{avg} = \frac{1}{A} \iint z \rho V\, dA.$$

In Fig. 4-17, $\dot{W}_{shear} = 0$ on the boundaries, because the boundaries of the control volume are not in contact with the fluid, except at the inlet and outlet where the flow is normal to the control surface. The flow work is included in the enthalpy term, and $\dot{W}_{shaft}$ represents the rate that shaft work is done by the system.

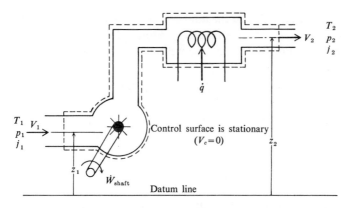

Fig. 4-17 Overall energy balance.

Instead of using average values such as those defined above, it is convenient to use the concept of a bulk property in conjunction with the mass average velocity defined by Eq. (4–11). A *bulk property* is defined such that the product of the bulk property and the mass flow rate over a given area normal to the flow is equal to the total flow rate of that property through the area. This definition will apply to specific properties, such as enthalpy per unit mass and kinetic energy per unit mass.

For example, the *bulk kinetic energy* per unit mass for a uniform density across a cross section A and with velocity u normal to A is defined by

$$\text{KE}_{\text{bulk}} = \frac{1}{u_{\text{avg}}A} \int_A \frac{u^3}{2} \, dA. \tag{4-56a}$$

It should be obvious from a study of Eq. (4–56a) that the bulk kinetic energy per unit mass of the flow at any cross section is *not* simply expressed as $\text{KE}_{\text{bulk}} = (u_{\text{avg}})^2/2$. The bulk kinetic energy at any cross section can be expressed in terms of the average velocity at that cross section by use of a *kinetic energy correction factor*, α:

$$\text{KE}_{\text{bulk}} = \alpha \frac{(u_{\text{avg}})^2}{2} \quad \text{where} \quad \alpha = \frac{1}{(u_{\text{avg}})^3 A} \int_A u^3 \, dA. \tag{4-56b}$$

For uniform velocity at a cross section, $\alpha = 1$; for laminar flow in a pipe, $\alpha = 2$ (see Example 4–7); and for turbulent flow in a pipe, α is equal to about 1.1.

The bulk enthalpy per unit mass of the fluid flowing j_{bulk} can be obtained from the following expression, where constant density is assumed:

$$j_{\text{bulk}} = \frac{1}{u_{\text{avg}}A} \int_A ju \, dA. \tag{4-56c}$$

The specific enthalpy of the fluid can often be approximated as a function of temperature only. Assuming a zero datum of enthalpy at the zero datum of temperature, Eq. (4–56c) can be written

$$(c_p T)_{\text{bulk}} = \frac{1}{u_{\text{avg}} A} \int_A c_p T u \, dA. \qquad (4\text{–}56\text{d})$$

In those cases where both the density and the specific heat at constant pressure can be assumed constant across the cross section, Eq. (4–56d) becomes

$$T_{\text{bulk}} = \frac{1}{u_{\text{avg}} A} \int_A T u \, dA. \qquad (4\text{–}56\text{e})$$

Equation (4–56e) defines the *bulk temperature* of a flow at any given cross section. This bulk temperature is sometimes called the *mixing cup temperature* since it represents the equilibrium temperature which would be obtained if the fluid flowing across a given section was directed immediately into an insulated chamber where it was mixed adiabatically. It is the temperature which defines the bulk enthalpy of the fluid at a cross section, and thus is important to the general flow problem.

Using bulk properties and assuming that elevation changes are small over both inlet and outlet areas, the system shown in Fig. 4–17 could be described for steady flow by eqn 6-21 (same)

$$(j_2 + (\text{KE})_2 + (\text{PE})_2) - (j_1 + (\text{KE})_1 + (\text{PE})_1) = \frac{\dot{q}}{\dot{m}} - \frac{\dot{W}_{\text{shaft}}}{\dot{m}}, \qquad (4\text{–}56\text{f})$$

where the enthalpy j and the kinetic energy are bulk properties defined above. Equation (4–56f) may be recognized as the *steady flow energy equation*, usually introduced in a basic course in thermodynamics. See Eq. (6–21).

Example 4–7. Calculate the kinetic energy correction factor for laminar flow in a round pipe. The velocity distribution is parabolic and is given by

$$V = V_{\text{max}} \left[1 - \left(\frac{r}{R}\right)^2\right].$$

Solution. The average velocity is

$$V_{\text{avg}} = \frac{1}{A} \iint V \, dA = \frac{1}{\pi R^2} \int_0^R V_{\text{max}} \left[1 - \left(\frac{r}{R}\right)^2\right] 2\pi r \, dr.$$

Therefore $V_{\text{avg}} = \tfrac{1}{2} V_{\text{max}}$, and

$$\tfrac{1}{2} \iint \rho V^3 \, dA = \tfrac{1}{2} \rho V_{\text{max}}^3 \int_0^R \left[1 - \left(\frac{r}{R}\right)^2\right]^3 2\pi r \, dr$$

$$= \pi \rho V_{\text{max}}^3 \int_0^R \left[1 - 3\left(\frac{r}{R}\right)^2 + 3\left(\frac{r}{R}\right)^4 - \left(\frac{r}{R}\right)^6\right] r \, dr = \frac{\pi R^2 \rho V_{\text{max}}^3}{8},$$

and, from Eq. (4-56b),
$$\alpha = \frac{\frac{1}{8}(\pi R^2 \rho V_{max}^3)}{\frac{1}{2}(\frac{1}{2}V_{max})^2(\frac{1}{2}\rho V_{max}\pi R^2)} = 2.$$

Example 4-8. A system has a mass flow rate of 5 lb_m/sec. The bulk enthalpy, velocity, and average height at the entrance are respectively 1000 Btu/lb_m, 100 ft/sec, and 100 ft. At exit these quantities are 1020 Btu/lb_m, 50 ft/sec, and 0 ft. Heat is transferred from the system at the rate of 50 Btu/sec. At what rate in horsepower must shaft work be supplied to this system? Neglect variation in the velocity over entrance and exit areas.

Solution.
$$\frac{V_2^2 - V_1^2}{2} + g(z_2 - z_1) + j_2 - j_1 = \frac{\dot{q}}{\dot{m}} - \frac{\dot{W}_{shaft}}{\dot{m}},$$

$$\frac{50^2 - 100^2}{2} + 32.2(0 - 100) + (1020 - 1000)(778)(g_c)$$

$$= \frac{-50(778)}{5/g_c} - \frac{\dot{W}_{shaft}}{5/g_c}.$$

(Notice that g_c and J, 778 ft lb_f/Btu, are included in various terms in order to balance the units.)

$$\dot{W}_{shaft} = -115,550 \frac{\text{ft lb}_f}{\text{sec}} \quad \text{(work is supplied to the system),}$$

power supplied = 115,550/550 = 210 hp.

The Differential Equation of Energy

The conservation of energy principle will now be applied to an infinitesimal element to derive a differential equation which is useful in the study of heat transfer and in many fluid flow problems. The integral conservation equation, Eq. (4-53), could be applied, and by use of Gauss' theorem a differential equation obtained directly. Again, however, the derivation will be carried out in a step-by-step manner. As for all derivations on infinitesimally small systems, it makes no difference whether the control volume viewpoint or the system viewpoint is utilized since equivalent equations would be derived.

This small element, Fig. 4-18, will be considered to be a system in which only heat and work are permitted to cross the boundaries. Thus, the same mass is observed throughout the analysis. The element observed is assumed to have the capability of storing energy *mechanically* as potential or kinetic energy and *thermally* as internal energy. Applying the first law of thermodynamics to the element:

rate of heat addition into element by conduction and radiation + rate of work done on element

= rate of change of stored energy of element   (4-57)

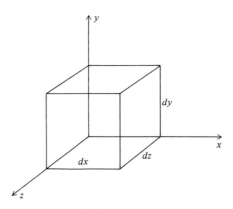

Fig. 4-18 Fluid element for energy balance.

In formulating the equation which describes conservation of energy, it will be useful to derive expressions describing the rates at which each of the energy terms above is changing in terms of either temperature, pressure, or velocity components.

The net rate of work done on the element is due to the surface forces. The net rate of work done due to the surface forces was derived in Section 3-4 as Eq. (3-24). The work term can then be written in the form of Eq. (3-24):

$$\text{rate work done on element per unit volume} = \frac{\partial}{\partial x}(u\tau_{xx} + v\tau_{xy} + w\tau_{xz})$$
$$+ \frac{\partial}{\partial y}(u\tau_{yx} + v\tau_{yy} + w\tau_{yz}) + \frac{\partial}{\partial z}(u\tau_{zx} + v\tau_{zy} + w\tau_{zz}).$$

Multiplying Eqs. (4-42a), (4-42b), and (4-42c) by u, v, and w respectively and then substituting into Eq. (3-24), along with Eqs. (3-19) and (3-20), one obtains the following equation, noting that $u(Du/Dt) = \frac{1}{2}(Du^2/Dt)$, etc.,

$$\text{rate of work done on element per unit volume} = \left(\rho\frac{D\text{KE}}{Dt} + \rho\frac{D\text{PE}}{Dt} - p\,\text{div}\,\mathbf{V} + \mu\Phi\right),$$

(4-58)

where

$$\text{KE} = \tfrac{1}{2}(u^2 + v^2 + w^2) = \text{kinetic energy per unit mass,}$$

$$\text{PE} = -\frac{1}{\rho}(xX + yY + zZ) = \text{potential energy per unit mass,}$$

$$p\,\text{div}\,\mathbf{V} = p\left(\frac{\partial u}{\partial x} + \frac{\partial v}{\partial y} + \frac{\partial w}{\partial z}\right)$$

$$= \text{reversible rate of compression work per unit volume}$$

and
$$\mu\Phi = \mu\left[2\left(\frac{\partial u}{\partial x}\right)^2 + 2\left(\frac{\partial v}{\partial y}\right)^2 + 2\left(\frac{\partial w}{\partial z}\right)^2 + \left(\frac{\partial u}{\partial y} + \frac{\partial v}{\partial x}\right)^2 \right.$$
$$\left. + \left(\frac{\partial v}{\partial z} + \frac{\partial w}{\partial y}\right)^2 + \left(\frac{\partial w}{\partial x} + \frac{\partial u}{\partial z}\right)^2 - \tfrac{2}{3}(\text{div } \mathbf{V})^2\right]$$

= irreversible rate of conversion of work to internal energy per unit volume.

Thus it can be seen that all of the work done on the element goes into changing the kinetic and potential energy of the element, compressing the element, or in changing the internal energy e of the element.

The term Φ is sometimes referred to as the *viscous dissipation function*, and it can be shown, by rearranging the above equation, that it is always positive. The product $\mu\Phi$ is the rate of work per unit volume being done by the stresses in shearing and distorting the fluid, and this work is dissipated within the fluid as *friction heat*. For many problems, particularly those in which shear rates are small, the viscous dissipation function may be neglected. In some cases, such as the flow of fluids through porous media, Φ can become very large, even for very low flow speeds.

The expression describing the net heat conduction across the boundaries will now be derived, using the Fourier law. Reference is again directed to Fig. 4–18. The heat conducted across the left-hand x face of the element per unit time is

$$\text{rate of heat conduction across left-hand } x \text{ face} = -k_x \frac{\partial T}{\partial x} \, dy \, dz,$$

where k_x is the thermal conductivity in the x direction.

The rate at which heat is conducted out of the element across the right-hand x face is

$$\text{rate of heat conduction across right-hand } x \text{ face} = -\left[k_x \frac{\partial T}{\partial x} \, dy \, dz + \frac{\partial}{\partial x}\left(k_x \frac{\partial T}{\partial x}\right) dx \, dy \, dz\right].$$

The net rate at which heat is conducted across the x faces into the element is, therefore,

$$\text{net rate of heat conduction across } x \text{ faces} = \frac{\partial}{\partial x}\left(k_x \frac{\partial T}{\partial x}\right) dx \, dy \, dz.$$

Likewise, equations for the rate of heat conduction in the y and z directions can be developed. The net rate of heat conduction across all of the boundaries into the element is

net rate of heat conduction across all faces

$$= \frac{\partial}{\partial x}\left(k_x \frac{\partial T}{\partial x}\right) dx \, dy \, dz + \frac{\partial}{\partial y}\left(k_y \frac{\partial T}{\partial y}\right) dx \, dy \, dz + \frac{\partial}{\partial z}\left(k_z \frac{\partial T}{\partial z}\right) dx \, dy \, dz.$$

Dividing by the volume of the element ($dx\,dy\,dz$) gives the rate of energy addition by conduction per unit volume,

$$\frac{\partial}{\partial x}\left(k_x\frac{\partial T}{\partial x}\right) + \frac{\partial}{\partial y}\left(k_y\frac{\partial T}{\partial y}\right) + \frac{\partial}{\partial z}\left(k_z\frac{\partial T}{\partial z}\right) = -\operatorname{div}\dot{\mathbf{q}}_c, \quad (4\text{-}59)$$

and

$$\dot{\mathbf{q}}_c = -\left(\mathbf{i}k_x\frac{\partial T}{\partial x} + \mathbf{j}k_y\frac{\partial T}{\partial y} + \mathbf{k}k_z\frac{\partial T}{\partial z}\right),$$

where $\dot{\mathbf{q}}_c$ is the heat conduction flux vector.

A similar analysis could be used to express the rate of energy addition by radiation per unit volume as $-\operatorname{div}\dot{\mathbf{q}}_r$, which is the net heat transferred per unit volume per unit time to the element by radiation. This is a complicated function of wavelength and temperature distribution in the fluid and boundaries surrounding the element. It will be discussed in more detail in Chapter 15. If the element in Fig. 4–18 is an interior element in a solid, then $-\operatorname{div}\dot{\mathbf{q}}_r$ is zero. Fortunately for many engineering problems, the temperatures are low enough so that $-\operatorname{div}\dot{\mathbf{q}}_r$ is negligible for cases when the fluid element is a gas or liquid.

The appropriate expressions may now be substituted in Eq. (4–57), and the following equation is obtained, representing the conservation of energy for any small element of fluid:

$$\overbrace{\frac{\partial}{\partial x}\left(k_x\frac{\partial T}{\partial x}\right) + \frac{\partial}{\partial y}\left(k_y\frac{\partial T}{\partial y}\right) + \frac{\partial}{\partial z}\left(k_z\frac{\partial T}{\partial z}\right) - \operatorname{div}\dot{\mathbf{q}}_r}^{\text{(rate of heat addition)}}$$

$$\underbrace{+\,\rho\frac{D(\mathrm{KE})}{Dt} + \rho\frac{D(\mathrm{PE})}{Dt} - p\left(\frac{\partial u}{\partial x} + \frac{\partial v}{\partial y} + \frac{\partial w}{\partial z}\right) + \mu\Phi}_{\text{(rate of work)}}$$

(rate of change of stored energy)

$$= \rho\frac{D(e)}{Dt} + \rho\frac{D(\mathrm{KE})}{Dt} + \rho\frac{D(\mathrm{PE})}{Dt}. \quad (4\text{-}60)$$

It can be seen in Eq. (4–60) that the kinetic and potential energy term will cancel from both sides of the equation so that

$$\rho\frac{De}{Dt} = -p\left(\frac{\partial u}{\partial x} + \frac{\partial v}{\partial y} + \frac{\partial w}{\partial z}\right) + \frac{\partial}{\partial x}\left(k_x\frac{\partial T}{\partial x}\right) + \frac{\partial}{\partial y}\left(k_y\frac{\partial T}{\partial y}\right)$$

$$+ \frac{\partial}{\partial z}\left(k_z\frac{\partial T}{\partial z}\right) + \mu\Phi + -\operatorname{div}\dot{\mathbf{q}}_r. \quad (4\text{-}61)$$

From thermodynamics, one can express

$$de = c_v\, dT + \left(\frac{\partial e}{\partial V}\right)_T dV. \tag{4-62}$$

If the variation of the internal energy with specific volume is neglible, then

$$\rho \frac{De}{Dt} = \rho c_v \frac{DT}{Dt}. \tag{4-63}$$

For mathematical simplicity, the thermal conductivity k is often assumed to be constant. The above assumptions allow the energy equation to be written as

$$\rho c_v \frac{DT}{Dt} = -p\left(\frac{\partial u}{\partial x} + \frac{\partial v}{\partial y} + \frac{\partial w}{\partial z}\right)$$
$$+ k\left(\frac{\partial^2 T}{\partial x^2} + \frac{\partial^2 T}{\partial y^2} + \frac{\partial^2 T}{\partial z^2}\right) + \mu\Phi - \operatorname{div} \dot{\mathbf{q}}_r. \tag{4-64}$$

For an incompressible substance, the divergence of the velocity vector is zero—see Eq. (4–17); hence, the first term on the right of Eq. (4–64) drops out, and $c_p = c_v$ so that the following equation is commonly used for this case:

$$\rho c_p \frac{DT}{Dt} = k\left(\frac{\partial^2 T}{\partial x^2} + \frac{\partial^2 T}{\partial y^2} + \frac{\partial^2 T}{\partial z^2}\right) + \mu\Phi - \operatorname{div} \dot{\mathbf{q}}_r. \tag{4-65}$$

The same final equations could have been obtained by assuming an element of fixed volume (a control volume) and fluid flowing across the boundaries. With such an assumption the volume would have been an open thermodynamic system. In such a system, stored energy is considered to be carried in and out of the element due to fluid motion. This energy carried in or out is sometimes referred to as the *convected energy*. For the control volume viewpoint, Eq. (4–65) is written by expanding the substantial derivative term and rearranging. The result is

$$\rho c_p \frac{\partial T}{\partial t} = -\rho c_p \left(u\frac{\partial T}{\partial x} + v\frac{\partial T}{\partial y} + w\frac{\partial T}{\partial z}\right)$$
$$+ k\left(\frac{\partial^2 T}{\partial x^2} + \frac{\partial^2 T}{\partial y^2} + \frac{\partial^2 T}{\partial z^2}\right) + \mu\Phi - \operatorname{div} \dot{\mathbf{q}}_r. \tag{4-66}$$

The term on the left is the rate of energy storage per unit volume for the control volume, and the first terms on the right are the *convective* terms, or rate at which energy is carried in by the fluid motion per unit volume.

Cylindrical Coordinates

The energy equation may be written in cylindrical coordinates for the case of constant ρ, μ, and k and zero radiation flux:

$$\rho c_p \left(\frac{\partial T}{\partial t} + u_r \frac{\partial T}{\partial r} + \frac{u_\theta}{r} \frac{\partial T}{\partial \theta} + u_z \frac{\partial T}{\partial z} \right)$$

$$= k \left[\frac{1}{r} \frac{\partial}{\partial r}\left(r \frac{\partial T}{\partial r} \right) + \frac{1}{r^2} \frac{\partial^2 T}{\partial \theta^2} + \frac{\partial^2 T}{\partial z^2} \right]$$

$$+ 2\mu \left\{ \left(\frac{\partial u_r}{\partial r} \right)^2 + \left[\frac{1}{r} \left(\frac{\partial u_\theta}{\partial \theta} + u_r \right) \right]^2 + \left(\frac{\partial u_z}{\partial z} \right)^2 \right\}$$

$$+ \mu \left\{ \left(\frac{\partial u_\theta}{\partial z} + \frac{1}{r} \frac{\partial u_z}{\partial \theta} \right)^2 + \left(\frac{\partial u_z}{\partial r} + \frac{\partial u_r}{\partial z} \right)^2 + \left[\frac{1}{r} \frac{\partial u_r}{\partial \theta} + r \frac{\partial}{\partial r}\left(\frac{u_\theta}{r} \right) \right]^2 \right\}. \tag{4-67}$$

Spherical Coordinates

The energy equation in spherical coordinates for constant ρ, μ, and k and zero radiation flux is:

$$\rho c_p \left(\frac{\partial T}{\partial t} + u_r \frac{\partial T}{\partial r} + \frac{u_\theta}{r} \frac{\partial T}{\partial \theta} + \frac{u_\varphi}{r \sin \theta} \frac{\partial T}{\partial \varphi} \right)$$

$$= k \left[\frac{1}{r^2} \frac{\partial}{\partial r}\left(r^2 \frac{\partial T}{\partial r} \right) + \frac{1}{r^2 \sin \theta} \frac{\partial}{\partial \theta}\left(\sin \theta \frac{\partial T}{\partial \theta} \right) + \frac{1}{r^2 \sin^2 \theta} \frac{\partial^2 T}{\partial \varphi^2} \right]$$

$$+ 2\mu \left[\left(\frac{\partial u_r}{\partial r} \right)^2 + \left(\frac{1}{r} \frac{\partial u_\theta}{\partial \theta} + \frac{u_r}{r} \right)^2 + \left(\frac{1}{r \sin \theta} \frac{\partial v_\varphi}{\partial \varphi} + \frac{u_r}{r} + \frac{u_\theta}{r} \cos \theta \right)^2 \right]$$

$$+ \mu \left\{ \left[r \frac{\partial}{\partial r}\left(\frac{u_\theta}{r} \right) + \frac{1}{r} \frac{\partial u_r}{\partial \theta} \right]^2 + \left[\frac{1}{r \sin \theta} \frac{\partial u_r}{\partial \varphi} + r \frac{\partial}{\partial r}\left(\frac{u_\varphi}{r} \right) \right]^2 \right.$$

$$\left. + \left[\frac{\sin \theta}{r} \frac{\partial}{\partial \theta}\left(\frac{u_\varphi}{\sin \theta} \right) + \frac{1}{r \sin \theta} \frac{\partial u_\theta}{\partial \varphi} \right]^2 \right\}. \tag{4-68}$$

REFERENCES

1. I. Shames, *Mechanics of Fluids*, McGraw-Hill, New York (1962).
2. V. L. Streeter, *Fluid Mechanics*, McGraw-Hill, New York (1962).
3. R. M. Olsen, *Essentials of Engineering Fluid Mechanics*, International Textbook Co., Scranton (1962).
4. J. F. Thorpe, "On the Momentum Theorem for a Continuous System of Variable Mass," *Amer. Physics J.* **30**, 637–640 (1962).

5. S. J. Kline, "Flow Visualization," National Committee for Fluid Mechanics Films, Enc. Brit. Educ. Corp. Chicago.
6. H. J. Schlichting, *Boundary Layer Theory*, 6th ed., McGraw-Hill, New York (1968).

PROBLEMS

1. What is the average velocity when 15 gallons of water per minute flow through a pipe with a 2-in. i.d. (inside diameter)? Assume the water temperature is 100°F.

2. An ideal gas at 80°F flows isothermally through a long pipe of constant diameter. At a section where the pressure is 300 psia, the average velocity is 110 ft/sec. At a section downstream, the pressure is found to be 260 psia. What is the average velocity of the gas at the second section?

3. An ideal gas initially at 140°F flows adiabatically through a pipe whose cross-sectional area varies with the length of the pipe as $A = A_0(1 + x/4)$, where x is the distance in feet. As the gas enters the pipe, it has a pressure of 300 psia and an average velocity of 100 ft/sec. Determine how far down the pipe the gas has traveled to a certain cross section where its pressure is 100 psia, its temperature is 40°F, and its average velocity is 50 fps.

4. Water at 100°F enters a pipe of uniform cross section at an average velocity of 0.3 fps, and is evaporated into steam as it moves through the pipe. The exit fluid is steam at 10 psia and 200°F. What is the average velocity of the exit stream?

5. Air flows through a constant area circular duct of 4-in. i.d. At a point along the flow where the pressure is 100 psia and the temperature is 80°F, the average velocity is 50 fps. Further downstream the pressure is 60 psia and the temperature is still 80°F. Determine the mass rate of flow through the pipe, in lb_m/hr and calculate the average velocity at the downstream point (where the pressure is 60 psia.) Assume a gas constant $R_g = 53.3$ ft $lb_f/lb_m°R$.

6. How many pounds of air will flow through a 3-in. i.d. pipe in 20 minutes if the density of the air is 0.060 lb_m/ft^3 and the average velocity of flow is 50 ft/sec?

7. Water at an average temperature of 100°F and at moderate pressure flows through a 1-in. i.d. pipe at the rate of 10 gal/min. Determine the mass rate of flow in lb_m/hr and the average fluid velocity in ft/sec ($V = \dot{m}/\rho A$).

8. An ideal gas flows through a pipe of constant diameter. At one cross section where its temperature is 80°F and pressure is 100 psia, the average velocity is 40 ft/sec. At a cross section downstream, the temperature is found to be 120°F and the pressure 60 psia. What is the average velocity at that cross section?

9. An ideal gas flows through a tube of uniform cross-sectional area. At one station along the tube where the temperature is 120°C and the pressure is 80 psia, the average velocity is 40 ft/sec. Further downstream the pressure is 60 psia and the temperature is 90°C. What is the average velocity at this station?

10. Air flows into a tank at a steady rate of 10 lb_m/hr. A leak in the tank lets out air at the rate of 4 lb_m/hr. The temperature of the air is essentially constant at 80°F. Estimate the rate of change of pressure with time if the volume of the tank is 12 ft^3.

11. An 8-in. i.d. pipe reduces into a 3-in. i.d. pipe. (a) If the flow rate is 200 gal/min, what are the average velocities in the two pipes? (b) If the fluid is water at 50°F, what is the mass flow rate?

12. Water flows through the branching pipe in the figure with average velocity of 7 ft/sec in the 6-in. i.d. pipe and 20 ft/sec in the 9-in. i.d. pipe. (a) What is the average velocity in the 12-in. i.d. pipe? (b) Would the answer change if the flow direction was reversed?

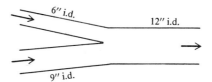

Problem 12

13. The average velocity of water entering one end of a 12-inch perforated pipe like that in the figure is 20 ft/sec. The perforations are arranged so that the velocity distribution through the pipe wall is linear. For a steady flow, calculate the average velocity V of the water leaving the pipe.

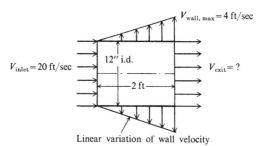

Linear variation of wall velocity

Problem 13

14. The average velocity of water at 50°F is 10 ft/sec in a 6-in. i.d. pipe. Determine the flow rate in ft^3/sec, slugs/sec, gal/min, and lb_m/min.

15. Derive the continuity equation in cylindrical coordinates by applying the principles of conservation of mass to a small element of volume.

16. Derive the continuity equation in spherical coordinates by applying the principle of conservation of mass to a small element of volume.

17. If distance is expressed in feet and velocity in ft/sec, what should be the units of the stream function ψ?

18. In a two-dimensional cylindrical coordinate system (defined by r and θ), show that the stream function would be defined by

$$\frac{1}{r}\frac{\partial \psi}{\partial \theta} = v_r \quad \text{and} \quad \frac{\partial \psi}{\partial r} = -v_\theta.$$

19. Find the forces R_x and R_y required to hold the plate in the figure in equilibrium if $\theta = 45°$, V_1 and $V_2 = 100$ ft/sec, and the flow rate is 0.50 ft³/sec. Neglect the weight of the plate and the water and assume one-dimensional flow.

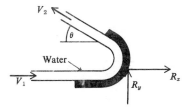

Problems 19 and 20

20. Solve Problem 19 for the case of $\theta = 0°$.

21. What is the force R and the orientation angle θ due to the water flowing through the nozzle in the diagram? The interior volume of the nozzle is 5 ft³. Assume one-dimensional flow.

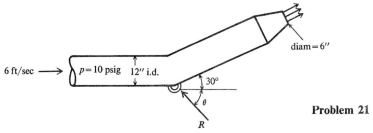

Problem 21

22. Water issues from the stationary nozzle in the figure at a velocity of 20 ft/sec relative to the nozzle. The jet then strikes a curved vane which deflects the flow upward. Find the thrust on the cart if the cart moves to the right with a uniform speed of 5 ft/sec.

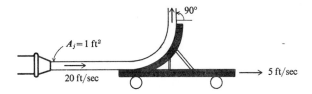

Problems 22, 23, and 24

23. Find the magnitude and direction of the velocity leaving the cart in Problem 22 relative to the stationary nozzle.

24. Work Problem 22, assuming that the cart is moving 5 ft/sec to the left.

25. The jet pump in the figure has a jet velocity V_j of 100 ft/sec and a jet area A_j of 0.04 ft². The jet entrains a secondary stream having an initial velocity V_s of 8 ft/sec in a constant-area pipe of 0.5 ft² area. The jet flow and secondary flow are thoroughly mixed at section 2. Neglect wall skin friction and assume one-

dimensional flow. (a) At section 2 what is the average velocity of the mixed flow? (b) Calculate the pressure difference $(p_2 - p_1)$. The pressure of the jet and secondary stream are the same at section 1. Both streams are water at 60°F.

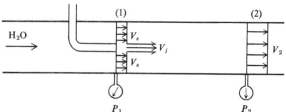

Problem 25

26. In the figure, water flows through a vertical 12-in. i.d. by 8-in. i.d. pipe bend at the rate of 10 ft³/sec. The volume of the bend is 3 ft³ and the entrance and exit pressures are 24.7 psia and 18.1 psia respectively. Calculate the forces on the bend due to the water. Assume one-dimensional flow.

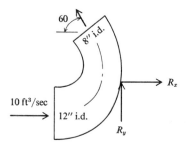

Problem 26

27. The velocity profile of a laminar boundary layer on a flat plate can be approximated by

$$u = V_\infty \left[\frac{3}{2}\frac{y}{\delta} - \frac{1}{2}\left(\frac{y}{\delta}\right)^3 \right].$$

Using the momentum theorem, find an expression for the drag on one side of a flat plate (see figure) as a function of ρ, V_∞, b, and δ at the rear of the plate (b is the width of the flat plate).

Problem 27

28. A wing tested in a wind tunnel is found to have its wake velocity distribution closely approximated by a cosine curve, as shown in the accompanying diagram. The minimum wake velocity is 55 percent of the free stream velocity. The wake width $w = 6$ in. and the free stream velocity is 100 fps. Find the drag on the

wing per foot of span. The pressure in the wake is equal to free stream pressure. $\rho = 0.002378$ slugs/ft³.

Problem 28

29. Derive an expression for the thrust of the jet engine in the figure in terms of the inlet and exit velocities (V_i and V_e), the mass flow of air ($\dot{m}_a$), the mass flow of fuel ($\dot{m}_f$), the exit pressure (p_e), and the exit area (A_e).

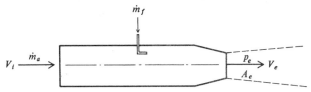

Problem 29

30. Derive Eq. (4-43a) from Eq. (4-42a).
31. Derive Eq. (4-43b) from Eq. (4-42b).
32. Derive Eq. (4-43c) from Eq. (4-42c).
33. Derive Eq. (4-44a) from Eq. (4-43a).
34. Derive Eq. (4-44b) from Eq. (4-43b).
35. Derive Eq. (4-44c) from Eq. (4-43c).
36. Does the fact that the density ρ appears outside the derivative in Eq. (4-39) and subsequent equations imply that ρ is constant?
37. Show that the dimensions of each term in Eqs. (4-45a-c) and (4-46a-c) can be expressed as force divided by length cubed.
38. Show that the left-hand side of Eq. (4-44a) can be written as

$$\rho \left[\frac{\partial u}{\partial t} + \frac{\partial u^2}{\partial x} + \frac{\partial (uv)}{\partial y} + \frac{\partial (uw)}{\partial z} \right].$$

Hint: Use the equation of continuity.

39. Air expands through a horizontal, adiabatic, converging nozzle. It enters the nozzle with a velocity of 20 ft/sec at a pressure of 100 psia and a temperature of 100°F. At the throat of the nozzle the air pressure is 52.8 psia and the temperature is 7°F. Find the velocity of the air in the throat of the nozzle. Assume uniform velocity at each cross section.

40. Using Newton's second law and the definition of work (work = force × distance), show that the amount of work expended on a mass m in increasing its velocity from zero velocity to velocity u is $\frac{1}{2}mu^2$. (Notice that $\frac{1}{2}mu^2$ is also its kinetic energy).

41. The inverse square law for the acceleration of gravity is

$$g = 32.2(r_0/r)^2,$$

where r_0 is the earth's radius, 3960 miles. (a) Using the definition of work, show that the work done in raising a mass, m, ΔZ distance above the earth is

$$m(32.2)r_0^2 \left[\frac{1}{r_0} - \frac{1}{r_0 + \Delta Z} \right].$$

(b) Show that if ΔZ is small compared with the earth's radius r_0, then the work is equal to $mg\Delta Z$, the change in potential energy.

42. Water is pumped in an insulated 3-in. pipe from a river to a nozzle in a manufacturing plant 80.0 ft above the river. The river temperature is 50°F and the rate of pumping is 150 gal/min. The pressure at the nozzle entrance is 2 psig and the power input to the pump is 15.0 hp. Determine the temperature of the water as it enters the nozzle. Assume uniform velocity at each cross section.

43. The pressure and velocity of water at the inlet of a pump is 15 psia and 20 ft/sec respectively. If the pump does work at the rate of 10 hp in pumping 1000 gal/min, find the pressure at the pump outlet. Assume no heat transfer, no change in internal energy, and flat velocity profiles. The inlet and outlet diameters are equal and at the same elevation.

44. Air at 200 psia and 120°F enters a rough, insulated 12-in. i.d. pipe at 40 ft/sec. At a downstream section, the velocity is 500 ft/sec. (a) What is the temperature at the downstream section? (b) What is the pressure at the downstream section? (c) What is the total wall shear force?

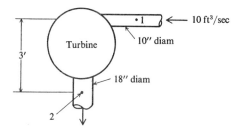

Problem 45

45. The inlet and outlet pressures of the turbine shown in the figure are 27 psig and −1 psig respectively. Water is flowing at the rate of 10 ft³/sec. Neglecting heat transfer and assuming flat velocity profiles, calculate the horsepower delivered to the turbine from the water.

46. An evacuated tank has a volume of 50 ft³. Find the temperature of air in the tank if a valve is opened and air rushes into the tank. Ambient air temperature

and pressure is 70°F and 14.7 psia. Assume that the tank is insulated so that no heat is transferred from the tank to the air, and neglect change in potential energy.

47. A small evacuated container is connected to a large air line that is at a pressure of 100 psia and a temperature of 140°F. If the container is filled with the air to the line pressure, what is the final temperature of the air in the container? (Assume heat losses and the energy storage of the container walls to be zero.)

48. A large reservoir contains air at 500°F and 150 psia. What is the maximum air velocity obtainable in a nozzle connected to this reservoir?

49. Prove that the velocity V_2 at some point for steady flow in a nozzle can be expressed as

$$V_2 = \left[1 - \left(\frac{A_2 \rho_2}{A_1 \rho_1}\right)^2\right]^{-1/2} \sqrt{2(j_1 - j_2)}.$$

Assume zero heat losses and flat velocity profiles.

50. An air and water mixture enters a dehumidifier at the rate of 800 lb_m/hr with an enthalpy of 50 Btu/lb_m. Water with an enthalpy of 15 Btu/lb_m drains out of the dehumidifier at the rate of 17 lb_m/hr. The air and water mixture leaving the dehumidifier has an enthalpy of 19 Btu/lb_m. Determine the heat removed in Btu/hr. Neglect kinetic and potential energy changes.

51. Show that for two-dimensional flow

$$-\left[\frac{\partial(\rho e u)}{\partial x} + \frac{\partial}{\partial y}(\rho e v)\right] = \frac{\partial}{\partial t}(\rho e) - \rho \frac{De}{Dt}.$$

52. Show that p (div V) can be expressed as

$$\rho \frac{D}{Dt}\left(\frac{p}{\rho}\right) - \frac{Dp}{Dt},$$

and, by assuming that the fluid is an ideal gas with specific heats which depend only on temperature, show that the Equation (4.64) can be written as

$$\rho c_p \frac{DT}{Dt} = \frac{Dp}{Dt} + k\left(\frac{\partial^2 T}{\partial x^2} + \frac{\partial^2 T}{\partial y^2} + \frac{\partial^2 T}{\partial z^2}\right) + \mu \Phi - \text{div } \dot{\mathbf{q}}_r.$$

CHAPTER 5

DIMENSIONAL ANALYSIS

The serious student of fluid mechanics and heat transfer takes a great step forward when he forms the habit of generalizing both analytical and experimental results. The nondimensionalizing technique is of great value in making such generalizations. If a quantity or equation is dimensionless or made up of dimensionless terms, it is applicable in all dimensional systems (Tables 1-1 and 5-1) and may be manipulated mathematically with great freedom. The procedure of nondimensionalizing will now be developed.

5-1 DIMENSIONLESS GROUPS

Dimensional systems and units have been discussed in Chapter 1. It was stated that in order to be *dimensionally homogeneous*, an equation must consist of terms having the same dimensions. Further, it has been shown by Bridgeman [1] that any dependent variable must have dimensions equal to a product of those of the independent variables, each raised to a power (positive, negative, or fractional); for example, this may be stated as

$$y = x_1^a \cdot x_2^b \cdot x_3^c, \ldots, x_n^n, \qquad (5\text{-}1)$$

where y is the dependent variable and the x's are the independent variables. This rule becomes important in dimensional analysis.

The *dimensionless group* is the key to all dimensional analysis. It is a grouping of variables arranged in such a way that the group has no dimensions. Such a group is called a pi (π) group:

$$\pi = x_1^a \cdot x_2^b \cdot x_3^c, \ldots, x_n^n.$$

For example, using an MLT system of dimensions,

$$\pi \triangleq M^0 L^0 t^0 \triangleq 1.$$

An example of a dimensionless group is the pressure coefficient $\Delta p/(\tfrac{1}{2}\rho V^2)$. Checking the dimensions, $\pi = \Delta p/(\tfrac{1}{2}\rho V^2)$,

$$\Delta p \triangleq \frac{M}{Lt^2} \qquad V^2 \triangleq \left(\frac{L}{t}\right)^2 \qquad \rho \triangleq \frac{M}{L^3}.$$

Therefore

$$\pi \triangleq \frac{(M/Lt^2)}{(M/L^3)(L/t)^2} \triangleq M^0L^0t^0 = 1,$$

which shows that it is indeed a pi group.

The general scheme of *dimensional analysis* will be to use one or more of several available techniques to replace the basic functional relationship that exists between the n variables by a relationship between fewer than n dimensionless groups.

Analysis of the behavior of any system can be made by varying the value of each of the dimensionless groups in turn. This results in fewer experiments and much improved possibilities for the presentation of data. For example, the following relationship may exist:

$$r = f(x,y).$$

This relationship may be represented graphically as a result of conducting experiments, holding y constant at several desired values (four in our example). If r is determined at four values of x, the results would appear as in Fig. 5–1.

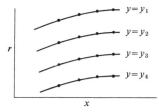

Fig. 5–1 A plot of r vs. x for fixed values of y.

This is a very satisfactory presentation of the data. However, if one more variable is added, $r = f(x,y,z)$, the presentation becomes more complex. For each of various values of z (assume four in this example), the experiments required in the case of two independent variables must be repeated; these results are shown in Fig. 5–2.

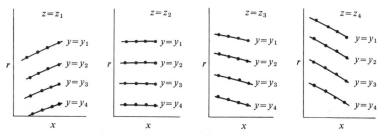

Fig. 5–2 A plot of r vs. x for fixed values of y and z.

Table 5-1 DIMENSIONS OF COMMON VARIABLES

Dimensional system	$MLTt$	$FLTt$	$FMLTtQ$*
Mass (m)	M	$\dfrac{Ft^2}{L}$	M
Force (f)	$\dfrac{ML}{t^2}$	F	F
Length (L)	L	L	L
Time (t)	t	t	t
Temperature (T)	T	T	T
Work (W)	$\dfrac{ML^2}{t^2}$	FL	FL
Heat (q)	$\dfrac{ML^2}{t^2}$	FL	Q
Acceleration (a)	$\dfrac{L}{t^2}$	$\dfrac{L}{t^2}$	$\dfrac{L}{t^2}$
Frequency (N)	$\dfrac{1}{t}$	$\dfrac{1}{t}$	$\dfrac{1}{t}$
Area (A)	L^2	L^2	L^2
Coefficient of thermal expansion (β)	$\dfrac{1}{T}$	$\dfrac{1}{T}$	$\dfrac{1}{T}$
Density (ρ)	$\dfrac{M}{L^3}$	$\dfrac{Ft^2}{L^4}$	$\dfrac{M}{L^3}$
Dimensional constant (g_c)	1.0	1.0	$\dfrac{ML}{t^2F}$
Specific heat at constant pressure (c_p); at constant volume (c_v)	$\dfrac{L^2}{t^2T}$	$\dfrac{L^2}{t^2T}$	$\dfrac{Q}{MT}$
Heat transfer coefficient (h); overall (U)	$\dfrac{M}{t^3T}$	$\dfrac{F}{tLT}$	$\dfrac{Q}{tL^2T}$

* This system, called the English Engineering System, is commonly used. Dimensional analysis may, however, be more conveniently carried out with one of the four-dimension systems, MLTt or FLTt.

Table 5-1 (*Continued*)

Dimensional system	MLTt	FLTt	FMLTtQ*
Work rate ($\dot{W}$)	$\dfrac{ML^2}{t^3}$	$\dfrac{FL}{t}$	$\dfrac{FL}{t}$
Heat flow rate ($\dot{q}$)	$\dfrac{ML^2}{t^3}$	$\dfrac{FL}{t}$	$\dfrac{FL}{t}$
Kinematic viscosity (ν)	$\dfrac{L^2}{t}$	$\dfrac{L^2}{t}$	$\dfrac{L^2}{t}$
Mass flow rate ($\dot{m}$)	$\dfrac{M}{t}$	$\dfrac{Ft}{L}$	$\dfrac{M}{t}$
Mechanical equivalent of heat (J)			$\dfrac{FL}{Q}$
Pressure (p)	$\dfrac{M}{Lt^2}$	$\dfrac{F}{L^2}$	$\dfrac{F}{L^2}$
Surface tension (σ)	$\dfrac{M}{t^2}$	$\dfrac{F}{L}$	$\dfrac{F}{L}$
Angular velocity (ω)	$\dfrac{1}{t}$	$\dfrac{1}{t}$	$\dfrac{1}{t}$
Volume flow rate ($\dot{m}/\rho$) = $\dot{Q}$	$\dfrac{L^3}{t}$	$\dfrac{L^3}{t}$	$\dfrac{L^3}{t}$
Thermal conductivity (k)	$\dfrac{ML}{t^3 T}$	$\dfrac{F}{tT}$	$\dfrac{Q}{LtT}$
Thermal diffusivity (α)	$\dfrac{L^2}{t}$	$\dfrac{L^2}{t}$	$\dfrac{L^2}{t}$
Velocity (V)	$\dfrac{L}{t}$	$\dfrac{L}{t}$	$\dfrac{L}{t}$
Viscosity, absolute (μ)	$\dfrac{M}{Lt}$	$\dfrac{Ft}{L^2}$	$\dfrac{M}{Lt}$
Volume (Vol)	L^3	L^3	L^3

Addition of yet another variable, w, would result in repeating Fig. 5-2 (and conducting experiments to provide the data) for as many separate values of w as one may wish to examine. For four values of w, a total of 16 separate graphs would result.

It should be noted that the use of four experimental determinations for each curve in Figs. 5-1 and 5-2 and the decision to examine the phenomenon at only four values of y in Fig. 5-2 and four values of w in the last example are perfectly arbitrary and, in many cases, not sufficient to make a complete determination. If the choice had been to use five values of z and w in the experimental determination in place of four in the above cases, then Fig. 5-2 would have had five separate graphs, and 25 separate graphs would have been required to account for three independent variables.

Thus, the problem of handling a greater number of variables multiplies the difficulty. Further, the investigator has more and more difficulty gaining real insight into the phenomenon with this mass of data.

Dimensional analysis can have great value if a reduction can be accomplished in the number of graphs needed and experiments required. The use of the concept of the dimensionless group will accomplish this. For example, it is entirely possible that the case of $r = f_1(x,y,z,w)$ may reduce to

$$\pi_1 = f_2(\pi_2).$$

The experimenter may then determine π_1 at various values of π_2 (four, for example), and the result may be plotted as shown in Fig. 5-3, which replaces the 16 figures previously required.

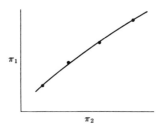

Fig. 5-3 A plot of the dimensionless functions π_1 and π_2.

In this situation the experimenter reduced his job to performing four tests which replaced the 64 tests required to produce the data previously required.

Once the graph of $\pi_1 = f_2(\pi_2)$ is obtained as in Fig. 5-3, information may be gained as to the nature of the functional relationship. If, for example, the data plots in a straight line which has a slope m and a π_1 intercept of $(\pi_1)_0$, then $\pi_1 = (\pi_1)_0 + m\pi_2$. Of course, this is only an assumed example and limitless other functional relationships may exist.

Dimensional Analysis by Nondimensionalizing the Basic Differential Equations

One technique of performing a dimensional analysis is to examine the governing differential equations and other pertinent relationships of a phenomenon using the tool of nondimensionalizing. This procedure will be developed using the following example.

A cylinder of diameter D is submerged in a flowing fluid with constant values of the free stream velocity V_∞, the density ρ, and the viscosity μ. Near the cylinder the velocity components in the x and y directions are u and v respectively. Figure 5-4 shows the arrangement, and all planes parallel to the plane of the figure are assumed identical (two-dimensional flow, infinitely long cylinder). Buoyancy is neglected. It is desired to calculate the force on this cylinder caused by the flowing fluid.

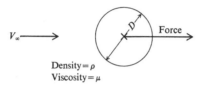

Fig. 5-4 Determination of drag on a cylinder in crossflow.

For this incompressible two-dimensional steady flow, the equation for conservation of mass is

$$\frac{\partial u}{\partial x} + \frac{\partial v}{\partial y} = 0, \qquad (5-2)$$

and the Navier-Stokes equations are:

$$\rho\left(u\frac{\partial u}{\partial x} + v\frac{\partial u}{\partial y}\right) = -\frac{\partial p}{\partial x} + \mu\left(\frac{\partial^2 u}{\partial x^2} + \frac{\partial^2 u}{\partial y^2}\right), \qquad (5-3)$$

$$\rho\left(u\frac{\partial v}{\partial x} + v\frac{\partial v}{\partial y}\right) = -\frac{\partial p}{\partial y} + \mu\left(\frac{\partial^2 v}{\partial x^2} + \frac{\partial^2 v}{\partial y^2}\right). \qquad (5-4)$$

In order to nondimensionalize Eqs. (5-2), (5-3), and (5-4), all lengths may be divided by an arbitrary reference length which will be taken as the diameter D of the cylinder. All velocities may be divided by an arbitrary reference velocity which, in this case, may be the free stream velocity V_∞, and pressures may be divided by ρV_∞^2, twice the free stream dynamic pressure. An asterisk is used to denote the nondimensionalized property; for example,

$$u^* = \frac{u}{V_\infty} \quad \text{and} \quad v^* = \frac{v}{V_\infty}. \qquad (5-5)$$

The following example shows how the technique is applied to one of the terms of an equation. Selecting the second term of Eq. (5-3) and multiplying

116   DIMENSIONAL ANALYSIS   5–1

by V_∞/V_∞ and D/D where appropriate,

$$v\frac{\partial u}{\partial y} = v\frac{V_\infty}{V_\infty}\frac{\partial\left(u\frac{V_\infty}{V_\infty}\right)}{\partial\left(y\frac{D}{D}\right)} = \frac{v}{V_\infty}V_\infty\frac{\partial\left(\frac{u}{V_\infty}\right)V_\infty}{\partial\left(\frac{y}{D}\right)D} = \frac{V_\infty^2}{D}v^*\frac{\partial u^*}{\partial y^*}.$$

Using this scheme and rearranging, one obtains:

$$\frac{\partial u^*}{\partial x^*} + \frac{\partial v^*}{\partial y^*} = 0, \tag{5-6}$$

$$u^*\left(\frac{\partial u^*}{\partial x^*}\right) + v^*\left(\frac{\partial u^*}{\partial y^*}\right) = -\frac{\partial p^*}{\partial x^*} + \frac{v}{V_\infty D}\left(\frac{\partial^2 u^*}{\partial x^{*2}} + \frac{\partial^2 u^*}{\partial y^{*2}}\right), \tag{5-7}$$

$$u^*\left(\frac{\partial v^*}{\partial x^*}\right) + v^*\left(\frac{\partial v^*}{\partial y^*}\right) = -\frac{\partial p^*}{\partial y^*} + \frac{v}{V_\infty D}\left(\frac{\partial^2 v^*}{\partial x^{*2}} + \frac{\partial^2 v^*}{\partial y^{*2}}\right). \tag{5-8}$$

The above equations are seen to contain only the dimensionless dependent variables u^*, v^*, and p^*, the dimensionless independent variables x^* and y^*, and the grouping of variables $v/(V_\infty D)$, which may be considered to be one parameter. A check of this grouping's dimensions shows that it is dimensionless:

$$\frac{v}{V_\infty D} \triangleq \frac{L^2/t}{L(L/t)} \triangleq 1.0.$$

The reciprocal of this group was proposed by Osborne Reynolds in 1883 as a similarity parameter for fluid flow. It is called the *Reynolds number* in his honor and is designated by the symbol Re:

$$\mathrm{Re} = \frac{V_\infty D}{v}. \tag{5-9}$$

A discussion of the concept of similarity in fluid flow is included later in this chapter.

In the example considered here, there are three unknowns: u^*, v^*, and p^*. Equations (5–6), (5–7), and (5–8) contain these unknowns. A solution of this set of equations would provide knowledge of these unknowns and allow for the calculation of u, v, p, and the force on the submerged body.

In the case described here, this force is the drag in the cylinder. *Drag* is defined as the force in the direction of the free stream velocity at infinity exerted on a submerged body by the flow. Similarly, the net force perpendicular to the free stream velocity at infinity is called the *lift*. In the case of the symmetrical body such as a circular cylinder, there is no net force in the direction perpendicular to the free stream velocity; the lift is zero.

The drag on a body consists of two parts: (1) The *skin friction drag* is the net component in the direction of the free stream velocity (velocity at a considerable distance from the body) of the frictional or shearing effect of a viscous fluid on the surface of the body; and (2) the *pressure drag*, or *form drag*, is the net of the pressure forces acting on the body in the direction of the free stream velocity.

Pressure drag has its origin in two phenomena: lift and flow separation. The pressure drag associated with lift is called *induced drag* and is caused by downward, induced velocities which arise from the vortices created by a lifting body such as an airfoil of finite span. In the case of the circular cylinder, there is no induced drag because no lift is created. The pressure drag which is associated with flow separation is called *wake drag*.

In all real fluids, because of frictional effects, the fluid layer just adjacent to the surface of a body does not continue to flow along the surface all the way to the rear of the body. At some point along the body this fluid layer will separate from the surface and a *wake* or eddying region is observed at the rear of the body. This is shown in Fig. 9–13, where the streamline represented by the surface of the body is seen to detach itself and leave the body surface at an angle. The pressures in the rear portion of the body, where the flow has separated, are lower than the pressures on the front portion of the body, and the wake drag occurs.

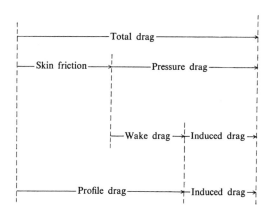

Fig. 5–5 Components of total drag.

The sum of the wake drag and skin friction drag is called *profile drag* because it is determined by the local shape (profile) of the body. Thus, there are two alternate ways of classifying drag, according to (1) whether drag comes from pressure or skin friction, or (2) whether drag depends upon lift or the profile of the body. These terms are clarified in Fig. 5–5.

The skin friction may be calculated by integrating the local shear stress at the wall over the entire surface area of the cylinder. Integration of the pressure over the surface will yield the pressure drag.

An analysis of this procedure follows.

An inspection of Eqs. (5–6), (5–7), and (5–8) indicates that the solution yields results of the form:

$$u^* = f_1(x^*, y^*, \text{Re}), \qquad (5\text{–}10)$$

$$v^* = f_2(x^*, y^*, \text{Re}), \qquad (5\text{–}11)$$

$$p^* = f_3(x^*, y^*, \text{Re}). \qquad (5\text{–}12)$$

As has been said, the solution may not be known but some insights can be obtained by examining the above relationships.

From Eqs. (5–10), (5–11), and (5–12) a conversion may be made to polar coordinates centered at the center line of the cylinder. Letting the dimensionless radius be defined by $r^* = r/R$, results in the following:

$$u_r^* = f_4(\theta, r^*, \text{Re}), \qquad (5\text{–}13)$$

$$u_\theta^* = f_5(\theta, r^*, \text{Re}), \qquad (5\text{–}14)$$

$$p^* = f_6(\theta, r^*, \text{Re}). \qquad (5\text{–}15)$$

Note: θ is of itself nondimensional.

In order to obtain the drag from the integration of shear stress and pressure, Newton's law of friction, Eq. (3–21), is used. The relationship for shear stress in polar coordinates is

$$\tau_w = \mu \left(\frac{\partial u_\theta}{\partial r} \right)_{r=R}.$$

Nondimensionalizing this equation (by $\frac{1}{2}\rho V_\infty^2$ for τ_w, by V_∞ for u_θ, and by R for r),

$$\tau_w^* = f_7 \left[\text{Re}, \left(\frac{\partial u_\theta^*}{\partial r^*} \right)_{r^*=1} \right].$$

But from Eq. (5–14),

$$\left(\frac{\partial u_\theta^*}{\partial r^*} \right)_{r^*=1} = f_8(\text{Re}, \theta),$$

so that

$$\tau_w^* = f_9(\text{Re}, \theta). \qquad (5\text{–}16)$$

Similarly,

$$(p^*)_{r^*=1} = f_{10}(\text{Re}, \theta) \qquad (5\text{–}17)$$

is the expression for the nondimensional pressure at the surface.

Equations (5–16) and (5–17) now provide the basis for integration over the surface.

The drag of a body is usually proportional to the size or area of the body. The area used is normally the cross-sectional area projected on the plane perpendicular to the direction of the free stream velocity. In the case of a circular cylinder of unit length this area is $(D)(1) = 2R$. Drag is also nondimensionalized and is given the symbol C_D, called the *drag coefficient**:

$$C_D = \frac{\text{drag}/A}{\frac{1}{2}\rho V_\infty^2} = \frac{\text{drag}/2R}{\frac{1}{2}\rho V_\infty^2} \quad \text{(for a cylinder of unit length)}. \quad (5\text{--}18)$$

If pressure drag is referred to as $(\text{drag})_p$ and skin friction as $(\text{drag})_f$, then, because of symmetry,

$$\text{drag}_p = 2\int_0^\pi p \cos\theta\, R\,d\theta \quad \text{or} \quad \frac{\text{drag}_p}{A} = \frac{\text{drag}_p}{2R} = \int_0^\pi p \cos\theta\, d\theta,$$

$$\text{drag}_f = 2\int_0^\pi \tau_w \sin\theta\, R\,d\theta \quad \text{or} \quad \frac{\text{drag}_f}{A} = \frac{\text{drag}_f}{2R} = \int_0^\pi \tau_w \sin\theta\, d\theta,$$

and

$$\frac{\text{drag}}{A} = \frac{\text{drag}}{2R} = \frac{\text{drag}_p}{2R} + \frac{\text{drag}_f}{2R} = \int_0^\pi p \cos\theta\, d\theta + \int_0^\pi \tau_w \sin\theta\, d\theta.$$

Dividing by $\frac{1}{2}\rho V_\infty^2$,

$$C_D = C_{D_p} + C_{D_f}$$

$$= \int_0^\pi p^* \cos\theta\, d\theta + \int_0^\pi \tau_w^* \sin\theta\, d\theta.$$

But using τ^* and p^* from Eq. (5–16) and (5–17),

$$C_D = \int_0^\pi f_{10}(\text{Re},\theta) \cos\theta\, d\theta + \int_0^\pi f_9(\text{Re},\theta) \sin\theta\, d\theta.$$

Although the functions $f_{10}(\text{Re},\theta)$ and $f_9(\text{Re},\theta)$ are not known, an examination shows that the evaluation of the two integrals above will yield an expression which is a function of the Reynolds number only, say $f_{11}(\text{Re})$. As the subscripts 1 through 11 have been used only to clarify the fact that these were different functions, this subscript may now be dropped. Therefore,

$$C_D = f(\text{Re}),$$

which is of the form $\pi_1 = f(\pi_2)$.

Thus, the dimensionless drag coefficient C_D of a cylinder is dependent only on the Reynolds number. Experiments which yield a plot of C_D vs. Re can be used for other cylinders of various diameters in fluids with a variety of values of density and viscosity and for a range of free stream velocity values.

---

* The use of the free stream dynamic pressure to nondimensionalize is customary, rather than the use of twice that pressure as in the example of Fig. 5–4. In a similar manner, lift per unit area may be nondimensionalized to give a *lift coefficient*, C_L:

$$C_L = \frac{\text{lift}/A}{\frac{1}{2}\rho V_\infty^2}. \quad (5\text{--}19)$$

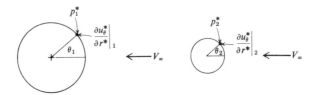

Fig. 5-6 Cylinders in similar flow situations.

This result is an introduction to the theory of similarity. The above result may be further clarified by reference to Fig. 5-6 and the discussion which follows.

Consider two cylinders in similar flow situations. At the (geometrically) corresponding points 1 and 2 on these cylinders, $r^* = 1$ and $\theta_1 = \theta_2$, so that, from Eqs. (5-16) and (5-17),

$$p^*\Big|_{\substack{r^*=1 \\ \theta=\theta_1}} = f_1(\text{Re}) \quad \text{and} \quad \tau^*\Big|_{\substack{r^*=1 \\ \theta=\theta_1=\theta_2}} = f_2(\text{Re}).$$

Therefore, for the same value of Reynolds number over the two cylinders, there is at points 1 and 2 the same nondimensional shear stress and pressure. The same equality exists between pressures and shear stresses at all sets of corresponding points. It follows that the nondimensional drag that depends only on these would be identical for the two cases. This leads again to the conclusion that the dimensionless drag is a function only of the Reynolds number. The result can be further generalized for all flows over all circular cylinders where the stated assumptions are valid (constant properties, flow normal to the axis of the cylinder). The relationship between C_D and Re for flow over cylinders is shown in Fig. 5-7.

It should be noted that the preceding discussion and the results shown in Fig. 5-7 are not valid when the flow speed becomes larger than roughly one-half of the speed of sound. When this occurs, another dimensionless parameter, the *Mach number*, M, (ratio of flow speed to speed of sound) comes into prominence, and then $C_D = f(\text{Re}, \text{M})$.

Unfortunately, solutions to the sets of equations such as (5-6), (5-7), and (5-8) are available for only a limited number of cases. The above analysis did yield some insights into the basic nature of the flow, however, including the development of the Reynolds number as an important dimensionless grouping of variables.

A more complex situation arises when the effects of temperature must be considered. This situation is described below.

Consider a body with a constant, uniform surface temperature T_w submerged in a flowing fluid (Fig. 5-8). The temperature of the fluid a large distance away from the surface in the free stream is T_∞ and is assumed constant.

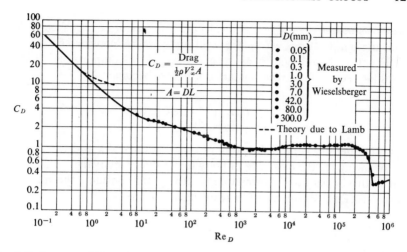

Fig. 5-7 Drag coefficient for circular cylinders. From *Boundary Layer Theory* by Schlichting. Copyright 1968, 1960, 1955 by McGraw-Hill Book Company, Inc. Used by permission of McGraw-Hill Book Company.

Consider that heat is being added to the fluid from the solid body, which might be a copper tube with a heating element inside. The heating element maintains the constant surface temperature T_w. The local temperature at each point in the fluid is constant with time since the stream temperature and velocity and the wall temperature are steady. The local fluid temperature near the body is T (a variable depending on location). The temperature differences of importance are $(T - T_\infty)$, and $(T_w - T_\infty)$. A dimensionless ratio may be constructed in this fashion:

$$T^* = \frac{T - T_\infty}{T_w - T_\infty}. \tag{5-20}$$

This dimensionless temperature (T^*) is the ratio of the fluid temperature above the free stream datum level (T_∞) to the temperature of the wall above that same datum. The ratio is the fractional part of the total possible fluid temperature change and has a value which varies from 0 (at large distances from the body) to 1.0 (on the surface of the body). For the present analysis,

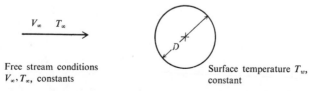

Fig. 5-8 Heat transfer from a cylinder to airstream.

interest is centered only in the local fluid temperature at each point in the fluid; the temperatures within the solid body are not considered.

For this two-dimensional steady flow of a compressible fluid, the continuity equation in rectangular coordinates becomes

$$\frac{\partial}{\partial x}(\rho u) + \frac{\partial}{\partial y}(\rho v) = 0. \tag{5-21}$$

Note that density is not a function of time but only of x and y.

Allowing for the existence of body forces due to buoyancy and assuming that both viscosity and thermal conductivity are constant, the momentum equations, Eqs. (4-44a-b), after rearrangement, are:

$$\rho\left(u\frac{\partial u}{\partial x} + v\frac{\partial u}{\partial y}\right) = -\frac{\partial p}{\partial x} + \rho g_x \beta(T - T_\infty)$$
$$+ \mu\left[\frac{\partial^2 u}{\partial x^2} + \frac{\partial^2 u}{\partial y^2} + \frac{1}{3}\frac{\partial}{\partial x}\left(\frac{\partial u}{\partial x} + \frac{\partial v}{\partial y}\right)\right], \tag{5-22}$$

$$\rho\left(u\frac{\partial v}{\partial x} + v\frac{\partial v}{\partial y}\right) = -\frac{\partial p}{\partial x} + \rho g_y \beta(T - T_\infty)$$
$$+ \mu\left[\frac{\partial^2 v}{\partial x^2} + \frac{\partial^2 v}{\partial y^2} + \frac{1}{3}\frac{\partial}{\partial y}\left(\frac{\partial u}{\partial x} + \frac{\partial v}{\partial y}\right)\right]. \tag{5-23}$$

In Eqs. (5-22) and (5-23), g_x and g_y represent the components of the gravitational acceleration vector. The term β is the coefficient of thermal expansion of the fluid (previously defined).

For the above conditions, the energy equation (see Problem 52, Chapter 4) is

$$\rho c_p\left(u\frac{\partial T}{\partial x} + v\frac{\partial T}{\partial y}\right) = k\left(\frac{\partial^2 T}{\partial x^2} + \frac{\partial^2 T}{\partial y^2}\right) + u\frac{\partial p}{\partial x} + v\frac{\partial p}{\partial y} + \mu\Phi, \tag{5-24}$$

where Φ, the two-dimensional dissipation function, is

$$\Phi = 2\left(\frac{\partial u}{\partial x}\right)^2 + 2\left(\frac{\partial v}{\partial y}\right)^2 + \left(\frac{\partial v}{\partial x} + \frac{\partial u}{\partial y}\right)^2 - \frac{2}{3}\left(\frac{\partial u}{\partial x} + \frac{\partial v}{\partial y}\right)^2. \tag{5-25}$$

A solution to the problem described by Eqs. (5-21) through (5-25) would give the temperature at each point in the fluid, and this information could be used with the Fourier equation to obtain the rates of heat transfer from the cylinder to the stream. An exact solution would be a simultaneous solution to the equations of continuity, momentum, and energy, and to an appropriate equation of state of the fluid. The unknown dependent variables in these equations are u, v, ρ, T, and p. These five variables are related by the five simultaneous equations. Application of the boundary conditions can be made provided a general solution of the differential equations is known. Instead, dimensional analysis will be employed to obtain some insight into the problem and perhaps to give information which will assist in correlation of available experimental data. As before, this dimensional analysis will be based upon the method of nondimensionalizing the differential equations that describe the problem.

In order to nondimensionalize Eqs. (5-21-25), the procedures described previously will be used. Since T_w and T_∞ are constant, the energy equation can be rewritten in terms of T^* to give

$$\rho c_p \left(u \frac{\partial T^*}{\partial x} + v \frac{\partial T^*}{\partial y} \right) = k \left(\frac{\partial^2 T^*}{\partial x^2} + \frac{\partial^2 T^*}{\partial y^2} \right)$$
$$+ \left(\frac{1}{T_w - T_\infty} \right) \left(u \frac{\partial p}{\partial x} + v \frac{\partial p}{\partial y} \right) + \frac{\mu \Phi}{(T_w - T_\infty)}. \quad (5\text{-}26)$$

The body force due to buoyancy will be assumed to act only in the y direction. The asterisk is used to denote all nondimensionalized properties. If we nondimensionalize lengths, velocities, and pressures (with D, V_∞, and ρV_∞^2 respectively) and rearrange, we get:

$$u^* \frac{\partial u^*}{\partial x^*} + v^* \frac{\partial u^*}{\partial y^*} = -\frac{\partial p^*}{\partial x^*} + \frac{\nu}{V_\infty D} \left[\frac{\partial^2 u^*}{\partial x^{*2}} + \frac{\partial^2 u^*}{\partial y^{*2}} + \frac{1}{3} \frac{\partial}{\partial x^*} \left(\frac{\partial u^*}{\partial x^*} + \frac{\partial v^*}{\partial y^*} \right) \right],$$
$$(5\text{-}27)$$

$$u^* \frac{\partial v^*}{\partial x^*} + v^* \frac{\partial v^*}{\partial y^*} = -\frac{\partial p^*}{\partial y^*} + \frac{g\beta D(T_w - T_\infty)T^*}{V_\infty^2}$$
$$+ \frac{\nu}{V_\infty D} \left[\frac{\partial^2 v^*}{\partial x^{*2}} + \frac{\partial^2 v^*}{\partial y^{*2}} + \frac{1}{3} \frac{\partial}{\partial y^*} \left(\frac{\partial u^*}{\partial x^*} + \frac{\partial v^*}{\partial y^*} \right) \right], \quad (5\text{-}28)$$

$$u^* \frac{\partial T^*}{\partial x^*} + v^* \frac{\partial T^*}{\partial y^*} = \frac{k}{\rho c_p V_\infty D} \left(\frac{\partial^2 T^*}{\partial x^{*2}} + \frac{\partial^2 T^*}{\partial y^{*2}} \right)$$
$$+ \frac{V_\infty^2}{c_p(T_w - T_\infty)} \left(u^* \frac{\partial p^*}{\partial x^*} + v^* \frac{\partial p^*}{\partial y^*} \right) + \frac{\mu V_\infty}{\rho c_p D(T_w - T_\infty)} \Phi^*. \quad (5\text{-}29)$$

Equations (5-27), (5-28), and (5-29) are the dimensionless or nondimensionalized equations of motion in the x direction and the y direction, and the energy equation. The continuity equation does not make any additional contribution to the dimensional analysis.

In addition to their providing starting points for solutions, the parameters $(V_\infty, D, T_w - T_\infty)$ used in the boundary conditions affect the solution of Eqs. (5–27), (5–28), and (5–29) through the coefficient terms of these equations. These coefficient terms are dimensionless. Also, as each term of the above equations is dimensionless and as the variables and their derivatives have been nondimensionalized, the coefficients *must* be dimensionless. In order of appearance in these equations, these dimensionless coefficients are:

$$\pi_1 = \frac{\nu}{V_\infty D},$$

$$\pi_2 = \frac{g\beta(T_w - T_\infty)D}{V_\infty^2},$$

$$\pi_3 = \frac{k}{\rho c_p V_\infty D},$$

$$\pi_4 = \frac{V_\infty^2}{c_p(T_w - T_\infty)},$$

$$\pi_5 = \frac{\nu V_\infty}{c_p D(T_w - T_\infty)}.$$

Since $\pi_5 = \pi_4 \pi_1$, there are only four independent coefficients or groupings of properties. Of these $(1/\pi_1)$ is the Reynolds number.

An examination of π_2 shows that it may be arranged thus:

$$\pi_2 = \frac{g\beta(T_w - T_\infty)D}{V_\infty^2} = \frac{g\beta(T_w - T_\infty)D^3}{\nu^2} \cdot \frac{\nu^2}{V_\infty^2 D^2}.$$

If we let

$$\text{Gr} = \frac{g\beta(T_w - T_\infty)D^3}{\nu^2},$$

then

$$\pi_2 = \frac{\text{Gr}}{\text{Re}^2}.$$

The quantity Gr is the *Grashof number*, which will prove to be of importance in certain classes of heat transfer problems where buoyancy effects are important.

Considering π_3 in a similar way yields

$$\pi_3 = \frac{k}{\rho c_p V_\infty D} = \frac{\alpha}{\nu} \frac{\nu}{V_\infty D} = \frac{1}{\text{Pr Re}},$$

where

$$\alpha = \frac{k}{\rho c_p}. \tag{5–30}$$

The quantity α is the *thermal diffusivity*. It contains only the thermophysical properties of the fluid so that it is a composite property of the fluid, and

$$\text{Pr} = \frac{\nu}{\alpha}. \tag{5-31}$$

The quantity Pr is the *Prandtl number*. Since it also contains only thermophysical properties of the fluid, it too may be considered to be a composite property.

The value of the Prandtl number is very important in almost all heat transfer computations involving fluid flow. In fact, basic assumptions used in deriving certain equations depend upon the magnitude of this parameter. The Prandtl number compares the rate of diffusion of momentum to the rate of diffusion of heat. It is usually important to specify whether the Prandtl number of the fluid of interest is "high," as for viscous oils, or "low," as for liquid metals, or "near unity," as for most gases. Some typical values of Prandtl number are shown below:

Fluid	Prandtl number	Designation
Mercury (200°F)	0.016	low
Air (100°F)	0.72	near unity
Glycerin (50°F)	31,000	high

The quantity π_4 has been given the name *Eckert number* (Ek):

$$\text{Ek} = \pi_4 = \frac{V_\infty^2}{c_p(T_w - T_\infty)}. \tag{5-32}$$

The Eckert number will prove to be of importance in describing the effects of compression and viscous dissipation in the flow of fluids.

All of these quantities, Re, Gr, Pr, and Ek, are dimensionless. When any *consistent* set of units is used, the same value of the dimensionless quantity will always exist for a given physical situation. For example the Prandtl number for air at normal temperature and pressure is approximately 0.7 and is independent of the units of k, μ, and c_p.

The value of most dimensionless groups will depend upon temperature and pressure since the physical properties appearing in these groups vary. Usually an average or representative dimensionless group is specified in which the physical properties are evaluated at the average temperature and pressure for the situation of interest. In some cases, *local* values of a dimensionless group are of interest. These would be the value of the group at a particular position in the fluid at a specified time. In such cases the local values of temperature and pressure are used to evaluate the fluid properties.

Equations (5–27), (5–28), and (5–29) can be written in terms of Re, Gr, Pr, and Ek:

$$u^* \frac{\partial u^*}{\partial x^*} + v^* \frac{\partial u^*}{\partial y^*} = -\frac{\partial p^*}{\partial x^*} + \frac{1}{\text{Re}}\left[\frac{\partial^2 u^*}{\partial x^{*2}} + \frac{\partial^2 u^*}{\partial y^{*2}} + \frac{1}{3}\frac{\partial}{\partial x^*}\left(\frac{\partial u^*}{\partial x^*} + \frac{\partial v^*}{\partial y^*}\right)\right], \quad (5\text{–}33)$$

$$u^* \frac{\partial v^*}{\partial x^*} + v^* \frac{\partial v^*}{\partial y^*} = -\frac{\partial p^*}{\partial y^*} + \frac{\text{Gr}}{\text{Re}^2} T^*$$
$$+ \frac{1}{\text{Re}}\left[\frac{\partial^2 v^*}{\partial x^{*2}} + \frac{\partial^2 v^*}{\partial y^{*2}} + \frac{1}{3}\frac{\partial}{\partial y^*}\left(\frac{\partial u^*}{\partial x^*} + \frac{\partial v^*}{\partial y^*}\right)\right], \quad (5\text{–}34)$$

$$u^* \frac{\partial T^*}{\partial x^*} + v^* \frac{\partial T^*}{\partial y^*} = \frac{1}{\text{Pr Re}}\left(\frac{\partial^2 T^*}{\partial x^{*2}} + \frac{\partial^2 T^*}{\partial y^{*2}}\right)$$
$$+ \text{Ek}\left(u^* \frac{\partial p^*}{\partial x^*} + v^* \frac{\partial p^*}{\partial y^*}\right) + \frac{\text{Ek}}{\text{Re}} \Phi^*. \quad (5\text{–}35)$$

These equations, together with the dimensionless conservation of mass equation and a proper equation of state, provide five equations for five unknowns, u, v, ρ, p, and T. Thus, the solution of these equations would provide information required to determine the temperature gradients. With the Fourier equation, a knowledge of the temperature gradients is sufficient for the calculation of heat transfer rates. The dimensionless temperature gradient evaluated at points on a solid surface is expressible as a dimensionless Nusselt number (see Problem 18 and Eq. (8–28)).

It is important to emphasize that only four dimensionless quantities (made up of fluid properties and boundary parameters) control the solution to the specified problem. With an arbitrary geometry and with given boundary conditions, a solution to any problem described by the five equations (for example, the local value of the dimensionless Nusselt number) can always be expressed (in dimensionless terms) as a function of only Re, Pr, Gr, and Ek, i.e., Nu = f(Re,Pr,Gr,Ek). This does not imply that an exact mathematical solution to the equations can be obtained for any given problem. The results do show that a dimensionless solution to this particular problem (and any problem described by these equations, whether the solution is exact or approximate) can be stated in terms of only four dimensionless parameters. When laboratory experiments become necessary to analyze such a problem in heat transfer, only these four parameters need to be varied, greatly reducing the number of measurements required.

The method of dimensional analysis illustrated here is useful where the differential equations describing the situation can be written. All parameters important to the problem are immediately obvious in such cases. The dimensional analysis arranges the parameters into the minimum number of groups that can be used to describe the problem. Other methods such as the pi theorem (see next section) can be used to obtain dimensionless groups when the differential equations are not known but pertinent parameters are specified. However, one of the weaknesses of the pi theorem method is that in many cases the important parameters are not apparent if the differential equations have not been written.

Dimensional Analysis Using the Pi Theorem

In addition to the use of an examination of the basic governing equations to obtain the dimensionless groups important in problems of fluid mechanics and heat transfer, other powerful methods are available. The most effective of these is the pi theorem (often called the Buckingham pi theorem).

This theorem may be developed in the following way. It is desired to determine the most effective way to present data and to perform experimental work by finding the minimum number of dimensionless groups that provide a relationship equivalent to the basic equation relating the variables known to affect the phenomenon under study.

The basic equation relating the variables may be written

$$f_1(x_1, x_2, x_3, x_4, \ldots, x_m) = 0,$$

where $x_1, x_2, x_3, \ldots, x_m$ are the m variables that have been determined to have influence in the problem to be considered.

The pi theorem states that a relationship may be found involving independent dimensionless groupings of variables called *pi groups*, which will have fewer terms than is the case for the basic equation and will represent it completely:

$$f_2(\pi_1, \pi_2, \pi_3, \ldots, \pi_n) = 0,$$

where $n < m$.

Further, the theorem states that

$$n = m - k, \qquad (5\text{--}36)$$

where k is the largest number of variables in the basic equation which will *not* combine into a dimensionless group.

A study of the pi theorem [2] indicates further that k is the number of variables whose units can be used to measure all of the variables. Also if i is the number of fundamental dimensions in the system chosen for the analysis, then k may not exceed i. In most cases $k = i$, so that the number of independent variables is usually reduced by an amount equal to the number of dimensions involved. These relationships are shown in Table 5–2 for convenience.

Table 5–2

i = number of fundamental dimensions
m = number of variables in the basic equation
$n = m - k$ = number of pi groups
k = the reduction factor $\leq i$

In order to use the above theorem in practice, Kline [3] notes that three conditions should be fulfilled:

1. The dependent variable and the independent variables of the basic equation must contain all variables which affect the phenomenon being investigated. Omission of any of the variables will cause errors in the results. Inclusion of a variable which really does not affect the result will cause additional pi groups to be obtained, which defeats the original purpose—reducing the complexity of the problem.

2. The pi groups formed should include, at least once, all of the original variables (x's).

3. The dimensions used in the analysis must be independent. If they are not independent, a provision must be made to account for this—for example, the use of g_c if both force and mass are to be included.

The process of obtaining the pi groups may involve the intuition of the experienced worker in the field based on intimate knowledge as to what type pi group will be most useful to him. It is possible to select pi groups by examination of the variables or by trial and error. However, a more formalized procedure is useful. In one such procedure, k of the variables (x's) are selected as repeating variables and are used in turn with the remaining variables in each successive pi group. Each of the repeating variables must have different dimensions, and the repeating variables collectively must contain all of the dimensions.

Thus, dimensional homogeneity demands that the following situation exist for an example where $m = 6$ and $k = 3$:

$$n = 6 - 3 = 3 \text{ groups.}$$

To obtain the pi groups, x_1, x_2, and x_3 are selected as the repeating variables with the following results:

$$\pi_1 = x_1^a \cdot x_2^b \cdot x_3^c \cdot x_4,$$
$$\pi_2 = x_1^d \cdot x_2^e \cdot x_3^f \cdot x_5,$$
$$\pi_3 = x_1^g \cdot x_2^h \cdot x_3^i \cdot x_6.$$

This format can be used for other values of m and k. The following examples serve to show how this procedure works:

Example 5–1. By using the pi theorem, find the dimensionless groups that relate the drag per unit length on a cylinder to the free stream velocity V_∞, the diameter D, the fluid density ρ, and the fluid viscosity μ.

Solution. The basic equation is

$$f_1(\text{drag}/L, \rho, V_\infty, D, \mu) = 0.$$

Note that drag/L is the drag per unit length of cylinder. An examination of this equation shows that, using the MLt system, $i = 3$ and $m = 5$; and

$$\text{drag}/L \triangleq \frac{ML}{t^2} \cdot \frac{1}{L} \qquad V_\infty \triangleq L/t,$$

$$\rho \triangleq \frac{M}{L^3} \qquad D \triangleq L \qquad \mu \triangleq M/Lt.$$

By trial and error it is seen that μ, ρ, and D may not form a dimensionless group, but addition of either drag or V_∞ would cause such a group to be formed; therefore, $k = 3$, and

$$n = m - k = 5 - 3 = 2 \text{ pi groups}.$$

Thus, $f_2(\pi_1, \pi_2) = 0$ is the relationship which may be used to replace the basic equation.

It is now only necessary to construct the groups according to the proposed rule. Intuition must play a role in selecting the best groups. Theory shows that the total number of groups possible by combining m variables, $k + 1$ at a time, is

$$\frac{m!}{(k+1)!\,(m-k-1)!},$$

which, for this situation, gives

$$\frac{5!}{(3+1)!\,(5-3-1)!} = 5 \text{ groups}.$$

Two of these must be selected.

The three repeating variables chosen are density, velocity, and the diameter, as this choice permits the drag to appear in only one group, and the important fluid property μ to appear in only one group:

$$\pi_1 = (V_\infty^a \cdot \rho^b \cdot D^c \cdot \mu),$$
$$\pi_2 = (V_\infty^d \cdot \rho^e \cdot D^f \cdot \text{drag}/L).$$

Substituting dimensions

for π_1: $\qquad M^0 L^0 t^0 \triangleq \left(\dfrac{L}{t}\right)^a \left(\dfrac{M}{L^3}\right)^b (L)^c \left(\dfrac{M}{Lt}\right).$

Equating exponents of M: $\quad 0 = b + 1$
of L: $\quad 0 = a - 3b + c - 1$
of t: $\quad 0 = -a - 1,$

from which

$$b = -1 \qquad a = -1 \qquad c = -1,$$

and

$$\pi_1 = \frac{\mu}{\rho V_\infty D} \qquad \text{(the reciprocal of the Reynolds number)}.$$

Substituting dimensions

for π_2 $\quad M^0L^0t^0 = \left(\dfrac{L}{t}\right)^d \left(\dfrac{M}{L^3}\right)^e (L)^f \left(\dfrac{M}{t^2}\right).$

Equating exponents of M: $\quad 0 = e + 1$
of L: $\quad 0 = d - 3e + f$
of t: $\quad 0 = -d - 2,$

from which

$$d = -2 \quad e = -1 \quad f = -1,$$

and

$$\pi_2 = \dfrac{\text{drag}}{LD\rho V_\infty^2},$$

or

$$f_2(\pi_1, \pi_2) = f_2\left(\dfrac{\text{drag}/LD}{\rho V_\infty^2}, \dfrac{\mu}{\rho V_\infty D}\right) = 0. \tag{5-37}$$

Note in π_2 that $(L)(D)$ is the projected area (A_p) of the cylinder. The drag coefficient is twice the value of the second group, π_2, arrived at above:

$$C_D = \dfrac{\text{drag}/A_p}{\tfrac{1}{2}\rho V_\infty^2}.$$

The dependent variable may be removed from Eq. (5-37) as follows*:

$$\dfrac{\text{drag}/DL}{\rho V_\infty^2} = f_3\left(\dfrac{\rho V_\infty D}{\mu}\right);$$

or, multiplying by 2,

$$\dfrac{\text{drag}/DL}{\tfrac{1}{2}\rho V_\infty^2} = C_D = f_4(\text{Re}),$$

which is the result obtained from an examination of the basic differential equation in the previous section.

Example 5–2. The surface coefficient of convective heat transfer is given the symbol h; it is also called the *film coefficient, unit thermal convective conductance,* and *convective heat transfer coefficient.* This coefficient, which will be more rigorously defined in a later chapter, is the ratio of the heat flux to a specified temperature difference. It is assumed that this film coefficient, h, is a function of the density ρ, the viscosity μ, the thermal conductivity k, and the heat capacity c_p, a characteristic dimension D (for example, the diameter of a sphere), and the free stream velocity V_∞. Find appropriate

---

* The reciprocal of π_1 is used for convenience. This fact may be accounted for in the function, f_3.

dimensionless groups which may be used to form a correct relationship describing the phenomenon.

Solution.

$$f_1(h,\rho,V_\infty,D,\mu,k,c_p) = 0.$$

$$h \triangleq \frac{M}{t^3 T}; \quad \rho \triangleq \frac{M}{L^3}; \quad V_\infty \triangleq \frac{L}{t}; \quad D \triangleq L;$$

$$\mu \triangleq \frac{M}{Lt}; \quad k \triangleq \frac{ML}{t^3 T}; \quad c_p \triangleq \frac{L^2}{t^2 T}.$$

From an examination of the above, there are seven variables and four basic dimensions, and four is the maximum number of variables which will not form a dimensionless group. So from Eq. (5-36), $m - k = 7 - 4 = 3$ dimensionless groups.

The variables μ, ρ, D, and k are chosen as the repeating variables. The pi groups may now be determined using the rules presented above:

$$\pi_1 \triangleq \mu^a \rho^b D^c k^d c_p,$$

$$\pi_1 \triangleq \left(\frac{M}{Lt}\right)^a \left(\frac{M}{L^3}\right)^b (L)^c \left(\frac{ML}{t^3 T}\right)^d \left(\frac{L^2}{t^2 T}\right).$$

Equating exponents of M: $0 = a + b + d$
of L: $0 = -a - 3b + c + d + 2$
of t: $0 = -a - 3d - 2$
of T: $0 = -d - 1.$

So
$$d = -1 \quad a = 1 \quad b = 0 \quad c = 0,$$
and
$$\pi_1 = \frac{\mu c_p}{k} = \text{Pr (the Prandtl number)}.$$

Similarly:
$$\pi_2 \triangleq \mu^f \rho^g D^i k^j V_\infty,$$

$$\triangleq \left(\frac{M}{Lt}\right)^f \left(\frac{M}{L^3}\right)^g (L)^i \left(\frac{ML}{t^3 T}\right)^j \left(\frac{L}{t}\right).$$

Equating exponents of M: $0 = f + g + j$
of L: $0 = -f - 3g + i + j + 1$
of t: $0 = -f - 3j - 1$
of T: $0 = j.$

So
$$j = 0 \quad f = -1 \quad g = +1 \quad i = 1,$$

and similarly,

$$\pi_2 = \frac{\rho V_\infty D}{\mu} = \text{Re (the Reynolds number)}.$$

Also,

$$\pi_3 \triangleq \mu^w \rho^x D^y k^z h,$$

$$\triangleq \left(\frac{M}{Lt}\right)^w \left(\frac{M}{L^3}\right)^x (L)^y \left(\frac{ML}{t^3 T}\right)^z \left(\frac{M}{t^3 T}\right).$$

Equating exponents of M: $\quad 0 = w + x + z + 1$
of L: $\quad 0 = -w - 3x + y + z$
of t: $\quad 0 = -w - 3z - 3$
of T: $\quad 0 = -z - 1.$

So
$$z = -1 \quad w = 0 \quad x = 0 \quad y = 1,$$
and

$$\pi_3 = \frac{hD}{k} = \text{Nu (the Nusselt number)}.$$

This allows the functional relationship between the groups to be written

$$f_2(\pi_1, \pi_2, \pi_3) = 0,$$
$$f_2(\text{Pr}, \text{Re}, \text{Nu}) = 0,$$

or, rearranging to get the dependent variable explicitly identified,

$$\text{Nu} = f_3(\text{Pr}, \text{Re}),$$

or

$$\frac{hD}{k} = f_3(\text{Pr}, \text{Re}). \tag{5-38}$$

The selection of the repeating and nonrepeating variables in this problem is the most crucial step in arriving at the final form of the equation. In this case, a knowledge of which groups would be most useful led to the selection. The use of h, V_∞, and c_p as nonrepeating variables assured that each would appear in only one group. This example indicates clearly that the dependent variable, in this case h, should be chosen as a nonrepeating variable so that it may be solved for easily in order to yield an explicit equation at the conclusion of the analysis—for example, Eq. (5-38).

Such insights as to the proper selection of the repeating (or nonrepeating) variables will vary with the experience of the individual and his knowledge of the phenomenon being investigated. It must be remembered that many other valid groups (in this case 21 total groups) may be obtained. They would appear to be equally good in the eyes of an inexperienced person but not nearly so useful as the ones obtained above.

5-2 SIMILITUDE

The discussion of dimensional analysis (Section 5-1) and of the value of nondimensional presentation of data led naturally to a brief mention of the concept of similarity in systems. This idea will now be developed in greater detail.

Engineers and scientists learned early that experiments were often required to provide information needed to construct devices and systems. It was apparent that it was not feasible in many cases to construct a full-scale device (prototype) upon which to perform these experiments because of the cost of the device and the test facilities required to perform the experimentation. On the other hand, if a model of the prototype could be used for experimental purposes and the results applied to the full-scale device, great savings in cost and time would obtain. The model is, in the great majority of cases, smaller than the prototype; however, occasionally a model larger than the prototype may be used if the small size of the prototype causes experimental difficulties.

A rule or theory for the application of the results of the model tests to the prototype is needed for these purposes. The theory of similitude performs this needed function. This theory was stated by Kline: "If two systems obey the same set of governing equations and conditions and if the values of all parameters in these equations and conditions are made the same, then the two systems must exhibit similar behavior provided that a unique solution to the set of equations and conditions exists." [3, p. 41.]

It will be remembered that the above conditions were met when it was argued that the dimensionless drag coefficient C_D could be applied to circular cylinders of other sizes and with different flow conditions than the ones for which C_D was experimentally determined. In the previous section it was pointed out (1) that the governing equations would be the same (in fact, identical in every respect) when presented in nondimensional form; (2) that the parameters in those equations, i.e., the Reynolds numbers, were made equal; and (3) that the geometric boundaries in the two cases were similar.

The theory of similarity applied to fluid mechanics and heat transfer can be used to establish three kinds of similarity:

1. Thermal similarity—Two flows are thermally similar if the absolute temperatures at corresponding points bear a constant fixed ratio to each other.

2. Kinematic similarity—Two flows are said to be kinematically similar if the two sets of streamlines are geometrically similar to each other. Geometric similarity infers that one geometric pattern may be obtained from the other by a constant linear ratio. (Inasmuch as the boundary of the system—for example, the surface of a cylinder—forms one of the streamlines, kinematic similarity includes the requirement that the two shapes under consideration be geometrically similar.) In other words, the motions of two kinematically similar systems are similar.

3. Dynamic similarity—It has been stated by Holt [4] that "two phenomena are dynamically similar if the dimensionless form of each physical variable has the same value at corresponding points." In two dynamically similar flows, each force (viscous force, inertia, etc.) at a point in one flow bears the same ratio to the same type force at the corresponding point in the other flow, and the two forces are parallel to each other. Also, whatever ratio is found to exist between two forces of the same type at a set of corresponding points in the two flows must be constant at all sets of corresponding points throughout the two flows for the ratios of all types of forces. Dynamic similarity requires, too, that there is a similarity of density (mass) distribution between the two flows. Thus, experimentally determined values of force, such as lift and drag, found by the use of a model may be applied to the prototype by use of the constant ratio of forces known to exist. An example of this follows.

In the case of drag on a cylinder, it was found that

$$C_D = \frac{\mathrm{drag}/2R}{\frac{1}{2}\rho V_\infty^2} = f(\mathrm{Re}).$$

If the result of a model test is to be applied to a full-scale device operating at the same Reynolds number, the following equations hold:

$$(C_D)_p = (C_D)_m,$$

where p refers to prototype, and m refers to model; and

$$\frac{\mathrm{drag}_p/2R_p}{\frac{1}{2}\rho_p V_{\infty p}^2} = \frac{\mathrm{drag}_m/2R_m}{\frac{1}{2}\rho_m V_{\infty m}^2}.$$

Rearranging,

$$\frac{D_p}{D_m} = \left(\frac{\rho_p}{\rho_m}\right)\cdot\left(\frac{R_p}{R_m}\right)\cdot\left(\frac{V_{\infty p}}{V_{\infty m}}\right)^2.$$

The ratios on the right side of the above equation are sometimes called scale factors (ratios); thus,

$$D_p = D_m(S_\rho)(S_r)(S_u)^2,$$

where

$$S_\rho = \frac{\rho_p}{\rho_m} \qquad S_r = \frac{R_p}{R_m} \qquad S_u = \frac{V_{\infty p}}{V_{\infty m}}.$$

It should be carefully noted that in the use of dimensionless groups for model analysis, even if the value of the group is equal in the model and in the prototype, the values of the individual variables in the groups are not necessarily equal. In fact, the scale factors will almost certainly not be equal to one.

Complete Similarity

Reference to the following pi group relationship shows the requirement for complete similarity:

$$\pi_1 = f(\pi_2, \pi_3, \ldots, \pi_n).$$

If it is decided to conduct model tests and apply the results to a prototype, then the following equations must apply:

$$(\pi_1)_p = (\pi_1)_m,$$

where p refers to prototype, and m refers to model; and

$$(\pi_1)_p = f[(\pi_2)_p, (\pi_3)_p, \ldots, (\pi_n)_p],$$
$$(\pi_1)_m = f[(\pi_2)_m, (\pi_3)_m, \ldots, (\pi_n)_m].$$

Thus, for insurance of equality of the two π_1 values,

$$(\pi_2)_p = (\pi_2)_m, \; (\pi_3)_p = (\pi_3)_m, \ldots, (\pi_n)_p = (\pi_n)_m,$$

or all corresponding π groups, π_2 through π_n, must be equal to insure that similarity exists.

In the actual case, it is often impossible to construct a model experiment in such a way that the above conditions are met completely. It must then be argued that certain of the pi groups "govern" or control the phenomenon under consideration. Fortunately, this is true in a large number of cases. It could, for example, be said that in high-speed flow, the effects of buoyancy, and thus of free convection, must be negligible, and no ill effects on the experiment would be expected, even if the Grashof numbers of model and prototype were unequal.

In other situations, the effect of one or more of the pi groups may just not be known. The experimenter must then hold model and prototype pi groups equal wherever possible and try to estimate by other means the errors caused by his inability to hold all pi groups equal.

It is difficult to overestimate the value of models in engineering. The *true model*, in which all of the pertinent pi groups are equal for model and prototype, is difficult to achieve. Other reasoning may lead to a deliberate distortion of the geometric scale in many cases. For example, models of rivers and harbors usually are *distorted models*. If a reasonable size for such models is maintained, the depth of the water would often be only a fraction of an inch and the effects of surface tension would be abnormally amplified. Also, if such models are used to study transport of solids (shifting of the river or harbor beds), the small depth would not allow a study to be carried out. Therefore, the depth is increased and the horizontal scale factor is greater than the vertical scale factor. Great care must be taken in applying such model results to the full-scale situation. Other cases of the distorted model occur where it is found that one or more variables will affect the model abnormally because of the geometric scale reduction.

5-3 INTERPRETATION OF DIMENSIONLESS GROUPS

It is sometimes possible to obtain some insight into the physical nature of dimensionless groups by application of some simple reasoning. However, such procedures are sometimes unsound and yield false or distorted ideas when applied to specific situations. To illustrate this, the following line of reasoning is developed:

It may be said that for similar flows, past a cylinder for example, the momentum equation yields the following for the inertia force*:

$$f_I \sim \rho u \frac{\partial u}{\partial x} + \rho v \frac{\partial u}{\partial y} \sim \rho \frac{V_\infty^2}{D}.$$

Similarly, the viscous force is

$$f_v \sim \mu \left(\frac{\partial^2 u}{\partial x^2} + \frac{\partial^2 u}{\partial y^2} \right) \sim \mu \frac{V_\infty}{D^2}.$$

The ratio of these forces is then

$$\frac{f_I}{f_v} \sim \frac{\rho V_\infty^2 / D}{\mu V_\infty / D^2} \sim \frac{\rho V_\infty D}{\mu},$$

which is the Reynolds number.

From this development it can be said that in similar flows, if the Reynolds numbers are identical, then the ratios of inertia to viscous forces are identical. However, great care should be taken in generalizing beyond this point, such as relating the value of the ratio of forces to the magnitude of this Reynolds number. Note that the magnitude of this Reynolds number is arbitrary because of the arbitrary selection of the reference length D in the above case. But the ratio of inertia to viscous forces at a point in a flow is certainly not arbitrary. Furthermore, the example of laminar flow through a long tube may be examined for the meaning of Reynolds number. Such flows are governed by pressure and viscous forces only. *Inertia forces* are everywhere zero, but a Reynolds number can be calculated, as values of the free stream velocity, density, viscosity, and a reference length are known. The calculation of the Reynolds number would yield a positive number, where any extension of the above argument would dictate that the Reynolds number should be zero, as the inertia force is zero.

It thus appears that the value of the ratio of inertia to viscous forces is not equal to the magnitude of the Reynolds number. It may still be useful, however, to use the value of the Reynolds number to indicate areas of

---

* The velocity v is proportional to the free stream velocity V_∞; $\partial u/\partial x$ and $\partial u/\partial y$ are proportional to V_∞/D; and $\partial^2 u/\partial x^2$ and $\partial^2 u/\partial y^2$ are proportional to V_∞/D^2.

Table 5-3

DIMENSIONLESS GROUPS OF FLUID MECHANICS AND HEAT TRANSFER

Name	Notation	Formula	Interpretation in terms of ratio
Biot number	Bo	$\dfrac{hL}{k_s}$	Surface conductance ÷ internal conduction of solid.
Cauchy number	Ca	$\dfrac{V^2}{B_s/\rho} = \dfrac{V^2}{a^2}$	Inertia force ÷ compressive force = (Mach number)^2.
Eckert number	Ek	$\dfrac{V^2}{c_p \Delta T}$	Temperature rise due to energy conversion ÷ temperature difference.
Euler number	Eu	$\dfrac{\Delta p}{\rho V^2}$	Pressure force ÷ inertia force.
Fourier number	Fo	$\dfrac{kt}{\rho c_p L^2} = \dfrac{\alpha t}{L^2}$	Rate of conduction of heat ÷ rate of storage of energy.
Froude number	Fr	$\dfrac{V^2}{gL}$	Inertia force ÷ gravity force.
Graetz number	Gz	$\dfrac{D}{L} \cdot \dfrac{V\rho c_p D}{k}$	Re Pr ÷ (L/d); heat transfer by convection in entrance region ÷ heat transfer by conduction.
Grashof number	Gr	$\dfrac{g\beta \Delta T L^3}{\nu^2}$	Buoyancy force ÷ viscous force.
Knudsen number	Kn	$\dfrac{\lambda}{L}$	Mean free path of molecules ÷ characteristic length of an object.
Lewis number	Le	$\dfrac{\alpha}{D_c}$	Thermal diffusivity ÷ molecular diffusivity.
Mach number	M	$\dfrac{V}{a}$	Macroscopic velocity ÷ speed of sound.

Table 5-3 (Continued)

Name	Notation	Formula	Interpretation in terms of ratio
Nusselt number	Nu	$\dfrac{hL}{k}$	Temperature gradient at wall ÷ overall temperature difference.
Péclet number	Pé	$\dfrac{V\rho c_p D}{k}$	(Re Pr); heat transfer by convection ÷ heat transfer by conduction.
Prandtl number	Pr	$\dfrac{\mu c_p}{k} = \dfrac{\nu}{\alpha}$	Diffusion of momentum ÷ diffusion of heat.
Reynolds number	Re	$\dfrac{\rho V L}{\mu} = \dfrac{VL}{\nu}$	Inertia force ÷ viscous force.
Schmidt number	Sc	$\dfrac{\mu}{\rho D_c} = \dfrac{\nu}{D_c}$	Diffusion of momentum ÷ diffusion of mass.
Sherwood number	Sh	$\dfrac{h_D L}{D_c}$	Mass diffusivity ÷ molecular diffusivity.
Stanton number	St	$\dfrac{h}{V\rho c_p} = \dfrac{h}{c_p G}$	Heat transfer at wall ÷ energy transported by stream.
Stokes number	Sk	$\dfrac{\Delta p L}{\mu V}$	Pressure force ÷ viscous force.
Strouhal number	Sl	$\dfrac{L}{tV}$	Frequency of vibration ÷ characteristic frequency.
Weber number	We	$\dfrac{\rho V^2 L}{\sigma}$	Inertia force ÷ surface tension force.

predominance of inertia forces (high Reynolds number) and areas of predominance of viscous forces (low Reynolds number). Further, in a given flow if inertia forces are found to predominate at a given Reynolds number, they would be expected to predominate even more at higher Reynolds numbers. Similarly, greater predominance of viscous force would be expected at Reynolds numbers lower than one at which viscous forces are known to predominate.

Considerable use is made in fluid mechanics and heat transfer of interpretations of a dimensionless group as a ratio of some pair of forces or other effects which, it may be argued, control the behavior of the system. For example, in addition to inertia forces and viscous forces, the following forces may be used in making such interpretation: pressure force, compressive force, surface tension force, gravity force, and electric and magnetic force.

Table 5-3 presents information about dimensionless groups of use in fluid mechanics and heat transfer. The use of these groups in generalizing data for ease of plotting and in empirical and analytical equations is widespread. The interpretations listed in the last column yield some insights into physical phenomena involved in fluid flow and heat transfer behavior. Great benefits accrue, also, from the use of these groups in guiding the design of experiments and in applying the results of modeling through the principle of similitude.

REFERENCES

1. P. W. Bridgeman, *Dimensional Analysis*, Yale University Press, New Haven (1922).
2. E. R. Van Driest, "On Dimensional Analysis and the Presentation of Data in Fluid Flow Problems," *J. Appl. Mech.* (March 1956).
3. S. J. Kline, *Similitude and Approximation Theory*, McGraw-Hill, New York (1965).
4. M. Holt, "Dimensional Analysis," *Handbook of Fluid Dynamics*, ed. by V. L. Streeter, McGraw-Hill, New York (1961), Section 15.
5. H. L. Langhaar, *Dimensional Analysis and the Theory of Models*, J. Wiley, New York (1951).
6. W. J. Duncan, *Physical Similarity and Dimensional Analysis*, Longmans, Green, New York (1953).
7. G. Murphy, *Similitude in Engineering*, Ronald Press, New York (1950).
8. L. I. Sedov, *Similarity and Dimensional Methods in Mechanics*, tr. by M. Hold, Academic Press, New York (1959).
9. D. C. Ipsen, *Units, Dimensions, and Dimensionless Numbers*, McGraw-Hill, New York (1960).
10. I. Shames, *Mechanics of Fluids*, McGraw-Hill, New York (1962).
11. V. L. Streeter, *Fluid Mechanics*, McGraw-Hill, New York (1962).
12. R. M. Olsen, *Essentials of Engineering Fluid Mechanics*, International Textbook Co., Scranton (1962).

PROBLEMS

1. Following the symbolism of Table 5-1, arrange the following variables into dimensionless groups.

   a) $D, \rho, \mu, V.$ c) $f, \omega, \mu, V.$ e) $f, \rho, \mu.$
   b) $f, \mu, N, D.$ d) $\Delta p, \rho, V.$ f) $\alpha, L, t.$

2. Air at a pressure of 14.7 psia and a temperature of 150°F flows past a 1-in. diameter sphere which has a surface temperature of 300°F. The free stream velocity is 60 miles per hour. Using properties evaluated at an average (film) temperature of 225°F, calculate (a) the Reynolds number, (b) the Prandtl number, and (c) the Grashof number based on difference between surface and stream temperature.

3. Water flows in a 6-in. diameter pipe at a pressure of 70 psi and a temperature of 210°F. The average velocity is 10 ft/sec. Calculate (a) the Reynolds number and (b) the Prandtl number.

4. The thrust force T of the propeller of an aircraft is a function of the axial velocity V, the diameter D, the rotational speed ω, the density of the fluid ρ, and the viscosity of the fluid μ. Find the independent pi groups needed to represent the function. Note: ω has the dimensions $1/t$.

In Problems 5–8 the variables are listed for a specific situation. In each case, find the required pi groups.

5. Fluid flows through an orifice. The velocity V of the jet is a function of the pressure drop Δp and the density of the fluid ρ.

6. An incompressible fluid flows through a pipe. The shear stress τ_w at the wall is a function of average velocity V_{avg}; pipe diameter is D; density of the fluid, ρ; viscosity of the fluid, μ; and the roughness e is expressed as average height of the wall roughness. Note: One dimensionless group will be e/D.

7. The velocity V of propagation of waves on the surface of a deep body of liquid depends only on the wave length L and the gravitational acceleration g.

8. The volume rate of flow $\dot{Q}$ of a fluid through a capillary tube is a function of the pressure drop per unit length $\Delta p/L$, the diameter D, and the viscosity μ.

9. The "Rayleigh method" of finding the dimensionless group is preferred by many engineers over the pi theorem. The problem solved in Example 5–2 would be approached as described below. Parts (c) and (d) below involve steps that are left for the student to perform.

a) Assume that the independent variables appear as a product of the variables raised to a power and are equated to the dependent variable by a constant of proportionality.

$$h = \text{constant } (V^a \rho^b \mu^c D^d k^e c_p^f).$$

b) Write the above dimensionally (use an $MLtT$ system here).

$$\frac{M}{t^3 T} \stackrel{\triangle}{=} \left(\frac{L}{t}\right)^a \left(\frac{M}{L^3}\right)^b \left(\frac{M}{Lt}\right)^c (L)^d \left(\frac{ML}{t^3 T}\right)^e \left(\frac{L^2}{t^2 T}\right)^f.$$

Equate powers of each dimension to assure dimensional homogeneity.

Powers of M: $1 = b + c + e$
      of L: $0 = a - 3b - c + d + e + 2f$
      of t: $-3 = -a - c - 3e - 2f$
      of T: $-1 = -e - f$.

c) Note that there are now four equations and six unknowns. Therefore, solve for four of the unknowns in terms of the other two. (Try solving in terms of a and f.) *Answer:* $a = a;\ b = a;\ c = f - a;\ d = a - 1;\ e = 1 - f;\ f = f$.

d) Write the original equation in (a) above in terms of the unknowns (a and f here). Rearrange into groups.

10. Solve Problem 9 in terms of the unknowns c and e.
11. Solve Problem 9 in terms of the unknowns b and f.
12. Solve Problem 4 using the Rayleigh method.
13. Solve Problem 5 using the Rayleigh method.
14. Solve Problem 6 using the Rayleigh method.
15. Show that the Froude number can be interpreted in terms of the ratio of inertia force to gravity force.
16. Show that the Cauchy number can be interpreted in terms of the ratio of inertia force to compressive force. What relationship exists between the Cauchy number and the Mach number?
17. Show that the Biot number can be interpreted in terms of the ratio of convective heat flow to heat flow by conduction.
18. The dimensionless temperature gradient evaluated at a surface, $(\partial T^*/\partial y^*)_{y^*=0}$, might be obtainable as a solution of Eqs. (5-33), (5-34), and (5-35), the continuity equation, the equation of state, and suitable boundary conditions. Show that this dimensionless temperature gradient is equivalent to the negative of a Nusselt number $(hL)/k$, where L is a characteristic length used to nondimensionalize y^*, and

$$h = \frac{(\dot{q}/A)_{\text{wall}}}{(T_w - T_\infty)}.$$

CHAPTER 6

IDEAL FLUID FLOW AND HEAT TRANSFER

Idealizations are often made in engineering to permit a better insight into a complex situation or to permit a solution to be obtained more readily or more easily. In the study of fluid flow and heat transfer, idealizations are quite common. The assumption that a fluid is free of viscous forces is a common idealization in the study of flow about airfoils and turbine blades or in short channels or pipes. In some situations, the assumption that fluid properties have constant values at any cross section in a pipe or channel is commonly made. This chapter will present some of the assumptions or idealizations made to simplify fluid studies. This should also be useful as background material for study of more complex problems in subsequent chapters.

6–1 INCOMPRESSIBLE FLUID IN FRICTIONLESS FLOW

In the study of certain fluid flow situations, it is convenient to consider the fluid to be free of viscous forces. For the relations in Chapter 3 describing Stokes' law of friction, it can be seen that the viscous forces in a fluid will be small when the velocity gradients are small. This situation often exists in fluids except near solid boundaries. If the effect of viscosity is neglected, the equations of motion (Eqs. 4–44a-c) become

$$\rho \frac{Du}{Dt} = X - \frac{\partial p}{\partial x}, \qquad (6\text{–}1a)$$

$$\rho \frac{Dv}{Dt} = Y - \frac{\partial p}{\partial y}, \qquad (6\text{–}1b)$$

$$\rho \frac{Dw}{Dt} = Z - \frac{\partial p}{\partial z}. \qquad (6\text{–}1c)$$

Equations (6–1a-c) describe the conservation of momentum for a *nonviscous* or *frictionless fluid*, sometimes referred to as an *ideal fluid*. They are called the *Euler equations*. Note that these equations are first order differential equations, whereas Eqs. (4–44a-c) are second order. Because the order of Eqs. (6–1a-c) is lower, these equations cannot be made to fit all of the boundary conditions that might be applied to Eqs. (4–44a-c). Specifically, Eqs. (6–1a-c) cannot be made to fit the boundary conditions of zero velocity at all points along any solid boundaries enclosing the fluid, but they are most useful in describing flow regions away from the wall or where the assumption of nonzero fluid velocity at the wall, commonly called *slip*, does not lead to great error.

Equations (6–1a-c) can be utilized to reveal some very interesting and useful information. This is done by use of the *velocity potential*, which is defined as a function whose partial derivative with respect to a direction coordinate is equal to the velocity component in that direction. Letting Φ stand for the velocity potential, the definition leads to the following:

$$\frac{\partial \Phi}{\partial x} = u \qquad \frac{\partial \Phi}{\partial y} = v \qquad \frac{\partial \Phi}{\partial z} = w. \tag{6-2}$$

According to Eq. (6–2), if an imaginary surface representing constant velocity potential exists in a fluid, then the flow will everywhere be normal to that surface. In a two-dimensional flow, the fluid motion (streamline) is everywhere normal to lines of constant velocity potential.

If it is assumed that the flow is steady and incompressible, the continuity equation is

$$\frac{\partial u}{\partial x} + \frac{\partial v}{\partial y} + \frac{\partial w}{\partial z} = 0. \tag{6-3}$$

Substituting Eq. (6–2) into (6–3),

$$\frac{\partial^2 \Phi}{\partial x^2} + \frac{\partial^2 \Phi}{\partial y^2} + \frac{\partial^2 \Phi}{\partial z^2} = 0. \tag{6-4}$$

Equation (6–4) is the *Laplace equation*, which is sometimes written with the Laplacian operator ∇^2:

$$\nabla^2 \Phi = 0 \qquad \text{where} \qquad \nabla^2 = \frac{\partial^2}{\partial x^2} + \frac{\partial^2}{\partial y^2} + \frac{\partial^2}{\partial z^2}. \tag{6-5}$$

A flow which satisfies Eq. (6–4), and thus satisfies Eq. (6–2), is referred to as *potential flow*. Differentiating the first equality of Eq. (6–2) with respect to y, and the second with respect to x, gives

$$\frac{\partial^2 \Phi}{\partial y\, \partial x} = \frac{\partial u}{\partial y} \qquad \frac{\partial^2 \Phi}{\partial x\, \partial y} = \frac{\partial v}{\partial x}. \tag{6-6}$$

Subtracting the two equations (noting that the order of differentiation of Φ is unimportant) gives

$$\frac{\partial v}{\partial x} - \frac{\partial u}{\partial y} = 0. \tag{6-7}$$

In a similar manner the following may be obtained:

$$\frac{\partial u}{\partial z} - \frac{\partial w}{\partial x} = 0, \tag{6-8}$$

$$\frac{\partial w}{\partial y} - \frac{\partial v}{\partial z} = 0. \tag{6-9}$$

Equations (6–7), (6–8), and (6–9) state that the fluid elements have no net rotation in space.* Such flows are referred to as *irrotational flows*. Thus, a flow described by Eq. (6–2)—a potential flow—is irrotational. In the absence of shear forces, the net resultant of forces acting on an element always passes through the center of mass of the element. Thus, there are no forces to cause rotation to occur.

Fig. 6–1 Irrotational and rotational motion of a fluid element. (a) Irrotational movement. (b) Rotational movement.

Rotation may be illustrated by reference to Fig. 6–1. In Fig. 6–1(a), a fluid element is shown as it moves along a pathline in steady irrotational flow. Although the particle direction changes, its orientation relative to a fixed axis does not. In Fig. 6–1(b), rotational flow is illustrated showing the particle orientation changing as it moves along the pathline. Particle distortion should not be confused with change of orientation or rotation. A particle may be distorted without rotating.

To show that the left side of Eq. (6–7) describes the rotation of an element, refer to Fig. 6–2. Assume that the velocity components of the element at the lower left-hand corner A are u and v. The y component of velocity at B is, therefore, $v + \partial v/\partial x(dx)$. The rotation of the element AB about the z axis passing through A, considering counterclockwise rotation to

---

* Equations (6–7), (6–8), and (6–9) can also be shown to be the components of twice the vector $(\nabla \times \mathbf{V})/2$, sometimes called the vorticity vector (see Problem 27).

The description of a flow as irrotational implies $\nabla \times \mathbf{V} = 0$ or curl $\mathbf{V} = 0$. In such cases the velocity $\mathbf{V}$ can always be represented by the gradient of a scalar function since the vector identity curl (grad Φ) = 0 can be easily verified. Thus, an irrotational flow is always potential flow.

be positive, is

$$\text{rate of rotation of } AB \text{ about } A \text{ in radians per unit time} = \frac{v + \frac{\partial v}{\partial x} dx - v}{dx} = \frac{\partial v}{\partial x}. \quad (6\text{--}10)$$

Likewise,

$$\text{rate of rotation of } AD \text{ about } A \text{ in radians per unit time} = \frac{-u - \frac{\partial u}{\partial y} dy + u}{dy} = -\frac{\partial u}{\partial y}. \quad (6\text{--}11)$$

The net rotation of the element about A is the average of the rotation of the sides, or

$$\text{rate of rotation of element about } A = \frac{1}{2}\left(\frac{\partial v}{\partial x} - \frac{\partial u}{\partial y}\right). \quad (6\text{--}12)$$

Referring to Eq. (6–7), $(\partial v/\partial x) - (\partial u/\partial y)$ is equal to zero and thus, the rate of rotation of the element about the z axis is zero. In potential flow, the rotation about all axes is zero; hence, the term irrotational flow.

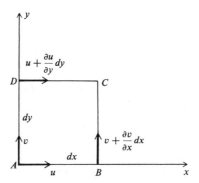

Fig. 6–2 Rotation of fluid element about A.

Sufficient information has now been presented to permit a solution of Eqs. (6–1a-c). Assuming that the only body force is that due to gravity, and letting the coordinate system be oriented so that the gravitational force is in the negative y direction, Eqs. (6–1a-c) become, for two dimensional flow:

$$\rho\left(\frac{\partial u}{\partial t} + u\frac{\partial u}{\partial x} + v\frac{\partial u}{\partial y}\right) = -\frac{\partial p}{\partial x}, \quad (6\text{--}13a)$$

$$\rho\left(\frac{\partial v}{\partial t} + u\frac{\partial v}{\partial x} + v\frac{\partial v}{\partial y}\right) = -g\rho - \frac{\partial p}{\partial y}. \quad (6\text{--}13b)$$

Using the relation for irrotationality from Eq. (6–7), Eqs. (6–13a-b) become:

$$\rho\left(\frac{\partial u}{\partial t} + u\frac{\partial u}{\partial x} + v\frac{\partial v}{\partial x}\right) = -\frac{\partial p}{\partial x}, \quad (6\text{–}14\text{a})$$

$$\rho\left(\frac{\partial v}{\partial t} + u\frac{\partial u}{\partial y} + v\frac{\partial v}{\partial y}\right) = -g\rho - \frac{\partial p}{\partial y}. \quad (6\text{–}14\text{b})$$

Equation (6–14a) may be multiplied by dx and Eq. (6–14b) by dy and added together.

It will also be assumed that $u = dx/dt$ and $v = dy/dt$. Therefore,

$$\frac{\partial u}{\partial t}dx = u\frac{\partial u}{\partial t}dt \quad \text{and} \quad \frac{\partial v}{\partial t}dy = v\frac{\partial v}{\partial t}dt.$$

The result is

$$\rho\left(u\frac{\partial u}{\partial t}dt + v\frac{\partial v}{\partial t}dt + u\frac{\partial u}{\partial x}dx + u\frac{\partial u}{\partial y}dy + v\frac{\partial v}{\partial x}dx + v\frac{\partial v}{\partial y}dy\right)$$

$$= -g\rho\,dy - \frac{\partial p}{\partial x}dx - \frac{\partial p}{\partial y}dy. \quad (6\text{–}15)$$

Expressions for the exact differentials du and dv can be factored from terms in Eq. (6–15). Adding $-(\partial p/\partial t)\,dt$ to each side of the equation, grouping terms, and simplifying, the following is obtained:

$$\rho(u\,du + v\,dv) - \frac{\partial p}{\partial t}dt = -g\rho\,dy - dp. \quad (6\text{–}16\text{a})$$

Noting that $u^2 + v^2 = V^2$, then $u\,du + v\,dv = V\,dV$, where V is the magnitude of the velocity. Then

$$\rho(V\,dV) - \frac{\partial p}{\partial t}dt = -g\rho\,dy - dp, \quad (6\text{–}16\text{b})$$

or

$$\frac{dp}{\rho} + g\,dy + V\,dV - \frac{1}{\rho}\frac{\partial p}{\partial t}dt = 0. \quad (6\text{–}16\text{c})$$

For the steady state case, $\partial p/\partial t = 0$, so

$$\frac{dp}{\rho} + g\,dy + V\,dV = 0. \quad (6\text{–}17\text{a})$$

Equation (6–17a) is usually referred to as the *steady-state Euler Equation*. For the case of an incompressible fluid ρ is constant and Eq. (6–17a) may be integrated to give

$$\frac{p}{\rho} + gy + \frac{V^2}{2} = \text{constant}. \quad (6\text{–}17\text{b})$$

Equation (6–17b) is the *steady incompressible flow Bernoulli equation*. It states that the sum of the specific kinetic energy, potential energy, and flow energy is constant throughout the fluid in irrotational flow.

It should be pointed out that even though the terms in Eq. (6–17b) are energy terms, the equation is in reality a momentum equation, since it resulted from an integration of mathematical statements of Newton's second law for an inviscid, steady, irrotational flow. It should also be emphasized that the sum of the terms on the left side of Eq. (6–17b), is constant throughout the entire flow field only if the flow field is irrotational. The kinetic energy term must be corrected if the velocity is not constant over the flow cross-section area, as discussed in Section 4–4.

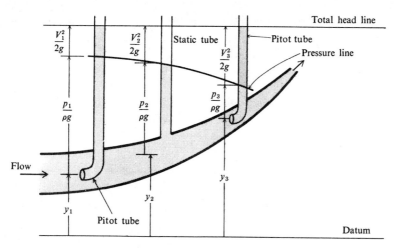

Fig. 6–3 Breakdown of Bernoulli equation terms.

The Bernoulli equation can be written in a different manner if Eq. (6–17b) is divided by g:

$$\frac{p}{\rho g} + \frac{V^2}{2g} + y = \text{constant},$$

$$\underset{\text{head}}{\text{pressure}} + \underset{\text{head}}{\text{velocity}} + \underset{\text{head}}{\text{elevation}} = \underset{\text{head}}{\text{total}}.$$

(6–17c)

Two graphical illustrations of Bernoulli's equations are shown in Figs. 6–3 and 6–4. Notice in Fig. 6–3 that the pressure head (gage pressure) is the actual height of liquid in the static tube and that the total energy line represents the height of liquid in a total head tube (or Pitot tube); see Problems 12, 13, and 14. In Fig. 6–4, it is seen that the portion of the pipe above the level of the reservoir has negative gage pressure (is below atmospheric pressure).

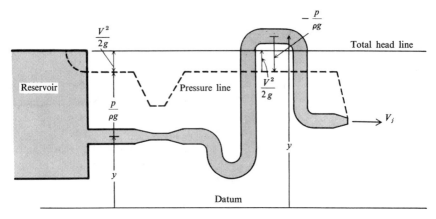

Fig. 6–4 Bernoulli terms for flow from a reservoir.

The derivation leading up to Eq. (6–17) was based upon the assumption that the flow was irrotational. It is possible to derive the Bernoulli equation along a streamline, regardless of whether the flow field is irrotational or rotational. However, in the case of a rotational flow, the constant in the Bernoulli equation will vary from streamline to streamline. The Eulerian equations for two-dimensional steady flow with the gravitational force in the negative y direction are

$$u\frac{\partial u}{\partial x} + v\frac{\partial u}{\partial y} = -\frac{1}{\rho}\frac{\partial p}{\partial x}, \tag{6–18a}$$

$$u\frac{\partial v}{\partial x} + v\frac{\partial v}{\partial y} = -g - \frac{1}{\rho}\frac{\partial p}{\partial y}. \tag{6–18b}$$

This set of equations can be easily integrated along a streamline after multiplying the first equation by dx and the second by dy. By virtue of the definition of a streamline,

$$\frac{v}{u} = \frac{dy}{dx}. \tag{6–19}$$

If v in Eq. (6–18a) and u in Eq. (6–18b) are eliminated by means of Eq. (6–19) and if the two equations are added, one obtains

$$\frac{1}{2}\frac{\partial}{\partial x}(u^2 + v^2)\,dx + \frac{1}{2}\frac{\partial}{\partial y}(u^2 + v^2)\,dy = -g\,dy - \frac{1}{\rho}\left(\frac{\partial p}{\partial x}\,dx + \frac{\partial p}{\partial y}\,dy\right). \tag{6–20a}$$

Noting that $u^2 + v^2 = V^2$, then

$$d\left(\frac{V^2}{2} + \frac{p}{\rho}\right) + g\,dy = 0, \tag{6–20b}$$

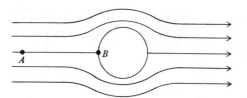

Fig. 6–5 Example 6–1.

since ρ is constant. Integrating Eq. (6–20b),

$$\frac{p}{\rho} + gy + \frac{V^2}{2} = \text{constant}. \tag{6–20c}$$

Eq. (6–20c) is valid even for a rotational flow field but the constant in Eq. (6–20c) is constant only along the same streamline. Only if the flow is irrotational will the constant be the same for other streamlines.

Example 6–1. Figure 6–5 shows water ($\rho = 62.0 \text{ lb}_m/\text{ft}^3$) flowing around a cylinder in steady flow. At point A the pressure of the water is 20 psia and the velocity is 50 ft/sec. At point B, which is on the same streamline and at the same elevation as point A, the velocity has been slowed to zero. What is the pressure at point B?

Solution. Equation (6–17b) can be written

$$\left[\frac{p}{\rho} + gy + \frac{V^2}{2}\right]_A = \left[\frac{p}{\rho} + gy + \frac{V^2}{2}\right]_B,$$

since the elevation y is the same at both A and B; and, since $V = 0$ at B,

$$\frac{p_A}{\rho} + \frac{V_A^2}{2} = \frac{p_B}{\rho},$$

or

$$p_B = p_A + \frac{\rho V_A^2}{2},$$

$$= 20 \frac{\text{lb}_f}{\text{in}^2} + \frac{(62.0)\dfrac{\text{lb}_m}{\text{ft}^3}(50)^2 \dfrac{\text{ft}^2}{\text{sec}^2}}{(2)(32.2)\dfrac{\text{lb}_m \text{ ft}}{\text{lb}_f \text{ sec}^2}(144)\dfrac{\text{in}^2}{\text{ft}^2}}$$

$$= (20 + 16.7)\frac{\text{lb}_f}{\text{in}^2} = 36.7 \text{ psi}.$$

The pressure at point B is referred to as the stagnation pressure, and is discussed later in this section.

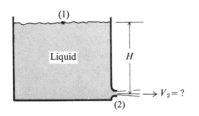
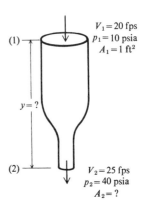

Fig. 6-6 Flow through a nozzle from reservoir.

Fig. 6-7 Flow in a reducing nipple.

Example 6-2. Using the Bernoulli equation, determine the velocity of the jet in Fig. 6-6.

Solution. Assuming the flow in the reservoir is inviscid, steady, and irrotational,

$$p_1 + \tfrac{1}{2}\rho V_1^2 + \rho g y_1 = p_2 + \tfrac{1}{2}\rho V_2^2 + \rho g y_2.$$

The pressure in a subsonic free jet is equal to the local atmospheric pressure, therefore, $p_1 = p_2$. Since the tank is large compared to the nozzle, $V_1 \ll V_2$, and may be assumed to equal zero.

$$V_2^2 = 2g(y_1 - y_2) = 2gH.$$

Hence

$$V_2 = \sqrt{2gH}.$$

This equation is known as the *Torricelli equation*.

Example 6-3. In Fig. 6-7, determine the vertical distance between station 1 and station 2 and also find the area A_2. Water is flowing in the pipe.

Solution. Assuming the flow is irrotational, steady and inviscid,

$$\frac{p_1}{\rho g} + \frac{V_1^2}{2g} + y_1 = \frac{p_2}{\rho g} + \frac{V_2^2}{2g} + y_2,$$

$$y_1 - y_2 = \frac{p_2 - p_1}{\rho g} + \frac{V_2^2 - V_1^2}{2g} = \frac{(40 - 10)(144)}{(62.4)(32.2)/32.2} + \frac{625 - 400}{2(32.2)}$$

$$= 69.3 + 3.5 = 72.8 \text{ ft}.$$

Using the equation for conservation of mass for this case,

$$\rho_1 A_1 V_1 = \rho_2 A_2 V_2, \quad \text{and} \quad \rho_1 = \rho_2.$$

Therefore

$$A_2 = \frac{V_1}{V_2} A_1 = \frac{20}{25}(1) = 0.8 \text{ ft}^2.$$

6-1 Comparison of the Bernoulli and Energy Equations

It is interesting to compare the Bernoulli equation (which was derived as a momentum equation) with the overall energy equation, Eq. (4–55) for steady, incompressible flow in a system consisting of solid boundaries except for an inlet (station 1) and an outlet (station 2), which allow the fluid to cross the control surface perpendicular to the surface (see Fig. 4–17). The steady flow energy equation can be written as

$$e_1 + \frac{p_1}{\rho} + gy_1 + \frac{V_1^2}{2} = e_2 + \frac{p_2}{\rho} + gy_2 + \frac{V_2^2}{2} - \frac{\dot{q}}{\dot{m}} + \frac{\dot{W}_{shaft}}{\dot{m}}, \quad (6\text{--}21)$$

where $\dot{W}_{shear}$ is zero for flow in the system above, because the fluid velocities are zero (no-slip condition) at the walls, and because the shear stresses are zero at the inlet and outlet planes (assuming uniform velocity profiles).

The Bernoulli equation for flow in the system between stations 1 and 2 can be written as

$$\frac{p_1}{\rho} + gy_1 + \frac{V_1^2}{2} = \frac{p_2}{\rho} + gy_2 + \frac{V_2^2}{2}. \quad (6\text{--}22)$$

Thus, it can be seen that the Bernoulli equation could have been derived as a very special form of the conservation of energy equation if one assumed that the fluid were adiabatic ($\dot{q} = 0$), isoenergetic ($e_1 = e_2$), and with zero shaft work.* Eq. (6–22) considers only the mechanical forms of stored energy (kinetic, potential, and flow energy).

A large viscous or frictional effect can cause significant changes in the internal energy of a fluid even when the flow is adiabatic. The irreversible conversion of mechanical energy to thermal energy can be expressed by the modified form of the energy equation, Eq. (6–21):

$$\frac{p_2 - p_1}{\rho} + \frac{V_2^2 - V_1^2}{2} + g(y_2 - y_1) = \frac{-\dot{W}_{shaft}}{\dot{m}} - \text{loss}. \quad (6\text{--}23)$$

The energy "loss" in Eq. (6–23) includes the irreversible conversion of energy resulting in net heat transferred out of the system and/or internal energy increase ($e_2 - e_1$). Rearranging Eq. (6–23), the term ($p_1/\rho + V_1^2/2 + gy_1$) might be considered as the "useful mechanical energy" input which is expended in providing (a) a shaft work output, (b) a useful mechanical energy output ($p_2/\rho + V_2^2/2 + gy_2$), and (c) an energy conversion leading to temperature rise and/or heat transfer.

Example 6–4. Water is flowing in a constant-diameter horizontal pipe that is insulated to prevent heat transfer. The pressure drop due to friction over a certain length is 2000 psi. Calculate the temperature rise of the water.

---

* Equation (6–22) would also be valid for the special case where

$$e_2 - e_1 = \frac{\dot{q}}{\dot{m}} - \frac{\dot{W}_{shaft}}{\dot{m}}$$

152 IDEAL FLUID FLOW AND HEAT TRANSFER

Solution. Assume the water is incompressible with a specific heat $c = 1.0$ Btu/lb$_m$ °F and a density of 62.4 lb$_m$/ft^3. From Eq. (6–23),

$$\text{loss} = 2000 \frac{\text{lb}_f}{\text{in.}^2} \frac{144 \text{ in.}^2 \text{ ft}^3}{\text{ft}^2 \, 62.4 \text{ lb}_m} = 4615 \frac{\text{lb}_f\text{-ft}}{\text{lb}_m},$$

but the loss is equal to $e_2 - e_1$, or

$$\text{loss} = c(T_2 - T_1).$$

Therefore,

$$T_2 - T_1 = \frac{4615 \text{ lb}_f\text{-ft}}{\text{lb}_m} \frac{\text{lb}_m \, °\text{F}}{1 \text{ Btu}} \frac{\text{Btu}}{778 \text{ ft lb}_f} = 5.93°\text{F}.$$

Dynamic and Total Pressures

Equation (6–22) may be rewritten by multiplying each term by the density ρ, which has been assumed to be constant:

$$\frac{\rho V_1^2}{2} + g\rho y_1 + p_1 = \frac{\rho V_2^2}{2} + g\rho y_2 + p_2. \tag{6–24}$$

Each term in Eq. (6–24) has the dimensions of pressure. The terms from left to right are given the names of *velocity pressure* (or *dynamic pressure*), *potential pressure*, and *static pressure*.

Equation (6–24) may be applied to any streamline in an incompressible, inviscid flow. It can be seen from Eq. (6–24) that as the fluid velocity decreases along a particular streamline with constant elevation, the static pressure increases. In the flow of fluid around a solid object there is always one streamline which intercepts the solid object at right angles to the surface (see Fig. 6–5). At the point of interception, called the *stagnation point*, the fluid velocity is zero. The static pressure at that point is called the *stagnation pressure*, or the *total pressure*, of the flow. The stagnation pressure is the maximum pressure which is obtained in an adiabatic flow where no work is done on the fluid. For a description of ways to measure total and static pressure, refer to Problems 12, 13, and 14.

The *local* stagnation pressure or local total pressure may be defined for the fluid at any point in the flow. If a uniform pressure exists across the cross section of a pipe flow, for example, but the velocity and elevation vary across that cross section, then the local stagnation pressure would vary across the cross section. If the local static pressure at a point located a distance y above a datum is p and the magnitude of the local fluid velocity at that point is V, then the local stagnation or local total pressure at that point is

$$p_T = p + \frac{\rho V^2}{2} + \rho g y. \tag{6–25}$$

The stagnation pressure is constant along a streamtube in the absence of heat and work.

6–1 INCOMPRESSIBLE FLUID IN FRICTIONLESS FLOW   153

Example 6–5. At a point in fluid flow where the velocity is 10.0 miles per hour, the static pressure is found to be 15.0 psia. The density of the fluid is 50 lb_m/ft^3. What is the local stagnation pressure, assuming that changes in elevation are neglected?

Solution.

$$p_T = \frac{\rho V^2}{2g_c} + p$$

$$= 50 \frac{lb_m}{ft^3} \frac{10.0^2 \frac{mile^2}{hr^2} \left(\frac{1}{3600}\right)^2 \frac{hr^2}{sec^2} 5280^2 \frac{ft^2}{mile^2}}{2\left(32.2 \frac{lb_m\, ft}{lb_f\, sec^2} 144 \frac{in.^2}{ft^2}\right)} + 15.0 \text{ psia}$$

$$= (1.14 + 15.0) \text{ psia} = 16.14 \text{ psia}.$$

Dynamic and Total Temperatures

Often there is interest in studying the flow of gases where a significant change in internal energy occurs. For a flow with zero heat transfer and zero work, the energy equation, Eq. (6–21), can be written as

$$j_1 + \frac{V_1^2}{2} + gy_1 = j_2 + \frac{V_2^2}{2} + gy_2, \qquad (6\text{--}26)$$

where j, the enthalpy, is defined by Eq. (4–54). For an ideal gas with constant specific heats, $\Delta j = c_p \Delta T$, therefore, the temperature change can be written as

$$T_2 - T_1 = \frac{V_1^2 - V_2^2}{2c_p} + \frac{g(y_1 - y_2)}{c_p}. \qquad (6\text{--}27)$$

Equation (6–27) can be applied to a control volume consisting of a section of a streamtube, as shown in Fig. 4–10. For a fluid element moving along the streamtube, in the absence of heat and work, the *stagnation temperature*, or *total temperature*, may be defined by

$$T_T = T + \frac{V^2}{2c_p}. \qquad (6\text{--}28)$$

The term T, which is the *static temperature*, is that temperature which would be sensed by a thermometer or thermocouple moving along at the stream velocity. The term $V^2/2c_p$, which has the dimension of temperature, is called the *dynamic temperature*. It is the temperature *rise* that would occur if a fluid stream were decreased to zero velocity adiabatically, with no change in elevation and no work. Equations (6–27) and (6–28) apply even when the decrease in velocity occurs with friction present, providing no heat is exchanged between fluid particles, and no work is performed on or by the fluid. If the decrease in velocity to zero were accomplished with no friction (no viscosity effect) and no heat transfer, then the rise in temperature is that which would occur between these two pressure levels during compression in a reversible adiabatic steady flow.

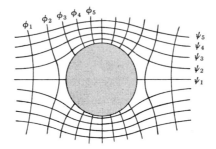

Fig. 6-8 Flow net around a cylinder in frictionless flow.

The Flow Net

Potential flow problems can often be solved by the use of the equations for the stream function and for the potential function. For the two-dimensional case, these are

$$\frac{\partial^2 \psi}{\partial x^2} + \frac{\partial^2 \psi}{\partial y^2} = 0, \qquad (6\text{-}29)$$

and

$$\frac{\partial^2 \Phi}{\partial x^2} + \frac{\partial^2 \Phi}{\partial y^2} = 0. \qquad (6\text{-}30)$$

Equations (6-29) and (6-30) are Laplace equations, with many known solutions. In solving them, suitable boundary conditions in terms of ψ or Φ must be specified. For example, solid boundaries must coincide with a streamline, or streamlines (lines of constant ψ), in the flow.

An x-y coordinate plot representing a solution of Eqs. (6-29) and (6-30) for a particular case will yield a family of lines, each with a constant value of ψ, and another family of lines, each with a constant value of Φ. The constant ψ lines will everywhere be orthogonal (intersecting at right angles) to the lines of constant Φ. This system of orthogonal lines makes up a *flow net*. A flow net around a circular cylinder is shown in Fig. 6-8.

The solution to Eqs. (6-29) and (6-30) for flow around an infinitely long cylinder (Fig. 6-8) is given by

$$\Phi = -V_\infty r \cos\theta - V_\infty R^2 \frac{\cos\theta}{r} = -V_\infty x - \frac{V_\infty x R^2}{x^2 + y^2}, \qquad (6\text{-}31)$$

and

$$\psi = -V_\infty r \sin\theta + V_\infty R^2 \frac{\sin\theta}{r} = -V_\infty y + \frac{V_\infty y R^2}{x^2 + y^2}, \qquad (6\text{-}32)$$

where V_∞ is the fluid velocity at a great distance, r is the distance from the axis of the cylinder, R is the radius of the cylinder, and θ is the angle measured from the positive x axis of the rectangular coordinate system with origin on the axis of the cylinder.

The solutions for Φ and ψ can be verified by substituting the expressions into Eqs. (6–29) and (6–30). The velocity components u and v at any point in the flow field can be obtained from Eqs. (6–2) or (4–30). The velocity vector can then be calculated from the components as

$$\mathbf{V} = \mathbf{i}u + \mathbf{j}v. \tag{6-33}$$

The magnitude of the velocity is, therefore, given by

$$V = \sqrt{u^2 + v^2} = \sqrt{\left(\frac{\partial \Phi}{\partial x}\right)^2 + \left(\frac{\partial \Phi}{\partial y}\right)^2}, \tag{6-34}$$

or

$$V = \sqrt{\left(\frac{\partial \psi}{\partial y}\right)^2 + \left(\frac{\partial \psi}{\partial x}\right)^2}, \tag{6-35}$$

and the velocity of the fluid along the surface of the cylinder (at $r = R$) is

$$V = 2V_\infty \sin \theta. \tag{6-36}$$

Notice that in frictionless flow there is a finite velocity of slip of the fluid along the surface, a situation which does not usually exist with real fluids.

The pressure distribution around the cylinder can be determined by Bernoulli's equation once the velocities are known. The body forces are usually small and are neglected.

Example 6–6. Determine the pressure distribution on the surface of the cylinder in the irrotational flow of an ideal fluid such as is shown in Fig. 6–8.

Solution. Neglecting changes in elevation, Eq. (6–17b) becomes

$$\frac{p_\infty}{\rho} + \frac{V_\infty^2}{2} = \frac{p}{\rho} + \frac{V^2}{2},$$

where p_∞ and V_∞ are the pressure and velocity of the undisturbed stream, a large distance away, and p and V are the local pressure and velocity along the surface, which are variable. Using Eq. (6–36),

$$\frac{p_\infty}{\rho} + \frac{V_\infty^2}{2} = \frac{p}{\rho} + \frac{4V_\infty^2 \sin^2 \theta}{2},$$

or

$$p = p_\infty + \frac{\rho V_\infty^2}{2}(1 - 4\sin^2 \theta).$$

156　IDEAL FLUID FLOW AND HEAT TRANSFER

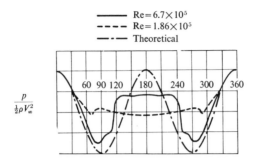

Fig. 6-9 Pressure distribution around a circular cylinder in the subcritical and supercritical range of Reynolds numbers. (From *Boundary Layer Theory* by Schlichting. Copyright 1968, 1960, 1955 by McGraw-Hill Book Company, Inc. Used by permission of McGraw-Hill Book Company.)

The dimensionless pressure ratio (or pressure coefficient) $(p - p_\infty)/\tfrac{1}{2}\rho V_\infty^2$ is plotted in Fig. 6-9, together with experimental results for a viscous fluid for two different Reynolds numbers. It can be seen that agreement between the potential theory, which assumes no viscosity, and experiments with a real, viscous fluid is very poor for both values of the Reynolds number. However, for a more streamlined body in flow of a viscous fluid the potential theory is very useful in estimating pressure distribution. This is seen for flow around an airfoil in Fig. 6-10. The reason for the poor agreement in the case

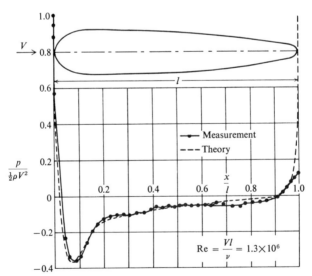

Fig. 6-10 Pressure distribution about a streamlined body of revolution. (From *Boundary Layer Theory* by Schlichting. Copyright 1968, 1960, 1955 by McGraw-Hill Book Company, Inc. Used by permission of McGraw-Hill Book Company.)

of the cylinder is due to formation of a wake caused by a phenomenon called *separation*, which occurs in real fluids. Separation is less significant in the case of streamlined objects. This was mentioned in Chapter 5 and will be discussed in detail in Chapter 9.

Source, Sink, and Free Vortex

In some types of potential flow problems, it is convenient to use cylindrical coordinates to describe the potential function and the stream function. In a two-dimensional cylindrical coordinate system, where r and θ are considered to be the coordinates of interest,

$$u_r = \frac{1}{r}\frac{\partial \psi}{\partial \theta} = \frac{\partial \Phi}{\partial r}, \tag{6-37}$$

$$u_\theta = -\frac{\partial \psi}{\partial r} = \frac{1}{r}\frac{\partial \Phi}{\partial \theta}. \tag{6-38}$$

The Laplace equation in cylindrical coordinates can be written for both Φ and ψ:

$$\nabla^2 \psi = \frac{\partial^2 \psi}{\partial r^2} + \frac{1}{r^2}\frac{\partial^2 \psi}{\partial \theta^2} + \frac{1}{r}\frac{\partial \psi}{\partial r} = 0, \tag{6-39}$$

and

$$\nabla^2 \Phi = \frac{\partial^2 \Phi}{\partial r^2} + \frac{1}{r^2}\frac{\partial^2 \Phi}{\partial \theta^2} + \frac{1}{r}\frac{\partial \Phi}{\partial r} = 0. \tag{6-40}$$

A *source* is defined as the steady, radial flow outward from a line. The streamlines and the lines of equipotential for a source are shown in Fig. 6–11. The solutions to Eqs. (6–39) and (6–40) for a source are

$$\Phi = A \ln r \quad \text{and} \quad \psi = A\theta. \tag{6-41}$$

where the *strength* of the source is defined as equal to A.

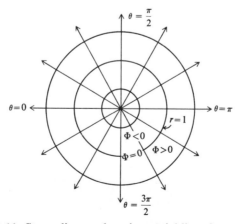

Fig. 6–11 Streamlines and equipotential lines for a source.

These solutions can be verified by substitution into the appropriate differential equations. The velocities can be obtained by use of Eqs. (6–37) and (6–38):

$$u_r = \frac{1}{r}\frac{\partial \psi}{\partial \theta} = \frac{A}{r}$$

and

$$u_\theta = -\frac{\partial \psi}{\partial r} = 0. \quad (6\text{–}42)$$

It is seen that the velocity has only a radial component and decreases as the distance r from the center point increases. At $r = 0$, the velocity is not defined (indeterminate mathematically). The rate of flow outward can be obtained by considering the flow across a cylindrical area at any radius r. For a unit depth normal to the $(r - \theta)$ plane, the mass flow rate is

$$\dot{m} = \rho 2\pi r u_r = \rho 2\pi A. \quad (6\text{–}43)$$

It can be seen that the mass flow rate outward is independent of r.

A *sink* is defined as the steady radial flow inward toward a line. The equations of a sink are obtained by multiplying the right side of Eq. (6–41) by (-1).

A free *vortex* is the steady irrotational flow of a fluid about a line. Solutions of Eqs. (6–39) and (6–40) for a vortex give

$$\psi = -B \ln r \quad \text{and} \quad \Phi = B\theta. \quad (6\text{–}44)$$

The velocities can be obtained from Eqs. (6–37) and (6–38):

$$u_r = \frac{\partial \Phi}{\partial r} = 0 \quad \text{and} \quad u_\theta = -\frac{\partial \psi}{\partial r} = \frac{B}{r}. \quad (6\text{–}45)$$

It can be seen that there is no radial velocity component, and the tangential velocity component decreases with distance from the line, or center, of the free vortex. This type of flow is approximated very closely in the pump-casing outside an impeller and in two-dimensional flow around bends, except very near the solid boundaries. The tornado is an example of a close approximation to free vortex flow in nature.

Only a few examples of potential flow solutions have been given. Because of the linearity of the Laplace equation, the principle of *superposition* of solutions is useful. These simple solutions may be combined with others to give flow solutions for more complex situations, such as the flow and pressure distribution around an airfoil. The potential solutions are particularly useful where effects due to viscosity are confined to a limited portion of the flow field.

6-1 INCOMPRESSIBLE FLUID IN FRICTIONLESS FLOW

Example 6-7. The solutions of a source flow and a uniform flow can be combined to produce flow around a half-body (Fig. 6-12). Using the principle of superposition, derive an equation for the pressure distribution around the half-body.

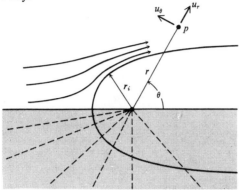

Fig. 6-12 Flow pattern past a half-body.

Solution. In polar coordinates the stream function of a uniform flow can be written as

$$(\psi)_{\text{uf}} = V_\infty r \sin \theta.$$

Adding the stream function of a uniform flow and a source—Eq. (6-41)—we get

$$\psi = V_\infty r \sin \theta + A\theta.$$

At the stagnation point, $u_r = 0$. Therefore,

$$u_r = \frac{1}{r}\frac{\partial \psi}{\partial \theta} = \left(V_\infty \cos \theta + \frac{A}{r}\right)_{\substack{\theta=\pi \\ r=r_s}} = 0.$$

Hence, $r_s = A/V_\infty$, which is the location of the stagnation point.

The equation of the streamline through the stagnation point is

$$\psi = V_\infty r_s \sin \pi + A\pi = V_\infty(r_s)(0) + A\pi = A\pi.$$

Therefore, the equation of the streamline describing the surface of the body, which includes the stagnation point, is

$$\psi = V_\infty r_b \sin \theta + A\theta = A\pi,$$

or

$$r_b = \frac{A(\pi - \theta)}{V_\infty \sin \theta}.$$

The function $r_b(\theta)$ describes the surface of the body. The velocity components at any point in the flow field are:

$$u_r = \frac{1}{r}\frac{\partial \psi}{\partial \theta} = V_\infty \cos\theta + \frac{A}{r},$$

$$u_\theta = -\frac{\partial \psi}{\partial r} = -V_\infty \sin\theta,$$

$$V^2 = u_r^2 + u_\theta^2 = V_\infty^2 + \frac{A^2}{r^2} + 2V_\infty \frac{A}{r}\cos\theta.$$

The pressure on the surface of the body can be determined from the Bernoulli equation, letting $r = r_b$:

$$p - p_\infty = \tfrac{1}{2}\rho V_\infty^2 \left(1 - \frac{V^2}{V_\infty^2}\right)$$

$$= -\tfrac{1}{2}\rho \left(\frac{A}{r_b}\right)\left(\frac{A}{r_b} + 2V_\infty \cos\theta\right).$$

The technique of superposition illustrated in the example above can be used to obtain other solutions, such as the flow around circular cylinders and the flow around rotating cylinders.

The techniques of complex variables are particularly useful in ideal fluid flow since they permit *conformal transformations*. Geometrical shapes for which solutions are not known can be transformed into shapes for which there are solutions available. The resulting solution can be transformed back into the original geometry leading to useful results not otherwise easily obtained. By this conformal transformation technique, for example, the flow net around a cylinder can be used to find the flow net around certain airfoil shapes.

REFERENCES

1. H. J. Schlichting, *Boundary Layer Theory*, 6th ed., McGraw-Hill, New York (1968).
2. L. M. Milne-Thomson, *Theoretical Hydrodynamics*, 3d ed., Macmillan, New York (1955).
3. V. L. Streeter, *Fluid Dynamics*, McGraw-Hill, New York (1948).
4. H. Rouse, *Elementary Mechanics of Fluids*, Wiley, New York (1946).
5. "Fluid Meters, Their Theory and Application," 5th ed., American Society of Mechanical Engineers, New York (1959).
6. H. S. Bean, "Flow Measurement, Part I—The Expanding Field of Fluid Metering," *Mech. Eng.* **36** (Apr. 1967).

7. A. H. Shapiro, "Pressure Fields and Fluid Acceleration," National Committee for Fluid Mechanics Films, Enc. Brit. Educ. Corp., Chicago.

8. A. H. Shapiro, "Vorticity," National Committee for Fluid Mechanics Films, Enc. Brit. Educ. Corp., Chicago.

9. H. Lamb, *Hydrodynamics*, Dover, New York (1955).

PROBLEMS

1. Water flows through a ¾-in. i.d. tube at an average velocity of 4 ft/sec. The energy input to the fluid at the tube wall surface is 5000 Btu/hr per ft² of tube inside surface area. Estimate the length of tube required for the bulk temperature of the water to be increased from 158°F to 162°F. Is this answer valid for both laminar and turbulent flow?

2. Light oil flows through a 2-in. i.d. tube. At the inlet the bulk temperature of the oil is 130°F and at the outlet the temperature is 170°F. The oil flows at an average velocity of 10 ft/sec. Calculate the rate (in watts) at which heat is transferred from the wall to the fluid.

3. Water flows through a 1.5-in. i.d. tube at an average velocity of 1.2 ft/sec. Estimate the length of tubing required for the water bulk temperature to be increased from 94°F to 106°F if the energy input at the tube wall is 0.4 kw per square foot of surface area.

4. Estimate the change in bulk temperature in degrees centigrade that will occur in each meter of length of tubing as water flows through a 3-cm i.d. tube at a mass flow rate of 100 gm/sec. Assume that heat is being added along the tube at a uniform rate of 0.5 kw per meter of length. Use fluid properties at 100°F as average values.

5. Hydraulic fluid ($\rho = 55$ lb_m/ft^3, $c_v = 0.48$ Btu/lb_m °F) flows through a device in steady, adiabatic flow. The pressure drop between inlet and outlet of the component is 800 psi. There is no work done on or by the component, and there is negligible change in kinetic and potential energy between inlet and outlet. Assuming no change in density or specific heat, estimate the bulk temperature change in the fluid between inlet and outlet.

6. Water flows through a 1-in. i.d. tube at the rate of 25 gal/min. The average bulk temperature of the water is 140°F and the heat addition at the wall is 1000 Btu/hr ft². Estimate the length of tube necessary for the bulk temperature of the water to be increased by 12°F.

7. In a round tube through which a fluid is flowing, the velocity and temperature distributions are

$$u(r) = 80\left(\frac{R-r}{R}\right)^{1/7} \quad \text{and} \quad T(r) = T_w + 40\left(\frac{R-r}{R}\right)^{1/5},$$

where r is the distance from the centerline in feet,
R is the inside radius of the tube in feet,
T is the temperature in °F,
T_w is the wall temperature (a constant) in °F, and
u is the velocity in feet per second.

Determine the value of the quantity $(T_B - T_w)$, where T_B is the bulk, or mixing-cup, temperature. *Hint:* Let $y = R - r$.

8. The velocity profile for turbulent flow in a pipe is given by

$$\frac{u}{u_{max}} = \left(\frac{y}{R}\right)^{1/n},$$

where u is the velocity at any radius, u_{max} is the maximum velocity, R is the radius of the pipe, y is the distance from the wall, and n is a quantity dependent on the value of the Reynolds number. Derive an expression for the average velocity in terms of u_{max} and n.

9. Show that the expression for the potential function for flow of an ideal fluid around a cylinder of radius R,

$$\Phi = -V_\infty r \cos\theta - V_\infty R^2 \frac{\cos\theta}{r},$$

satisfies the Laplace equation for Φ:

$$\frac{\partial^2 \Phi}{\partial r^2} + \frac{1}{r^2} \frac{\partial^2 \Phi}{\partial \theta^2} + \frac{1}{r} \frac{\partial \Phi}{\partial r} = 0.$$

10. Air at 30 psia pressure and 100°F enters a heater at the rate of 50,000 ft³/hr. The exit air temperature from the heater is 200°F. Estimate the rate at which heat must be supplied in the heater. Neglect any changes in kinetic or potential energy.

11. The velocity profile in flow through a 4-in. i.d. pipe is given by

$$\frac{u}{u_{max}} = \left(\frac{y}{R}\right)^{1/6},$$

where u is the local velocity at a particular value of y measured from the pipe wall, and R is the pipe radius. If the fluid has a density of 50 lb_m/ft^3 and if u_{max} is 20 ft/sec, determine the mass rate of flow through the pipe in lb_m/hr.

12. Prove that the velocity measured by the Pitot tube in the figure on page 163 is

$$V = \sqrt{\frac{2g(\rho_m - \rho)\Delta h}{\rho}},$$

where ρ_m is the density of the manometer fluid, ρ is the density of the flowing fluid, and Δh is the height differential of the manometer fluid. *Note:* The right side of the above equation is usually multiplied by an empirical factor C to take into account deviations from an ideal fluid; C is usually close to unity.

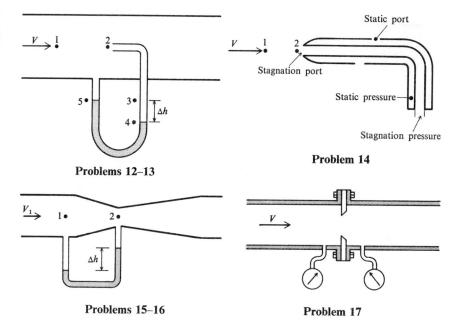

Problems 12–13

Problem 14

Problems 15–16

Problem 17

13. For the Pitot tube of problem 12, if $\Delta h = 2$ in., the manometer fluid is oil (specific gravity $= 0.83$), and the flowing fluid is air ($\rho = 0.0765 \text{ lb}_m/\text{ft}^3$), what is the velocity V? Assume $C = 0.995$.

14. A *Pitot-static tube* (see figure above) is a combined Pitot tube and static tube. If the pressure difference between the static port and the stagnation port is 0.1 psi for air flowing, what is the velocity V, if $\rho = 0.07 \text{ lb}_m/\text{ft}^3$? Assume $C = 1.0$.

15. Prove that the volume flow rate measured by a *venturi meter* like that in the figure above is

$$\dot{Q} = A_1 V_1 = A_1 \left[\frac{2(p_1 - p_2)}{\rho[(A_1^2/A_2^2) - 1]} \right]^{1/2},$$

where A_1 and A_2 are the areas at stations 1 and 2 respectively. *Note:* The right side of the above equation is usually multiplied by the empirical discharge coefficient C to take into account deviations from an ideal fluid; C usually ranges from 0.96 to 0.98.

16. For the venturi meter of Problem 15, what is the volume flow rate of water ($\rho = 62.4 \text{ lb}_m/\text{ft}^3$) if the manometer fluid is mercury ($\rho_m = 848 \text{ lb}_m/\text{ft}^3$ and $\Delta h = 6$ in.)? The diameters at station 1 and 2 are 4 and 1 in. respectively. Assume $C = 0.97$.

17. An *orifice meter* (see figure) operates on the same principle as a venturi meter, but it has a lower cost of installation. However, the discharge coefficient C is much lower than that of a venturi meter. C usually ranges from 0.55 to 0.75. If water is flowing through an orifice meter with a pressure difference

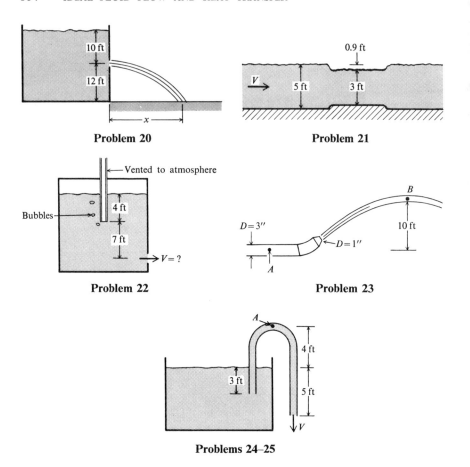

Problem 20

Problem 21

Problem 22

Problem 23

Problems 24–25

of 4 psi, what is the volume flow rate? The pipe diameter is 12 in. and the orifice diameter 6 in. Assume $C = 0.625$.

18. A laminar jet of water falling from a faucet has a velocity of 2 ft/sec and a diameter of 1 in. at an elevation of 2.0 ft. Calculate the velocity and diameter at an elevation of 0.0 ft.

19. The diameters of a vertical laminar jet of water are $\frac{7}{16}$ in. and $\frac{3}{16}$ in., measured 1.0 ft apart. What is the flow rate of water in gal/sec?

20. Water flows from an orifice in the side of the tank in the figure above. Calculate the distance x.

21. Water flows through a rectangular channel like that in the diagram, 3 ft wide. Find the flow rate. Neglect viscous effects.

22. The accompanying figure shows a constant head tank. (a) What is the efflux velocity V? (b) What is the efflux velocity if the water level is 2 ft above the end of the vent?

23. The velocity at B in the figure on page 164 is 20 ft/sec. What is the pressure at A?
24. The figure represents a *siphon*. (a) Find the velocity of the water leaving the siphon. (b) What is the pressure at A?
25. In the siphon of Problem 24, how high above the discharge end of the siphon could point A be before the absolute pressure goes to zero?
26. Use a plotter with Eqs. (6–31) and (6–32) to generate a set of curves such as those shown in Fig. 6–8.
27. Show that the vector cross product $\nabla \times \mathbf{V}$ is a vector with components equal to the left-hand terms in Eqs. (6–7), (6–8), and (6–9).

CHAPTER 7

FUNDAMENTALS OF FLUID FLOW AND HEAT TRANSFER IN VISCOUS FLUIDS

In the previous chapter, idealizations of fluid behavior were made in order to simplify the solution of problems. In some cases, such as the prediction of the pressure distribution around a streamlined object, the results obtained by neglecting viscous forces are useful and closely approximate the actual situation. In other situations, however, such as the prediction of drag on an object, or in the study of flows where large velocity gradients exist, the neglecting of viscous forces can lead to very erroneous results.

7–1 THE BOUNDARY LAYER CONCEPT

Even the apparently simple case of steady, incompressible flow of a fluid along a flat wall is quite complex to analyze with viscous effects present. Consider such a flow, shown in Fig. 7–1, with the two-dimensional rectangular coordinate system having an origin at the leading edge of the plate. The plate is assumed to be infinitely thin and the flow approaching the leading edge of the plate is uniform and parallel to the plate. The undisturbed, or free stream, velocity is designated V_∞ and is assumed to be constant. In the vicinity of the plate the local velocity components u and v, parallel and normal to the plate, are assumed to be variable with position.

If the additional assumptions are made that body forces are absent and the properties of the fluid (except temperature and pressure) are constant, then the fundamental equations—continuity, Eq. (4–16); momentum, Eqs. (4–44a-c); and energy, Eq. (4–65)—become:

$$\frac{\partial u}{\partial x} + \frac{\partial v}{\partial y} = 0, \tag{7–1}$$

$$u\frac{\partial u}{\partial x} + v\frac{\partial u}{\partial y} = -\frac{1}{\rho}\frac{\partial p}{\partial x} + \nu\left(\frac{\partial^2 u}{\partial x^2} + \frac{\partial^2 u}{\partial y^2}\right), \tag{7–2}$$

$$u\frac{\partial v}{\partial x} + v\frac{\partial v}{\partial y} = -\frac{1}{\rho}\frac{\partial p}{\partial y} + \nu\left(\frac{\partial^2 v}{\partial x^2} + \frac{\partial^2 v}{\partial y^2}\right), \tag{7–3}$$

$$u\frac{\partial T}{\partial x} + v\frac{\partial T}{\partial y} = \frac{k}{\rho c_p}\left(\frac{\partial^2 T}{\partial x^2} + \frac{\partial^2 T}{\partial y^2}\right) + \frac{\mu}{\rho c_p}\Phi. \tag{7–4}$$

7-1 THE BOUNDARY LAYER CONCEPT

The solution of the continuity equation, Eq. (7–1), and the momentum equations, Eqs. (7–2) and (7–3), can be sought independent of Eq. (7–4), since properties are assumed to be constant and temperature does not appear in the first three equations. The three unknowns in the continuity and momentum equations above are u, v, and p. The boundary conditions on Eqs. (7–1), (7–2), and (7–3) are:

$$u = v = 0 \quad \text{at} \quad y = 0,$$
$$u = V_\infty \quad \text{at} \quad x = 0,$$
$$u \to V_\infty \quad \text{as} \quad y \to \infty \quad \text{for} \quad x > 0.$$

The closed-form solution of Eqs. (7–1), (7–2), and (7–3), with the boundary conditions for this most simple viscous flow problem, has challenged mathematicians for decades and is not known even today.

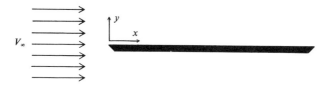

Fig. 7–1 Flow along a thin flat plate.

In 1904 L. Prandtl [1] proposed an approximate method for solution of the momentum and continuity equations given above. Prandtl suggested that, under certain conditions, the flow field could be divided into two regions. In the region near the wall, the viscous forces are significant and must be considered. This region near the wall is referred to as the *boundary layer*. The other region, outside the boundary layer, is a region where the viscous forces are insignificantly small and can be ignored.

This boundary layer concept and its significance will now be considered for the flow of fluid in the vicinity of the solid surface or wall, such as that shown in Fig. 7–1.

In the boundary layer the velocity of the fluid decreases from the free stream value, V_∞, to a value of zero at the wall. At the outer edge of the boundary layer the velocity, V_∞, is assumed to be parallel to the surface and constant for all positions x along the surface. Viscous effects cause the fluid at the surface to cling to this surface and thus have a velocity of zero; this is referred to as a "no slip" condition and it holds for a continuum flow. Intuition suggests that the local velocity u approaches the free stream value asymptotically. This is verified by experiment.

A typical plot of u versus y at a given x is shown in Fig. 7–2. Such a plot is referred to as a velocity profile. Since the velocity approaches the free stream value almost asymptotically, a distance greater than the boundary layer thickness must be traversed to reach a point where the velocity u actually equals the free stream velocity V_∞. However, an arbitrary distance from the wall can be selected at which the local velocity u is practically the same as V_∞. Usually this point is selected so that $u = 0.99 V_\infty$. The distance y from the wall to that arbitrary point is referred to as the *boundary layer thickness*, δ. Inside the boundary layer ($y < \delta$) the viscous forces are significant since there is a significant velocity gradient; see Stokes' law, Eq. (3–20). Outside the boundary layer ($y > \delta$) the velocity gradients are small, and therefore the viscous forces are negligible. Two important results follow from this boundary layer concept of Prandtl.

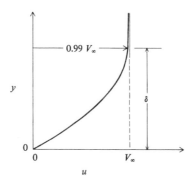

Fig. 7–2 Typical velocity profile near a wall.

For the flow *outside* the boundary layer (the free stream), the terms involving viscosity in Eqs. (7–2) and (7–3) can be neglected in comparison with the remaining inertia and pressure terms. This reduction in the order of the equations leads to no difficulty since the boundary conditions at $y = \delta$ can be satisfied. The resulting equations, referred to as the Euler equations, have been discussed in Chapter 6. The usefulness of these equations for flow along solid surfaces is that they give the relationship between the pressure distribution and the local velocity at the edge of the boundary layer ($y = \delta$). This is important in obtaining a solution to the fundamental equations describing the flow and the temperature inside the boundary layer.

The equations describing the flow *inside* the boundary layer can be simplified for cases where the boundary layer thickness δ is relatively small compared to the distance x from the front or leading edge of the solid body. This condition has been found experimentally to exist at large values of the local Reynolds number, $(V_\infty x)/\nu$, based on the distance x from the leading edge. For typical conditions encountered by an aircraft flying in the atmosphere, for example, the boundary layer thickness δ is small compared to the distance x from the leading edge of the surface, except when x is very near the leading edge.

7–1 THE BOUNDARY LAYER CONCEPT

The assumption that the boundary layer thickness is small permits simplification of Eqs. (7–2) and (7–3). This will be done by an order of magnitude analysis of terms. Let x and u be considered arbitrarily to be of order (1) and let y be considered to be of order (δ)—since we are restricting our analysis to the boundary layer, and no value of y greater than δ will occur—where $\delta \ll 1$. The order of magnitude of the derivatives of the velocity components u can be written as follows, where the symbol O stands for "the order of":

$$\frac{\partial u}{\partial x} = O(1), \quad \frac{\partial u}{\partial y} = O\left(\frac{1}{\delta}\right), \quad \frac{\partial^2 u}{\partial y^2} = O\left(\frac{1}{\delta^2}\right), \quad \frac{\partial^2 u}{\partial x^2} = O(1).$$

From the continuity equation, Eq. (7–1), $\partial v/\partial y$ cannot be greater than of order (1), since $\partial u/\partial x$ is of order (1). Therefore, v must be no greater than order (δ). Hence the following statements can be expressed:

$$v = O(\delta), \quad \frac{\partial v}{\partial y} = O(1), \quad \frac{\partial^2 v}{\partial y^2} = O\left(\frac{1}{\delta}\right), \quad \frac{\partial v}{\partial x} = O(\delta), \quad \frac{\partial^2 v}{\partial x^2} = O(\delta).$$

Equations (7–1) to (7–3), rewritten with the appropriate order of magnitude for each term, will have the form

$$\overset{1}{\frac{\partial u}{\partial x}} + \overset{1}{\frac{\partial v}{\partial y}} = 0, \tag{7–5}$$

$$\overset{1\cdot 1}{u\frac{\partial u}{\partial x}} + \overset{\delta\cdot(1/\delta)}{v\frac{\partial u}{\partial y}} = -\frac{1}{\rho}\frac{\partial p}{\partial x} + \nu\left(\overset{1}{\frac{\partial^2 u}{\partial x^2}} + \overset{1/\delta^2}{\frac{\partial^2 u}{\partial y^2}}\right), \tag{7–6}$$

$$\overset{1\cdot\delta}{u\frac{\partial v}{\partial x}} + \overset{\delta\cdot 1}{v\frac{\partial v}{\partial y}} = -\frac{1}{\rho}\frac{\partial p}{\partial y} + \nu\left(\overset{\delta}{\frac{\partial^2 v}{\partial x^2}} + \overset{1/\delta}{\frac{\partial^2 v}{\partial y^2}}\right). \tag{7–7}$$

From Eq. (7–6), it is evident that the term $\partial^2 u/\partial x^2$ is very small compared with $\partial^2 u/\partial y^2$ and may be neglected. Since no single term in any one equation may be of greater order than the greatest of other terms, we can see from Eq. (7–6) that the kinematic viscosity must be of order (δ^2), or $\delta = O(\nu^{1/2})$. But since both u and x are of order (1), then it is valid to state:

$$\frac{\delta}{x} = O\left(\frac{\nu}{ux}\right)^{1/2} \quad \text{or} \quad \frac{\delta}{x} = O(\text{Re}^{-1/2}). \tag{7–8}$$

From Eq. (7–8) it can be seen that for a "rough rule of thumb," the boundary layer thickness δ can be considered relatively thin if the Reynolds number is at least of order (1000).

In examining Eq. (7–7), it is obvious that all terms are of order (δ) or smaller and hence they are of lesser order than the largest in Eq. (7–6). The pressure gradient $(1/\rho)(\partial p/\partial y)$ must be of order (δ) or smaller, and hence the pressure at the solid boundary cannot differ appreciably from that at the edge of the boundary layer. Therefore, under the condition of large Reynolds number, the equations of momentum, Eqs. (7–2) and (7–3), for the boundary layer become:

$$u \frac{\partial u}{\partial x} + v \frac{\partial u}{\partial y} = -\frac{1}{\rho} \frac{dp}{dx} + v \frac{\partial^2 u}{\partial y^2}, \qquad (7\text{–}9)$$

$$\frac{\partial p}{\partial y} = 0. \qquad (7\text{–}10)$$

The continuity equation, Eq. (7–1), remains unchanged and is valid inside or outside the boundary layer:

$$\frac{\partial u}{\partial x} + \frac{\partial v}{\partial y} = 0. \qquad (7\text{–}11)$$

Equations (7–9), (7–10), and (7–11) are known as *Prandtl's boundary layer equations*. Their usefulness will be seen from the fact that solutions to these equations can be obtained for cases where solutions to the momentum equations, Eqs. (7–2) and (7–3), are not known.

Equation (7–9) is a partial differential equation of the parabolic type. Note that the order of the differential equation has not been reduced in dropping the $\partial^2 u/\partial x^2$. A general closed-form solution to these simplified equations is not known. However, problems of practical importance such as the flat plate problem of Fig. 7–1 can be solved using this equation and the equation of continuity. These solutions are particularly valuable since they give insight into other problems for which solutions are not available. Equations (7–9) and (7–10) are approximations to the momentum equations, Eqs. (7–2) and (7–3), and are valid for high values of Reynolds number. An exact solution to the boundary layer equations can be considered an approximate solution to the momentum equations, valid inside the boundary layer.

Equation (7–10) is significant because it implies that the pressure at the outer edge of the boundary layer is the same as that at all positions y inside the layer, for a given location x. Thus the pressure on the wall can be predicted from the Euler equations describing the free stream flow, after making a small correction to account for the boundary layer thickness. Methods of correcting the results involve a *displacement thickness*.* Free stream solutions are obtained by the use of potential theory, which has been discussed in Chapter 6, and by other methods.

---

* The displacement thickness δ_D indicates the distance by which streamlines outside the boundary layer are shifted outward due to formation of the boundary layer. It is defined by

$$\delta_D = \frac{1}{V_\infty} \int_0^\infty (V_\infty - u)\, dy$$

Boundary Layer Transition

The equations discussed so far in this chapter are valid only for laminar flow, since no variation of velocity with time, and thus no turbulence, is assumed to exist.

Boundary layers may be either laminar or turbulent depending upon the flow conditions. The boundary layer is often illustrated by means of a picture showing the boundary layer thickness at locations along a surface. It is convenient to discuss the flow along a very thin flat plate, such as was shown in Fig. 7–1. Figure 7–3 illustrates how a plate of finite thickness with a sharp leading edge is used to approximate an infinitesimally thin plate. The sharp leading edge permits a splitting of the flow so that a portion above the plate approximates the thin plate flow conditions.

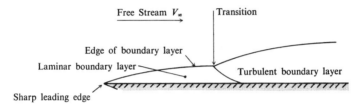

Fig. 7–3 Boundary layer along a flat plate.

The vertical scale in Fig. 7–3 is enlarged, since the boundary layer thickness δ is usually small compared to the length of the flat plate. It can be seen in the figure that the laminar boundary layer (that part of the boundary layer in which the flow is laminar) starts from zero thickness and grows in thickness with increasing distance from the leading edge.

At some distance back from the leading edge, the laminar boundary layer undergoes a change or transition, and the boundary layer becomes turbulent. Experiments have shown that the transition will normally occur at a distance x, measured from the leading edge, where the value of a *local Reynolds number based on x* $[Re_x = (V_\infty x)/\nu]$ has a value equal to approximately 320,000. This is an approximation, or rule of thumb, since the actual transition Reynolds number depends on intensity of turbulence, surface roughness, pressure gradient, and temperature difference between the free stream and the plate.

Figure 7–4 shows the effect of intensity of turbulence on the critical Reynolds number for transition on a flat plate. Decreasing intensity results in increasing values of the critical Reynolds number, up to an upper limit of Reynolds number of about 3×10^6. With Reynolds numbers greater than that value, decreasing intensity no longer has any effect. As yet no analytical methods have been developed for accurately predicting transition Reynolds

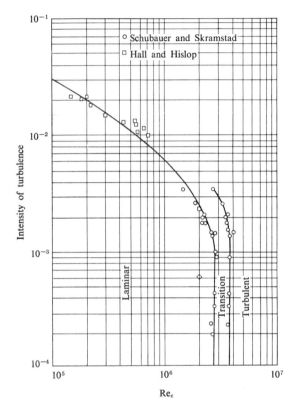

Fig. 7-4 Critical Reynolds number Re_c for transition to turbulence of a boundary layer on a flat plate as function of the stream turbulence [2, 3]. (From *Heat and Mass Transfer* by Eckert and Drake. Copyright 1959 by McGraw-Hill Book Company, Inc. Used by permission of McGraw-Hill Book Company.)

numbers. If the leading edge of the flat plate were not sharp but consisted of a flat edge normal to the free stream, then a small, disorganized flow, turbulent in nature, might occur near the leading edge before the laminar boundary layer commences. In such a case, the laminar boundary layer could commence with a nonzero thickness. In a body with a rounded leading edge (such as an airfoil) the laminar boundary commences with a nonzero thickness at the stagnation point.

The Thermal Boundary Layer

It is necessary to know the local values of u and v before Eq. (7–4) can be solved. Boundary layer concepts have permitted the calculation of these velocity components, for certain cases, in the region near the wall. The concept of a boundary layer may also be applied to the local temperature in the fluid, permitting a simplification of Eq. (7–4). This is done by defining a

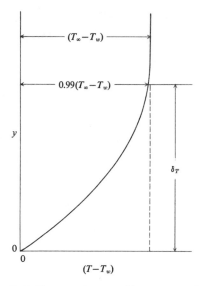

Fig. 7-5 Temperature profile near a wall.

temperature boundary layer, or *thermal boundary layer*, in which there is a significant variation of temperature.

In many fluid flow/heat transfer problems, the region in which such a significant temperature variation exists is confined to the fluid near the wall, or in approximately the same region as the velocity boundary layer. The thermal boundary layer thickness δ_T may be defined as that distance from the surface where difference between the temperature and the surface temperature $(T - T_w)$ is 99 percent of the difference between the free stream temperature and the surface temperature $(T_\infty - T_w)$. This is illustrated in Fig. 7-5, where it can be seen that the free stream temperature is approached asymptotically.

Using an order of magnitude analysis (assuming $\delta_T \ll x$) similar to that which led to Eqs. (7-9) and (7-10), the energy equation—Eq. (7-4)—for flow inside the boundary layer becomes:

$$u\frac{\partial T}{\partial x} + v\frac{\partial T}{\partial y} = \frac{k}{\rho c_p}\left(\frac{\partial^2 T}{\partial y^2}\right) + \frac{\mu}{\rho c_p}\left(\frac{\partial u}{\partial y}\right)^2. \tag{7-12}$$

Equation (7-12) could also have been derived by assuming that conduction of heat in the x direction is negligible compared to that in the y direction, and velocity gradients with respect to y are much larger than with respect to x.

It is sometimes convenient to discuss the velocity and temperature profiles within the boundary layer in dimensionless terms. The distance y

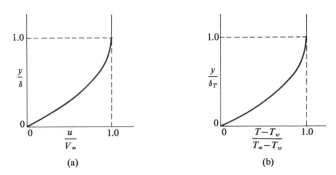

Fig. 7-6 Dimensionless profiles. (a) Velocity profiles. (b) Temperature profiles.

normal to the wall is conveniently nondimensionalized with the local boundary layer thickness, δ or δ_T. The local velocity component u, for example, is nondimensionalized with the free stream velocity V_∞. The temperature excess $(T - T_w)$ is nondimensionalized by the difference $(T_\infty - T_w)$. The dimensionless velocity profile is shown in Fig. 7-6(a), and the dimensionless temperature profile is shown in Fig. 7-6(b).

Notice that in both cases the dimensionless values range, conveniently, from 0 to 1.0 inside the boundary layer.

7-2 THE ENTRANCE REGION IN CONDUITS

As fluid enters a conduit from a large reservoir, a boundary layer like that described in the previous section will be formed along the conduit wall. The fluid close to the conduit wall adheres to the walls, and the viscous effects cause the velocity of adjoining fluid layers to be reduced successively from the free stream value. The free stream value is the velocity in the core of Fig. 7-7. Figure 7-7 shows fluid leaving a reservoir or region of very low fluid velocity and flowing through a channel consisting of two semi-infinite, parallel flat plates. This type of flow closely approximates flow through channels with a small height/width ratio.

Near each wall is a region of retarded velocity, or a boundary layer. The velocity in this boundary layer varies from zero at the wall to a "free

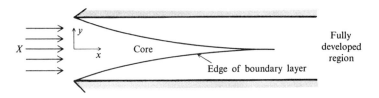

Fig. 7-7 Development of boundary layer in the entrance of a conduit.

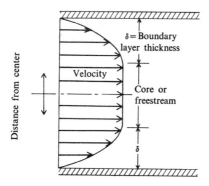

Fig. 7–8 Typical velocity profile in entrance region.

stream" value at the edge of the boundary layer. The edge of the boundary layer is again defined as that place where the local velocity is 99 percent of the free stream value.

Although the velocity at the inlet to the channel is not exactly parallel or uniform, such an assumption is often made to permit a more simple mathematical analysis. A further assumption is that the fluid outside the boundary layers (in the core or free stream) is uniform in velocity at any cross section along the flow. Thus, a typical velocity profile at some cross section of the channel would appear as shown in Fig. 7–8.

Deceleration of the fluid in the boundary layer has two important effects. It causes a y component of velocity to develop, which can be seen from the continuity equation, Eq. (7–11). The fluid in the channel outside the boundary layer accelerates to meet the requirements of overall mass conservation in the channel. Simultaneous with the deceleration of the axial velocity near the wall and the acceleration of the axial velocity near the center of the channel, there will be a steady growth in the thickness of each boundary layer. If the axial dimension of the channel is large enough, the two boundary layers will eventually intersect. This growing together of the two boundary layers occurs asymptotically, and it is, therefore, impossible to describe a definite location where the layers combine. At some finite distance from the entrance, the change in axial velocity with x at all values of y will be so slight that it is insignificant. The velocity profile is then said to be *fully developed*. For the example discussed in Fig. 7–7, the velocity profile has the shape of a parabola when fully developed.

Transition to Turbulence

In the case of a wide channel (large spacing between plates), with a high inlet velocity, or with a fluid of relatively low kinematic viscosity, it is possible that transition to turbulence may occur before the boundary layers grow together. This is shown in Fig. 7–9.

176   FLUID FLOW AND HEAT TRANSFER IN VISCOUS FLUIDS

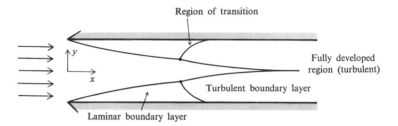

Fig. 7–9 Development of a turbulent velocity profile.

In this case it is turbulent boundary layers that grow together. As a result, the fully developed profile downstream is a turbulent velocity profile and differs considerably from the parabolic laminar profile. Turbulent flow is more apt to occur at high values of the Reynolds number ($V_{\text{avg}}l/\nu$), based on the plate spacing l. Turbulent channel flow will be discussed in more detail later.

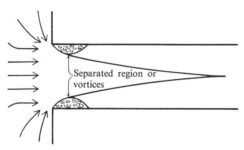

Fig. 7–10 Vortices formed due to sharp-edged entrance.

The situations discussed above are idealized concepts of what usually occurs at the entrance of such channels. As shown in Fig. 7–10, the flow entering a channel is often of nonuniform velocity across the channel and usually has some y component of velocity. If the y component of velocity at the entrance edges is large enough, a *separated region* (or wake region of relatively low velocity) will form near the inlet (Fig. 7–10). The wave region has a swirling motion, and the swirls are sometimes referred to as *vortices*. The vortices may cause instant transition to turbulence in some cases, or they may merely cause some effect on the establishment of a laminar boundary layer. The shape of the inlet region is very important. For example, the separation can be prevented by rounding the entrance to the channel.

The location of the point of transition depends to an extent on the intensity of the turbulence existing in the fluid at the entrance. It has been found experimentally that, by reducing the turbulence of the inlet fluid, by rounding

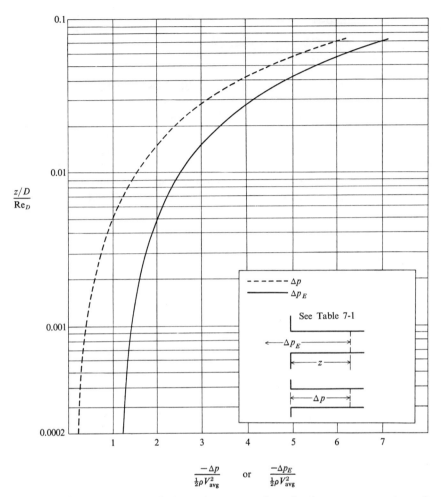

Fig. 7–11 Langhaar's prediction of pressure drop in the entrance region of a tube in laminar flow. [From H. L. Langhaar, *Trans. ASME* **66**, 671 (1944).]

the entrance to the channel, and by eliminating any vibration of the channel walls, laminar flow can be maintained into the fully developed region for very large values of the Reynolds number.

The preceding discussion for parallel-wall channels can also be applied to flow in round tubes. In this case, the boundary layer forms around the entire tube and grows toward the axis of the tube.

The length required for the formation of a fully developed laminar profile in a round tube can be estimated by use of the theoretical equation of Langhaar [4]. Defining an entrance length, L_e, as that distance required for the centerline velocity to reach 99 percent of the fully developed value,

Langhaar's equation is

$$\frac{L_e}{D} = 0.0575 \frac{V_{\text{avg}} D}{\nu} = 0.0575 \, \text{Re}_D. \tag{7-13}$$

Equation (7-13) holds for Reynolds numbers based on diameter of about 2000 and less. For Reynolds numbers above that value, transition to a turbulent boundary layer will occur. The entrance length decreases sharply from that predicted by Eq. (7-13) when transition to turbulence occurs and then increases with increasing Reynolds number. An exact prediction of entrance length in turbulent flow cannot be made with any great reliability. The shape of the entrance is very important, with much shorter entrance lengths occurring for square-edged entrances than for rounded ones. A distance of at least 50 diameters is normally required to establish a fully developed turbulent velocity profile in the case of rounded entrances.

The entrance effect is important because the pressure drop in the region prior to the establishment of the fully developed flow is much greater than for an equivalent length in the region of fully developed flow. The pressure drop, Δp, in the entrance region may be estimated by the use of Fig. 7-11 taken from the results of Langhaar [4]. An additional pressure drop upstream from the entrance, due to fluid acceleration in the reservoir, is included in the term Δp_E.

Table 7-1
ANALYTICAL RESULTS OF LANGHAAR

$\dfrac{z/D}{\text{Re}_D}$	$\dfrac{-2\Delta p g_c}{\rho u_{\text{avg}}^2}$	$\dfrac{-2\Delta p_E g_c}{\rho u_{\text{avg}}^2}$
0.000205	0.2120	1.2122
0.00083	0.3860	1.3866
0.001805	0.5652	1.5732
0.003575	0.8300	1.8300
0.00535	1.0420	2.0442
0.00838	1.360	2.3614
0.01373	1.844	2.8474
0.01788	2.188	3.1898
0.02368	2.636	3.6354
0.0341	3.380	4.3840
0.04488	4.112	5.1146
0.06198	5.232	6.2326
0.0760	6.152	7.1478

Note: When flow is not horizontal, replace $(-\Delta p)/\rho$ by $[(-\Delta p)/\rho - \Delta z(g/g_c)]$. Δp is the pressure drop between the entrance ($z = 0$) and a point at a distance z from the entrance. Δp_E is the pressure drop between the reservoir, where the velocity is zero, and point z measured from the tube entrance. This takes into account an *entrance effect* outside the tube, not included in Δp. From H. L. Langhaar, *Trans. ASME* **64**, A-55 (1942).

7-3 DETERMINATION OF PRESSURE DROP IN CONDUIT FLOW

In engineering analysis and design related to flowing fluid systems in closed conduits, there often is a need to know something about the pressure drop which may occur between inlet and outlet, where the flow rate is specified. In some cases the inverse problem may be of interest; that is, a given pressure drop is available between two specified locations in a system, and knowledge of the flow rate is desired. Knowledge of the power required when a specified pressure drop and flow rate exist is also important in many engineering situations. A convenient method of relating pressure drop to mass flow rate, for a variety of engineering situations, is obviously desirable. This need has led to the use of a dimensionless quantity called the *friction factor*.

The friction factor is actually a dimensionless pressure gradient. Letting the symbol f stand for the friction factor,

$$f = -\frac{dp^*}{dz^*}, \tag{7-14}$$

where

$$p^* = \frac{p}{\frac{1}{2}\rho V_{avg}^2} \quad \text{and} \quad z^* = \frac{z}{D},$$

and z is an axial dimension in the direction of the flow. D is a characteristic dimension of the duct or conduit, such as diameter.

Equation (7–14) may be rewritten to give

in fully developed
① or ⑦
laminar turbulent

$$f = \frac{-\frac{dp}{dz} D}{\frac{1}{2}\rho V_{avg}^2}. \tag{7-15}$$

In the case of a constant pressure gradient dp/dz between two reference sections 1 and 2 in the flow, Eq. (7–14) can be written

$$f = \frac{(p_1 - p_2)D}{\frac{1}{2}\rho V_{avg}^2 L}, \tag{7-16}$$

where L is the distance between the two reference sections. Equation (7–16) could be used to predict the pressure drop between two reference sections in a conduit of specified size if the fluid density and the average velocity are specified or if the mass rate of flow is known. The use of Eq. (7–16) would require knowledge of a proper value of the friction factor f. Values of f can be determined in some cases by theoretical means; in other cases experimental data can be utilized along with dimensional analysis to provide the needed values.

Theoretical values of f are obtained from knowledge of velocity profiles and the use of Stokes' law of friction. One such development is as follows.

Three factors will normally contribute to the pressure drop occurring along a conduit as fluid flows through it. They are:

1. Changes of elevation in conjunction with gravity;

2. Changes of velocity (acceleration) caused by changes in density, mass flow rate, or area; and
3. Viscous friction, or shear stress at the conduit wall.

It should be pointed out that the pressure does not always *drop* in the flow direction; for example, in vertical downflow, or in flow with heat removal from the fluid, the pressure may actually increase in the direction of flow.

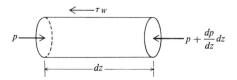

Fig. 7–12 Force balance on a fluid element.

In the case of steady, incompressible flow of a fluid through a conduit of constant cross-sectional area, the average velocity of flow is a constant along the conduit. At sufficient distance from an inlet (or any area change), the flow will be fully developed and non-accelerating. If the conduit is also horizontal then there will be no pressure change due to elevation changes. Consider such a flow in a round tube. The only force acting on the fluid to cause a pressure drop is the viscous force. A typical fluid element in such a situation is shown in Fig. 7–12. Neglecting the pressure difference between the top and bottom of the tube due to elevation, and considering the pressure to be uniform across a cross section, conservation of momentum (net pressure force − net viscous force = 0) gives

$$-\frac{\pi D^2}{4}\frac{dp}{dz}dz - \pi D\, dz\, \tau_w = 0, \tag{7-17}$$

or

$$\frac{dp}{dz} = -\frac{4\tau_w}{D}. \tag{7-18}$$

Equation (7–18) shows the relationship between the pressure gradient and the shear stress at the wall.

The dimensionless wall shear stress τ_w^*, which is usually referred to as the skin friction coefficient C_f, may be related to the friction factor f by use of Eqs. (7–18) and (7–15):

$$C_f = \tau_w^* = \frac{\tau_w}{\tfrac{1}{2}\rho V_{\text{avg}}^2} = \frac{-\dfrac{D}{4}\dfrac{dp}{dz}}{\tfrac{1}{2}\rho V_{\text{avg}}^2} = \frac{f}{4}. \tag{7-19}$$

7-3 DETERMINATION OF PRESSURE DROP IN CONDUIT FLOW

The friction factor f defined by Eq. (7–15) is called the *Moody friction factor* [5], or the *Darcy friction factor* [6]. The dimensionless wall shear stress, or coefficient of friction C_f, is sometimes called the *Fanning friction factor* [7], and the symbol f is also often used for this term. Care should be taken to check which friction factor is intended when equations involving these terms are used. It will be the practice of this book to use the symbol f for the *Moody friction factor*, as in Eqs. (7–15) and (7–16).

From Stokes' law of friction (in cylindrical coordinates),

$$\tau_w = -\mu \frac{\partial u_z}{\partial r}\bigg|_{\text{wall}}, \qquad (7\text{--}20)$$

where u_z is the velocity of the fluid in the axial direction. The negative sign in Eq. (7–20) appears because a positive r is measured outward from the tube axis toward the wall. Comparison of Eqs. (7–18) and (7–20) shows that the pressure gradient can be determined once the velocity profile is known. Equation (7–18) is valid where acceleration or elevation changes are present, but it gives only that portion of the pressure gradient due to friction. For fully developed flow, where the velocity profile does not change with z, the pressure gradient is constant, and therefore pressure drop between two specified points can be calculated.

Example 7–1. For laminar, fully developed flow through a round tube, the velocity profile is given by

$$u_z = 2V_{\text{avg}}\left(1 - \frac{r^2}{R^2}\right),$$

where r is the distance from the tube axis, and R is the tube radius. It is desired to have an expression for the friction factor f in terms of fluid kinematic viscosity ν, tube diameter D, and average velocity of flow V_{avg}.

Solution. Combining Eqs. (7–18) and (7–20),

$$\frac{dp}{dz} = \frac{4\mu}{D}\frac{\partial u_z}{\partial r}\bigg|_{\text{wall}}.$$

The velocity gradient at the wall can be evaluated from the given velocity profile,

$$\frac{\partial u_z}{\partial r}\bigg|_{\text{wall}} = \frac{-4V_{\text{avg}}r}{R^2}\bigg|_{r=R} = \frac{-4V_{\text{avg}}}{R}.$$

Thus,

$$\frac{dp}{dz} = \frac{4\mu}{D}\left(\frac{-4V_{\text{avg}}}{R}\right) = \frac{-16\mu V_{\text{avg}}}{D^2/2}$$

$$= \frac{-32\mu V_{\text{avg}}}{D^2}.$$

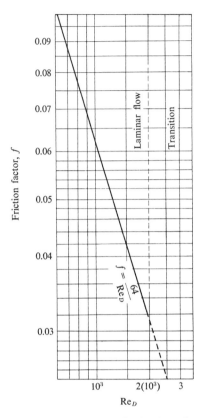

Fig. 7-13 Friction factors for laminar flow, from Eq. (7-21).

Using Eq. (7-15),

$$f = \frac{32\mu V_{\text{avg}} D}{D^2 \frac{1}{2} \rho V_{\text{avg}}^2} = \frac{64\mu}{D\rho V_{\text{avg}}}.$$

Since the Reynolds number based on tube diameter is

$$\text{Re}_D = \frac{D\rho V_{\text{avg}}}{\mu} = \frac{DV_{\text{avg}}}{\nu},$$

the solution to the example can be written as

$$f = \frac{64}{\text{Re}_D}. \tag{7-21}$$

Equation (7-21) is shown plotted in Fig. 7-13. Since the assumption of laminar flow was made in the derivation of Eq. (7-21), it is restricted to use at Re_D below 2000. Experimental data have shown the validity of Eq. (7-21) for laminar flow in tubes.

REFERENCES

1. L. Prandtl, "Motion of Fluids with Very Little Friction," *Proc. Third International Math. Conf., Heidelberg, 1904* (NACA TM 452, 1928).
2. G. B. Schubauer and H. K. Skramstad, *J. Res.* (National Bureau of Standards) **38**, 28 (1947).
3. A. A. Hall and G. S. Hislop, Aeronautical Research Committee, Report and Memorandum, No. 1843 (1938).
4. H. L. Langhaar, *Trans. ASME* **64**, A-55 (1942)
5. L. F. Moody, *Trans. ASME* **66**, 671 (1944).
6. V. L. Streeter, *Handbook of Fluid Dynamics*, McGraw-Hill, New York (1961).
7. J. G. Knudsen and D. Katz, *Fluid Dynamics and Heat Transfer*, McGraw-Hill, New York (1958).

PROBLEMS

1. For a flat moving plate with a velocity relative to air of 300 ft/sec, estimate the distance from the leading edge to the point of transition for (a) sea level, (b) 50,000, and (c) 100,000 ft altitude. Assume a critical Reynolds number of 2.6×10^6. The value of the reciprocal of kinematic viscosity $1/\nu$ at these altitudes, in sec/ft², is sea level, 6380; 50,000 ft, 1210; 100,000 ft, 110.
2. Work Problem 1 for a critical Reynolds number of 320,000.
3. Air at one atmospheric pressure and 70°F flows through a wind tunnel past a sharp-edged flat plate. The free stream air velocity is 100 ft/sec. Transition occurs 9 in. back from the leading edge. Estimate the intensity of turbulence in the tunnel.
4. Calculate the entrance length for flow of water at 100°F from a large reservoir into a 0.25-in. i.d. round tube at an average velocity of 0.5 ft/sec.
5. For Problem 4, calculate the pressure drop (a) from the entrance to a point 1 ft downstream from the entrance, and (b) from the reservoir upstream (where the velocity is zero) to a point 1 ft downstream from inlet.
6. Water flows from a large reservoir vertically upward through a $\frac{1}{8}$-in. i.d. tube at an average velocity of 0.5 ft/sec. The water temperature is 150°F and the reservoir pressure is 50 psia. Estimate the pressure in the tube at a point 8 in. from the inlet.
7. For 100°F water flowing through a 1-in. i.d. pipe, calculate the velocity at which the flow would tend to undergo transition to turbulence. Assume a critical Reynolds number, based on diameter, of 2000. Repeat the problem for air at 100°F and 1 atmospheric pressure.

CHAPTER 8

LAMINAR FLOW AND HEAT TRANSFER

The technique for analyzing fluid flow for the cases where the effect of fluid viscosity could be ignored has been illustrated in Chapter 6. Although the results of such an assumption are useful in obtaining information concerning certain types of flow, there are many situations where the effect of viscosity cannot be ignored. Some techniques for obtaining theoretical solutions to fluid flow problems with and without heat transfer, will now be illustrated.

8-1 LAMINAR FLOW AND HEAT TRANSFER IN TUBES

The situation which is to be studied in this section is that of a fluid flowing through a tube of constant diameter D at a constant mass rate of flow. It is known that the Reynolds number is low so that laminar flow may be assumed and the product of the Eckert and Prandtl numbers is small, allowing for the assumption that the viscous dissipation is negligible. Radiation heat transfer will be neglected. Further, body forces are assumed to be absent and the values of μ, c_p, ρ, and k are constant. The fluid is, also, assumed to flow parallel to the axis of the tube, a situation which would exist at a large distance from any change of area, and in the absence of swirl or buoyancy effects. The flow is thus assumed to be *fully developed* laminar flow.

Equations which represent the flow are often expressed quite adequately in rectangular coordinates. In this case (the flow through a long round tube) the use of cylindrical coordinates is desirable since that system leads to the most simple form of the basic equations of mass, momentum, and energy conservation. The incompressible continuity equation and the momentum equations in cylindrical coordinates have already been presented in Eqs. (4–22) and (4–45). The energy equation for the incompressible case in cylindrical coordinates can be obtained from Eq. (4–67). The result is

$$\rho c_p \left(\frac{\partial T}{\partial t} + u_r \frac{\partial T}{\partial r} + \frac{u_\theta}{r} \frac{\partial T}{\partial \theta} + u_z \frac{\partial T}{\partial z} \right)$$
$$= k \left[\frac{1}{r} \frac{\partial}{\partial r} \left(r \frac{\partial T}{\partial r} \right) + \frac{1}{r^2} \frac{\partial^2 T}{\partial \theta^2} + \frac{\partial^2 T}{\partial z^2} \right] + \mu \Phi_{\text{cyl}}, \quad (8\text{–}1)$$

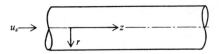

Fig. 8-1 Orientation of coordinate system for tube flow.

in which Φ_{cyl} is the viscous dissipation function in cylindrical coordinates:

$$\Phi_{\text{cyl}} = 2\left[\left(\frac{\partial u_r}{\partial r}\right)^2 + \left(\frac{1}{r}\left(\frac{\partial u_\theta}{\partial \theta} + u_r\right)\right)^2 + \left(\frac{\partial u_z}{\partial z}\right)^2\right]$$
$$+ \left[\left(\frac{\partial u_\theta}{\partial z} + \frac{1}{r}\frac{\partial u_z}{\partial \theta}\right)^2 + \left(\frac{\partial u_z}{\partial r} + \frac{\partial u_r}{\partial z}\right)^2 + \left(\frac{1}{r}\frac{\partial u_r}{\partial \theta} + r\frac{\partial}{\partial r}\left(\frac{u_\theta}{r}\right)\right)^2\right].$$
(8-2)

The cylindrical coordinate system is oriented for this case with the z axis aligned along the tube axis (Fig. 8-1). There will be a z component of velocity, u_z, but the r and θ components, u_r and u_θ, are zero.

Velocity Distribution

The fluid velocity distribution will be determined first. The velocity profile can be determined from the momentum equation without the use of the energy equation, since the properties of the fluid are assumed constant. The continuity equation, Eq. (4-22), reduces to

$$\frac{\partial u_z}{\partial z} = 0. \qquad (8\text{-}3)$$

This implies that u_z is a function of r and θ only. Flow will be assumed symmetrical about the axis of the tube so that u_z is not a function of θ but only of r, the radial distance from the tube axis. The momentum equation, Eqs. (4-45a-c) reduces to

$$\frac{\partial p}{\partial r} = 0, \quad \frac{\partial p}{\partial \theta} = 0, \quad \frac{\partial p}{\partial z} = \mu\left(\frac{\partial^2 u_z}{\partial r^2} + \frac{1}{r}\frac{\partial u_z}{\partial r}\right). \qquad (8\text{-}4)$$

Therefore, the pressure p will depend only on z, the distance along the axis. Equation (8-4) can be rewritten as an ordinary differential equation:

$$\frac{dp}{dz} = \frac{\mu}{r}\frac{d}{dr}\left(r\frac{du_z}{dr}\right). \qquad (8\text{-}5)$$

This equation could be valid only if $dp/dz = $ constant, since the left side is a function only of z and the right side is a function only of r. Considering dp/dz as a constant, Eq. (8-5) can be integrated twice to obtain a general solution. Using the boundary conditions,

$$u_z = 0 \quad \text{at} \quad r = R \quad \text{and} \quad u_z = u_{\max} \quad \text{at} \quad r = 0,$$

where R is the tube radius, the solution is

$$u_z = \frac{R^2}{4\mu}\frac{dp}{dz}\left[\left(\frac{r}{R}\right)^2 - 1\right] = -\frac{dp/dz}{4\mu}(R^2 - r^2). \tag{8-6}$$

The velocity profile is seen to be parabolic (Fig. 8–2) and, for a given tube radius, depends only on the fluid viscosity and the pressure gradient.

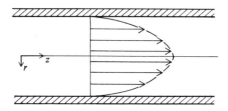

Fig. 8–2 The parabolic velocity profile in laminar tube flow.

This type of flow is variously referred to as *Hagen-Poiseuille* flow*, *Poiseuille flow*, or *capillary flow* (since laminar flow usually occurs only with very small tube diameters). It can be seen that for this type of flow, the velocity at the tube centerline ($r = 0$) is a maximum, and it can be shown that $u_{max} = 2V_{avg}$.

The equation for the velocity profile can be written

$$u_z = 2V_{avg}\left[1 - \left(\frac{r}{R}\right)^2\right]. \tag{8-7}$$

The shear stress at the wall is found by the use of the Stokes relation for the cylindrical case with no radial velocity component (Newton's law of viscosity). This was done in Section 7–3. The result is

$$C_f = \frac{8\mu}{R\rho V_{avg}} = \frac{16}{Re_D} = \frac{f}{4}, \tag{8-8}$$

where

$$Re_D = \frac{\rho V_{avg} D}{\mu}. \tag{8-9}$$

Thus the dimensionless wall shear stress C_f, or the friction factor f, depends only upon the Reynolds number in this type of flow.

The Reynolds number defined by Eq. (8–9) can be used as the criterion to establish whether the assumption of laminar flow is valid. Generally speaking, for tube flow, if the value of the Reynolds number based on tube diameter, Eq. (8–9), is less than 2000, the flow may be assumed to be laminar.

---

* Pronounced "pwah zŏo′yah", with the "ŏo" pronounced as in book.

The Temperature Profile

Once the velocity profile in the tube is known, the energy equation can be investigated. For the assumptions stated thus far (ρ = constant; k = constant; $\Phi_{\text{cyl}} = 0$; $u_r = u_\theta = 0$; steady flow), the energy equation reduces to

$$\rho c_p \left(\frac{\partial T}{\partial t} + u_z \frac{\partial T}{\partial z} \right) = k \left[\frac{1}{r} \frac{\partial}{\partial r} \left(r \frac{\partial T}{\partial r} \right) + \frac{1}{r^2} \frac{\partial^2 T}{\partial \theta^2} + \frac{\partial^2 T}{\partial z^2} \right]. \quad (8\text{--}10)$$

The general solution to this partial differential equation is not available. Some of the terms become zero, however, by specification of certain boundary conditions on temperature. A solution can be obtained for those particular boundary conditions. The assumption that the temperature at each point on the wall is constant with time, for example, will eliminate the dependency on time of the temperature at each location in the fluid, and the first term in Eq. (8–10) becomes zero. The assumption of no temperature variation around the tube, for any fixed z, will eliminate the dependency of temperature in the fluid on θ provided there are no buoyancy forces of significance. Buoyancy effects can be ignored when the Grashof number is much less than the square of the Reynolds number. Neglecting conduction in the z direction, compared to that of the r direction, would permit elimination of the term $\partial^2 T / \partial z^2$. Equation (8–10) then becomes

$$\rho c_p u_z \frac{\partial T}{\partial z} = k \left[\frac{1}{r} \frac{\partial}{\partial r} \left(r \frac{\partial T}{\partial r} \right) \right]. \quad (8\text{--}11)$$

Equation (8–11) is also difficult to solve since the temperature depends on both r and z. However, a special case where the heat flux at the wall is uniform and heating begins a very long distance upstream yields useful results. The requirement that heating begins some long distance upstream is to assure that the temperature profiles are fully developed, or, in other words, that the temperature boundary layer has grown to the center of the tube. The case of constant wall heat flux is an important one in engineering because it arises quite often in such problems as nuclear heating, electric resistance heating, counterflow heat exchangers, and radiant heating.

Change in bulk specific enthalpy of the fluid between any two cross sections a distance Δz apart is reflected in a difference of the bulk temperature between the two points. If a uniform amount of heat is added at the tube wall at all positions, then the change in the bulk temperature will be linear with distance (since the mass flow rate and the specific heat are assumed constant).

With constant thermal conductivity k and uniform heat flux at the wall, the Fourier-Biot law requires that for all z

$$\left(\frac{\partial T}{\partial r} \right)_{r=R} = \frac{\dot{q}}{kA} = \text{constant}. \quad (8\text{--}12)$$

188 LAMINAR FLOW AND HEAT TRANSFER   8-1

Since the change in bulk temperature is linear with distance z, and the temperature gradient at the wall is constant, it seems likely that the temperature T at any fixed r would also vary linearly with z at the same rate.

The temperature of the wall T_w also varies linearly with z at that same rate and, at all positions z, the temperature difference $(T_w - T_B)$ has a constant value. Such a variation of temperatures indicates that the dimensionless temperature profile $(T - T_w)/(T_B - T_w)$ vs. r is the same for all values of z.

Notice that Eq. (8–11) has the following boundary conditions:

$$\text{at } r = 0, \frac{\partial T}{\partial r} = 0; \text{ at } r = R, T = T_w;$$

$$\text{at } 0 < r < R, \frac{\partial T}{\partial z} = \frac{dT_B}{dz} = \text{constant}. \tag{8-13}$$

For an infinitely long tube, the local velocity u_z can be expressed in terms of the average velocity, V_{avg}, using the previous solution, given in Eq. (8–7):

$$u_z = 2V_{\text{avg}}\left[1 - \left(\frac{r}{R}\right)^2\right], \tag{8-14}$$

where

$$V_{\text{avg}} = -\frac{1}{8\mu}\left(\frac{dp}{dz}\right)R^2. \tag{8-15}$$

Equation (8–11) can be nondimensionalized by letting

$$T^* = \frac{T - T_w}{T_B - T_w}, \quad u^* = \frac{u_z}{u_{\max}} = \frac{u_z}{2V_{\text{avg}}}, \quad r^* = \frac{r}{D}. \tag{8-16}$$

Letting

$$\alpha = \frac{k}{\rho c_p} \quad \text{and} \quad \frac{\partial T}{\partial z} = \frac{dT_B}{dz},$$

and noting that T^* is a function only of r^*, Eq. (8–11) becomes

$$u^*\frac{dT_B}{dz} = \left[\frac{\alpha(T_B - T_w)}{2V_{\text{avg}}D^2}\right]\frac{1}{r^*}\frac{d}{dr^*}\left(r^*\frac{dT^*}{dr^*}\right). \tag{8-17}$$

From an energy balance on the fluid between two cross sections a distance dz apart, it can be shown that

$$\frac{dT_B}{dz} = \frac{4\alpha \dot{q}''}{kDV_{\text{avg}}}, \tag{8-18}$$

where $\dot{q}''$ is the heat flux (heat flow per unit area per unit time) at the tube wall. Substituting in Eq. (8–16) for u^*, from Eq. (8–15), and for dT_B/dz, from Eq. (8–17), gives

$$\frac{8D\dot{q}''}{k(T_B - T_w)} [1 - (2r^*)^2] = \frac{1}{r^*} \frac{d}{dr^*}\left(r^* \frac{dT^*}{dr^*}\right). \quad (8\text{–}19)$$

Equation (8–19) can be solved by separation of variables and integration. The first integration yields

$$\frac{8D\dot{q}''}{k(T_B - T_w)}\left(\frac{r^{*2}}{2} - r^{*4}\right) = r^* \frac{dT^*}{dr^*} + C_1. \quad (8\text{–}20)$$

Since
$$dT^*/dr^* = 0 \quad \text{at} \quad r^* = 0, \quad C_1 = 0.$$

A second integration yields

$$\frac{8D\dot{q}''}{k(T_B - T_w)}\left(\frac{r^{*2}}{4} - \frac{r^{*4}}{4}\right) = T^* + C_2. \quad (8\text{–}21)$$

From Eq. (8–16), $T = T_w$ and $T^* = 0$ when $r^* = \frac{1}{2}$. So

$$\frac{8D\dot{q}''}{k(T_B - T_w)}\left(\frac{1}{16} - \frac{1}{64}\right) = C_2. \quad (8\text{–}22)$$

Substituting for C_2, from Eq. (8–22), into Eq. (8–21), gives

$$T^* = \frac{8D\dot{q}''}{k(T_B - T_w)}\left(\frac{r^{*2}}{4} - \frac{r^{*4}}{4} - \frac{3}{64}\right). \quad (8\text{–}23)$$

Equation (8–23) gives the dimensionless temperature as a function of dimensionless radius and the dimensionless coefficient $(8D\dot{q}'')/[k(T_B - T_w)]$. Since $(T_B - T_w)$ appears on both sides of Eq. (8–23), it does not need to be known in order to compute values of temperature. The local temperatures may be computed by knowing the local wall temperature, the tube radius, the fluid thermal conductivity, and the heat flux at the tube wall. Since the local wall temperatures are not normally known, it is desirable to change Eq. (8–23) to a different form. From Eq. (8–23), using the definition of the bulk temperature—Eq. (4–56e)—and Eq. (8–14), and noting that T_w is not dependent on r, it can be shown that

$$(T_B - T_w) = -\frac{22}{96} \frac{\dot{q}'' D}{k}. \quad (8\text{–}24)$$

Equation (8–24) shows the constant difference between the bulk and wall temperature at any distance z for a fixed value of heat flux.

The Heat Transfer Coefficient

The ratio of the heat flux at the wall to the temperature difference $(T_w - T_B)$ is defined as the *heat transfer coefficient*, h. It is variously called the unit thermal convective conductance, the film coefficient, and the film heat transfer factor:

$$h = \frac{\dot{q}''}{(T_w - T_B)}. \tag{8-25}$$

Equation (8-25) is referred to as *Newton's Law of Cooling*.

The heat transfer coefficient h is merely a convenience for simplifying the extremely complex phenomenon of convection heat transfer. It has no meaningful physical significance, may be arbitrarily defined for a given problem, and can rarely be predicted with exactness in real engineering problems. Estimates of h are often utilized in fact, simply because the solution to a problem is not obtainable.

Using Eqs. (8-25) and (8-24),

$$h = \frac{96k}{22D}, \tag{8-26}$$

or, in terms of the dimensionless Nusselt number,

$$\mathrm{Nu}_D = \left(\frac{hD}{k}\right) = 4.36. \tag{8-27}$$

Thus, the Nusselt number is a constant for this case. Equation (8-27) is useful for predicting the wall temperature–bulk temperature difference for a given, constant heat flux or for predicting the heat flux for a specified temperature difference. It is limited to laminar, fully developed flow in constant-area tubes with temperature profiles fully developed. In actual situations the effect caused by buoyancy forces and the effects of variation of viscosity with temperature may cause large discrepancies between the predicted—(Eq. 8-27)—and the actual values.

Just as the friction factor f was shown to be a dimensionless pressure drop related to the value of the velocity gradient at the wall, the Nusselt number is equal to a dimensionless temperature gradient at the wall, as shown below.

$$\frac{\dot{q}''}{(T_w - T_B)} = h = \frac{k\left(\frac{\partial T}{\partial r}\right)_{\mathrm{wall}}}{(T_w - T_B)} = \frac{-k\left(\frac{\partial T^*}{\partial r^*}\right)_{\mathrm{wall}}}{D},$$

and

$$\mathrm{Nu}_D = \frac{hD}{k} = -\left(\frac{\partial T^*}{\partial r^*}\right)_{\mathrm{wall}}. \tag{8-28}$$

The problem solution represented by Eqs. (8–26) and (8–27) is an extremely rare case in which the local heat transfer coefficient is constant over a surface. Usually the local coefficient varies with position, and an average value, $\bar{h}$, is specified for convenience.

8–2 THE ENTRANCE REGION IN LAMINAR FLOW

The equations for tube flow which were developed in the previous section assumed that the velocity and temperature profiles were fully developed or, in other words, were functions of the radius r only, and that the radial component of velocity u_r was zero. This is descriptive of the conditions actually existing in a tube at a large distance from the entrance or from any sudden change in the channel dimensions. In the entrance region of any channel the flow is not parallel, the radial velocity u_r is not zero, and the axial velocity u_z is a function of both spatial coordinates z and r. The reason for this was explained in Section 7–2 (Fig. 7–7).

It was stated in Section 8–1, for the case of constant wall heat flux in round tubes, that the dimensionless temperature profile $(T - T_w)/(T_B - T_w)$ becomes independent of z at large distances downstream from the tube entrance and at large distances from the start of heating. This implied that the dimensionless temperature profile must be established over some starting length, during which the profile depends on z. The solution of the temperature equation is more difficult in such cases since the dimensionless temperature then depends upon axial distance z as well as upon r. The problem is even more complex when the velocity profile is also not fully established at the start of heating. This would be the situation when heating of the fluid begins at the tube entrance.

Only the case where the velocity is assumed to be fully developed and parallel to the tube axis will be considered at this time. This profile might be arbitrary but is assumed to be constant with z. Equation (8–11) is valid for such a case. In any case where the fluid properties are assumed to be constant, knowledge of the temperature profiles is not necessary to obtain a solution for the velocity profile. Assuming that a solution for the velocity profile is available, then the energy or temperature equation, Eq. (8–11), is the only equation left to be solved. For most boundary conditions the solution to this equation is difficult to obtain, and, therefore, dimensional reasoning might be utilized to gain some insight into the probable form of the solution.

Equation (8–11) is nondimensionalized by letting

$$V^* = \frac{u_z}{V_{\text{avg}}}, \quad T^* = \frac{T}{T_0}, \quad r^* = \frac{r}{D}, \quad z^* = \frac{z}{z_{\text{max}}}, \quad (8\text{–}29)$$

where z_{max} is any arbitrary fixed value of z, usually the largest possible value. T_0 is any constant temperature such as the inlet temperature. Substituting the above into Eq. (8–11), gives

$$\left[\frac{\rho c_p D^2 V_{avg}}{k z_{max}}\right] V^* \frac{\partial T^*}{\partial z^*} = \frac{1}{r^*}\frac{\partial}{\partial r^*}\left(r^* \frac{\partial T^*}{\partial r^*}\right). \tag{8–30}$$

The coefficient appearing in the brackets is the dimensionless *Graetz number*, Gz. It may be written in terms of any arbitrary value of z:

$$\text{Gz} = \frac{\mu c_p}{k}\frac{\rho D V_{avg}}{\mu}\frac{D}{z} = \left(\text{Pr Re}_D \frac{D}{z}\right). \tag{8–31}$$

The dimensionless temperature, and therefore the dimensionless temperature gradients at the wall, depend only upon the Graetz number. The dimensionless temperature at the wall can be related to the Nusselt number (Section 8–1). Therefore, it is expected that the solution to various problems of this type could be expressed in terms of Nusselt number–Graetz number correlations, or in other words,

$$\text{Nu}_D = f(\text{Gz}) = f\left(\text{Pr Re}_D \frac{D}{z}\right). \tag{8–32}$$

Equation (8–32) gives the general functional relationship that should be expected in the theoretical solution or in the correlation of experimental data for the situation being discussed.

In 1885 Graetz [1] presented an analytical solution for the entrance problem, assuming that an established parabolic velocity profile existed at the start of heating and assuming that the wall of the tube had a uniform temperature. Various investigators have made improvements on Graetz's solution and others have solved the problem for various profiles and boundary conditions [18].

A generalized procedure, based on a superposition principle was presented by Sellars, Tribus, and Klein [2] and is useful for a variety of boundary conditions and for rectangular as well as circular ducts. The solutions for the local Nusselt number based on diameter vs. the local Graetz number for the case of fully developed laminar flow in a round tube are shown in Fig. 8–3, for both constant wall temperature and constant wall heat flux. In the figure, the curve for constant wall heat flux approaches a constant value of Nusselt number equal to 4.36 as the value of z becomes very large. This agrees with the result found in the previous section for fully developed temperature profiles. The curve for constant wall temperature approaches a value of Nusselt number equal to 3.66 at large values of z. This is the value found by Graetz for the fully developed temperature profile for the case of constant wall temperature.

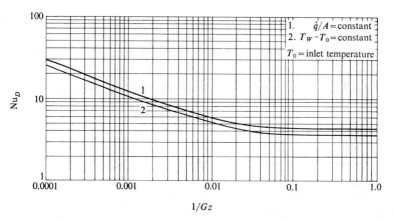

Fig. 8-3 The local Nusselt number near the start of heating for fully developed laminar flow in a tube.

The establishment of a constant value of Nusselt number, and therefore a constant value of the heat transfer coefficient, might be surprising in view of the fact that for constant wall temperature the bulk fluid temperature continues to approach the wall temperature as z increases. In the limit, for very large values of z the fluid bulk temperature becomes equal to the wall temperature. This is explained by the fact that beyond the entrance region, as the temperature difference between the tube wall and the fluid decreases, the heat flux at the wall decreases at the same rate. Their ratio, which is defined as the heat transfer coefficient, therefore does not change with z. This might also be explained in terms of the temperature gradient at the wall, which becomes constant downstream from the thermal inlet region. Note that for both cases the local Nusselt number is very large near the start of heating and decreases with distance downstream, approaching the constant values asymptotically.

From Fig. 8-3, it can be seen that the distance required to reach fully developed temperature conditions can be approximated as that distance of z that makes the reciprocal of Gz equal to about 0.05. Thus the thermal entrance length $L_{e,t}$ can be approximated by the equation

$$\frac{L_{e,t}}{D} = 0.05(\text{Re}_D)(\text{Pr}). \tag{8-33}$$

Equation (8-33) for thermal entrance length should be compared with Eq. (7-13) for the velocity entrance length. For fluids having a Prandtl number near 1.0, it can be seen that the thermal and velocity entrance lengths are approximately the same. Fluids having large Prandtl numbers, such as viscous oils, have very large thermal entrance lengths compared to velocity entrance lengths, and the assumption of a fully developed velocity profile is a good approximation even near tube inlets. For fluids having very low Prandtl numbers, such as liquid metals, the thermal entrance length is very short compared to the distance needed to establish the velocity profiles, and therefore an assumption that nearly uniform velocity exists across the tube section is a good approximation in solving the temperature inlet problem. This assumption of uniform velocity across the tube is sometimes referred to as *rod flow* or *slug flow*.*

In practical engineering design, it is necessary to know values of the average heat transfer coefficient over some heated length so that overall rates of heat transfer may be calculated. An empirical equation, valid for constant wall temperature and fully developed velocity profiles in a tube, has been developed by Hausen [3] for such purposes. It expresses the average Nusselt number based on tube diameter $\overline{\mathrm{Nu}}_D$ to the Graetz number based on the total length over which heating is occurring, $\overline{\mathrm{Gz}}$:

$$\overline{\mathrm{Nu}}_D = \left(3.66 + \frac{0.0668\,\overline{\mathrm{Gz}}}{1 + 0.04(\overline{\mathrm{Gz}})^{2/3}}\right)\left(\frac{\mu_B}{\mu_w}\right)^{0.14}. \qquad (8\text{--}34)$$

where μ_B is the viscosity evaluated at the mean bulk temperature and μ_w is the viscosity evaluated at the wall temperature.

For situations where the properties, particularly the viscosity, do not vary too much across the fluid, the viscosity ratio may be neglected and the fluid properties evaluated at a mean film temperature:

$$T_f = \left(\frac{T_w + T_{B,\mathrm{avg}}}{2}\right).$$

This situation normally exists for gases. $T_{B,\mathrm{avg}}$ is the average bulk temperature between inlet and outlet:

$$T_{B,\mathrm{avg}} = \frac{T_{B,1} + T_{B,2}}{2}.$$

For the case of liquids, particularly viscous oils, the viscosity ratio term is needed to account for the viscosity variation. For such cases, properties in the Graetz number and Nusselt number should be evaluated at the average bulk temperature between inlet and outlet.

The average heat transfer coefficient in the Nusselt number of Eq. (8–34) is used to obtain the total heat transfer by the defining relation for $\bar{h}$, which is similar to Eq. (8–25):

$$\dot{q} = \bar{h}A(T_w - T_{B,\mathrm{avg}}). \qquad (8\text{--}35)$$

---

* Not to be confused with *slug flow* in two-phase flow (Section 12–3).

8-2 THE ENTRANCE REGION IN LAMINAR FLOW

In Section 13-2 it will be shown that the proper temperature difference in Eq. (8-35) is a logarithmic one, although a simple arithmetic mean, $(T_w - T_{B,\text{avg}})$, is usually sufficiently accurate.

The use of Eq. (8-34) in the solutions of heat transfer problems is often complicated by the fact that trial and error is involved, since the unknown quantity cannot be expressed explicitly. This is illustrated in the following example.

Example 8-1. Estimate the length z of a tube of 0.05 ft inside diameter necessary to raise the bulk temperature of a light oil from 95° to 105°F if the tube wall temperature is constant beyond some location ($z = 0$) and is equal to 150°F, and the average flow velocity is 2 ft/sec. Ignore buoyancy and variable property effects. Assume that the velocity profile is fully developed at the start of heating.

Solution. The properties of light oil at the average bulk temperature of

$$T_{B,\text{avg}} = \frac{T_{B,1} + T_{B,2}}{2} = \frac{95 + 105}{2} = 100°F,$$

are found to be

$$\rho = 56.0 \text{ lb}_m/\text{ft}^3$$
$$c_p = 0.46 \text{ Btu/lb}_m \text{ °F}$$
$$\mu = 1.53 \times 10^{-2} \text{ lb}_m/\text{ft-sec}$$
$$\nu = 27.4 \times 10^{-5} \text{ ft}^2/\text{sec}$$
$$k = 0.076 \text{ Btu/hr-ft °F}$$
$$\text{Pr} = 340$$
$$\mu_w = 0.53 \times 10^{-2} \text{ lb}_m/\text{ft-sec at } T_w = 150°F.$$

The Reynolds number is then

$$\text{Re}_D = \frac{V_{\text{avg}} D}{\nu} = \frac{(2 \text{ ft/sec})(0.05 \text{ ft})}{27.4 \times 10^{-5} \text{ ft}^2/\text{sec}} = 365,$$

which indicates that the flow is laminar. Hence Eq. (8-34) can be applied. Considering the equation, one observes that two quantities are unknown, $\overline{\text{Nu}_D}$ and z. Therefore, another equation must be found which will enable these quantities to be evaluated. Performing an energy balance on the fluid within the tube:

$$\dot{q} = c_p(\rho V_{\text{avg}} A_{\text{cross}}) \Delta T_{\text{bulk}} = -\bar{h} A_{\text{surf}}(T_{B,\text{avg}} - T_{\text{wall}}),$$

$$c_p \rho V_{\text{avg}} \frac{\pi D^2}{4}(105 - 95) = \bar{h} \pi D z (150 - 100).$$

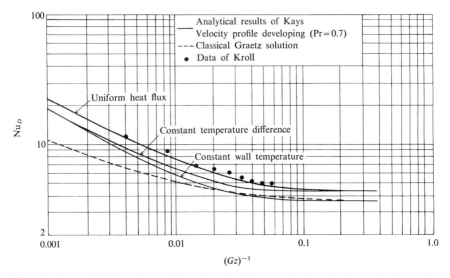

Fig. 8–4 Local Nusselt number for simultaneously developing velocity and temperature profiles (Pr = 0.7). (From *Fluid Dynamics and Heat Transfer* by Knudsen and Katz. Copyright 1958 by McGraw-Hill Book Company, Inc. Used with permission of McGraw-Hill Book Company.)

Dividing both sides of this equation by the thermal conductivity k, solving for the Nusselt number, and then substituting in the known quantities,

$$\overline{\text{Nu}}_D = \frac{305}{z},$$

where z is in feet. This allows Eq. (8–34) to be written as

$$\overline{\text{Nu}}_D = \frac{305}{z} = \left\{3.66 + \left[\frac{0.0668(\overline{\text{Gz}})}{1 + 0.04(\overline{\text{Gz}})^{2/3}}\right]\right\}\left(\frac{\mu_B}{\mu_w}\right)^{0.14}.$$

which can now be solved by trial and error to yield the desired value of z. Rearranging this, and substituting the values of the fluid properties, yields

$$\frac{305}{z} = \left\{3.66 + \left[\frac{\frac{414}{z}}{1 + 0.04\left(\frac{6200}{z}\right)^{2/3}}\right]\right\}(1.16).$$

After several trials a value of $z = 29.7$ is found to satisfy the above equation, thus 29.7 ft of heated length is required to change the bulk temperature by the specified amount.

The very complicated problem of simultaneously developing velocity and temperature profiles has been solved numerically by Kays [4] for the special case of Pr = 0.7. This solution is useful for air and most common gases since their Prandtl numbers are very close to that value. The solution is presented in terms of the local Nusselt number based on diameter and the local Graetz number and is shown in Fig. 8–4. Also shown in the figure for comparison is the curve for constant wall temperature, assuming fully established velocity profile. It can be seen that higher heat transfer coefficients are predicted for the case where the profiles develop simultaneously.

The same warnings are necessary here as in the previous section. The effects of property variations and buoyancy may cause serious deviations from the results predicted by the curves and equations. Care should be taken in using these results for very close estimation of laminar heat transfer situations.

8–3 SIMILARITY METHODS IN LAMINAR BOUNDARY FLOW

The boundary layer simplifications made by Prandtl and discussed in Section 7–1 represented a significant advancement in the study of fluid flow and heat transfer. Whereas the momentum equations have defied analytical solution for all but the most simple flow situations, the boundary layer equations of Prandtl have been solved for problems that are fairly complex. The solutions of the boundary layer equations give values of the local velocity components and local temperatures in the vicinity of a wall, and these may be used to predict such important information as the viscous drag on the wall, the formation of wake, and heat transfer rates.

The Flat Plate

In order to illustrate a very important method of solving boundary layer problems, the flat plate in Fig. 8–5 will be analyzed. It is assumed that the surrounding fluid is infinite in extent and flows with steady velocity parallel to the plate. The fluid properties are assumed to be constant and body forces as well as viscous dissipation are assumed to be negligible. The Reynolds number based on the length of the plate is assumed to be large enough for boundary layer assumptions to apply, and yet small enough that the flow in the boundary layer is laminar.

Because of the infinite extent of the fluid and the thinness of the plate and the boundary layer, the pressure gradient in the free stream parallel to the

Fig. 8–5 Flow along a thin flat plate.

plate is assumed to be zero. Since this free stream pressure is transmitted unchanged across the boundary layer—Eq. (7–10)—the pressure gradient throughout the boundary layer is also zero. It will be assumed for this special case that the x component of velocity u has a similar profile at all positions x along the plate for any stream velocity and any fluid. In other words the dimensionless velocity profile shown in Fig. 7–6(a) is assumed to be the same for all laminar flows along a flat plate and for all positions along the plate. This similarity of profiles implies that a unique function exists such that

$$\frac{u}{V_\infty} = f\left(\frac{y}{\delta}\right), \tag{8-36}$$

where δ is the local boundary layer thickness at any value of x.

With the similarity expressed by Eq. (8–36), the variable (u/V_∞) would be a function only of the *single* variable (y/δ). If a differential equation exists that relates the two variables, then that equation would be an ordinary differential equation, which generally can be solved in a more direct manner than partial differential equations such as the boundary layer equations.

The boundary layer equations may be used to obtain the ordinary differential equation relating (u/V_∞) and (y/δ). The solution of that equation must yield an expression for boundary layer thickness at each position x in terms of known quantities as well as the desired functional relationship between (u/V_∞) and (y/δ).

Since fluid properties have been assumed to be constant, the continuity and momentum boundary layer equations can be examined for a solution independent of the temperature or energy equation. For the case of zero pressure gradient, the boundary layer equations—Eqs. (7–11) and (7–9)—may be used:

$$\frac{\partial u}{\partial x} + \frac{\partial v}{\partial y} = 0, \tag{8-37}$$

$$u\frac{\partial u}{\partial x} + v\frac{\partial u}{\partial y} = \nu \frac{\partial^2 u}{\partial y^2}. \tag{8-38}$$

These equations can be solved by utilizing the stream function (see Section 4–2), which, by its very definition, satisfies Eq. (8–37), the continuity equation, or

$$u = \frac{\partial \psi}{\partial y} \quad \text{and} \quad v = -\frac{\partial \psi}{\partial x}. \tag{4-30}$$

Now by substituting for u and v from Eq. (4–30), the boundary layer momentum equation becomes a partial differential equation for ψ in terms of the independent variables x and y:

$$\frac{\partial \psi}{\partial y}\frac{\partial^2 \psi}{\partial x\, \partial y} - \frac{\partial \psi}{\partial x}\frac{\partial^2 \psi}{\partial y^2} = \nu \frac{\partial^3 \psi}{\partial y^3}. \tag{8-38a}$$

Equation (8–38a) is a single equation which satisfies both the requirements of conservation of momentum and conservation of mass. The dependent variable ψ is expressed as a function of two independent variables, x and y. The equation is a *partial* differential equation and a straightforward integration is not possible.

The concept of similar velocity profiles is utilized to transform this equation into an *ordinary* differential equation which can be solved by numerical integration. If similarity exists, the dimensionless velocity (u/V_∞) is a function of a single dimensionless distance such as (y/δ), as expressed in Eq. (8–36). Since the combined continuity-momentum equation is expressed in terms of the stream function, Eq. (8–38a) might be utilized to find an expression for a dimensionless stream function in terms of a single dimensionless distance.

The dimensionless distance (y/δ) seems to be an obvious choice except for the fact that the boundary layer thickness is an unknown function of x, the distance from the leading edge of the plate. The boundary layer thickness can be assumed to depend upon x and the parameters ν and V_∞:

$$\delta = \delta(x, \nu, V_\infty).$$

Dimensional reasoning leads to the conclusion that the boundary layer thickness δ is proportional to $\sqrt{(\nu x)/V_\infty}$, or

$$\delta = C\sqrt{(\nu x)/V_\infty}, \qquad (8\text{–}39)$$

where C is an unknown constant of proportionality. Thus a dimensionless distance y^* might be defined by

$$y^* = \left(\frac{y}{\delta}\right) = \frac{y}{C\sqrt{(\nu x)/V_\infty}}. \qquad (8\text{–}40)$$

Since the dimensionless distance y^* contains both x and y, it permits defining any point (x,y) in the boundary layer in terms of a single variable y^*. Hopefully a dimensionless stream function can now be found that will depend only on the value of y^*:

$$\psi^* = f(y^*) \text{ only}.$$

Since the stream function ψ is defined by Eq. (4–30), $\partial \psi / \partial y = u$, an attempt is made to find a dimensionless stream function such that

$$\frac{d\psi^*}{\partial y^*} = \left(\frac{u}{V_\infty}\right).$$

Using the parameters x, ν, and V_∞, trial and error leads to the following possibility:

$$\psi^* = \frac{\psi}{\sqrt{\nu x V_\infty}}. \qquad (8\text{–}41)$$

Equations (8–40) and (8–41) contain the dependent variable ψ and independent variables x and y. These two equations will now be used as transformation equations, with the hope that an ordinary differential equation might be obtained from Eq. (8–38a). In order to carry out this transformation, the following rule of chain differentiation is utilized:

If ψ^* is a function of y^* only, and if y^* is a function of two independent variables x and y, then

$$\frac{\partial \psi^*}{\partial x} = \frac{d\psi^*}{dy^*} \frac{\partial y^*}{\partial x} \quad \text{and} \quad \frac{\partial \psi^*}{\partial y} = \frac{d\psi^*}{dy^*} \frac{\partial y^*}{\partial y}.$$

Thus, from Eqs. (8–40) and (8–41) it can be shown that

$$\frac{d\psi^*}{dy^*} = C\frac{u}{V_\infty}. \tag{8-42}$$

Equation (8–42) shows that if the dimensionless derivative $d\psi^*/dy^*$ can be obtained as a function of the dimensionless distance y^*, then the local velocity component u can be determined at any position x and y. This is done by nondimensionalizing the continuity-momentum equation, Eq. (8–38a), by the use of Eqs. (8–40) and (8–41). The result is

$$\psi^* \frac{d^2\psi^*}{dy^{*2}} + \frac{2}{C}\frac{d^3\psi^*}{dy^{*3}} = 0. \tag{8-43}$$

The boundary conditions $u = v = 0$ at $y = 0$ and $u \to V_\infty$ as $y \to \infty$ become

$$\psi^* = \frac{d\psi^*}{dy^*} = 0 \text{ at } y^* = 0 \quad \text{and} \quad \frac{d\psi^*}{dy^*} \to C \text{ as } y^* \to \infty. \tag{8-44}$$

The three boundary conditions required to solve the third order differential equation, Eq. (8–43), are given in Eq. (8–44). The constant C can be arbitrary and is determined by requiring (u/V_∞) to be equal to 0.99 at $y^* = 1.0$. The transformation can be considered successful up to this point since the equation obtained, Eq. (8–43), is an ordinary differential equation, and the boundary conditions for the transformed equation are simple.

The solution to Eq. (8–43) is obtainable by a series expansion or by numerical methods. With the use of a digital computer the solution is not difficult (see Problem 32). Because a solution was originally obtained in 1908 by H. Blasius [5], the solution is often referred to as the *Blasius solution*.

The solution consists of values of the velocity components u and v for all x and y. These are shown in Fig. 8–6. The value of $C = 5$ leads to a u velocity component approximately equal to the free stream velocity ($u = 0.99V_\infty$) at the edge of the boundary layer ($y^* = 1$).

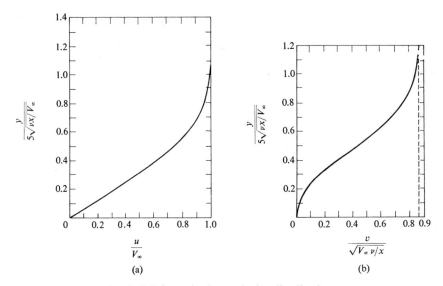

Fig. 8-6 Dimensionless velocity distribution.

The solution for the y velocity component v is obtained by use of Eq. (8-41) and the definition of the stream function, Eq. (4-30). It is interesting to note that the velocity component v does not become zero at the edge of the boundary layer but approaches a constant value for each position x. This flow outward is necessitated by the deceleration of the fluid in the boundary layer and the requirements of continuity. Note the similarity between Figs. 8-6a and 7-6a, where

$$\delta = 5\sqrt{(\nu x)/V_\infty}, \tag{8-45}$$

or

$$\frac{\delta}{x} = \frac{5}{\sqrt{\mathrm{Re}_x}}. \tag{8-46}$$

A large amount of experimental data has been gathered which shows the validity of the Balsius solution. One such set of data is shown in Fig. 8-7. These data justify the boundary layer assumptions made in Section 7-1 and the important assumption of similarity—Eq. (8-36)—made in transforming the boundary layer equations to an ordinary differential equation. An exception exists very near the leading edge of the plate, where the second derivative $\partial^2 u/\partial x^2$ is not small compared to $\partial^2 u/\partial y^2$. In this region the boundary layer assumptions do not apply, and the solutions obtained in this section do not agree closely with experiments. Because the region in which this is true is usually very small compared to the size of the entire plate, this discrepancy can usually be ignored when considering average values over the entire plate. This will be done in the analysis on page 203.

LAMINAR FLOW AND HEAT TRANSFER

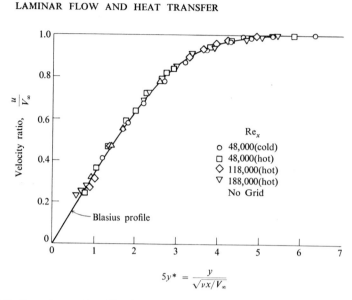

Fig. 8–7 Experimental verification of the Blasius solution [7]. (Used with permission of the American Society of Mechanical Engineers.)

The solution in Fig. 8–6 can be used to obtain the viscous shear stress at the wall at each x. Along the wall there is no v component of velocity and $\partial v/\partial x = 0$. Thus the local shear stress at the wall can be expressed from Eq. (3–20) or Eq. (3–21):

$$(\tau_{xy})_{\text{wall}} = \tau_w(x) = \mu \left(\frac{\partial u}{\partial y}\right)_{y=0}. \tag{8–47}$$

The velocity gradient at the wall can be obtained from the slope $\partial u^*/\partial y^*$ measured at $y^* = 0$ in Fig. 8–6:

$$\frac{\partial u^*}{\partial y^*} = C \frac{\frac{1}{V_\infty}}{\sqrt{V_\infty/(\nu x)}} \frac{\partial u}{\partial y} = C \sqrt{\frac{\nu x}{V_\infty^3}} \frac{\partial u}{\partial y},$$

or

$$\frac{\partial u}{\partial y} = \frac{1}{C} \sqrt{V_\infty^3/(\nu x)} \frac{\partial u^*}{\partial y^*}. \tag{8–48}$$

The slope of $\partial u^*/\partial y^*$ at the wall is equal to 1.66, and C is equal to 5.0. Therefore, the local shear stress at the wall, from Eq. (8–47), is

$$\tau_w(x) = 0.332\mu\sqrt{V_\infty^3/(\nu x)}. \tag{8–49}$$

It is desirable to express the shear stress at the wall in dimensionless terms as has been done previously. We define the coefficient of drag as before, but, since drag is due only to wall shear, this coefficient will be called

8-3 SIMILARITY METHODS IN LAMINAR BOUNDARY FLOW

the coefficient of friction and designated C_f:

$$C_f = \frac{\text{drag/area}}{\frac{1}{2}\rho V_\infty^2} = \frac{\tau_w(x)}{\frac{1}{2}\rho V_\infty^2} = \frac{0.332}{\frac{1}{2}\rho V_\infty^2}\mu\sqrt{V_\infty^3/(\nu x)},$$

or

$$C_f = \frac{0.664}{\sqrt{\text{Re}_x}}. \qquad (8\text{-}50)$$

Equation (8–50) expresses the *local friction coefficient* C_f (the dimensionless wall shear stress) at each position x along the plate. It is usually desirable to have an expression for the total drag on the plate or a total friction coefficient. This is obtained by integration of the local drag over the entire plate of length L and unit width:

$$\overline{C}_f = \frac{1}{L}\int_0^L C_f(x)\,dx \qquad (8\text{-}51)$$

$$\overline{C}_f = \frac{1}{L}\int_0^L \frac{0.664}{\sqrt{V_\infty/\nu}}x^{-1/2}\,dx = \frac{0.664}{L\sqrt{V_\infty/\nu}}\frac{x^{1/2}}{\frac{1}{2}}\bigg|_0^L = \frac{1.328}{\sqrt{V_\infty L/\nu}}$$

$$= \frac{1.328}{\sqrt{\text{Re}_L}}. \qquad (8\text{-}52)$$

Equation (8–52) is the expression for the friction coefficient in terms of the Reynolds number based on the plate length. It can be used to obtain the total drag on a thin flat plate in steady, laminar, parallel flow. Its use should be limited to the range of Reynolds number of approximately

$$10^4 < \text{Re}_L < 10^6. \qquad (8\text{-}53)$$

At Reynolds numbers below 10^4 the boundary layer assumptions are not valid. At Reynolds numbers above 10^6 transition to turbulence is most likely to occur along the plate (Section 7–1).

Laminar flow is most likely to occur over an entire plate of moderate size at low velocities or with gases at low pressures where the kinematic viscosity is fairly large.

Example 8–2. To assure laminar flow at 100,000 feet altitude, what is the maximum length of thin flat plate, assuming that the velocity relative to the air is 400 ft/sec and that transition occurs at $\text{Re}_L = 5 \times 10^5$? What is the drag on both sides of that plate per foot of width? At 100,000 feet altitude the kinematic viscosity of air is 0.0091 ft²/sec and the density is 1.062×10^{-3} lb$_m$/ft³.

Solution.

$$\text{Re}_L = \frac{V_\infty L}{\nu} = 500{,}000.$$

$$L = \frac{500{,}000\nu}{V_\infty} = \frac{(500{,}000)(0.0091)\text{ ft}^2/\text{sec}}{(400)\text{ ft/sec}} = 11.37\text{ ft.}$$

The friction coefficient is

$$\overline{C}_f = \frac{1.328}{\sqrt{\mathrm{Re}_L}} = \frac{1.328}{\sqrt{500{,}000}} = 1.89 \times 10^{-3}.$$

The drag on the plate per foot of width is

$$\begin{aligned}\mathrm{drag} &= 2\overline{C}_f \tfrac{1}{2}\rho V_\infty^2 L \\ &= \frac{(2)(1.89 \times 10^{-3})(\tfrac{1}{2})(1.062 \times 10^{-3})\dfrac{\mathrm{lb}_m}{\mathrm{ft}^3}(400)^2 \dfrac{\mathrm{ft}^2}{\mathrm{sec}^2}(11.37\ \mathrm{ft})}{32.2 \dfrac{\mathrm{lb}_m\ \mathrm{ft}}{\mathrm{lb}_f\ \mathrm{sec}^2}} \\ &= 0.1135\ \mathrm{lb}_f/\mathrm{ft}\ \text{of width.}\end{aligned}$$

Similarity solutions such as the one described above can also be obtained for a number of other constant property laminar boundary layer situations, including the cases where the free stream velocity V varies as a power of the distance from the leading edge [17]:

$$V = Cx^m.$$

For the special case of the flat plate, V is a constant for all x, therefore $m = 0$ and $C = V_\infty$.

The Temperature Equation

The temperature or energy boundary layer equation can also be solved for the case of the flat plate with constant wall temperature T_w, using the assumption of similarity for the temperature profiles as well as for the velocity profiles. For the temperature profiles this assumption would be

$$T^* = \frac{T - T_w}{T_\infty - T_w} = f\left(\frac{y}{\delta_T}\right), \tag{8-54}$$

where δ_T is the thermal boundary layer thickness.

For the case of a thin flat plate in steady laminar flow, assuming constant properties and neglecting viscous dissipation the energy equation for the boundary layer equation, Eq. (7-12), is

$$u\frac{\partial T}{\partial x} + v\frac{\partial T}{\partial y} = \alpha \frac{\partial^2 T}{\partial y^2}, \tag{8-55}$$

where $\alpha = \dfrac{k}{\rho c_p}$, the thermal diffusivity.

Equation (8-55) describes the temperature field inside the thermal boundary layer ($y < \delta_T$). Because of the assumption made in its derivation that

$$\frac{\partial^2 T}{\partial x^2} \ll \frac{\partial^2 T}{\partial y^2},$$

the equation is valid only where the thermal boundary layer thickness is much less than the distance from the start of heating, in this case the leading edge of the plate. Equation (8–55) can be rewritten by the substitution of the expression for the dimensionless temperature T^*, recalling that both T_w and T_∞ are constant parameters:

$$u\frac{\partial T^*}{\partial x} + v\frac{\partial T^*}{\partial y} = \alpha\frac{\partial^2 T^*}{\partial y^2}. \qquad (8\text{–}56)$$

The question might now arise as to what length should be used to nondimensionalize distance, since the thermal boundary layer thickness has been introduced. Because u and v appear in Eq. (8–56) and must be determined by use of the Blasius solution (Fig. 8–6) prior to solving for temperature, it is best to use the same length δ for nondimensionalizing as was used previously. Thus, as before,

$$y^* = \frac{y}{C\sqrt{(\nu x)/V_\infty}} \quad \text{and} \quad \psi^* = \frac{\psi}{\sqrt{\nu x V_\infty}}. \qquad (8\text{–}57)$$

After introduction of the stream function and Eq. (8–57), Eq. (8–56) becomes

$$\frac{d^2 T^*}{dy^{*2}} + \frac{C\,\text{Pr}}{2}\psi^*\frac{dT^*}{dy^*} = 0. \qquad (8\text{–}58)$$

The boundary conditions of Eq. (8–56) become

$$T^* \to 1 \text{ as } y^* \to \infty, \quad \text{and} \quad T^* = 0 \text{ at } y^* = 0.$$

The parameter Pr appearing in Eq. (8–58) is the Prandtl number, and the constant C has the same value of 5.0 as before. Solutions can be obtained for arbitrary values of the Prandtl number; that is, a solution exists for each value of Pr. Pohlhausen [6] first obtained solutions for this problem in 1921. The results are shown in Fig. 8–8 for values of Pr from 1000 to 0.6. In more recent times, the Prandtl number range has been extended to higher and lower values than those shown in the figure.

The thickness of the thermal boundary layer δ_T is the distance from the wall to the point where the temperature change $(T - T_w)$ is 99 percent of the total change $(T_\infty - T_w)$. The effect of the Prandtl number on the temperature profile and upon the thermal boundary layer thickness is apparent from the solutions to Eq. (8–58) shown in Fig. 8–8. The thermal boundary layer is thinner than the velocity boundary layer for values of Prandtl numbers greater than 1.0, since the thickness of the velocity boundary layer corresponds to a value of $y^* = 1.0$. For Pr less than 1.0, the opposite is true. For Pr = 1.0, the boundary layer thicknesses are not only the same but the

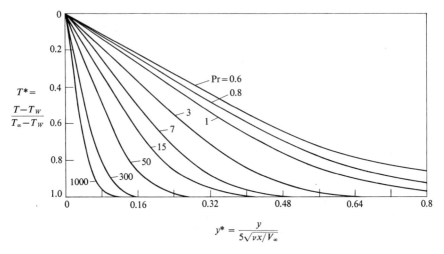

Fig. 8-8 Temperature profiles in laminar flow along a flat plate. [From E. Pohlhausen, *Z. angew. Math. Mech.* **1**, 115 (1921). Used with permission.]

velocity and temperature profiles are identical; that is, the same function relates both

$$\frac{u}{V_\infty} \quad \text{and} \quad \frac{T - T_w}{T_\infty - T_w} \quad \text{to} \quad \frac{y}{5\sqrt{(\nu x)/V_\infty}}.$$

From Fig. 8-8, the thickness ratio of the velocity boundary layer to the thermal boundary layer thickness can be estimated. This can be given by

$$\frac{\delta}{\delta_T} \approx (\text{Pr})^{1/3}. \tag{8-59}$$

The local heat transfer coefficient as defined previously relates the local rate of heat transfer per unit area at any position x to a temperature difference, in this case $T_w - T_\infty$:

$$h = \frac{\dot{q}/A}{T_w - T_\infty}. \tag{8-60}$$

Values of the local heat transfer coefficient for a given Prandtl number can be obtained from Fig. 8-8. The local heat flux is related to the temperature derivative at the wall by the Fourier law:

$$\left(\frac{\dot{q}}{A}\right)_{\text{wall}} = -k\left(\frac{\partial T}{\partial y}\right)_{y=0} = h(T_w - T_\infty).$$

Since

$$\left(\frac{\partial T}{\partial y}\right)_{y=0} = -(T_w - T_\infty)\frac{\partial y^*}{\partial y}\left(\frac{dT^*}{dy^*}\right)_{y^*=0} = -(T_w - T_\infty)\frac{1}{C}\sqrt{\frac{V_\infty}{\nu x}}\left(\frac{dT^*}{dy^*}\right)_{y^*=0},$$

then it can be seen that

$$k(T_w - T_\infty)\frac{1}{C}\sqrt{\frac{V_\infty}{\nu x}}\left(\frac{dT^*}{dy^*}\right)_{y^*=0} = h(T_w - T_\infty).$$

By rearranging and solving for the local Nusselt number,

$$\mathrm{Nu}_x = \frac{hx}{k} = \frac{1}{C}\left(\frac{dT^*}{dy^*}\right)_{y^*=0}\sqrt{\mathrm{Re}_x}.$$

The derivative $(dT^*/dy^*)_{y^*=0}$ depends on the value of the Prandtl number (Fig. 8–8).

Pohlhausen showed that, in the range of Prandtl numbers $0.6 < \mathrm{Pr} < 15$, the numerical value of the derivative could be approximated by

$$\left(\frac{dT^*}{dy^*}\right)_{y^*=0} = 1.660\sqrt[3]{\mathrm{Pr}}. \tag{8-61}$$

Thus the local Nusselt number can be expressed as a function of the Prandtl number and the local Reynolds number:

$$\mathrm{Nu}_x = \frac{hx}{k} = 0.332\sqrt[3]{\mathrm{Pr}}\sqrt{\mathrm{Re}_x}. \tag{8-62}$$

The assumptions made in the derivation of Eq. (8–62) should be summarized. They include:

1. Steady laminar boundary layer flow along a flat plate,
2. Constant properties μ, c_p, k, and ρ,
3. Negligible body forces,
4. Negligible viscous dissipation,
5. Constant wall and free stream temperatures, and
6. $0.6 < \mathrm{Pr} < 15$.

Since the temperature of the fluid varies from T_w to T_∞, there will be some variation of the properties which were assumed to be constant. For small temperature differences this variation will not be large and can be taken into account by evaluating the properties at an average temperature, sometimes called the *film temperature:*

$$T_f = \frac{T_w + T_\infty}{2}. \tag{8-63}$$

The average heat transfer coefficient over the entire plate of length L can be obtained from Eq. (8–62) by considering all fluid properties to be constant and by using the following definition:

$$\bar{h} = \frac{1}{L}\int_0^L h\, dx. \tag{8-64}$$

The result is an expression for the average Nusselt number $\overline{\mathrm{Nu}}$ based on $\bar{h}$ and L:

$$\overline{\mathrm{Nu}} = \frac{\bar{h}L}{k} = 0.664 \sqrt[3]{\mathrm{Pr}}\, \sqrt{\mathrm{Re}_L}. \tag{8-65}$$

Equation (8-65) can be used to obtain the total heat loss from a flat plate in laminar flow, where the same restrictions apply as in Eq. (8-62).

8-4 INTEGRAL METHODS FOR BOUNDARY LAYER PROBLEMS

The solution to the boundary layer problem discussed in the previous section is sometimes referred to as an *exact* solution to the boundary layer equations. This is so in spite of the fact that closed solutions to the ordinary differential Eqs. (8-43) and (8-58) are not known. The similarity solutions are valuable for the insight that they give into the problems of fluid flow and heat transfer along a flat surface. However, the similarity method has serious limitations that restrict its use in general types of problems. An approximate method of solution of the boundary layer equations is often more convenient to use than the exact method. An example using an approximate method will now be presented, following closely the developments of Pohlhausen [10] and von Kármán [11].

The flat plate in steady laminar flow with negligible body forces and negligible viscous heating will again be considered. As before, the temperature of the plate will be assumed constant and equal to T_w. The free stream velocity and temperature are constant and equal to V_∞ and T_∞ respectively. The differential equations and boundary conditions describing the flow are the same as those developed in the previous sections in Eqs. (8-37), (8-38), and (8-55):

$$\frac{\partial u}{\partial x} + \frac{\partial v}{\partial y} = 0, \tag{8-66}$$

$$u\frac{\partial u}{\partial x} + v\frac{\partial u}{\partial y} = \nu \frac{\partial^2 u}{\partial y^2}, \tag{8-67}$$

$$u\frac{\partial T}{\partial x} + v\frac{\partial T}{\partial y} = \alpha \frac{\partial^2 T}{\partial y^2}. \tag{8-68}$$

At $y = 0$, $u = v = 0$ and $T = T_w$.

As $y \to \infty$, $u \to V_\infty$, $\dfrac{\partial u}{\partial y} \to 0$, and $T \to T_\infty$.

The continuity equation, Eq. (8-66), can be rewritten in the following form, since $v = 0$ at $y = 0$:

$$v = -\int_0^y \frac{\partial u}{\partial x}\, dy. \tag{8-69}$$

Using Eq. (8–69), the velocity component v can be eliminated from the momentum equation, Eq. (8–67):

$$u\frac{\partial u}{\partial x} - \frac{\partial u}{\partial y}\int_0^y \frac{\partial u}{\partial x}dy = \nu\frac{\partial^2 u}{\partial y^2}. \qquad (8\text{–}70)$$

For any arbitrary but fixed value of x, the distance from the leading edge, Eq. (8–70) is valid at any point y measured from the wall. Equation (8–70) may, therefore, be multiplied by dy and integrated over the boundary layer thickness, assuming that x is held constant for any integration:

$$\int_0^\delta u\frac{\partial u}{\partial x}dy - \int_0^\delta \left(\frac{\partial u}{\partial y}\int_0^y \frac{\partial u}{\partial x}dy\right)dy = \int_0^\delta \nu\frac{\partial^2 u}{\partial y^2}dy. \qquad (8\text{–}71)$$

The right-hand term of Eq. (8–71) may be simplified:

$$\int_0^\delta \nu\frac{\partial^2 u}{\partial y^2}dy = \nu\frac{\partial u}{\partial y}\bigg|_{y=\delta} - \nu\frac{\partial u}{\partial y}\bigg|_{y=0} = -\nu\frac{\partial u}{\partial y}\bigg|_{y=0}.$$

The derivative $\partial u/\partial y = 0$ at $y = \delta$, the edge of the boundary layer. Integration of the left side of Eq. (8–71) requires use of integration by parts. Recalling from calculus that $d(AB) = A\,dB + B\,dA$,

$$AB = \int A\,dB + \int B\,dA + \text{constant}, \quad \int A\,dB = AB - \int B\,dA + \text{constant},$$

or, with limits,

$$\int_{B_1}^{B_2} A\,dB = AB\bigg|_{B_1}^{B_2} - \int_{B_1}^{B_2} B\,dA.$$

The term $\int_0^\delta \left(\frac{\partial u}{\partial y}\int_0^y \frac{\partial u}{\partial x}dy\right)dy$ in Eq. (8–71) is divided so that

$$A = \int_0^y \frac{\partial u}{\partial x}dy, \qquad dB = \frac{\partial u}{\partial y}dy,$$

$$dA = \frac{\partial u}{\partial x}dy, \qquad B = \int_0^y \frac{\partial u}{\partial y}dy = u,$$

$$A\bigg|_0^\delta = \int_0^\delta \frac{\partial u}{\partial x}dy, \qquad B\bigg|_0^\delta = V_\infty.$$

Therefore

$$\int_0^\delta \left(\frac{\partial u}{\partial y}\int_0^y \frac{\partial u}{\partial x}dy\right)dy = V_\infty \int_0^\delta \frac{\partial u}{\partial x}dy - \int_0^\delta u\frac{\partial u}{\partial x}dy,$$

and Eq. (8–71) becomes

$$\int_0^\delta u\frac{\partial u}{\partial x}dy - V_\infty \int_0^\delta \frac{\partial u}{\partial x}dy + \int_0^\delta u\frac{\partial u}{\partial x}dy = -\nu\frac{\partial u}{\partial y}\bigg|_{y=0}. \qquad (8\text{–}72)$$

Remembering that V_∞ is constant, that the order of integration or differentiation is interchangeable, and that the integrals are evaluated at zero and at δ, which depends only on x, Eq. (8–72) can be rewritten to give

$$\frac{d}{dx}\int_0^\delta (V_\infty - u)u\, dy = \nu \frac{\partial u}{\partial y}\bigg|_{y=0}. \tag{8–73}$$

In a similar manner the energy equation, Eq. (8–68) may be combined with Eq. (8–66) and integrated to give

$$\frac{d}{dx}\int_0^{\delta_T} (T_\infty - T)u\, dy = \alpha \left(\frac{dT}{dy}\right)_{\text{wall}}. \tag{8–74}$$

Equations (8–73) and (8–74) are useful forms of the boundary layer equations for the flat plate because they can be integrated, once u and T are stated as functions of y. Actually u and T are unknown functions. *Approximate* expressions for the functions can be obtained by intuition, however, and by use of the boundary conditions that have been given.

Assume that the velocity *and* temperature profiles at all points x along the plate are *similar* for all flow conditions, that is,

$$\frac{u}{V_\infty} = f_1\left(\frac{y}{\delta}\right), \tag{8–75}$$

and

$$\frac{T - T_w}{T_\infty - T_w} = f_2\left(\frac{y}{\delta_T}\right). \tag{8–76}$$

The unknown functions will be approximated by means of the polynomials:

$$\frac{u}{V_\infty} = a + b\left(\frac{y}{\delta}\right) + c\left(\frac{y}{\delta}\right)^2 + d\left(\frac{y}{\delta}\right)^3, \tag{8–77}$$

and

$$\frac{T - T_w}{T_\infty - T_w} = e + f\left(\frac{y}{\delta_T}\right) + g\left(\frac{y}{\delta_T}\right)^2 + h\left(\frac{y}{\delta_T}\right)^3, \tag{8–78}$$

where a, b, c, d, e, f, g, and h are constants to be determined from the boundary conditions.

The boundary conditions that have been given for Eqs. (8–66), (8–67), and (8–68) are used, assuming that free stream values are attained at $y = \delta$ and $y = \delta_T$ instead of at ∞ (see the problems at end of chapter):

$$\frac{u}{V_\infty} = \frac{3}{2}\left(\frac{y}{\delta}\right) - \frac{1}{2}\left(\frac{y}{\delta}\right)^3, \tag{8–79}$$

and

$$\frac{T - T_w}{T_\infty - T_w} = \frac{3}{2}\left(\frac{y}{\delta_T}\right) - \frac{1}{2}\left(\frac{y}{\delta_T}\right)^3. \tag{8–80}$$

Equations (8–79) and (8–80) can be substituted into Eqs. (8–73) and (8–74) to give expressions for the local thermal and velocity boundary layer thicknesses, δ_T and δ. Equation (8–73) becomes

$$\frac{d}{dx}\left(\frac{39}{280} V_\infty^2 \delta\right) = \frac{3}{2}\frac{\nu V_\infty}{\delta},$$

which can be solved by separation of variables to give

$$\frac{\delta^2}{2} = \frac{140}{13}\frac{\nu x}{V_\infty} + \text{constant.}$$

The boundary layer thickness δ is assumed to be zero at $x = 0$; therefore the constant of integration is zero and

$$\delta = 4.64\sqrt{(\nu x)/V_\infty}. \tag{8–81}$$

For the cases where δ_T is always smaller than δ (Pr $>$ 1), the following expression is obtained for the thermal boundary layer thickness:

$$\frac{\delta_T}{\delta} = \frac{0.976}{\sqrt[3]{\text{Pr}}}. \tag{8–82}$$

Equation (8–82) is approximately correct for gases in which Pr $\sim$ 1. For fluids having Pr much less than 1.0 (the liquid metals) Eq. (8–82) is not valid.

Equations (8–81) and (8–82) can be combined with (8–79) and (8–80) to give the local temperature and velocity at every point in the boundary layers. Knowledge of the local temperatures permits computation of local Nusselt numbers along the surface. The result is

$$\text{Nu}_x = 0.332\sqrt[3]{\text{Pr}}\sqrt{\text{Re}_x}. \tag{8–83}$$

Note the similarity between Eq. (8–83) and Eq. (8–62), the "exact" solution.

This good agreement in the solutions obtained by the two methods does not always occur, but sufficiently good agreement obtained in a number of cases has encouraged the use of this method.

Figure 8–9 shows another situation which has been analyzed successfully by the integral method.

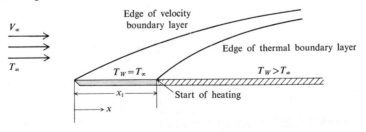

Fig. 8–9 Velocity and thermal boundary layer formation.

Flow along a flat plate is again assumed under the same conditions as before, except heating is assumed to begin at some distance x_t, measured from the leading edge. In this situation the thermal boundary layer does not begin to grow until the start of heating and would lie entirely within the velocity boundary layer (for $\text{Pr} > 1$). Using the same assumed forms of the velocity and temperature profiles as before—Eqs. (8–79) and (8–80)—the result is an expression for δ_T:

$$\frac{\delta_T}{\delta} = \frac{0.976}{\sqrt[3]{\text{Pr}}} \sqrt[3]{1 - (x_t/x)^{3/4}}, \tag{8-84}$$

where δ is given by Eq. (8–81). The resulting expression for the local Nusselt number is

$$\text{Nu}_x = (0.332 \sqrt[3]{\text{Pr}} \sqrt{\text{Re}_x}) \div \sqrt[3]{1 - (x_t/x)^{3/4}}. \tag{8-85}$$

Notice that Eqs. (8–84) and (8–85) reduce to Eqs. (8–82) and (8–83) for $x_t = 0$.

The integral method has been used to solve a variety of problems including those where pressure gradients along the surface exist [12], turbulent boundary layers [13], flows with magnetic body forces present [14], and flows induced by heating [15].

8–5 REYNOLDS ANALOGY

Analogies were discussed briefly in Section 3–6 and were described as a similarity in some respects between things otherwise unlike. An analogy or similarity can be shown to exist, under certain conditions, between the transport of momentum and the transport of thermal energy (or heat) in a fluid. A similar analogy involving the transport of mass will be discussed later.

In the laminar boundary layer on a flat plate (assuming constant properties, no viscous dissipation, and no buoyancy forces), the dimensionless momentum and energy equations are

$$u^* \frac{\partial u^*}{\partial x^*} + v^* \frac{\partial u^*}{\partial y^*} = \frac{1}{\text{Re}} \frac{\partial^2 u^*}{\partial y^{*2}}, \tag{8-86}$$

$$u^* \frac{\partial T^*}{\partial x^*} + v^* \frac{\partial T^*}{\partial y^*} = \frac{1}{\text{Pr Re}} \frac{\partial^2 T^*}{\partial y^{*2}}, \tag{8-87}$$

where

$$u^* = \frac{u}{V_\infty} \quad \text{and} \quad T^* = \frac{T - T_w}{T_\infty - T_w}. \tag{8-88}$$

The boundary conditions are assumed to be

at $y = 0$, $u^* = T^* = 0$; at $y = \infty$, $u^* = T^* = 1$.

Notice that identical boundary conditions exist for u^* and T^*. If, in addition, the Prandtl number $\Pr = 1$, then the same differential equation describes both u^* and T^*. This means that the solution to Eq. (8–86) is identical to Eq. (8–87), and plots of dimensionless velocity vs. dimensionless distance will be identical to plots of dimensionless temperature vs. dimensionless distance. This was shown to be true in the exact solutions of Section 8–3 for $\Pr = 1$ (or $\nu = \alpha$). If the velocity profile is identical to the temperature profile, then the velocity gradient at the wall will have the same value as the temperature gradient at the wall:

$$\left(\frac{\partial T^*}{\partial y^*}\right)_{\text{wall}} = \left(\frac{\partial u^*}{\partial y^*}\right)_{\text{wall}} \quad \text{or} \quad \frac{l}{T_\infty - T_w}\left(\frac{\partial T}{\partial y}\right)_{\text{wall}} = \frac{l}{V_\infty}\left(\frac{\partial u}{\partial y}\right)_{\text{wall}}.$$

Now since

$$\dot{q}/A = h(T_w - T_\infty) = -k\left(\frac{\partial T}{\partial y}\right)_{\text{wall}},$$

and

$$\tau_{\text{wall}} = \mu\left(\frac{\partial u}{\partial y}\right)_{\text{wall}},$$

then

$$\frac{h}{k} = \frac{\tau_w}{\mu V_\infty}. \tag{8–89}$$

Defining the dimensionless coefficient of friction as

$$C_f = \frac{\tau_w}{\frac{1}{2}\rho V_\infty^2} = \tau_w^*, \tag{8–90}$$

substituting Eq. (8–90) into (8–89) to solve for h, and multiplying the numerator and denominator by c_p, gives

$$h = \frac{\frac{1}{2}C_f \rho V_\infty k}{\mu}\left(\frac{c_p}{c_p}\right). \tag{8–91}$$

Since

$$\Pr = \frac{\mu c_p}{k} = 1,$$

then

$$\frac{h}{\rho V_\infty c_p} = \frac{C_f}{2}. \tag{8–92}$$

Equation (8–92), although not expressed in the form originally put forth by Osborne Reynolds [16], is referred to as the *Reynolds analogy*. The dimensionless group $h/(\rho V_\infty c_p)$ is called the *Stanton number* and given the symbol St. The Stanton number is equal to the Nusselt number divided by the product of the Reynolds and Prandtl numbers. Thus Reynolds analogy can be written

$$\text{St}_x = \frac{\text{Nu}_x}{\text{Pr Re}_x} = \frac{C_f}{2} \quad (\text{Pr} = 1) \tag{8-93}$$

Equation (8-93) permits the calculation of the local heat transfer coefficient when the local friction coefficient is known. It is valid for flow over a flat plate within the assumptions listed above.

The real usefulness of analogies is not in laminar flows but is in turbulent heat transfer where theoretical analysis is usually absent but where experimental data on friction coefficients or friction factors are available. This will be developed in more detail in the next chapter.

REFERENCES

1. L. Graetz, *Ann. Physik* **18**, 79 (1883); **25**, 337 (1885).
2. J. R. Sellars, M. Tribus, and J. S. Klein, *Trans. ASME* **78**, 441 (1956).
3. H. Hausen, *Z. ver. deut. Ing. Beih, Verfarenstech* **4**, 91 (1943).
4. W. M. Kays, *Trans. ASME* **77**, 1265 (1955).
5. H. Blasius, *Z. f. Math. u. Physik* **56** (1908); or NACA TM No. 1256 (1950).
6. E. Pohlhausen, *Z. angew. Math. Mech.* **1**, 115 (1921).
7. G. H. Junkhan and G. K. Serovy, "Effects of Free Stream Turbulence and Pressure Gradient on Flat Plate Boundary Layer Velocity Profiles and on Heat Transfer," ASME Paper 66-WA/HT-4 (Nov. 1966).
8. F. D. Fisher and J. G. Knudsen, "Heat Transfer from Isothermal Plates," Chemical Engineering Progress Symposium, Series 29.
9. E. M. Sparrow and J. L. Gregg, *J. Aeronautical Sci.* **24**, 852 (1957).
10. K. Pohlhausen, *Z. angew. Math. Mech.* **1**, 252 (1921).
11. T. von Kármán, *Z. angew Math. Mech.* **1**, 232 (1921).
12. H. Holstein and J. Bohlen, *Lilienthal-Bericht* **S10**, 5 (1940).
13. A. Buri, Thesis, Zurich (1931).
14. R. Hugelman, Ph.D. Thesis (M.E.), Oklahoma State University (1964).
15. E. R. G. Eckert and J. W. Jackson, NACA Report 1015 (1951).
16. O. Reynolds, *Trans. Royal Society* (London) **174A**, 935 (1883).
17. V. M. Falkner and S. W. Skan, *Phil. Mag.* **12**, 865 (1931).
18. J. R. Whiteman and W. B. Drake, "Heat Transfer to Flow in a Round Tube with Arbitrary Velocity Distribution," ASME Paper 57-HT-1 (1957).

PROBLEMS

1. Use Eq. (8–6) to obtain expressions for the maximum velocity and the average velocity in fully developed laminar tube flow.

2. Calculate the rate in gal/hr at which water (average temperature of 70°F) will flow through a $\frac{1}{16}$ in. diameter tube if the pressure drop between two points 30 ft apart along the tube is 8 psi. Check the Reynolds number based on diameter, Re_D, to verify that the flow is laminar.

3. Use the Reynolds number calculated for Problem 2 to calculate the wall shear stress along the tube described in that problem. Show that a force balance on the fluid between the two pressure taps will give the pressure drop stated for the problem.

4. Calculate the velocity at which the flow ceases to be laminar ($Re_D = 2000$) for the following fluids (assume the fluids are all at an average temperature of 100°F and flow through a $\frac{1}{2}$-in. tube): (a) Water; (b) Air; (c) Light oil; (d) isopropyl alcohol.

5. Derive Eq. (8–18).

6. Air flows through a tube having an inside diameter (i.d.) of 3 cm. Heat is being added at the tube wall at the rate of 0.1 kw per square meter. Assuming that the velocity and temperature profiles are fully developed, estimate the wall temperature at a point where the bulk temperature of the air is 100°F. Evaluate the fluid properties at 100°F. Assume $p = 1$ atmosphere.

7. Air flows under laminar conditions through a 2-in. i.d. round tube at the rate of 5 lb_m/hr. Heat is added to the air through the tube wall at the rate of 1.0 watt per foot of tube length. Assuming that the section of the tube being considered is sufficiently far from the inlet so that entrance effects are overcome (i.e., the flow is fully developed), estimate the length of tube required to change the bulk temperature of the air from 90°F to 110°F. Assume fluid properties to be constant and equal to their values at 100°F, the average bulk temperature of the air within the section in question. Calculate the tube wall temperature at the cross sections of the tube where the bulk temperature of the air is 90°F and 110°F.

8. Water flows through a 0.1-in. tube at an average velocity of 0.4 ft/sec in fully developed laminar flow. Estimate the value of the heat transfer coefficient when the fluid receives a uniform heat flux from the wall of the tube. (Assume the thermal conductivity is that for water at 100°F). What would be the wall temperature at the place where the fluid bulk temperature is 100°F, if heat is being added uniformly at the rate of 5000 Btu/hr ft² at the tube wall?

9. Use Eq. (8–27) to calculate the length of a $\frac{1}{2}$-in. i.d. tube required to change the bulk temperature of air from 80°F to 84°F. The tube wall temperature is maintained at a value such that $(T_w - T_B) = 36°F$ at all positions along the tube. The air flows at a rate of 0.0003 lb_m/sec.

10. Calculate the local heat transfer coefficient h at a distance of 20 diameters from the start of heating of light oil flowing in fully developed laminar flow through a $\frac{1}{4}$-in. tube at a Reynolds number $Re_D = 150$. Assume a film temperature of 100°F and constant wall heat flux.

11. For the problem in Example 8–1 on page 195, calculate the average heat transfer coefficient $\bar{h}$ over the calculated length.

12. Calculate the rate at which heat is added to the oil in Example 8–1.

13. Calculate the pressure drop over the calculated length in Example 8–1.

14. Calculate the logarithmic mean temperature difference for Example 8–1 and compare with the temperature difference that was actually used.

15. Calculate the entrance length that is required for the fully developed velocity profile assumed in Example 8–1.

16. Using Fig. 8–3, calculate the local heat transfer coefficient h at 5-ft intervals from the start of heating in Example 8–1.

17. For Example 8–1, *estimate* the distance from the start of heating required to raise the bulk temperature from 95°F to 100°F. Assume that the properties given in the example are constant over the temperature range.

18. Use Eq. (8–40) and (8–41) to derive Eq. (8–43).

19. Derive Eq. (8–58).

20. Using the definition of the stream function, ψ, and Eqs. (8–40) and (8–41), determine an expression for velocity component v in the y direction for laminar flow along a flat plate.

21. Air at a mean film temperature of 100°F flows with a constant free stream velocity of 200 ft/min over a flat plate 10 ft long held at a uniform temperature. At a point located at $y = 0.5$ in. from the center ($x = 5$ ft) of the plate, determine the velocity components u and v. What is the local heat transfer coefficient h at the center of the plate ($x = 5$ ft)?

22. For the same air flow as described in the problem above, use the definition of a boundary layer to determine its thickness at $x = 5$ ft.

23. Derive the expression for the average convective heat transfer coefficient and obtain Eq. (8–65), starting with the expression for the local coefficient, Eq. (8–62).

24. Air at a mean film temperature of 100°F flows along a flat plate in steady laminar flow at a free stream velocity of 15 mph. At a position $x = 2$ ft from the leading edge, determine:
    a) the velocity in the x direction (u), 0.05 in. from the wall,
    b) the local heat transfer coefficient,
    c) the local boundary layer thickness, and
    d) the local coefficient of friction.

25. Estimate the local rate of heat loss from a flat plate at a point 3 in. from the leading edge when the plate is exposed to an air stream having a static temperature of 480°F and a pressure of 1 atmosphere. The free stream velocity relative

to the plate is 400 ft/sec. The plate is maintained at a surface temperature of 520°F. Use the average film temperature as a reference temperature for evaluation of fluid properties.

26. Estimate the rate of heat loss from a flat plate 3 ft wide by 3 ft long maintained at a constant wall temperature of 250°F when air at 150°F flows parallel to the plate with a free stream velocity of 30 ft/sec.

27. Estimate the total rate of heat loss from a flat plate, held at a constant temperature of 140°F exposed on one side to an air stream at 60°F flowing with a free stream velocity of 180 ft/min. The size of the plate is 4 ft wide by 6 ft long (in the direction of air flow).

28. Calculate the local temperature in a laminar boundary layer along a flat plate and at a point 3.0 in. back from the leading edge and 0.01 in. out from the wall. Assume the fluid is air at a free stream temperature of 80°F and atmospheric pressure. The free stream velocity is 100 mph and the flat plate temperature is 120°F.

29. Using Fig. 8–8, derive an expression to predict the heat transfer coefficient for a fluid having a Prandtl number Pr = 300. Put your answer in the form of a Nusselt number–Reynolds number relationship.

30. Derive an expression for $\partial^2 u / \partial y^2$ in terms of ψ^*, y^*, and some or all of the parameters V_∞, x, y, ν, and C using Eqs. (8–40) and (8–41).

31. Determine the coefficient C_T below:

$$\frac{\partial^2 T^*}{\partial y^2} = C_T \frac{\partial^2 T^*}{\partial y^{*2}} \quad \text{if} \quad y^* = \frac{y}{5\sqrt{(\nu x)/V_\infty}}.$$

32. Solve the Blasius problem, Eqs. (8–43) and (8–44), by use of a digital computer and obtain the results shown in Fig. 8–6. This can be done by solving Eq. (8–43) for the third derivative, differentiating this twice to obtain expressions for the fourth and fifth derivatives, and writing truncated Taylor series expansions for

$$\psi^*(y^* + h), \quad \frac{d}{dy^*}[\psi^*(y^* + h)], \quad \text{and} \quad \frac{d^2}{dy^{*2}}[\psi^*(y^* + h)],$$

where h is an increment in y^*. $\psi^*(0)$ and $[d\psi^*(0)]/dy^*$ are given as boundary conditions. The value of $d^2\psi^*/dy^{*2}$ at $y^* = 0$ is needed to start the integration and must be guessed. A solution is obtained when a guessed value leads to a fit of the boundary condition $\psi^*(\infty) = 1$ after integration across the boundary layer thickness. A Newton-Lagrangian interpolation scheme may be used to speed up the solution.

33. Estimate the value of the fluid velocity components v and u normal and parallel to a flat plate at a point 2 ft from the leading edge and 0.04 in. out from the wall. The fluid flowing along the plate is air with a free stream velocity of 80 mph. Assume a mean film temperature of 100°F and a pressure of 1 atmosphere.

34. Calculate the local temperature in a laminar boundary layer along a flat plate and at a point 4.0 in. from the leading edge and 0.02 in. out from the wall.

218   LAMINAR FLOW AND HEAT TRANSFER

Assume the fluid is air at a free stream temperature of 70°F and atmospheric pressure. The free stream velocity is 100 mph and the flat plate temperature is 130°F.

35. Air at a mean film temperature of 100°F flows along a flat plate in steady laminar flow at a free stream velocity of 10 mph. At a position $x = 2$ ft from the leading edge of the plate, determine:
    a) the x component of velocity u, 0.1 in. from the wall,
    b) the local heat transfer coefficient h_x,
    c) the local boundary layer thickness δ, and
    d) the local friction coefficient C_f.

36. Air flows with a constant free stream velocity of 4 ft/sec along a flat plate of constant wall temperature. The film temperature of the air is 200°F. For a distance of 4 ft from the leading edge, calculate (using the exact solution):
    a) the boundary layer thickness,
    b) the local shear stress at the wall,
    c) the local heat transfer coefficient, and
    d) the velocity component u at a point 0.1 in. from the wall.

37. Estimate the total drag in lb_f on one side of a sharp-edged flat plate which is placed in an air stream with the plate surface parallel to the flow. The air is at 40°F and 1 atmosphere and has a free stream velocity relative to the plate of 12 ft/sec. The plate dimensions are 4 ft (in the flow direction) by 8 ft wide.

38. Using Eq. (8–77) and the following boundary conditions:

$$\text{at } y = 0,\ u = 0 \quad \text{and} \quad \frac{\partial^2 u}{\partial y^2} = 0;$$

$$\text{at } y = \delta,\ u = U_\infty \quad \text{and} \quad \frac{\partial u}{\partial y} = 0$$

evaluate the constants a, b, c, and d and obtain Eq. (8–79).

39. Using the following boundary conditions:

$$\text{at } y = 0,\ T = T_w \quad \text{and} \quad \frac{\partial^2 T}{\partial y^2} = 0;$$

$$\text{at } y = \delta_T,\ T = T_\infty \quad \text{and} \quad \frac{\partial T}{\partial y} = 0$$

evaluate the constants e, f, g, and h in Eq. (8–78) and obtain Eq. (8–80).

40. Use Eqs. (8–73) and (8–79) to obtain the differential equation for $\delta(x)$.

41. Substituting both Eqs. (8–79) and (8–80) into Eq. (8–74) will lead to a differential equation. Obtain the equation and solve it to obtain Eq. (8–82).

42. Use Eq. (8–79) to obtain an expression for the local coefficient of friction:

$$C_f = \frac{\tau_w}{\tfrac{1}{2}\rho V_\infty^2}$$

for the flat plate. *Hint:* $\cdots \tau_w = \mu(\partial u/\partial y)_{y=0}$.

43. Use Eqs. (8–80), (8–81), and (8–82) to obtain Eq. (8–83).
44. Derive Eq. (8–84).
45. Derive Eq. (8–85) using Eqs. (8–84), (8–81), and the definition of the Nusselt number.
46. Assume that the velocity profile on a flat plate can be approximated by the following polynomial:

$$\frac{u}{V_\infty} = ay + by^2 + cy^3 + dy^4.$$

Evaluate the constants and obtain an expression for the dimensionless velocity u/V_∞ as a function of dimensionless distance y/δ. Use the following boundary conditions:

$$\frac{\partial^2 u}{\partial y^2} = 0 \text{ at } y = 0; \quad u = V_\infty \quad \frac{\partial u}{\partial y} = 0, \quad \frac{\partial^2 u}{\partial y^2} = 0 \text{ at } y = \delta.$$

47. Use the results of Problem 46 to obtain an expression for the dimensionless local coefficient of friction (Problem 42).
48. Assume that the temperature profile on a flat plate can be approximated by the following polynomial:

$$\frac{T - T_w}{T_\infty - T_w} = a\left(\frac{y}{\delta_T}\right) + b\left(\frac{y}{\delta_T}\right)^2 + c\left(\frac{y}{\delta_T}\right)^3 + d\left(\frac{y}{\delta_T}\right)^4.$$

Evaluate the constants and obtain an expression for the dimensionless temperature in terms of dimensionless distance. Use the following boundary conditions:

$$\frac{\partial^2 T}{\partial y^2} = 0 \text{ at } y = 0; \quad T = T_\infty, \quad \frac{\partial T}{\partial y} = \frac{\partial^2 T}{\partial y^2} = 0 \text{ at } y = \delta_T.$$

49. Use the results of Problem 48 to derive an expression for the local Nusselt number along the plate.
50. Show that the boundary conditions used in Problems 46 and 48 involving second derivatives at the wall are valid by using the momentum and energy equations, Eqs. (8–67) and (8–68) evaluated at the wall.
51. Assume that the velocity profile in the boundary layer is approximated by:

$$\frac{u}{V_\infty} = a + b\left(\frac{y}{\delta}\right) + c\left(\frac{y}{\delta}\right)^2 + d\left(\frac{y}{\delta}\right)^3 + e\left(\frac{y}{\delta}\right)^4,$$

with the following boundary conditions:

$$u = 0, \quad \frac{\partial^2 u}{\partial y^2} = 0 \quad \text{at} \quad y = 0;$$

$$u = V_\infty, \quad \frac{\partial u}{\partial y} = 0, \quad \frac{\partial^2 u}{\partial y^2} = 0 \quad \text{at} \quad y = \delta.$$

Determine the constant coefficients a, b, c, d, and e.

220 LAMINAR FLOW AND HEAT TRANSFER

52. The temperature profile in a laminar boundary layer on a flat plate can be approximated by

$$\frac{T - T_w}{T_\infty - T_w} = 2\left(\frac{y}{\delta_T}\right) - 2\left(\frac{y}{\delta_T}\right)^3 + \left(\frac{y}{\delta_T}\right)^4.$$

If

$$\delta_T = \frac{\delta}{\sqrt[3]{Pr}} \quad \text{and} \quad \delta = 5.8\sqrt{\frac{\nu x}{V_\infty}},$$

derive an expression for the Nusselt number Nu_x as a function of the Reynolds number Re_x and the Prandtl number Pr.

53. Verify Reynolds analogy for $Pr = 1$ for the case of steady laminar flow along a flat plate, with constant properties and no viscous dissipation, by comparing Eqs. (8–50), (8–62), and (8–93).

54. Derive an expression for the average Stanton number, $\overline{St} = \bar{h}/\rho V_\infty c_p$ in terms of the average coefficient of friction over the entire length of plate.

55. Calculate a value of the local Stanton number for Problem 25. Assume that the Prandtl number of the air is close enough to 1.0 for Reynolds analogy to be valid.

56. If the Prandtl number of a fluid were greater than 1.0, discuss the relationship that would exist between the Stanton number and the coefficient of friction. Use the results of Section 8–3 or 8–4 to arrive at some qualitative conclusion.

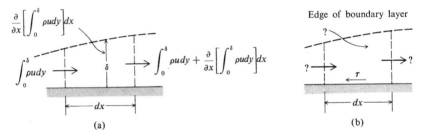

Problem 57 (a) Conservation of mass. (b) Conservation of momentum.

57. Derive the von Kármán momentum integral equation (Eq. 8–73) by applying the momentum theorem to the control volume around a portion of the boundary layer (see figure). Assume a zero pressure gradient, $dp/dx = 0$, and an incompressible fluid.

58. By dimensional reasoning, show that

$$\delta = f\left(\sqrt{\frac{\nu x}{V_\infty}}\right).$$

59. Use Eqs. (8–28) and (8–23) to derive Eq. (8–27).

CHAPTER 9

TURBULENT FLOW AND HEAT TRANSFER

The equations describing the conservation of mass, momentum, and energy, which were described in Chapter 4, are valid for turbulent as well as laminar flow. It must be noted, however, that the quantities such as velocity, pressure, and temperature in these equations are instantaneous values. In some cases the properties such as density, viscosity, specific heat, and thermal conductivity may be variable and must also be considered to be instantaneous values. In turbulent flow the instantaneous values always vary with time (see Fig. 3-15) and the derivatives of velocity and temperature with respect to time in the conservation equations are nonzero. The variation with time of a dependent property, such as a velocity component, is completely random in turbulent flow. This leads to seemingly insurmountable difficulties in the solution of these equations for most problems in turbulent flow. The lack of success in applying the theoretical approach to problems of turbulent flow and heat transfer has encouraged the development of alternate methods such as dimensional analysis, analogies, and semiempirical equations. Some of the methods in common use in the study and prediction of turbulent behavior will be presented in this chapter.

9–1 TIME-SMOOTHING

The technique known as time-smoothing is useful for the insight that it gives into turbulent behavior. The technique points out just what quantities must be known or approximated in order to make predictions. Demonstration of the technique will be made for flow with no body forces and with constant properties, but it can be developed for more complex situations. The momentum equations which apply in this case may be obtained from Eqs. (4–44a-c). Only the x component equation will be demonstrated:

$$\frac{\partial u}{\partial t} + u\frac{\partial u}{\partial x} + v\frac{\partial u}{\partial y} + w\frac{\partial u}{\partial z} = -\frac{1}{\rho}\frac{\partial p}{\partial x} + \nu\left(\frac{\partial^2 u}{\partial x^2} + \frac{\partial^2 u}{\partial y^2} + \frac{\partial^2 u}{\partial z^2}\right). \quad (9\text{–}1)$$

Equation (9–1) may be combined with the equation of continuity for incompressible flow to give the following form (Problem 38, Chapter 4):

$$\frac{\partial u}{\partial t} + \frac{\partial u^2}{\partial x} + \frac{\partial (uv)}{\partial y} + \frac{\partial (uw)}{\partial z} = -\frac{1}{\rho}\frac{\partial p}{\partial x} + \nu\left(\frac{\partial^2 u}{\partial x^2} + \frac{\partial^2 u}{\partial y^2} + \frac{\partial^2 u}{\partial z^2}\right). \quad (9\text{–}1a)$$

Since each value of velocity and pressure in Eq. (9–1a) is an instantaneous value, substitution may be made from Eqs. (3–25) and (3–29):

$$\frac{\partial}{\partial t}(\bar{u} + u') + \frac{\partial}{\partial x}(\bar{u} + u')(\bar{u} + u')$$

$$+ \frac{\partial}{\partial y}(\bar{u} + u')(\bar{v} + v') + \frac{\partial}{\partial z}(\bar{u} + u')(\bar{w} + w')$$

$$= -\frac{1}{\rho}\frac{\partial}{\partial x}(\bar{p} + p') + \nu\left[\frac{\partial^2}{\partial x^2}(\bar{u} + u') + \frac{\partial^2}{\partial y^2}(\bar{u} + u') + \frac{\partial^2}{\partial z^2}(\bar{u} + u')\right]. \quad (9\text{–}2)$$

Equation (9–2) may be simplified to give

$$\frac{\partial}{\partial t}(\bar{u} + u') + \frac{\partial}{\partial x}[(\bar{u})^2 + 2\bar{u}u' + (u')^2] + \frac{\partial}{\partial y}[\bar{u}\bar{v} + u'\bar{v} + v'\bar{u} + u'v']$$

$$+ \frac{\partial}{\partial z}[\bar{u}\bar{w} + \bar{u}w' + u'\bar{w} + u'w']$$

$$= -\frac{1}{\rho}\frac{\partial}{\partial x}(\bar{p} + p') + \nu\left[\frac{\partial^2}{\partial x^2}(\bar{u} + u') + \frac{\partial^2}{\partial y^2}(\bar{u} + u') + \frac{\partial^2}{\partial z^2}(\bar{u} + u')\right].$$

$$(9\text{–}2a)$$

Time-smoothing is accomplished by taking the time average of each term in Eq. (9–2a). The average is assumed to be taken over a sufficiently long period of time to give a steady or unchanging value for the term. It is assumed, therefore, that the only variation that occurs with time is that due to the random turbulent variation, and the average values do not change with time. A time average is indicated by use of a bar over a term.

The following rules will be helpful in taking time averages of the terms in Eq. (9–2a). If f_1 and f_2 are fluctuating variables (such as velocity or pressure) expressed as functions of some independent variable (such as position x), then the time average of the time average is the time average:

$$\bar{\bar{f}}_1 = \bar{f}_1, \quad \bar{\bar{f}}_2 = \bar{f}_2. \quad (9\text{–}3a)$$

The time average of the sum is the sum of the time averages:

$$\overline{f_1 + f_2} = \bar{f}_1 + \bar{f}_2. \quad (9\text{–}3b)$$

The time average of the product is the product of the time average:

$$\overline{\bar{f}_1 \bar{f}_2} = \bar{f}_1 \bar{f}_2. \quad (9\text{–}3c)$$

The time average of the derivative is the derivative of the time average:

$$\overline{\frac{\partial f_1}{\partial x}} = \frac{\partial \bar{f}_1}{\partial x}, \quad \overline{\frac{\partial f_2}{\partial x}} = \frac{\partial \bar{f}_2}{\partial x}. \quad (9\text{–}3d)$$

The time average of each derivative in Eq. (9–2a) becomes the derivative of the time average in each case according to Eq. (9–3d). Also the time average of the sum of terms becomes the sum of the time averages, according to Eq. (9–3b). For example,

$$\overline{\bar{p} + p'} = \bar{p} + \overline{p'} = \bar{p}. \tag{9–4}$$

The time average of a term involving the product of an average value and a fluctuating value is zero according to Eq. (9–3c). For example,

$$\overline{(\bar{u}u')} = 0. \tag{9–5}$$

The time average of the terms involving the product of a fluctuating quantity and the quantity itself is nonzero, however, since the product is always positive. Thus

$$\overline{(u')^2} \neq 0. \tag{9–6}$$

Likewise the product of any two fluctuating quantities is not necessarily zero since a *correlation* normally exists between each of these quantities. For example, at some fixed point in a fluid in "steady" turbulent flow, a positive value of the fluctuating component u' might always occur with a negative value of the fluctuating component v'. In such a case a negative value of u' would always occur with a positive value of v'. In any case the product of u' and v' would always be negative and therefore, the time average would be nonzero. That such correlations between fluctuating components do exist can be argued on theoretical grounds and can be demonstrated experimentally.

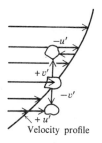

Fig. 9–1 Demonstration of correlation between fluctuating velocity components.

Consider, for example, a lump or particle of fluid located in a steady turbulent flow field where there is some arbitrary velocity profile (Fig. 9–1). If the fluid particle has a downward fluctuation of velocity (negative v') at some instant of time, it will arrive at a position where the local mean velocity in the flow direction is less than the particle's component of velocity in that direction. The particle will, therefore, tend to increase the local fluid velocity giving a positive fluctuation u' to the local velocity. In a similar manner an upward velocity (positive v') leads to a decrease in the local stream velocity (negative u'). Thus the product $u'v'$ is usually negative. For example,

$$\overline{(u'v')} \neq 0. \tag{9-7}$$

The time average of fluctuating components $\overline{u'}$ and $\overline{p'}$ are zero by definition. The derivative of $\bar{u}$ with respect to time is zero for "steady" flow. Using the above reasoning, Eq. (9-2a) becomes, after rearranging,

$$\rho\left(\bar{u}\frac{\partial \bar{u}}{\partial x} + \bar{v}\frac{\partial \bar{u}}{\partial y} + \bar{w}\frac{\partial \bar{u}}{\partial z}\right)$$
$$= -\frac{\partial \bar{p}}{\partial x} + \mu\left(\frac{\partial^2 \bar{u}}{\partial x^2} + \frac{\partial^2 \bar{u}}{\partial y^2} + \frac{\partial^2 \bar{u}}{\partial z^2}\right) - \rho\left(\frac{\partial}{\partial x}\overline{u'u'} + \frac{\partial}{\partial y}\overline{u'v'} + \frac{\partial}{\partial z}\overline{u'w'}\right). \tag{9-8}$$

In a similar manner the y and z component equations would be

$$\rho\left(\bar{u}\frac{\partial \bar{v}}{\partial x} + \bar{v}\frac{\partial \bar{v}}{\partial y} + \bar{w}\frac{\partial \bar{v}}{\partial z}\right)$$
$$= -\frac{\partial \bar{p}}{\partial y} + \mu\left(\frac{\partial^2 \bar{v}}{\partial x^2} + \frac{\partial^2 \bar{v}}{\partial y^2} + \frac{\partial^2 \bar{v}}{\partial z^2}\right) - \rho\left(\frac{\partial}{\partial x}\overline{u'v'} + \frac{\partial}{\partial y}\overline{v'v'} + \frac{\partial}{\partial z}\overline{w'v'}\right); \tag{9-9}$$

$$\rho\left(\bar{u}\frac{\partial \bar{w}}{\partial x} + \bar{v}\frac{\partial \bar{w}}{\partial y} + \bar{w}\frac{\partial \bar{w}}{\partial z}\right)$$
$$= -\frac{\partial \bar{p}}{\partial z} + \mu\left(\frac{\partial^2 \bar{w}}{\partial x^2} + \frac{\partial^2 \bar{w}}{\partial y^2} + \frac{\partial^2 \bar{w}}{\partial z^2}\right) - \rho\left(\frac{\partial}{\partial x}\overline{u'w'} + \frac{\partial}{\partial y}\overline{v'w'} + \frac{\partial}{\partial z}\overline{w'w'}\right). \tag{9-10}$$

The Eqs. (9-8), (9-9), and (9-10) have the same form as the steady flow momentum equations which have been applied to laminar flow, except for the fact that these equations have additional terms on the extreme right caused by the presence of the fluctuating components of velocity. These "extra" terms might be thought of as describing the additional effect on the flow due to the turbulence. In laminar flow these terms would be zero and the equations would reduce to the proper form. The additional stresses due to the fluctuating components have been given the name *Reynolds stresses* after Osborne Reynolds [1] who first presented this technique.

These additional stresses due to turbulence are also called *apparent* or *virtual stresses*. The *total stress* is the sum of the apparent stress and the viscous or molecular stress. For example, the shear stress τ_{xy} can be

expressed as
$$(\tau_{xy})_{\text{total}} = (\tau_{xy})_{\text{molecular}} + (\tau_{xy})_{\text{apparent}}. \tag{9-11}$$

Expressions for the stresses may be obtained by a comparison of Eqs. (4-42a-c), (3-20), (9-8), (9-9), and (9-10):

$$(\tau_{xy})_{\text{total}} = \mu\left(\frac{\partial \bar{u}}{\partial y} + \frac{\partial \bar{v}}{\partial x}\right) - \rho \overline{u'v'}. \tag{9-12}$$

For parallel flow in the x direction, $\bar{v} = 0$, and the expression for molecular shear stress contains only the derivative $\partial \bar{u}/\partial y$ (Newton's law of viscosity). At any point in the flow the magnitude of $\overline{u'v'}$ is related to the magnitude of $\partial \bar{u}/\partial y$. This suggests the definition of a proportionality between these two quantities:

$$-\overline{u'v'} = \frac{(\tau_{xy})_{\text{apparent}}}{\rho} = \epsilon_M \frac{\partial \bar{u}}{\partial y}. \tag{9-13}$$

The quantity ϵ_M is called the *eddy diffusivity of momentum*. In parallel flow the total stress would be given by

$$(\tau_{xy})_{\text{total}} = (\mu + \epsilon_M \rho)\frac{\partial \bar{u}}{\partial y}, \tag{9-14}$$

or

$$\frac{(\tau_{xy})_{\text{total}}}{\rho} = (\nu + \epsilon_M)\frac{\partial \bar{u}}{\partial y}. \tag{9-15}$$

Equation (9-15) shows the proportionality between the shear stress and the average velocity gradient at any point in a turbulent flow. For purely laminar flow the eddy diffusivity of momentum becomes zero, and Eq. (9-15) reduces to Newtons law of viscosity. In a thin layer adjacent to solid walls the turbulent eddies become small and ϵ_M becomes much smaller than the kinematic viscosity ν, and again Newtons law of viscosity applies. Away from the wall, however, the turbulent eddies are usually large enough that $\epsilon_M \gg \nu$, and the total stress is very closely approximated by the apparent stress.

Definition of the eddy viscosity and the resulting equations, such as Eq. (9-15), do not really lead to theoretical solutions in turbulent flow. This is so because the eddy diffusivity is a parameter of the fluid motion and is not a fluid property, as is viscosity. A large amount of study has been given to the eddy diffusivity and the quantities such as $\overline{u'v'}$, but no general method of predicting these has resulted. The concept of an eddy diffusivity is justified by its usefulness in the method of analogies which will be discussed in detail later.

The equations describing conservation of mass and energy may also be transformed by time-smoothing.

Using the time-smoothing technique, the continuity equation for incompressible flow becomes

$$\frac{\partial \bar{u}}{\partial x} + \frac{\partial \bar{v}}{\partial y} + \frac{\partial \bar{w}}{\partial z} = 0, \qquad (9\text{--}16)$$

which can be used to derive the following relationship:

$$\frac{\partial u'}{\partial x} + \frac{\partial v'}{\partial y} + \frac{\partial w'}{\partial z} = 0. \qquad (9\text{--}17)$$

Equation (9–16) shows that mass must be conserved "on the average," and Eq. (9–17) shows that mass must also be conserved at any instant of time.

The energy equation for the case of constant properties, and no viscous dissipation becomes, after time-smoothing,

$$\rho c_p \left(\bar{u} \frac{\partial \bar{T}}{\partial x} + \bar{v} \frac{\partial \bar{T}}{\partial y} + \bar{w} \frac{\partial \bar{T}}{\partial z} \right)$$
$$= k \left(\frac{\partial^2 \bar{T}}{\partial x^2} + \frac{\partial^2 \bar{T}}{\partial y^2} + \frac{\partial^2 \bar{T}}{\partial z^2} \right) - \rho c_p \left(\frac{\partial}{\partial x} \overline{u'T'} + \frac{\partial}{\partial y} \overline{v'T'} + \frac{\partial}{\partial z} \overline{w'T'} \right). \qquad (9\text{--}18)$$

An *eddy diffusivity of heat*, ϵ_H, may be defined in a manner similar to that for ϵ_M. For heat transfer in the y direction,

$$\frac{(\dot{q}/A)_{\text{total}}}{\rho c_p} = -(\alpha + \epsilon_H) \frac{\partial \bar{T}}{\partial y}, \qquad (9\text{--}19)$$

where α is the molecular thermal diffusivity, $\alpha = k/(\rho c_p)$, and ϵ_H is given by

$$\overline{v'T'} = -\epsilon_H \frac{\partial \bar{T}}{\partial y}, \qquad (9\text{--}20)$$

and

$$\left(\frac{\dot{q}}{A} \right)_{\text{apparent}} = +\rho c_p \overline{v'T'} = -\epsilon_H \rho c_p \frac{\partial \bar{T}}{\partial y}. \qquad (9\text{--}21)$$

The eddy diffusivity of heat is also a flow-related quantity and not a fluid property. As with ϵ_M, no generally useful theory for predicting values of ϵ_H is known.

The usefulness of the eddy diffusivity concept in analogies will be demonstrated later in this chapter.

9-2 FLOW ALONG A FLAT PLATE

The turbulent boundary layer on a flat plate is of paramount importance in many practical problems in fluid flow and heat transfer. The formation of a turbulent boundary layer was discussed in Section 7–1 and illustrated in Fig. 7–3. The method of analogy will now be demonstrated for this case. Making boundary layer assumptions, and further assuming that properties are constant, that body forces are negligible, and that the flow is "steady" with no viscous dissipation, the conservation equations can be written in terms of average quantities.

$$\frac{\partial \bar{u}^*}{\partial x^*} + \frac{\partial \bar{v}^*}{\partial y^*} = 0, \tag{9-22}$$

$$\bar{u}^* \frac{\partial \bar{u}^*}{\partial x^*} + \bar{v}^* \frac{\partial \bar{u}^*}{\partial y^*} = \frac{\partial}{\partial y^*}\left[\left(\frac{\nu + \epsilon_M}{V_\infty L}\right)\frac{\partial \bar{u}^*}{\partial y^*}\right], \tag{9-23}$$

$$\bar{u}^* \frac{\partial \bar{T}^*}{\partial x^*} + \bar{v}^* \frac{\partial \bar{T}^*}{\partial y^*} = \frac{\partial}{\partial y^*}\left[\left(\frac{\alpha + \epsilon_H}{V_\infty L}\right)\frac{\partial \bar{T}^*}{\partial y^*}\right]. \tag{9-24}$$

The energy equation, Eq. (9–24), is written in terms of the dimensionless temperature $\bar{T}^* = (\bar{T} - T_w)/(T_\infty - T_w)$, where T_∞ and T_w are constant and the momentum equation, Eq. (9–23), is written in terms of the dimensionless velocity $\bar{u}^* = \bar{u}/V_\infty$ so that the boundary conditions on u^* and T^* are identical at the wall ($y = 0$) and in the free stream ($y \to \infty$). Equations (9–23) and (9–24) are of the same form and would, therefore, have identical solutions for $\bar{T}^*$ and $\bar{u}^*$ if the following were true at every point in the boundary layer:

$$(\nu + \epsilon_M) = (\alpha + \epsilon_H). \tag{9-25}$$

Since the eddy terms ϵ_M and ϵ_H are much larger than the molecular terms ν and α away from the wall, and since the opposite is true near the wall, then for the solutions to be identical it is required that both the following be true:

$$\nu = \alpha \quad (\text{or } \Pr = 1), \tag{9-26}$$

and

$$\epsilon_M = \epsilon_H \quad (\text{or } \Pr_t = 1), \tag{9-27}$$

where $\Pr_t = \epsilon_M/\epsilon_H$ and is called the *turbulent Prandtl number*. In other words, for the average velocity and temperature profiles to be identical in the stated problem, both the molecular Prandtl number and the turbulent Prandtl number must be equal to 1.0. If this condition is met, then it is not difficult to show, by the reasoning of Section 7–5, that Reynolds analogy [1] is valid for this turbulent case, and

$$\text{St} = \frac{C_f}{2}. \quad (\text{Pr and } \Pr_t \text{ equal to } 1.0) \tag{9-28}$$

The assumption $\epsilon_M = \epsilon_H$ is not exactly true in most cases, but is generally a good approximation, except for liquid metals. Most of the early analogies have been based on the assumption that $\epsilon_M = \epsilon_H$ and represent various methods that allow for the variation of molecular Prandtl number [2, 3, 4]. One of these analogies, due to von Kármán [3], is given below as representative of these methods:

$$\frac{\mathrm{Nu}_x}{\mathrm{Re}_x \, \mathrm{Pr}} = \mathrm{St}_x = \frac{C_f/2}{1 + 5\sqrt{C_f/2}\,\{(\mathrm{Pr} - 1) + \ln[1 + \tfrac{5}{6}(\mathrm{Pr} - 1)]\}}. \qquad (9\text{--}29)$$

Note that the von Kármán analogy, Eq. (9–29), reduces to the Reynolds analogy, for Pr = 1. It also should be noted that for turbulent flow the friction coefficient C_f cannot be determined purely by analytical methods, and experimentation or semiempirical methods are needed. The analogy is useful primarily because values of C_f can be determined more easily by experimentation than can values of Nusselt number or Stanton number.

Colburn [5] suggested that most experimental data can be approximated by simply introducing a correction term to allow for the effect of Prandtl number variation. His suggestion led to the following relation:

$$\mathrm{St}_x \, \mathrm{Pr}^{2/3} = \frac{C_f}{2}. \qquad \text{(Colburn's analogy)} \qquad (9\text{--}30)$$

The left-hand term has been given the name *Colburn j factor*, or simply *j factor*.

The local coefficient of friction C_f has been determined experimentally for turbulent flow along a flat plate. For ($5 \times 10^5 < \mathrm{Re}_x < 10^7$), it has been found to be given approximately by

$$C_f = 0.0576 \, \mathrm{Re}_x^{-0.2}. \qquad (9\text{--}31)$$

If the expression for C_f, Eq. (9–31), is substituted into Eq. (9–30), the following results:

$$\mathrm{St}_x \, \mathrm{Pr}^{2/3} = 0.0288 \, \mathrm{Re}_x^{-0.2}. \qquad (9\text{--}32)$$

Multiplying Eq. (9–32) by $\mathrm{Re}_x \, \mathrm{Pr}^{1/3}$ leads to

$$\mathrm{Nu}_x = \frac{hx}{k} = 0.0288 \, \mathrm{Pr}^{1/3} \, \mathrm{Re}_x^{0.8}. \qquad (9\text{--}33)$$

Equations (9–32) and (9–33) are valid only in the turbulent region downstream of the transition point. To determine an average heat transfer coefficient $\bar{h}$ over the entire plate, an average coefficient for the turbulent region must be obtained and combined with the average coefficient for the laminar region:

$$\bar{h} = \frac{\displaystyle\int_0^{x\mathrm{crit}} h \, dx + \int_{x\mathrm{crit}}^{L} h \, dx}{L}. \qquad (9\text{--}34)$$

For the laminar region, from Eq. (8–62):

$$h = \frac{0.332k}{x} \text{Pr}^{1/3} \text{Re}_x^{1/2}. \tag{9-35}$$

Equations (9–33) and (9–35) can now be combined by assuming $x_{\text{crit}} = 500{,}000\nu/V_\infty$ (see Fig. 7–4) to give:

$$\overline{\text{Nu}} = \frac{\bar{h}L}{k} = 0.036 \, \text{Pr}^{1/3}(\text{Re}_L^{0.8} - 23{,}200). \tag{9-36}$$

Equation (9–36) can be used to obtain the average heat transfer coefficient for flow over a flat plate when heating is started at the leading edge. The film temperature $T_f = \tfrac{1}{2}(T_w + T_\infty)$ should be used to evaluate the fluid properties in Eqs. (9–33) and (9–36).

Example 9-1. Calculate the heat loss from one side of a plate to air at 1 atmosphere of pressure. The plate is 5 ft wide by 6 ft long in the direction of flow. The free stream velocity relative to the plate is 200 ft/sec and the free stream temperature is 60°F. The plate temperature is uniform and equal to 140°F.

Solution. The film temperature is

$$T_f = \frac{60 + 140}{2} = 100°\text{F}.$$

and

$$\text{Pr} = 0.706$$
$$\text{Pr}^{1/3} = 0.890$$
$$\nu = 0.18 \times 10^{-3} \text{ ft}^2/\text{sec}.$$
$$k = 0.0156 \, \frac{\text{Btu}}{\text{hr ft °F}}.$$
$$\text{Re}_L = \frac{V_\infty L}{\nu} = \frac{(200)(6)}{0.18 \times 10^{-3}} = 6{,}660{,}000.$$

Using Eq. (9–36),

$$\bar{h} = \frac{0.036k}{L} \text{Pr}^{1/3}(\text{Re}_L^{0.8} - 23{,}200)$$

$$= \frac{(0.036)(0.0156)(0.890)}{6}(288{,}000 - 23{,}200) = 22 \, \frac{\text{Btu}}{\text{hr ft}^2 \text{ °F}}.$$

$$\dot{q} = \bar{h}A(T_w - T_\infty)$$
$$= (22)(6 \times 5)(140 - 60) = 52{,}800 \text{ Btu/hr}.$$

9-3 TURBULENT FLOW IN CONDUITS

The transition to turbulent flow in a conduit such as a tube or pipe has been discussed briefly in Section 7–2. It was shown that if the Reynolds number based on conduit diameter (or equivalent diameter) is large enough, the boundary layer in the entrance region will become unstable and undergo transition to turbulence at some place downstream from the entrance (see Fig. 7–9). At some distance further downstream, the turbulent boundary layers will grow together and a fully developed velocity profile will exist from that point downstream. This assumes constant fluid properties and constant diameter.

The Reynolds number at which this transition occurs is so small ($Re_D = 2000$) that turbulent flow usually occurs in the majority of practical conduit flow problems. In most situations of practical interest, the transition occurs near the entrance. The turbulent boundary layer grows much more rapidly with axial distance than does the laminar boundary layer. This results in a fully developed flow occurring nearer the entrance than for the laminar case. Since the point of transition is highly dependent upon the entrance conditions, there is no general equation for predicting the entrance length required for the development of fully developed velocity profiles. However, the local friction factor in turbulent channel flow reaches a nearly constant value in a much shorter distance than that required for the development of a constant velocity profile. For a circular tube, the dimensionless distance required for the friction factor to become constant is given by an equation of Latzko [6]:

$$\frac{L}{D} = 0.623(Re_D)^{1/4}. \tag{9-37}$$

For a Reynolds number based on D of 20,000, for example, the friction factor becomes constant within about 7.5 diameters from the entrance. This fully developed value of the friction factor for turbulent flow in conduits is presented in Fig. 9–2 as a function of Reynolds number and pipe roughness. Such plots are essentially empirical in nature because there is at present no suitable analytical approach to the prediction of turbulent friction factors. A variety of equations for prediction of friction factors in turbulent flow may be found in the literature. The curve for smooth pipes shown in Fig. 9–2 is based on an equation of Nikuradse [7], which he obtained from correlation of experimental data covering the range $4000 < Re_D < 3,240,000$:

$$\frac{1}{\sqrt{f}} = 2.0 \log_{10}(Re_D\sqrt{f}) - 0.80. \tag{9-38}$$

Equation (9–38) is difficult to use in computation since the friction factor f cannot be expressed explicitly. This difficulty is not found in the

9-3 TURBULENT FLOW IN CONDUITS 231

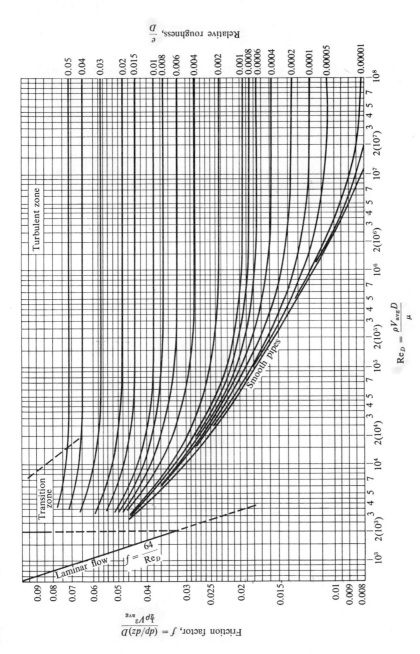

Fig. 9-2 Friction factors for pipe flow according to Moody [33]. (From Giedt's *Principles of Engineering Heat Transfer.* Copyright 1957, D. Van Nostrand Company, Inc., Princeton, New Jersey. Used with permission.)

empirical equation of Drew et al. [8], valid in the range $3000 < \text{Re}_D < 3{,}000{,}000$:

$$f = 0.0056 + 0.5(\text{Re}_D)^{-0.32}. \tag{9-39}$$

When the Reynolds number range is restricted to turbulent flow with $\text{Re}_D < 100{,}000$, the simple equation given below has been frequently used:

$$f = 0.184(\text{Re}_D)^{-0.2}. \tag{9-40}$$

Equations (9–38), (9–39), and (9–40) are for smooth tubes and do not give proper values of the friction factor for most commercial pipes. For such cases, the *relative roughness* of the pipe (e/D) is important. Values of relative roughness for different types of pipe material were given in Fig. 9–3. An equation which gives the friction factor for tubes of varying roughness is

$$\frac{1}{\sqrt{f}} = -2 \log_{10}\left(\frac{e/D}{3.7} + \frac{2.5}{\text{Re}_D\sqrt{f}}\right). \tag{9-41}$$

The curves for rough pipe shown in Fig. 9–2 approach asymptotic values at high values of Reynolds number showing that the friction factor becomes independent of Reynolds number. These asymptotic values may be approximated by the equation:

$$\frac{1}{\sqrt{f}} = 2 \log_{10}\left(\frac{D}{e}\right) + 1.74. \tag{9-42}$$

Example 9–2. Water at 100°F flows through a 2-in. i.d. galvanized iron pipe at an average velocity of 15 ft/sec. Determine the pressure drop due to friction per foot of pipe length.

Solution. The Reynolds number is first calculated:

$$\text{Re}_D = \frac{V_{\text{avg}} D}{\nu} = \frac{(15) \text{ ft/sec } (2/12) \text{ ft}}{(0.74 \times 10^{-5}) \text{ ft}^2/\text{sec}} = 3.38 \times 10^5.$$

The flow is turbulent. Using Fig. 9–3 for 2-in. i.d. galvanized iron pipe, the relative roughness is $e/D = 0.0030$. From Fig. 9–2, the value of f is 0.0265. The pressure drop per foot is obtained from Eq. (7–15):

$$\frac{dp}{dz} = \frac{-f\rho V_{\text{avg}}^2}{2g_c D} = \frac{-(0.0265)(62)\frac{\text{lb}_m}{\text{ft}^3}(15)^2 \frac{\text{ft}^2}{\text{sec}^2}}{(2)(32.2)\frac{\text{lb}_m\text{-ft}}{\text{lb}_f\text{-sec}^2}\left(\frac{2}{12}\right)\text{ft}}$$

$$= -34.4 \frac{\text{lb}_f}{\text{ft}^2\text{-ft}} = -0.239 \frac{\text{psi}}{\text{ft}}.$$

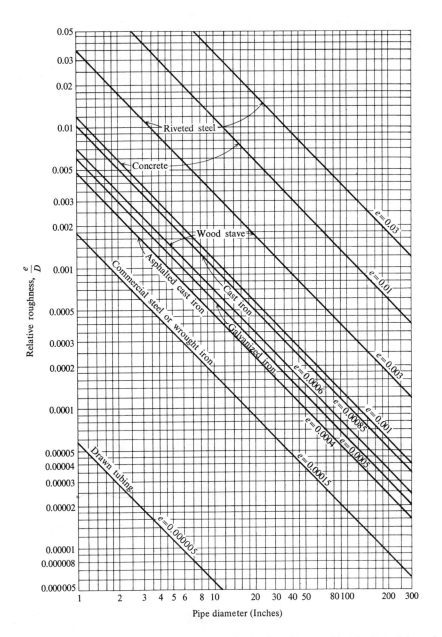

Fig. 9-3 Relative roughness of pipes (*e* in feet). (From Giedt's *Principles of Engineering Heat Transfer*. Copyright 1957, D. Van Nostrand Company, Inc., Princeton, New Jersey. Used with permission.)

Nonisothermal Flow

In nonisothermal flow, or flow where temperature variations exist, there will be an effect on the velocity profile due to variations of viscosity. This in turn will have an effect on the friction factor at a particular value of Reynolds number. Maurer and Le Tourneau [30] discuss this subject in detail. Good engineering approximations can be made by using the rules given below.

In using Fig. 9–2 for nonisothermal flow of gases, the kinematic viscosity ν should be evaluated at a mean film temperature T_f, where

$$T_f = \frac{T_B + T_w}{2},$$

and T_B and T_w are the bulk and wall temperatures respectively.

For liquids the friction factor should be evaluated at the mean bulk temperature and then multiplied by the factor $(\mu_w/\mu_B)^{0.14}$, where μ_w and μ_B are the viscosities at the wall and bulk temperatures respectively.

Figure 9–2 will be very useful in the studies of turbulent flow to be presented in later chapters.

Pressure Drop in Noncircular Ducts

It is often necessary to determine pressure drop for a situation where the duct or conduit is noncircular in cross section. In such situations a question arises as to the proper length to use to determine the Reynolds number for calculation of the friction factor. We might also now ask ourselves whether Fig. 9–2 and Eqs. (7–15) or (7–16) give proper values of pressure drop in such situations. Experience has shown that the concept of an equivalent diameter, usually called the hydraulic diameter, is useful for engineering calculations. The *hydraulic diameter* is defined as four times the cross-sectional area of the duct divided by the wetted perimeter:

$$D_H = \frac{4A}{WP}. \tag{9-43}$$

The hydraulic diameter may be substituted for the diameter D in Eqs. (7–15) and (7–16) and in calculation of the Reynolds number for use in Fig. 9–2.

Pressure Drop in Components

The pressure drop due to friction for fluids flowing through long straight tubes can be readily calculated by the use of the friction factor chart, Fig. 9–2, for either laminar or turbulent flow. Additional pressure drops usually exist in actual piping due to entrance effects, exit effects, or the presence of components such as valves, elbows, tees, and bends. In many cases the

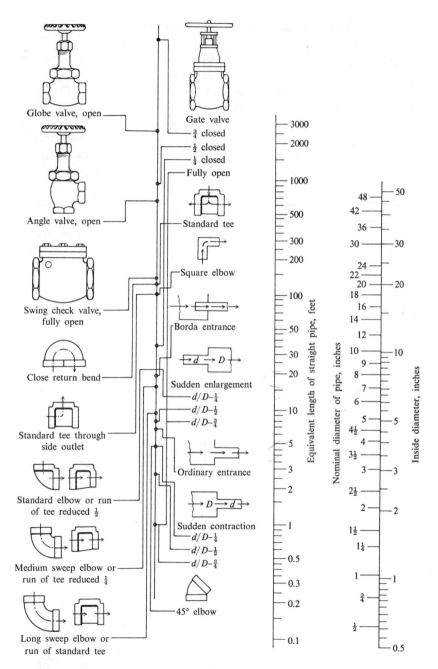

Fig. 9–4 Resistance of valves and fittings (courtesy of Crane Company).

pressure drop due to these factors is larger than that due to friction in the tubing or conduit.

Theoretical methods of determining pressure drop across components are complex and not suitable for engineering calculations. Instead, it has been common practice to use empirical information presented in the form of curves, tables, or nomographs. A widely used nomograph of this type is shown in Fig. 9–4. For various types of components the pressure drop across that component is presented in terms of an equivalent length of straight standard steel pipe. The actual pressure drop across that component is determined by the use of Eq. (7–16), making the assumption that the pressure drop across the component is the same as that across an equivalent length of pipe, where the friction factor is again determined from Fig. 9–2. The equivalent length of each component through which the fluid might flow in series can simply be added to the actual length of piping through which the fluid flows to give a total equivalent length for an entire system in the case of systems using standard commercial steel pipe or pipe of equivalent roughness.

Example 9–3. Determine the pressure drop across a fully open globe valve which is installed in a piping system having a 2-in. nominal diameter (2.067 in. i.d.). Water at 100°F flows through the system at the rate of 80 gal/min.

Solution: From Fig. 9–4, the equivalent length is 47 ft. The Reynolds number for the flow is

$$\text{Re}_D = \frac{V_{\text{avg}} D}{\nu}$$

$$= \frac{(4)(80) \text{ gal/min} (0.1337) \text{ ft}^3/\text{gal} (12) \text{ in./ft}}{(\pi)(2.067) \text{ in.} (0.74 \times 10^{-5}) \text{ ft}^2/\text{sec} (60) \text{ sec/min}} = 1.79 \times 10^5,$$

$$V_{\text{avg}} = \frac{\dot{m}}{\rho A} = \frac{(80) \text{ gal/min} (0.1337) \text{ ft}^3/\text{gal} \ 144 \text{ in}^2/\text{ft}^2}{\frac{\pi (2.067)^2}{4} \text{ in}^2 (60) \text{ sec/min}} = 7.68 \text{ ft/sec}.$$

From Fig. 9–2, for commercial steel pipe ($e/D = 0.0009$) the friction factor at that Reynolds number is 0.0208. Using Eq. (7–16), the pressure drop is determined:

$$p_1 - p_2 = \frac{\frac{1}{2}\rho V_{\text{avg}}^2 L f}{g_c D} = \frac{(\frac{1}{2})(62.0) \text{ lb}_m/\text{ft}^3 (7.68)^2 \text{ ft}^2/\text{sec}^2 (47) \text{ ft } (0.0208)}{(32.2) \text{ lb}_m \text{ ft/lb}_f \text{ sec}^2 \left(\frac{2.067}{12}\right) \text{ ft}}$$

$$= 322 \frac{\text{lb}_f}{\text{ft}^2} = 2.23 \text{ psi.}$$

9-3 TURBULENT FLOW IN CONDUITS

Another common method of calculating pressure loss in components, entrances, and so forth involves the use of an empirical loss coefficient K_L. Since at high values of Reynolds number the pressure drop in most instances appears to be proportional to the square of the average velocity in the conduit, the loss coefficient is defined by the equation

$$\Delta p = K_L \frac{\rho V_{\text{avg}}^2}{2}.$$

Values of K_L for various situations may be found in the literature. A table of loss coefficients for commercial pipe fittings is given in Table 9-1.

Table 9-1

LOSS COEFFICIENTS FOR COMMERCIAL PIPE FITTINGS (CRANE COMPANY)

Globe valve, wide open	10.0
Angle valve, wide open	5.0
Gate valve	
wide open	0.19
$\frac{3}{4}$ open	1.15
$\frac{1}{2}$ open	5.6
$\frac{1}{4}$ open	24.0
Return bend	2.2
90° elbow	0.90
45° elbow	0.42

Lost Work Due to Friction

The energy balance equation for incompressible steady flow in a pipe was given by Eq. (6-23) as

$$\frac{p_2 - p_1}{\rho} + \frac{V_2^2 - V_1^2}{2} + g(y_2 - y_1) = \frac{-\dot{W}_{\text{shaft}}}{\dot{m}} - \text{loss}.$$

The energy "loss" includes the net heat transferred out of the system plus the increase in internal energy. However, the sum of these two terms equals the internal conversion of energy due to viscosity or friction, which is given the symbol l_f:

$$l_f = \frac{-\dot{q}}{\dot{m}} + (e_2 - e_1).$$

From Eqs. (7-18) and (7-19), it can be shown that the lost work due to friction in a pipe is

$$l_f = \frac{(\Delta p)}{\rho} = \frac{L f V_{\text{avg}}^2}{2D}.$$

For a pipe system composed of several sections of pipe of various lengths and diameters and of components of various equivalent lengths,

Eq. (6–23) can be rewritten as

$$\frac{p_2 - p_1}{\rho} + \frac{V_2^2 - V_1^2}{2} + g(y_2 - y_1) + \frac{\dot{W}_{shaft}}{\dot{m}} + \sum \frac{LfV_{avg}^2}{2D} = 0.$$

The summation sign is used to indicate a summation over all pipe section lengths and component equivalent lengths.

Example 9–4. Water at 70°F flows from a large reservoir through a 2-in. i.d. commercial steel pipe as shown in Fig. 9–5. If the pump is rated at 2 hp with an efficiency of 85 percent, what is the flow rate?

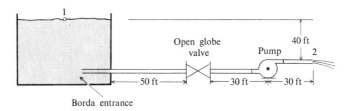

Fig. 9–5 Flow diagram for Example 9–4.

Solution. From Fig. 9–4 the equivalent length of the Borda entrance is 5 ft while the equivalent length of the open globe value is 55 ft. Therefore, the equivalent length of the 2-in. pipe is 5 + 50 + 55 + 30 + 30, or 170 feet.

$$\frac{\dot{W}_{shaft}}{\dot{m}} = \frac{-(0.85)(2)(550)}{V(62.4)(\pi/4)(2/12)^2/32.2} = \frac{-22{,}100}{V} \text{ ft}^2/\text{sec}^2.$$

From Fig. 9–3, $e/D = 0.0009$.

Since the velocity is unknown, a Reynolds number must be assumed in order to get a first approximation to a friction factor. This assumption must be checked after a velocity is calculated.

Assuming $Re > 10^6$, from Fig. 9–2, $f = 0.019$, and

$$l_f = \frac{(170)(0.019)V^2}{(2)(2/12)} = 9.7V^2.$$

The equation at the top of the page between points (1) and (2) is

$$0 + \frac{V^2 - 0}{2} + (32.2)(-40) - \frac{22{,}100}{V} + 9.7V^2 = 0,$$

or

$$V^3 - 126.2V = 2190.$$

9–3   TURBULENT FLOW IN CONDUITS   239

Solving for V by trial and error, $V = 16.2$ ft/sec, and

$$\text{Re} = \frac{VD}{\nu} = \frac{(16.2)(2/12)}{10^{-5}} = 2.7 \times 10^5.$$

From Fig. 9–2, the friction factor, $f = 0.0204$. Using this new value of f, the recomputed velocity is

$$V = 15.7 \text{ ft/sec}.$$

Subsequent computations do not improve the answer significantly. Therefore,

$$\dot{Q} = (\pi/4)(2/12)^2(15.7) = 0.342 \text{ ft}^3/\text{sec}.$$

Example 9–5. Asphalted 6, 9, and 12 in. diameter cast-iron pipes, 1000, 1800, and 2000 ft in length respectively, are connected in parallel (Fig. 9–6). If the pressure at branch 1 is 700 psig, calculate the pressure at branch 2. Because of the length of the pipes, neglect the pressure losses due to elbows, fittings, etc.

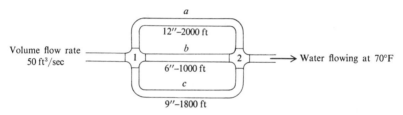

Fig. 9–6 Flow diagram for Example 9–5.

Solution. Equation (6–34) for each pipe can be written as below, assuming the same pressure drop exists in each branch:

$$\frac{p_2 - p_1}{\rho} + 0 + g(y_2 - y_1) = -\left(\frac{LfV^2}{2D}\right)_a = -\left(\frac{LfV^2}{2D}\right)_b = -\left(\frac{LfV^2}{2D}\right)_c.$$

From Fig. 9–3, the relative roughness of pipes a, b, and c are seen to be 0.0004, 0.0008, and 0.00055 respectively. For large Reynolds numbers the corresponding friction factors from Fig. 9–2 are 0.0158, 0.0171, and 0.0187 respectively. Therefore,

$$\frac{V_a}{V_b} = \left(\frac{f_b L_b D_a}{f_a L_a D_b}\right)^{1/2} = \left(\frac{(0.0171)(1000)(12)}{(0.0158)(2000)(6)}\right)^{1/2} = 1.04;$$

$$\frac{V_a}{V_c} = \left(\frac{f_c L_c D_a}{f_a L_a D_c}\right)^{1/2} = \left(\frac{(0.0187)(1800)(12)}{(0.0158)(200)(9)}\right)^{1/2} = 1.19.$$

For mass to be conserved, since density has been assumed constant:

(total volume flow rate = 50 ft³/sec)

$$\therefore \left(\frac{\pi}{4}\right)(1)^2 V_a + \left(\frac{\pi}{4}\right)\left(\frac{6}{12}\right)^2 \frac{V_a}{1.04}$$

$$+ \left(\frac{\pi}{4}\right)\left(\frac{9}{12}\right)^2 \frac{V_a}{1.19} = 50 \text{ ft}^3/\text{sec}.$$

$$\therefore V_a = 37.2 \text{ ft/sec},$$
$$V_b = 37.2/1.04 = 35.8 \text{ ft/sec},$$
$$V_c = 37.2/1.19 = 31.2 \text{ ft/sec}.$$

The Reynolds numbers for pipes a, b, and c are 3.72×10^6, 1.79×10^6, and 2.34×10^6 respectively. At these Reynolds numbers, $f_a = 0.0158$, $f_b = 0.0172$, and $f_c = 0.0187$. These f values are approximately equal to our original estimates, hence there is no need to recompute the velocities. The loss energy due to friction is

$$l_{fa} = \left(\frac{LfV^2}{2D}\right)_a = \frac{(2000)(0.0158)(37.2)^2}{2(1)} = 21{,}850 \text{ ft}^2/\text{sec}^2.$$

If nodes (1) and (2) are at the same elevation, then

$$\frac{p_2 - p_1}{\rho} = -l_{fa} = -21{,}850 \text{ ft}^2/\text{sec}^2.$$

$$p_2 = p_1 - \rho l_{fa} = 700 - (62.4)(21{,}850)/(32.2)(144)$$
$$= 700 - 295 = 405 \text{ psig}.$$

Universal Velocity Profiles

Friction factors are not difficult to measure in turbulent flow of an incompressible fluid. Many of the semiempirical theories of turbulent flow involve the use of the friction factor as a parameter which is assumed to be a known quantity. For example, in turbulent pipe flow, it might be assumed that the local axial velocity u_z depends only on the value of the average velocity V_{avg}, the distance from the wall y, the fluid kinematic viscosity ν, and the friction factor f. In other words,

$$u_z = f_1(V_{avg}, y, \nu, f).$$

Dimensional analysis leads to the following possible dimensionless relationship:

$$\left(\frac{u_z}{V_{avg}\sqrt{f/8}}\right) = f_2\left(\frac{yV_{avg}}{\nu}\sqrt{f/8}\right). \tag{9-44}$$

9-3 TURBULENT FLOW IN CONDUITS

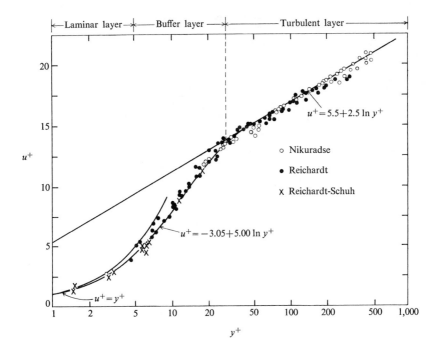

Fig. 9-7 Universal velocity distribution for turbulent flow in pipes. (From H. Reichardt, NACA TM1047, 1943. Courtesy NASA.)

The integer 8 which appears is arbitrary, and is introduced to follow the literature in which the Fanning friction factor or dimensionless wall shear stress has been commonly used (see Section 7-3). Equation (9-44) may be written in the following format, which is common to the literature:

$$u^+ = f(y^+)$$

where

$$u^+ = \frac{u_z}{u_f}, \qquad u_f = V_{\text{avg}}\sqrt{f/8}, \qquad \text{and} \qquad y^+ = \frac{y u_f}{\nu}.$$

The quantity

$$u_f = V_{\text{avg}}\sqrt{f/8}$$

is called the *friction velocity*, since it is related to the value of the friction factor.

Equation (9-44) is significant because, if it is valid, it would imply that all turbulent velocity profiles could be plotted in dimensionless terms *on a single curve*. This would be a *universal velocity profile for turbulent flow*. Experimental data confirms the existence of such a universal profile. Figure 9-7 shows such a plot.

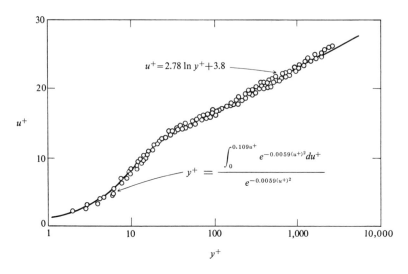

Fig. 9-8 Universal velocity profile of Deissler [10]. (Used with permission of the American Society of Mechanical Engineers.)

Von Kármán [9] suggested that the equations shown in Fig. 9-7 imply the existence of three distinct regions in the turbulent boundary layer:

1. A laminar sublayer ($y^+ < 5$) where $u^+ = y^+$.
2. A buffer layer ($5 < y^+ < 30$) where $u^+ = -3.05 + 5 \ln y^+$.
3. A turbulent core ($y^+ > 30$) where $u^+ = 5.5 + 2.5 \ln y^+$.

The division of the turbulent boundary layer into three distinct regions is not entirely realistic, but the concept has been widely used and has led to some useful results. The implication that the laminar sublayer is a region free of turbulent eddies is not correct, since these eddies gradually become zero as the wall is approached [28].

Deissler [10] suggested an equation for a universal profile which is more realistic from a physical standpoint, and which uses a single equation from the wall to $y^+ = 26$. He also suggested an equation for the turbulent core. His equations and some experimental data (Fig. 9-8) show excellent agreement.

The universal velocity profiles that have been discussed are valid only near the wall and lead to an unrealistic situation at the centerline of the tube, namely, a nonzero velocity gradient. They are quite useful, however, since both pressure drop and heat transfer are closely tied to the happenings near the wall. In many computations, however, it is desirable to have an expression for the velocity profile that is valid over most of the cross section. This would be true, for example, in the determination of bulk temperatures by

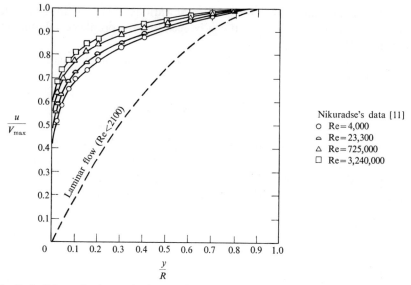

Fig. 9-9 Dimensionless velocity profiles at various values of Reynolds number. (From *Fluid Dynamics and Heat Transfer* by Knudsen and Katz. Copyright 1958 by McGraw-Hill Book Company, Inc. Used with permission of McGraw-Hill Book Company.)

the use of Eq. (4–56e). Experimental data of Nikuradse [11] suggests the following velocity profile for turbulent flow:

$$\frac{u_z}{V_{\max}} = \left(\frac{y}{R}\right)^{1/n}, \qquad (9\text{--}45)$$

where the n in the exponent varies with Reynolds number as shown in Table 9–2.

Equation (9–45) appears to fit most experimental data well. It does not predict the correct velocity gradient at the wall, however, but rather predicts an infinite value.

Table 9-2

$\mathrm{Re}_D = \dfrac{V_{\mathrm{avg}} D}{\nu}$	n
4,000	6
23,000	6.6
110,000	7
1,100,000	8.8
2,000,000	10
3,240,000	10

The dimensionless velocity profiles at various values of the Reynolds number are shown in Fig. 9–9. It can be seen that the profile gets flatter with increasing Reynolds number.

The profiles shown in Fig. 9–9 can be plotted on a single curve by use of an equation suggested by Prandtl [12]:

$$\frac{V_{max} - u_z}{u_f} = -5.75 \log_{10} \frac{y}{R}. \tag{9-46}$$

Equation (9–46) is a universal velocity distribution function for turbulent pipe flow and is sometimes called the *velocity defect law*. It does not predict a proper value of the velocity gradient at the wall or at the centerline but does accurately predict the average velocity across a tube cross section.

Heat Transfer

In most heat transfer applications, the heating of the fluid begins at the tube entrance and the temperature and velocity profiles develop simultaneously. As a rule of thumb, in turbulent conduit flow, a distance of about 40 diameters from the entrance is required for the local heat transfer coefficient h to approach a constant value. The coefficient decreases very rapidly near the entrance and approaches the constant value asymptotically. The average heat transfer coefficient will not depend upon the length of the tube so long as the length is greater than 60 diameters.

For tubes of sufficient length, an expression for the heat transfer coefficient might be obtained by use of analogy, as was done for the case of the flat plate in turbulent flow. In fact, analogies were first formulated for pipe flow and later applied to external flows such as over the flat plate. Colburn [5] suggested an analogy for turbulent flow in round tubes. In terms of the Moody friction factor (Fig. 9–2), his analogy may be written

$$\overline{St} \; Pr^{2/3} = \frac{f}{8}, \tag{9-47}$$

where the average Stanton number is

$$\overline{St} = \frac{\bar{h}}{\rho V_{avg} c_p}, \tag{9-48}$$

and the friction factor f is defined by Eq. (7–15). An expression for the friction factor in turbulent tube flow is obtained from Eq. (9–40). The result is:

$$\overline{St} \; Pr^{2/3} = 0.023 \; Re_D^{-0.2},$$

or

$$\overline{St} = 0.023 \; Re_D^{-0.2} \; Pr^{-2/3}. \tag{9-49}$$

Equation (9–49) is usually referred to as the *Colburn equation*. Experimental data has confirmed the validity of Eq. (9–49) for the following range of conditions:

$$Re_D > 10{,}000 \quad 0.7 < Pr < 160 \quad L/D > 60.$$

In Eq. (9–49), the specific heat should be evaluated at the average bulk temperature and all other properties at the average film temperature. L/D is the ratio of total tube length to diameter.

Equation (9–49) can be rewritten multiplying both sides by the product of the Reynolds and Prandtl number:

$$\overline{St}\, Re_D\, Pr = \overline{Nu}_D = 0.023\, Re_D^{0.8}\, Pr^{1/3}. \tag{9-50}$$

Equation (9–50) is of the form

$$\overline{Nu} = f(Re_D, Pr).$$

Most of the early work involving turbulent heat transfer in tubes used this form of correlation of data, and its use is quite popular even today.

Probably the most widely used heat transfer correlation for flow through a tube with constant wall temperature is the modified Dittus-Boelter [13] equation:

$$\overline{Nu}_D = \frac{\bar{h}D}{k} = 0.023\, Re_D^{0.8}\, Pr^n, \quad \begin{array}{l} n = 0.4 \\ n = 0.3 \end{array} \quad \begin{array}{l} T_w > T_B \text{ (heating),} \\ T_w < T_B \text{ (cooling).} \end{array} \tag{9-51}$$

Equation (9–51) is applicable under the conditions that $Re_D > 10{,}000$, and $0.7 < Pr < 100$, and $L/D > 60$. All fluid properties in Eq. (9–51) should be evaluated at the arithmetic mean bulk temperature of the fluid. Since the average bulk temperature is unknown in many practical problems, any computation may involve a trial-and-error procedure. Note the similarity between Eqs. (9–51) and (9–50).

An extension of Eq. (9–51) for higher Prandtl numbers was proposed by Sieder and Tate [14]. Their extension involved the use of a viscosity ratio to allow for large variations of that property which occurs across the tube section for high Prandtl number fluids. The most widely accepted form of this equation is

$$\frac{\bar{h}D}{k} = 0.023\, Re_D^{0.8}\, Pr^{1/3} \left(\frac{\mu_B}{\mu_w}\right)^{0.14}. \tag{9-51a}$$

All properties except μ_w are evaluated at the average bulk temperature. This equation is useful for fluids where $0.7 < Pr < 16{,}700$. The restrictions that $Re_D > 10{,}000$ and $L/D > 60$ remain.

Equations of the form of Eq. (9–49) are widely used because the Stanton number can be obtained directly from experimental measurement. An energy balance on the fluid flowing through a round tube between stations 1 and 2 gives

heat added at wall = change of fluid enthalpy,

$$\bar{h}\pi DL(T_w - T_B)_{\text{avg}} = \rho V_{\text{avg}} \frac{\pi D^2}{4} c_p(T_{B_2} - T_{B_1}), \qquad (9\text{--}52)$$

or

$$\overline{\text{St}} = \frac{D}{4L} \frac{(T_{B_2} - T_{B_1})}{(T_w - T_B)_{\text{avg}}}. \qquad (9\text{--}52a)$$

No fluid properties need to be known to determine the Stanton number, according to Eq. (9–52a). In contrast, the expression for the Nusselt number from Eq. (9–52) is

$$\overline{\text{Nu}} = \frac{D^2}{4L} \frac{\rho V_{\text{avg}} c_p}{k} \frac{(T_{B_2} - T_{B_1})}{(T_w - T_B)_{\text{avg}}}. \qquad (9\text{--}52b)$$

It can be seen that experimental determination of the Nusselt number requires knowledge of the specific heat, the thermal conductivity, and the mass velocity of the fluid (ρV_{avg}).

With constant Prandtl number, the Colburn form, Eq. (9–49), can be plotted with a smaller range of ordinate than the Dittus-Boelter form, Eq. (9–51), since the Stanton number varies as $\text{Re}^{-0.2}$, rather than as $\text{Re}^{0.8}$ like the Nusselt number.

The equations in this section for Nusselt number and Stanton number are useful in approximate calculations for flow through noncircular conduits. In such cases the diameter is merely replaced by the hydraulic diameter D_H, defined by Eq. (9–43).

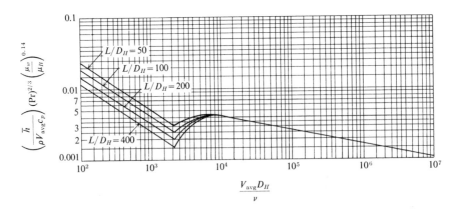

Fig. 9–10 Heat transfer correlation for transition region [14]. (Reprinted from *Industrial Engineering Chemistry*, Vol. **28**, 1936, Page 1434. Copyright 1936 by the American Chemical Society. Reprinted by permission.)

9-3 TURBULENT FLOW IN CONDUITS

Sieder and Tate [14] recommended a set of curves for prediction of the heat transfer coefficient in the transition region ($2100 < \text{Re}_D < 10{,}000$) for conduit flow. These are shown in Fig. 9–10. The heat transfer rates in this transition region fluctuate because of the unstable nature of the flow so that the assumption of a steady set of values for properties and parameters necessary to use Fig. 9–10 will yield a fixed value of this heat transfer coefficient. This does not allow for the reality of the unstable nature of the flow. Therefore, great faith cannot be placed in the use of Fig. 9–10 in the transition range.

Example 9–6. Determine the length of a 2-in. i.d. tube necessary for the bulk temperature of light oil to be raised from 90°F to 110°F if the tube wall temperature is constant and equal to 150°F. The oil flows through the tube with a mean velocity of 20 ft/sec.

Solution. Evaluating the fluid properties at the mean bulk temperature,

$$T_{B,M} = \left(\frac{90+110}{2}\right)°F = 100°F,$$

indicates that

$$\rho = 56.0 \text{ lb}_m/\text{ft}^3,$$
$$c_p = 0.46 \text{ Btu/lb}_m \, °F,$$
$$\nu = 0.982 \text{ ft}^2/\text{hr} = 27.3 \times 10^{-5} \text{ ft}^2/\text{sec},$$
$$k = 0.076 \text{ Btu/hr ft } °F,$$
$$\text{Pr} = \frac{\mu c_p}{k} = 330.$$

Hence, for this particular problem, the Reynolds number is

$$\text{Re}_D = \frac{V_{\text{avg}} D}{\nu} = \frac{(20 \text{ ft/sec})(2/12 \text{ ft})}{(27.3)(10^{-5}) \text{ ft}^2/\text{sec}}$$
$$= 1.22 \times 10^4.$$

The Reynolds number is sufficiently large for the flow to be assumed to be turbulent. The problem is now reduced to that of choosing an equation for the heat transfer coefficient for (a) turbulent pipe flow, (b) constant wall temperature, and (c) $\text{Pr} = 330$.

Only Eq. (9–51a) is valid for condition (c) and is, therefore, to be employed in the solution.

Observing that all properties except μ_w are to be evaluated at the average bulk temperature,

$$\frac{hD}{k} = 0.023 \, \text{Re}_D^{0.8} \, \text{Pr}^{1/3} \left(\frac{\mu_B}{\mu_w}\right)^{0.14}.$$

Substituting,

$$h = \frac{(0.076) \text{ Btu/hr ft °F}}{(2/12) \text{ ft}} (0.023)(12{,}200)^{0.8}(330)^{1/3} \left(\frac{55}{32}\right)^{0.14}$$

$$= 146 \text{ Btu/hr ft}^2 \text{ °F}.$$

Making an energy balance on the fluid, letting the heat added at the wall equal the change in enthalpy of the fluid,

$$\dot{q} = h\pi DL(T_w - T_{B,\text{avg}}) = \frac{\pi}{4} D^2 \rho V_{\text{avg}} c_p \Delta T_B,$$

so that

$$L = \frac{\rho V_{\text{avg}} D c_p \Delta T_B}{4(T_w - T_{B,\text{avg}})(\bar{h})}$$

$$= \frac{(56 \text{ lb}_m/\text{ft}^3)(20 \text{ ft/sec})(3600 \text{ sec/hr})(2/12 \text{ ft})(0.46 \text{ Btu/lb}_m \text{ °F})(20)}{(4)(50)(146 \text{ Btu/hr ft}^2 \text{ °F})}$$

$$= 213.0 \text{ ft.}$$

Notice that the L/D ratio is sufficient to satisfy the requirement specified for Eq. (9–52).

9–4 FLOW AROUND BODIES

Many engineering situations exist where a fluid must flow around a submerged body. In such situations, the flow is accelerated and/or decelerated in the region near the solid body. This was shown in Fig. 6–8 for a nonviscous fluid, where the streamlines near a cylinder are shown converging and then diverging. An inspection of the momentum equations for an ideal fluid, Eq. (6–1), indicates that any acceleration or deceleration of the fluid in a particular direction will cause a pressure gradient to exist in that direction. These accelerations and decelerations and the accompanying pressure gradients also exist in the flow of viscous fluids around bodies, although the behavior of viscous and nonviscous fluids is quite different for a given body.

The pressure gradients which exist in a viscous fluid flowing around a body have an important effect on the velocity profiles which exist in the boundary layer next to the body. The effect of pressure gradient on the velocity profile has already been seen for the case of laminar tube flow, Eq. (8–6).

The pressure gradient along an exterior surface affects the velocity profile in the boundary layer. This effect can be shown for the case of an airfoil by use of Fig. 9–11.

In the flow around an airfoil, such as that shown in Fig. 9–11, the coordinate system is defined so that the x direction is always parallel to the solid surface nearest that position and the y direction is normal to that surface. Thus, the coordinate system changes with position for a curved surface. It is not difficult to show that the equations derived for a flat surface, such as Eq. (7–9), are valid in this type of coordinate system, providing the radius of curvature of the surface is large [15].

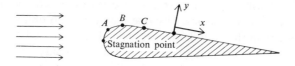

Fig. 9–11 Flow around a streamlined object.

Equation (7–9) may be written for the special case of $y = 0$, the condition at the wall. At $y = 0$, both the x and y velocity components u and v are zero and Eq. (7–9) becomes

$$\frac{dp}{dx} = \mu \left(\frac{\partial^2 u}{\partial y^2}\right)_{y=0}. \tag{9–53}$$

Equation (9–53) shows that the second derivative of velocity with respect to y evaluated at the wall has the same sign as the pressure gradient. That is, a positive pressure gradient dp/dx requires that $(\partial^2 u/\partial y^2)_{y=0}$ be positive, and a negative pressure gradient requires that the derivative be negative. This permits a qualitative description of the velocity profile in the boundary layer at various positions along the airfoil surface.

The pressure can be assumed to be constant across the boundary layer at any position x, Eq. (7–10). Therefore, the pressure on the airfoil surface in Fig. 9–11 is the same as the pressure at the outer edge of the boundary layer, in the stream. The stream velocity near the airfoil increases from the stagnation point to the point of maximum thickness, B. In this region the fluid decreases in pressure as it moves from the stagnation point to B. The pressure gradient dp/dx in this region is negative. At point B, the fluid velocity in the stream reaches a maximum and the pressure gradient is equal to zero. Downstream from point B, the free stream fluid velocity decreases and the pressure increases with increasing x. The pressure gradient dp/dx is then positive. The effect of the pressure gradient is shown in Fig. 9–12 where the velocity profiles at points A, B, and C are shown for comparison.

Curve B, where dp/dx is zero, is similar to the profile on a flat plate. The second derivative of velocity at the wall is zero in keeping with Eq. (9–53). Curve A, for negative dp/dx, is seen to have higher velocities near the wall than curve B, due to the "push" given by the pressure gradient. Curve C for the location with a positive pressure gradient has lower velocities near the wall and is seen to be quite different from curve B, notably in that it displays an inflection point or change in curvature from negative to positive. The shape of these curves can again be seen to agree with Eq. (9–53).

The three typical curves for the airfoil shown in Fig. 9–12 are not similar to each other, and therefore the similarity methods of Section 8–3 could not be expected to apply directly. Several schemes for solving flow problems of this type, where the boundary layer remains laminar, have been proposed [16, 17, 18] but will not be discussed here due to their complexity.

There are several reasons why one might wish to obtain solutions for the local velocity in the boundary layer around an immersed body such as the airfoil of Fig. 9–11. Two of these reasons have already been discussed for the flat plate case—the need to know the shear stresses acting on the surface, and the need for knowledge of local velocities in order to solve the energy equations. Another reason might be the requirement to predict the location where transition from a laminar to a turbulent boundary layer occurs. Methods of making such predictions will not be discussed but are described in some detail by Schlichting [15]. As with the flat plate, the transition is highly dependent on the intensity of the free stream turbulence.

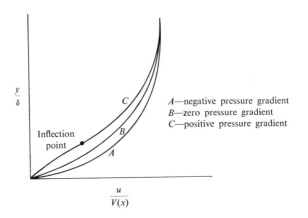

Fig. 9–12 Comparison of velocity profiles for positive, negative, and zero pressure gradients. A—negative; B—zero; C—positive.

Separation

A very important reason for wishing to know the velocity profile can be seen by referring to the profile C (Fig. 9–12) for the positive pressure gradient case. If the pressure gradient is severe enough, the velocity profile could conceivably exhibit a zero or even a negative first derivative at the wall, $(du/dy)_{y=0} < 0$. This is an indication of flow stoppage or flow reversal of the fluid in that region. In such cases, the slow moving fluid near the wall, coming into the region of increasing pressure, has insufficient momentum to overcome the adverse pressure forces. The fluid cannot flow back upstream because of the oncoming fluid and is hindered from flowing downstream

Fig. 9-13 Flow with boundary layer separation around a circular cylinder.

because of the pressure gradient; therefore, it flows away from the wall, or separates from the wall. This is called *flow separation*, or *boundary layer separation*. Downstream from the region of separation, a turbulent wake or region of disorderly flow exists, filling the void left by the separated fluid. Separation and the resulting wake can be seen in Fig. 9-13 for flow around a circular cylinder.

Typical velocity profiles along a surface where separation occurs are shown in Fig. 9-14.

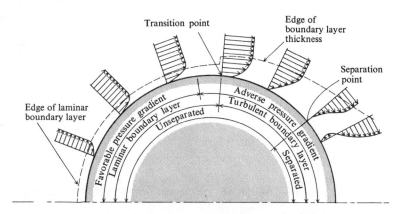

Fig. 9-14 Diagrammatic representation of boundary layer growth on a cylinder.

The *point* or *line of separation* is that point (or line) on the surface where the first derivative of the velocity with respect to y at the wall is zero:

$$\left(\frac{\partial u}{\partial y}\right)_{y=0} = 0 \quad \text{(point of separation)}. \tag{9-54}$$

At the separation point, the shear stress at the wall is zero, according to Stokes' law.

The profiles in Fig. 9-14 can be seen to change shape as the pressure gradient changes from negative to positive. At the position along the surface where the first derivative of velocity with respect to y taken at the wall is equal to zero, separation occurs; the flow leaves the body and is replaced downstream of that point by the backflow of the wake. It can be seen from the above discussion that separation occurs only in regions of increasing pressure. In flows around bodies, this is normally on the downstream side of the body. *Streamlining* consists of tapering the rear shape more slowly, reducing the magnitude of the pressure gradient in the flow direction, and preventing or retarding separation. Separation may occur on one side of a streamline body, such as the airfoil in Fig. 9-11, if the *angle of attack* is made sufficiently large. If separation occurs far enough forward of the trailing edge, it can result in reduction of lift (stalling).

Separation plays a significant role in the drag which exists on non-streamline bodies, such as a circular cylinder in flow perpendicular to its axis. The pressure distribution around such a cylinder was shown in Fig. 6-9. According to the theoretical curve, there should be no net pressure force on the cylinder for the case of an ideal fluid, or fluid with no viscosity. The two experimental curves, however, show a marked difference in the pressure distribution, in both cases giving a net pressure component in the direction of mainstream flow. In other words, the average pressure over the front half of the cylinder is greater than that over the rear half. This net pressure component which occurs in real fluids contributes to the total drag existing on the cylinder. The drag due to pressure (or *form drag*) plus the drag due to viscous shear (*friction* or *viscous drag*) make up the total drag on the cylinder. This was discussed previously and was shown in Fig. 5-5.

A curve of drag coefficients for the circular cylinder was presented in Fig. 5-7. The drag coefficient C_D, or dimensionless drag, was shown to be dependent only on the value of the Reynolds number for flow that is incompressible. Curves such as the one in Fig. 5-7 are obtained by use of experimental data. Similar curves for spheres and disks are shown in Fig. 9-15. Similar curves for other geometries may be found in the literature.

The drag coefficient curves which appear in Figs. 5-7 and 9-15 both have a sudden dip in the region of Reynolds number between 10^5 and 10^6. This sudden decrease in the drag coefficient in both cases is explained by the effect of boundary layer transition on separation. The experimental pressure

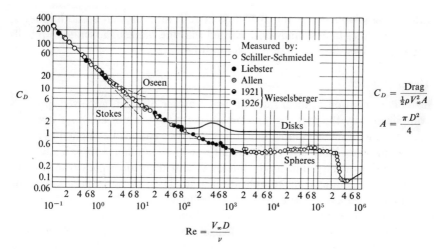

Fig. 9-15 Drag coefficient for spheres and disks as a function of the Reynolds number [15]. (From *Boundary Layer Theory* by Schlichting. Copyright 1968, 1960, 1955 by McGraw-Hill Book Company, Inc. Used by permission of McGraw-Hill Book Company.)

distributions around a cylinder at a value of Reynolds number above and at a Reynolds number below the dip in Fig. 5-7 were given in Fig. 6-9. At a Reynolds number of 1.86×10^5 the boundary layer on the cylinder is laminar up to the point of separation (about 80° from the stagnation point). The wake region is quite large and a relatively large net pressure difference exists between the front and rear half of the cylinder, leading to a large value of drag coefficient C_D.

At the higher Reynolds number of 6.7×10^5, the boundary layer has undergone a transition to turbulence and the wake has shifted back to about 140° from the stagnation point. The smaller wake leads to a reduced value of drag coefficient due to the decreased form drag. The shift in separation point caused by transition to turbulence is explained by the fact that the turbulent boundary layer has higher velocities near the wall than the laminar boundary layer, and, therefore, has greater ability to overcome the adverse pressure gradient which leads to separation. The value of Reynolds number at which the sudden decrease in C_D occurs is dependent upon the intensity of turbulence of the free stream. This characteristic is used as a means of determining the intensity of turbulence in wind tunnels by observing the value of the Reynolds number at which the drop in drag coefficient of a sphere occurs.

The transition to a turbulent boundary layer and the resulting decrease in C_D can also be brought about by the introduction of roughness on the solid surface. This leads to the anomaly that drag on an object can sometimes be reduced by making the object rougher. An example of the use of this anomaly is the manufactured roughness in the dimpled cover of golf balls. It is interesting to note that in the early days of golf, the balls were smooth spheres, but the caddies in Scotland soon learned that the old rough, scuffed-up balls could be driven farther than new smooth balls.

Heat Transfer

The determination of local rates of heat transfer over an immersed object is complicated by the fact that the pressure gradient is not constant over the surface, and a single, simple method of analysis cannot be applied over the entire surface. Normally the boundary layer is laminar approximately up to the point of maximum body thickness and then undergoes transition to turbulence, again causing complication in analysis. When separation of the boundary layer occurs, the methods of boundary layer theory can no longer be applied and empirical methods must often be employed.

Although any number of shapes of immersed bodies are of practical interest, the circular cylinder will be discussed as a typical shape since it is a simple geometry and the one which has been most widely investigated. Consider the flow of a fluid normal to the axis of a infinite circular cylinder as shown in Fig. 9–16.

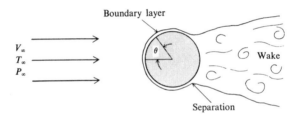

Fig. 9–16 Flow around a circular cylinder.

The fluid at a large distance from the cylinder has a static temperature of T_∞ and a static pressure of p_∞. The velocity of the fluid a large distance away and relative to the cylinder is V_∞. As in the case of the flat plate, a boundary layer forms near the cylinder as the fluid is slowed down by the effect of viscosity. With the cylinder at a temperature different from the free stream temperature, a thermal boundary layer will also be formed.

In the case of the circular cylinder the separation phenomenon will occur, except at very low values of the Reynolds number. In the region downstream from the place where the boundary layer separates from the wall, a turbulent wake region occurs.

The variation in the flow field around a cylinder with changing Reynolds number is indicated by the changes which occur in C_D (Fig. 5–7). It can be seen that no one simple equation could be expected to predict heat transfer over the surface for all arbitrary Reynolds numbers.

The effect of boundary layer growth, separation, and transition on the local Nusselt number on a cylinder in cross-flow can be seen in Fig. 9–17 from the measurements of Giedt [19] and Zapp [20].

The three bottom curves of Zapp [20] are for a laminar boundary layer with no transition to turbulence. The minimum in these three curves occurs at the point where the laminar boundary layer separates (approximately 80°). At this point the turbulent wake commences and the Nusselt number is seen to increase above the values for the forward portion of the cylinder.

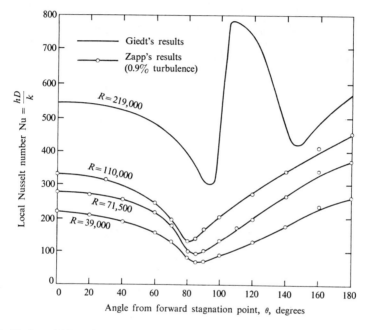

Fig. 9–17 Local Nusselt number on a circular cylinder [19]. [From W. H. Giedt, *Trans. ASME* **71**, 375 (1949). Reproduced by permission.]

The upper curve of Giedt [19] for $Re_D = 219{,}000$ is typical of behavior at higher Reynolds numbers $(V_\infty D)/\nu$ where the laminar boundary layer changes to a turbulent boundary layer (approximately 95°) and then separation of the turbulent boundary layer occurs (at approximately 140°). The sudden shift in the separation point from 80° back to 140° resulting from this transition causes the wake to exist over a much smaller area at the rear of the cylinder than at lower Reynolds numbers, leading to the decrease in C_D discussed previously.

An alternate way of presenting data for flow around a cylinder is by use of a polar plot such as the one shown in Fig. 9–18, presented by Schmidt and Wenner (23). A similar type of plot for flow at low Reynolds numbers is

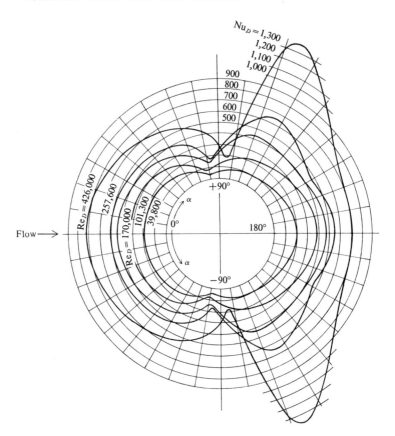

Fig. 9-18 Local heat transfer coefficients around a cylinder in flow of air normal to its axis [23]. [From E. M. Schmidt and K. Wenner, *Forsch. Gebiete Ing.* **12**, 65 (1933).]

given in Fig. 9-19 [22]. The large variation in local heat transfer coefficient with Reynolds number is apparent in these figures. It can be seen that at low values of Reynolds number, the local Nusselt number (and therefore the local heat transfer coefficient) is largest on the upstream side of the cylinder. In contrast, at higher values of Reynolds number the local Nusselt number is highest on the downstream side. At the highest values of Reynolds number, the Nusselt number is a maximum just downstream from the point of transition to turbulence.

In order to obtain average values of the heat transfer coefficient for the entire cylinder, it would be necessary to average the values from an appropriate curve shown in Fig. 9-17. As has been stated previously, there is no simple analytical solution to this problem. However, a correlation of the same form as predicted for other heat transfer problems (with negligible Eckert and Grashof numbers) appears to be useful. That is, one would expect

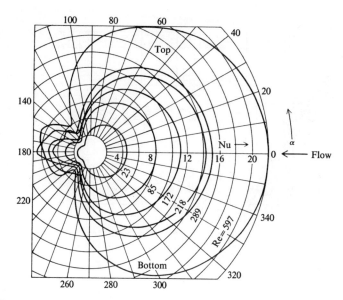

Fig. 9-19 Local heat transfer coefficients on a cylinder in cross-flow at low Reynolds number[22]. (Used with permission of the American Society of Mechanical Engineers.)

the average Nusselt number to be some function of the Prandtl number and the Reynolds number based on the diameter of the cylinder.

An experimental plot of data for air in cross-flow (flow normal to the axis) over a cylinder is given in Fig. 9-20, with the average Nusselt number plotted against the Reynolds number based on diameter. The Prandtl number is assumed to be constant for all conditions. It can be seen that a single curve correlates the data well. However, since the curve is not a straight line, it cannot be approximated by a single simple equation. It is convenient for computational purposes to approximate the curve in Fig. 9-20 by several straight-line segments, each of which is described by the equation

$$\overline{\mathrm{Nu}_D} = C_1 \, \mathrm{Re}_D^n. \tag{9-55}$$

The values of the constant C_1 and the exponent n in Eq. (9-55) are given in Table 9-3 below for various ranges of the Reynolds number. All fluid properties should be evaluated at the mean film temperature $T_f = \frac{1}{2}(T_\infty + T_w)$. The pressure is assumed to be approximately one atmosphere.

Equation (9-55) for a constant value of Pr = 0.7 can be empirically modified for use with fluids having other values of Prandtl number. The result is

$$\overline{\mathrm{Nu}_D} = C_2 \, \mathrm{Re}_D^n \, \mathrm{Pr}^{1/3}. \tag{9-56}$$

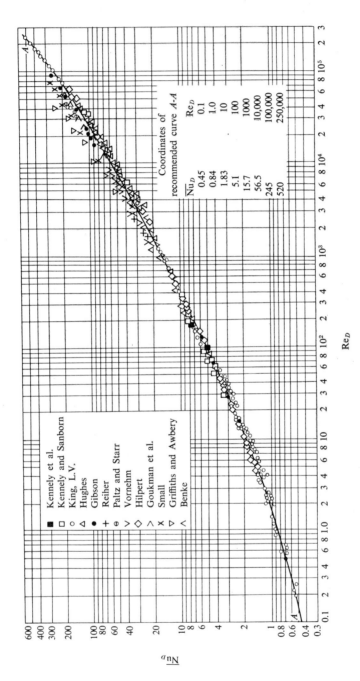

Fig. 9-20 Data for heating and cooling air flowing normal to single cylinder, corrected for radiation to surroundings [24]. (From *Heat Transmission* by McAdams. Copyright 1954 by William H. McAdams. Used by permission of McGraw-Hill Book Company.)

Table 9-3
COEFFICIENTS FOR EQUATIONS
(9-55) AND (9-56)

$Re_D = \dfrac{V_\infty D}{\nu}$	C_1	C_2	n
1–4	0.891	0.989	0.330
4–40	0.821	0.911	0.385
40–4000	0.615	0.683	0.466
4000–40,000	0.174	0.193	0.618
40,000–250,000	0.0239	0.0266	0.805

The value of the exponent n is unchanged and both n and C_2 can be obtained from Table 9-3. Equation (9-56) is useful for most liquids except liquid metals and highly viscous oils.

The values given in Fig. 9-20 and Table 9-3 are for a turbulence intensity of approximately 1 to 2 percent. Higher intensities of turbulence will give values of n larger than predicted above. For the case of the tube banks where the wake from one cylinder flows across the next cylinder downstream, corrections can be made by the use of empirical equations, which will be discussed in a later section.

Example 9-7. Determine the heat transferred per foot of length from a 1-in. o.d. (outside diameter) cylinder with a surface temperature of 212°F to air flowing normal to the cylinder at 50 ft/sec and 188°F.

Solution. The properties ν and k should be evaluated at the film temperature $T_f = 200°F$ (1 atmosphere assumed for pressure):

$$\nu = 0.239 \times 10^{-3} \text{ ft}^2/\text{sec}, \quad k = 0.0174 \text{ Btu/hr ft °F}.$$

$$Re_D = \frac{V_\infty D}{\nu} = \frac{(50)(1)}{(12)(2.39 \times 10^{-4})} = 17{,}400,$$

$$C_1 = 0.174, \quad n = 0.618.$$

$$\bar{h} = \frac{(0.174)k}{D} Re_D^{0.618} = \frac{(0.174)(0.0174)(12)(415)}{1} = 15.1 \frac{\text{Btu}}{\text{hr ft}^2 \text{ °F}}.$$

$$\frac{\dot{q}}{A} = \bar{h}(T_w - T_\infty),$$

$$\dot{q} = (15.1)\frac{\pi(1)(1)(212 - 188)}{12} = 95 \text{ Btu/hr}.$$

Spheres

For flow of gases over a sphere the following equation has been recommended for $25 < \text{Re}_D < 100{,}000$:

$$\overline{\text{Nu}_D} = 0.37 \, \text{Re}_D^{0.6}.$$

Properties should be evaluated at the film temperature.

9–5 FLOW ACROSS TUBE BANKS

The equations described in Section 9–4 for flow around a cylinder assume that the cylinder is immersed in an infinite fluid; that is, the cylinder is a large distance away from any other solid object. In such cases the flow field and the heat transfer are not affected by the presence of other surfaces. In most heat transfer equipment, the heat exchange takes place between one fluid which flows inside the tubes and another fluid which flows around the tubes. Often the tubes are arranged fairly close together and there may be a large number of tubes in a matrix. Such a set of tubes is referred to as a *tube bundle* or *tube bank*. A picture of a typical tube bank is shown in Fig. 9–21. In common situations where the fluid flow outside the tubes is normal or nearly normal to the axes of the tubes, most of the tubes are totally or partially in the wake of the tubes upstream. In such situations the equations of Section 9–4 could not be expected to give reliable predictions. This section will present methods for analysis of fluid flow and heat transfer in tube banks.

In the flow around a single cylinder in an infinite fluid, the pressure drop of the stream is of little concern since it is negligibly small between two points, one upstream and one downstream of the cylinder. In tube banks,

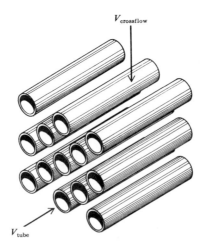

Fig. 9–21 A typical bank of tubes in cross-flow.

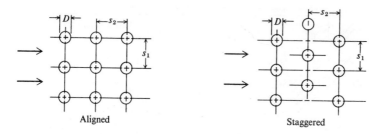

Fig. 9-22 Arrangement of tubes in a bank.

however, where the flow is usually contained within a vessel, the pressure drop experienced by the fluid in passing through the tube bank is important. This pressure drop determines the work required to move a given quantity of fluid through the tube bank and thus contributes to the economic considerations in the design of the equipment.

Tube banks may be arranged with respect to the external fluid flow in either an aligned or a staggered arrangement (Fig. 9-22).

Charts for the prediction of pressure drop through tube banks for both staggered and aligned tubes are common in the literature. These charts usually give values of the friction factor for tube banks, f_{TB}, as a function of Reynolds number for various tube spacings and arrangements. The friction factor f_{TB} may be defined in any one of several ways. In this section, the following definition will be used:

$$f_{TB} = \frac{(\Delta p)\rho}{2nG_{\max}^2} \qquad (9\text{-}57)$$

where n is the number of rows of tubes in the flow direction and $G_{\max}$ is the mass flow velocity (i.e., mass per unit time and area) between the tubes. Values of f_{TB} for aligned tubes are shown in Fig. 9-23 from Giedt [19].

The dip in the curves for f_{TB} in the transition zone for in-line (aligned) tubes does not occur for the staggered tubes. This is thought to be due to the manner in which transition occurs in staggered tubes, turbulence beginning at the exit of the tube bank and gradually moving upstream as the Reynolds number is increased. In aligned tubes, transition appears to occur simultaneously throughout the tube bank, as it does in a long pipe or tube. This transition from laminar to turbulent flow in aligned tubes at some definite, critical Reynolds number causes the jump in curves 3 and 5 of Fig. 9-23.

The friction factors given in Fig. 9-23 are for banks having more than 10 tube rows in the direction of the flow. For banks having less than 10 rows, the values of pressure drop given by Fig. 9-23 will be low. In any case, it should be remembered that the pressure drop obtained from Fig. 9-23 does not include effects that might be caused by entrance or exit effects or the presence of baffles and other obstructions in the tube channels.

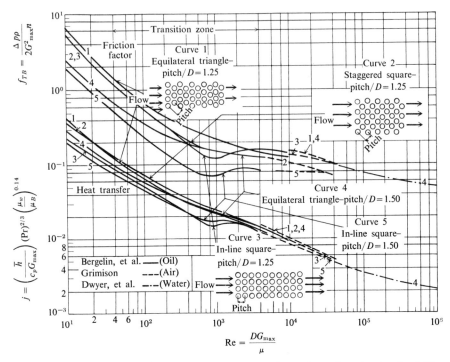

Fig. 9-23 Heat transfer and pressure drop for flow through tube banks [19]. (From Giedt's *Principles of Engineering Heat Transfer*. Copyright 1957, D. Van Nostrand Company, Inc., Princeton, New Jersey.)

Values of heat transfer coefficient can also be obtained from Fig. 9-23. Note again that a dip occurs in the curves for aligned tubes in the transition region. For values of Reynolds number above about 5000, the curves for all of the tube banks investigated appear to be very close to each other and a single predictive equation is useful, where high accuracy is not required. It may conveniently be expressed in terms of the Nusselt number:

$$\overline{\mathrm{Nu}_D} = \frac{\bar{h}D}{k} = 0.33 \left(\frac{G_{\max}D}{\mu}\right)^{0.6} \mathrm{Pr}^{0.3}. \tag{9-58}$$

In using Fig. 9-23 and Eq. (9-58), the fluid properties (except in the viscosity ratio term) should be evaluated at the mean film temperature. Note the similarity between Eq. (9-58) and Eq. (9-56), which was for flow over a single tube. Grimison [25] suggested a method of utilizing an equation of the form of Eq. (9-55) for airflow over tube banks by using values of C_1 and n which depend upon the tube arrangement and spacing. His suggested values for a Reynolds number range of 2000-40,000 are given in Table 9-4. For liquids, Eq. (9-56) may be used where C_2 is obtained by multiplying C_1 by 1.11.

Table 9-4

COEFFICIENTS FOR PREDICTION OF HEAT TRANSFER COEFFICIENTS IN TUBE BANKS 10 OR MORE ROWS DEEP
$Re_D = 2000\text{--}40,000$.

$a = S_1/D$, and $b = S_2/D$ in Fig. 9-22.

		$a = 1.25$		$a = 1.5$		$a = 2$		$a = 3$	
	b	C_1	n	C_1	n	C_1	n	C_1	n
Aligned	1.25	0.348	0.592	0.275	0.608	0.100	0.704	0.0633	0.752
tubes	1.5	0.367	0.586	0.250	0.620	0.101	0.702	0.0678	0.744
	2	0.418	0.570	0.299	0.602	0.229	0.632	0.198	0.648
	3	0.290	0.601	0.357	0.584	0.374	0.581	0.286	0.608
Staggered	0.6							0.213	0.636
tubes	0.9					0.446	0.571	0.401	0.581
	1			0.497	0.558				
	1.125					0.478	0.565	0.518	0.560
	1.25	0.518	0.556	0.505	0.554	0.519	0.556	0.522	0.562
	1.5	0.451	0.568	0.460	0.562	0.452	0.568	0.488	0.568
	2	0.404	0.572	0.416	0.568	0.482	0.556	0.449	0.570
	3	0.310	0.592	0.356	0.580	0.440	0.562	0.421	0.574

Table 9-4 (as well as Fig. 9-23) is valid only if there are more than 10 tubes in the direction of flow. For less than 10 tubes, Table 9-5 may be used to correct values of $\bar{h}$ obtained by use of Fig. 9-23, or Eqs. (9-55) and (9-56) with Table 9-4.

Table 9-5

RATIO OF $\bar{h}$ FOR n ROWS DEEP TO $\bar{h}$ FOR 10 ROWS DEEP*

$n =$	1	2	3	4	5	6	7	8	9	10
Staggered tubes	0.68	0.75	0.83	0.89	0.92	0.95	0.97	0.98	0.99	1.0
In-line tubes	0.64	0.80	0.87	0.90	0.92	0.94	0.96	0.98	0.99	1.0

* From W. H. McAdams, *Heat Transmission*, 3rd ed., McGraw-Hill, New York, 1954. Reproduced by permission of the publisher.

REFERENCES

1. O. Reynolds, *Trans. Roy. Soc.* (London) **174A**, 935 (1883).
2. L. Prandtl, *Physik, Z.* **11**, 1072 (1910).
3. T. von Kármán, *Trans. ASME* **61**, 705 (1939).
4. R. C. Martinelli, *Trans. ASME* **69**, 947 (1947).
5. A. P. Colburn, *Trans. AIChE* **29**, 174 (1933).
6. H. Latzko, *Z. angew. Math. Mech.*, **1**, 268 (1921), cited in R. G. Deissler, *Trans. ASME* **77**, 1221 (1955).
7. J. Nikuradse, "Strömungsgesetze in rauhen Rohren," VDI-Forschungsheft 361, Berlin (1933).
8. J. B. Drew, E. C. Koo, and W. H. McAdams, *Trans. AIChE* **28**, 56 (1932).
9. T. von Kármán, *J. Aeronautical Sci.* **1**, 1 (1934).
10. R. G. Deissler, *Trans. ASME* **73**, 101 (1951).
11. J. Nikuradse, "Gesetzmässigkeiten der turbulenten Strömung in glatten Rohren," VDI-Forschungsheft 356, Berlin (1932).
12. L. Prandtl, "Über die ausgebildete Turbulenz," ZAMM **5**, 136 (1925), and *Proc. 2nd Inter. Congress Applied Mech.*, Zurich (1926).
13. F. W. Dittus, and L. M. K. Boelter, University of California at Berkeley, *Publs. Eng.* **2**, 443 (1930).
14. E. N. Sieder and G. E. Tate, *Indus. Eng. Chem.* **28**, 1429 (1936).
15. H. Schlichting, *Boundary Layer Theory*, 4th ed., McGraw-Hill, New York (1960).
16. K. Pohlhausen, *Z. angew. Math. Mech.* **1**, 252 (1921); see also H. L. Dryden, NACA Report 497 (1934).
17. L. Haworth, ARC R&M 1632 (1935).
18. H. Geortler, *J. Math. Mech.* **6**, 1 (1957).
19. W. H. Giedt, *Trans. ASME* **71**, 375 (1949).
20. G. M. Zapp, M.S. Thesis, Oregon State University (1950).
21. R. Hilpert, *Forsch. Gebiete Ing.* **4**, 215 (1933).
22. E. R. G. Eckert and E. Soehngen, *Trans. ASME* **74**, 343 (1952).
23. E. N. Schmidt and K. Wenner, *Forsch. Gebiete Ing.* **12**, 65 (1933).
24. W. H. McAdams, *Heat Transmission*, 3d ed. McGraw-Hill, New York (1954).
25. E. D. Grimison, *Trans. ASME* **59**, 583 (1937).
26. O. P. Bergelin, G. A. Brown and S. C. Doberstein, *Trans. ASME* **74**, 953 (1952).
27. O. E. Dwyer, T. V. Sheean, J. Weisman, F. L. Horn, and R. T. Schomer, *Indus. Eng. Chem.* **48**, 1836 (1956).
28. S. J. Kline et al. "The Structure of Turbulent Boundary Layers," *J. Fluid Mech.* **30**, 741 (1967).

29. H. Rouse, *Fluid Mechanics for Hydraulic Engineers*, Dover, New York (1961), p. 215.
30. G. W. Maurer and B. W. Le Tourneau, "Friction Factors for Fully Developed Turbulent Flow in Ducts with and without Heat Transfer," *Trans. ASME* **86**, 627 (1964).
31. R. Gilmont, "A Falling-Ball Viscometer," *Instrument and Control Systems* **36** (9), 121 (1963).
32. A. H. Shapiro, "The Fluid Dynamics of Drag," National Committee for Fluid Mechanics Films.
33. L. F. Moody, *Trans. ASME* **66**, 671 (1944).

PROBLEMS

1. Derive Eq. (9–8) from Eq. (9–2a).
2. Derive Eq. (9–9) from Eq. (4–44b).
3. Derive Eq. (9–10) from Eq. (4–44c).
4. Derive Eq. (9–18) from the energy equation given in Eq. (4–65). Assume no viscous dissipation and no radiant heat transfer.
5. Derive Eq. (9–17) from Eq. (9–16) and the instantaneous conservation of mass equation.
6. If the z component of average velocity were zero (two-dimensional turbulent flow) would the fluctuating component w' necessarily be zero? Discuss.
7. Using Eqs. (9–33), (9–34), and (9–35) derive Eq. (9–36).
8. Derive an equation similar to Eq. (9–36) but assume that the critical Reynolds number for transition to turbulence is (a) 320,000, (b) 10^6.
9. Estimate the total rate of heat loss from one side of a flat plate placed with surface parallel to the fluid flowing past it. The plate is 4 ft wide by 5 ft in the flow direction. The fluid is light oil at a temperature of 130°F, flowing at a velocity of 30 ft/sec. The plate temperature is uniform and equal to 170°F.
10. Calculate the fraction of the surface area of the plate covered by a laminar boundary layer in the example problem of Section 9–2. Calculate the fraction of the total heat loss transferred through the laminar boundary layer.
11. What error would have been made in the example of Section 9–2 if it had been assumed that the turbulent boundary layer began at the leading edge of the plate ($x = 0$)? Would such an assumption be good for approximation purposes under any set of conditions?
12. A 10-psi pressure difference is imposed across a 400-ft length of 1-in. i.d. commercial steel pipe through which water at 60°F is flowing. Estimate the flow rate of the water in gal/min.
13. Estimate the pressure drop for saturated water at 100°F flowing at 40 gal/min through a 2-in. i.d. steel pipe system. The total system contains 600 ft of pipe, 3 standard elbows, 2 open globe valves, and turns through one standard tee.

14. Compare the pressure drop per 100 ft of pipe for water 60°F, flowing at 2000 gal/min through 10-in. i.d. pipe for (a) concrete pipe, $e = 0.01$ feet, and (b) commercial steel pipe.

15. Water at an average bulk temperature of 225°F flows through a rectangular channel 2-in. by 3 in. at a rate of 400 gal/min. Assume that the flow is isothermal and the walls of the channel are very smooth. Estimate the pressure drop (in psi) that will occur over 50 ft of the channel. Neglect entrance or exit effects.

16. Assuming isothermal flow of water at a temperature of 200°F at the rate of 2.0 gal/min through a 0.05-ft i.d. smooth pipe, estimate the pressure drop over a length of 400 ft.

17. Calculate the rate at which benzene (average temperature 100°F) will flow through a 0.1-in. i.d. tube if the pressure drop between two points 30 in. apart is 0.1 psi. Check the Reynolds number based on diameter (Re_D) to verify that the flow is laminar or turbulent. $\rho = 54$ lb_m/ft^3, $\nu = 0.65 \times 10^{-5}$ ft^2/sec.

18. Water at 50°F flows through 500 feet of straight 4-in. i.d. cast-iron pipe at the rate of 5000 lb_m/hr. Determine the frictional pressure drop, neglecting entrance or exit losses.

19. Using the loss coefficient, estimate the rate at which 60°F water is flowing through a half-opened 4-in. gate valve if the measured pressure drop across the valve is 10 psi. Express your answer in gal/min.

20. Estimate the rate of flow, in liters per hour, of 20°C water through 40 meters of 5-cm i.d. smooth tube if the pressure drop over the tube length is 80 mm of mercury.

21. Calculate the hydraulic diameter, in inches, of the following cross section:
    a) An equilateral triangle 4 in. on a side.
    b) An ellipse (major axis 6 in., minor axis 3 in.).
    c) An isosceles triangle (base 8 in., height 12 in.).
    d) A rectangle, 8 by 4 in.

22. Compare Table 9–1 and Fig. 9–14 for the case of a $\frac{1}{4}$-closed gate valve. Use water at 50°F, assume a 2-in. i.d. pipe, and an average velocity of 12 ft/sec. Select a different diameter of the pipe and a different velocity and compare the results with the above.

23. Water (50°F) flows at the rate of 1000 gal/min through an ordinary entrance into a 300-ft length of 6 in. i.d. commercial steel pipe. In the pipe run are three standard elbows, one open angle valve, and one turn through the side outlet of a standard tee. Estimate the pressure drop.

24. Water at 70°F is being pumped from the lower to the higher reservoir (see figure) through a 1-in. i.d. commercial steel pipe. The pump is rated at 1 hp and has an efficiency of 80 percent. Find the volumetric flow rate.

25. Crude oil (specific gravity $= 0.855$ and $\nu = 0.0001$ ft^2/sec) flows through a 6-in. i.d. cast-iron pipe at the rate of 500 gal/min. Calculate the head loss ($h_L = l_f/g$) per 1000 feet of pipe line.

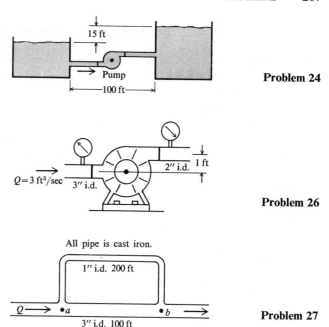

Problem 24

Problem 26

Problem 27

26. How much usable horsepower does the pump in the illustration deliver to water flowing through it at the rate of 3.0 ft³/sec if the pressure on the suction side is 6-in. of mercury vacuum, and on the discharge side 20 in. of mercury positive?

27. Determine the flow rate of water at 70°F if the pressure at point a in the figure is 80 psig and at point b, 30 psig. Neglect the minor losses due to fittings, etc. All pipe is cast iron.

28. If the flow rate in Example 9–5 had been 75 ft³/sec, what would be the pressure at point 2?

29. In Problem 27, if the flow rate were 5 ft³/sec, what would be the value of the pressure drop $p_a - p_b$?

30. Solve Example 9–4 by digital computer, assuming the pipe is 1-in. i.d. commercial steel and the reservoir height above the pipe discharge is 20 ft. Curve-fit the appropriate friction factor curve of Fig. 7–13 by several straight lines.

31. Solve Example 9–5 by digital computer, assuming that pipes a, b, and c in Fig. 9–5 are 1, 2, and 3 in. i.d. respectively, and assuming $Q = 0.1$ ft³/sec. Curve-fit each of the three appropriate friction factor curves of Fig. 9–2 by several straight lines.

32. Water (75°F) flows through a smooth 2-in. i.d. tube at the rate of 8000 gal/hr. Determine the thickness of the so-called laminar sublayer (in inches) assuming fully developed turbulent pipe flow.

268   TURBULENT FLOW AND HEAT TRANSFER

33. Estimate the distance from the wall to the edge of the turbulent core for Problem 32 above.
34. Estimate the number of diameters from inlet required for the friction factor to become constant for tube flow at Reynolds numbers based on diameters of 10^4, 10^5, 10^6, and 10^7.
35. Determine the friction factor f for flow of 100°F water through a smooth 3-in. pipe at 20 ft/sec, using Eq. (9–38). Compare with the value of f read from Fig. 9–2.
36. Determine the percent error in calculating f by Eq. (9–40) at a Reynolds number $Re_D = 3 \times 10^6$, assuming that Eq. (9–39) gives a "correct" value of f.
37. Plot Eq. (9–41) on Fig. 9–2.
38. Using Eq. (9–45), derive an expression for the ratio of the average to the maximum velocity in turbulent pipe flow in terms of the exponent n only.
39. Water at 50°F flows with an average velocity of 12 ft/sec through a smooth 2-in. i.d. pipe. What is the maximum velocity and what is the velocity $\frac{1}{8}$ in. from the wall?
40. Air at 100°F flows through a smooth 4-in. i.d. round pipe at an average velocity of 60 ft/sec. Determine the ratio of average to maximum velocity.
41. Calculate velocities at 10 evenly spaced points between the center of the tube and the wall for the flow conditions described in Problems 39 and 40 above. Plot both sets of points on a single graph using the coordinates suggested by Eq. (9–46) and compare with a plot of that equation.
42. Ammonia (liquid) flows through a 0.05-ft diameter round tube at the rate of 0.5 lb_m/sec. The inside wall temperature of the tube is maintained at 40°F. The average bulk temperature of ammonia is 80°F. Estimate the heat transfer from the ammonia per foot of tube length.
43. Water flows through a 0.1-ft diameter tube at the rate of 200 lb/min. It is desired to change the temperature of the fluid by heating. The wall is maintained at a constant temperature of 150°F. The inlet bulk temperature of the water is 90°F. Determine the length of pipe necessary to heat the water 20°F.
44. Water flows through a 2-in. diameter tube at the rate of 500.0 gal/hr. Assuming that the average bulk temperature is 100°F and the wall temperature is 180°F, estimate the change of bulk temperature which would occur in 20 ft of pipe.
45. Calculate the length of $\frac{1}{2}$-in. i.d. tube required to change the bulk temperature of air from 80°F to 84°F. The tube wall temperature is maintained at a constant value of 118°F. The air flows at a rate of 0.03 lb_m/sec.
46. Water at an average temperature of 150°F flows through a 1-in. i.d. tube at a flow rate of 12 gal/min. Estimate the change in bulk temperature which the water will undergo in each foot of pipe length if the average wall temperature of the tube is 200°F.
47. Water at an average temperature of 125°F flows through a 4-in. by 8-in. smooth rectangular duct at a rate of 1000 gal/min. The walls are maintained at an

average temperature of 175°F. Determine values of the friction factor and the average heat transfer coefficient.

48. Estimate the density of a 6-in. diameter sphere which is observed to fall at a constant velocity of 4 ft/sec in 75°F water.

49. Determine the terminal velocity of rise of a 1-in. o.d. (outside diameter) Ping-Pong ball in water at 75°F, assuming that $g = 32.174$ ft/sec^2 and neglecting the weight of the ball and the air inside compared to the weight of the water it displaces.

50. Carbon dioxide at 125°F and 1 atmosphere flows around a circular tube having an o.d. of 2 in. The relative velocity of the cylinder and the free stream is 60 ft/sec. If the outside surface temperature of the cylinder is 75°F, estimate the length of tube required to exchange 500 Btu/hr.

51. Air at 14.7 psia and 200°F flows around a 1.5-ft diameter sphere with a free stream velocity of 8 mph. Determine the drag on the sphere.

52. Show that an equation suggested by Eckert for flow around a cylinder,

$$\overline{\mathrm{Nu}_D} = 0.43 + 0.48\ \mathrm{Re}_D^{0.5},$$

for the Reynolds number range $1 < \mathrm{Re}_D < 4{,}000$, agrees with Eq. (9–55) and Table 9–2 within 5 percent.

53. Air at 1 atmosphere of pressure and 150°F flows normally past a circular cylinder which has an outside diameter of 2 in. The relative velocity of the cylinder and the free stream is 150 ft/sec. If the surface of the cylinder is maintained at 250°F, estimate the heat loss from the cylinder per foot of length.

54. Determine the drag exerted on the cylinder in Problem 53.

55. Water at 60°F flows past a 1-in. o.d. cylinder at a relative velocity of 20 ft/sec. The surface temperature of the cylinder is constant and equal to 140°F. Estimate the average heat transfer coefficient h and the drag coefficient C_D.

56. For air flowing past a cylinder at a Reynolds number of 39,000, compare the average Nusselt number obtained by integrating the curve in Figure 9–17 with the value calculated by Eq. (9–55).

57. Estimate the rate of heat loss from a 1.0-in. diameter sphere which moves through 80°F still air at a velocity of 100 mph. The surface temperature of the sphere is 120°F.

58. Use Eq. (9–55) to derive an expression for the free stream velocity as a function of the heat loss per unit surface area of the cylinder and the temperature difference between the cylinder and the free stream. Describe how this might be used as a velocity measuring device (hot-wire anemometer).

59. Assume that air with a mean bulk temperature of 50°F and a pressure of 1 atmosphere flows at the rate of 1500 lb$_m$/hr through a tube bank. The tubes are 0.5-in. o.d. spaced square, in line with a pitch of 0.625. The tube bank consists of 10 rows of tubes with 20 tubes in each row. The tubes are 6 ft long and have an outside surface temperature of 212°F. Calculate the heat transferred to the air and the pressure drop through the bank. Use Fig. 9–23.

60. A 20 × 20 bank of tubes (20 rows deep and 20 rows high), 1-in. o.d., and 8 ft long are arranged in a staggered arrangement with the tubes in each row 1.5 in. between centers and the rows 2 in. between centers. Air at 200 psia and 75°F flows over the tubes at the rate of 25,000 lb_m/hr. The outside surface temperature of the tubes is 325°F. Estimate the change in bulk temperature of the air. Use Tables 9–4 and 9–5.

CHAPTER 10

FREE CONVECTION

The body forces created by density variations in a fluid have been discussed previously. Such body forces can cause a significant effect on the motion of a fluid and, therefore, affect the energy transport within the fluid. Whether this effect is significant or not depends upon the relative magnitude of the body force and the other forces acting upon the fluid.

From Table 5–3, it can be seen that the dimensionless ratio of the Grashof number to the square of the Reynolds number describes the relative magnitude of the buoyancy forces and the inertia forces. With this ratio, all heat transfer problems in fluids can be classified into one of three categories:

$$\frac{Gr}{Re^2} \ll 1 \quad \text{forced convection,}$$

$$\frac{Gr}{Re^2} \approx 1 \quad \text{mixed convection,}$$

$$\frac{Gr}{Re^2} \gg 1 \quad \text{free convection,}$$

where

$$Gr = \frac{g\beta(T_w - T_\infty)L^3}{\nu^2} \quad \text{and} \quad Re = \frac{V_\infty L}{\nu}.$$

In forced convection, such as was considered in Chapters 7, 8, and 9, it was assumed that the buoyancy forces were negligible and $Re^2 \gg Gr$. As a result, the Grashof number did not appear in the equations which were derived. In free convection the motion of the fluid is due entirely to buoyancy forces, usually confined to a thin layer near the heated or cooled surface. The surrounding stationary fluid exerts a viscous drag on this layer of moving fluid, and the resulting inertia forces in the fluid layer are usually small. In such cases $Re^2 \ll Gr$, and the Reynolds number will not appear in the equations which will be derived. Free convection is often referred to as *natural convection*.

Mixed convection problems often confront the engineer but will not be discussed except to point out that in such cases both the Grashof and Reynolds numbers will appear in any equations developed [12, 13, 14].

10-1 FREE CONVECTION ON VERTICAL SURFACES

An important, but simple problem in natural convection is the case of a heated or cooled vertical wall [37]. The case of a heated wall is shown in Fig. 10–1. The coordinate system is located with the origin at the lower end of the wall, with the x axis parallel and the y axis normal to the wall.

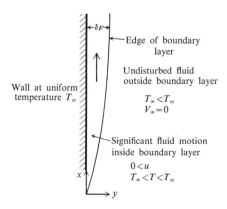

Fig. 10–1 The formation of a free convection boundary layer on a heated wall.

Heat transferred to the fluid from the wall creates density differences in the fluid. The body force due to gravity causes the lighter fluid near the wall to rise. The fluid beyond this less dense layer of fluid is assumed to be stationary. The region near the wall where there is a significant velocity will be considered as a *free convection boundary layer*. Boundary layer assumptions similar to those made for forced convection (Chapter 7) can be shown to give good approximations to the fluid behavior at high Grashof numbers ($\text{Gr} > 10^4$). The velocity profile in this boundary layer will be quite different from that in forced convection, however, since the surrounding fluid velocity in free convection is zero. The fluid velocity in free convection must, therefore, be zero at the solid surface, increase to a maximum value, and then decrease asymptotically toward zero at the outer edge of the boundary layer. It might be arbitrarily assumed that the edge of the boundary layer is that point where the velocity is less than 1 percent of the maximum value.

An alternate definition might be that point where the dimensionless temperature $(T - T_\infty)/(T_w - T_\infty)$ is less than some arbitrary value, such as 0.02 [5]. The flow field may then be divided into two regions: the boundary layer region and the undisturbed fluid region. The thickness of the boundary layer in free convection will be designated δ_F.

The continuity, momentum, and energy equations valid within the boundary layer can now be obtained for this case. The same type of reasoning

10-1   FREE CONVECTION ON VERTICAL SURFACES

used in obtaining the forced convection boundary layer equation is followed. If body forces are assumed to be significant only in the x direction, parallel to the wall, then the momentum equation for steady laminar boundary flow becomes

$$\rho\left(u\frac{\partial u}{\partial x} + v\frac{\partial u}{\partial y}\right) = -\rho g - \frac{dp}{dx} + \mu \frac{\partial^2 u}{\partial y^2}. \tag{10-1a}$$

The body force per unit volume due to the gravity force in the negative x direction is given by the term $-\rho g$. It is assumed that there are no significant y momentum changes, and therefore there is no pressure gradient normal to the wall. The pressure gradient dp/dx parallel to the wall is the static pressure variation due to the effect of gravity on the undisturbed fluid.

For the case of the vertical flat plate with no fluid motion outside the boundary layer, $V_\infty = 0$, Eq. (10–1a) can be evaluated at the edge of the boundary layer to give

$$\rho_\infty g = -\frac{dp}{dx}, \tag{10-1b}$$

where ρ_∞ is the density of the fluid outside the boundary layer and is a constant if the fluid in the free stream is considered incompressible. Equation (10–1b) is simply a statement that pressure variation in the x direction is caused only by the static elevation head. Since the pressure gradient is the same inside the boundary layer, Eq. (10–1b) may be substituted into Eq. (10–1a) to give

$$\rho\left(u\frac{\partial u}{\partial x} + v\frac{\partial u}{\partial y}\right) = g(\rho_\infty - \rho) + \mu\frac{\partial^2 u}{\partial y^2}, \tag{10-1c}$$

or

$$u\frac{\partial u}{\partial x} + v\frac{\partial u}{\partial y} = g\left(\frac{\rho_\infty}{\rho} - 1\right) + \nu\frac{\partial^2 u}{\partial y^2}. \tag{10-1d}$$

Assuming that the coefficient of thermal expansion is constant and the change in density is small,

$$\frac{\rho_\infty}{\rho} = 1 + \beta(T - T_\infty), \tag{10-1e}$$

and Eq. (10–1d) becomes

$$u\frac{\partial u}{\partial x} + v\frac{\partial u}{\partial y} = g\beta(T - T_\infty) + \nu\frac{\partial^2 u}{\partial y^2}. \tag{10-1f}$$

The continuity and energy equations are the same as the forced convection case for boundary layer flow over a flat plate. Thus, the following equations are valid within the free convection boundary layer for the laminar steady flow case on a vertical wall:

$$\frac{\partial u}{\partial x} + \frac{\partial v}{\partial y} = 0, \tag{10-2a}$$

$$u\frac{\partial u}{\partial x} + v\frac{\partial u}{\partial y} = \nu\frac{\partial^2 u}{\partial y^2} + g\beta(T - T_\infty). \tag{10-2b}$$

$$u\frac{\partial T}{\partial x} + v\frac{\partial T}{\partial y} = \alpha\frac{\partial^2 T}{\partial y^2}. \tag{10-2c}$$

The boundary conditions are:

$v = 0$ at $y = 0$
$u = 0$ at $y = 0$ and $y = \infty$
$T = T_w$ at $y = 0$
$T = T_\infty$ at $y = \infty$.

Since the dependence on time has been neglected, the equations can be used to describe steady flow only. Solutions to Eqs. (10–2a-c) can be used only for steady *laminar* free convection boundary layers. Experience has shown that a free convective boundary layer on a vertical wall will ordinarily remain laminar for values of the product of the Prandtl and Grashof numbers less than 10^9. This product (Pr Gr) is sometimes referred to in the literature as the *Rayleigh number*. At the point x, measured from the start of the boundary layer, where the value of Pr Gr becomes equal to 10^9, the transition from laminar to turbulent boundary layer will ordinarily occur:

$$\begin{aligned}(Pr\ Gr) &< 10^9 \quad \text{(laminar flow)}\\ (Pr\ Gr) &> 10^9 \quad \text{(turbulent flow)}\end{aligned} \tag{10-3}$$

Initially, only solutions to Eqs. (10–2a-c), valid for the laminar case (Pr Gr $< 10^9$) will be considered. Equations valid for the turbulent case will be presented later.

It should be noted that the momentum equation, Eq. (10–2b), involves the local temperature. This means that a solution to the momentum equation cannot be obtained independently of the energy equation as was done in the forced convection equations. Both the energy and the momentum equations must be solved simultaneously.

A method of solution of Eqs. (10–2a-c) was first presented by Pohlhausen from a study of the experimental data of Schmidt and Beckmann [1, 2]. This data is shown in Fig. 10–2. Because of the similarity of the velocity profiles and the similarity of the temperature profiles at various positions on the plate, Pohlhausen was able to obtain parameters which would convert the equations describing this case, Eq. (10–2), into a pair of ordinary differential

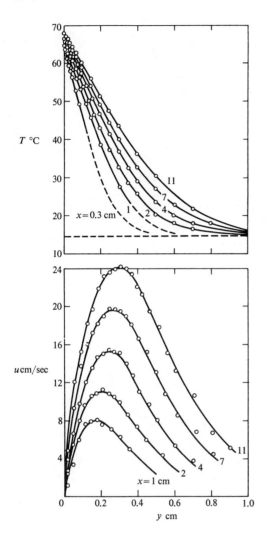

Fig. 10–2 Experimental free convection data of Schmidt and Beckmann [1, 2]. Vertical plate, $T_w = 66°C$ in air at 15°C. The distance x is distance from lower edge of plate. (From E. Schmidt, "Heat Transfer by Natural Convection," paper presented at 1961 International Heat Transfer Conference, Boulder, Colo.)

equations. The method is similar to that used in the Blasius solution except for the parameters. The stream function is again used to satisfy the continuity equation. The combined continuity-momentum equation and the energy equation, Eqs. (10–2a-c), become,

$$\frac{\partial \psi}{\partial y}\frac{\partial^2 \psi}{\partial x \partial y} - \frac{\partial \psi}{\partial x}\frac{\partial^2 \psi}{\partial y^2} = \nu \frac{\partial^3 \psi}{\partial y^3} + g\beta(T - T_\infty), \quad (10\text{–}4a)$$

$$\frac{\partial \psi}{\partial y}\frac{\partial T^*}{\partial x} - \frac{\partial \psi}{\partial x}\frac{\partial T^*}{\partial y} = \alpha \frac{\partial^2 T^*}{\partial y^2}, \quad (10\text{–}4b)$$

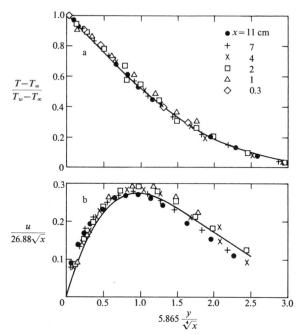

Fig 10–3 The data of Fig. 10–2 plotted for comparison with Pohlhausen's solution. The constants involve the properties of air. [From E. Schmidt, "Heat Transfer by Natural Convection," paper presented at 1961 International Heat Transfer Conference, Boulder, Colo.]

where
$$T^* = \frac{T - T_\infty}{T_w - T_\infty}.$$

For a given value of Prandtl number it can be shown that the free convective boundary layer thickness is proportional to $x/(\mathrm{Gr}_x)^{1/4}$. Pohlhausen suggested the following transformations:

$$y^* = \frac{y}{x/C_1} \qquad \psi^* = \frac{\psi}{4\nu C_1}, \quad (10\text{–}5\text{a},\text{b})$$

where

$$C_1 = \left(\frac{\mathrm{Gr}_x}{4}\right)^{1/4}. \quad (10\text{–}5\text{c})$$

The transformations convert Eqs. (10–4a-b) into

$$\frac{d^3\psi^*}{dy^{*3}} + 3\psi^* \frac{d^2\psi^*}{dy^{*2}} - 2\left(\frac{d\psi^*}{dy^*}\right)^2 + T^* = 0, \quad (10\text{–}6\text{a})$$

$$\frac{d^2 T^*}{dy^{*2}} + 3\Pr \psi^* \frac{dT^*}{dy^*} = 0, \quad (10\text{–}6\text{b})$$

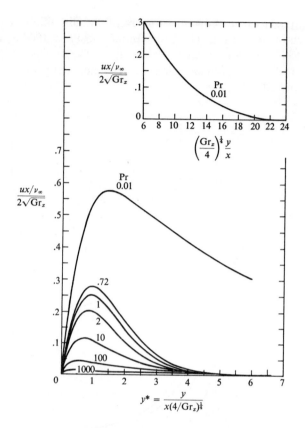

Fig. 10-4 Dimensionless velocity distributions for various Prandtl numbers [4]. [S. Ostrach, NACA Rept. 1111 (1953). Courtesy NASA.]

with the boundary conditions

$$\psi^* = \frac{d\psi^*}{dy^*} = 0 \quad T^* = 1 \quad y^* = 0$$

$$\frac{d\psi^*}{dy^*} = 0 \quad T^* = 0 \quad y^* \to \infty.$$

This pair of ordinary differential equations with the boundary conditions can be solved by any one of a number of methods. The use of an analog computer for a solution is described at the end of this chapter.

Pohlhausen solved the equations for air with a Prandtl number of 0.733. The data of Fig. 10-2 is shown in Fig. 10-3 along with Pohlhausen's solution. This confirms the idea that the velocity profiles are similar and that the temperature profiles are similar at various points along the plate.

Shuh [3] later calculated values for several Prandtl numbers greater than 1.0. Ostrach [4] solved Eqs. (10–6a-b) for $0.01 < \text{Pr} < 1000$ by the use of a numerical technique and a digital computer. The results of Ostrach are shown in the form of dimensionless velocity and temperature profiles in Figs. 10–4 and 10–5.

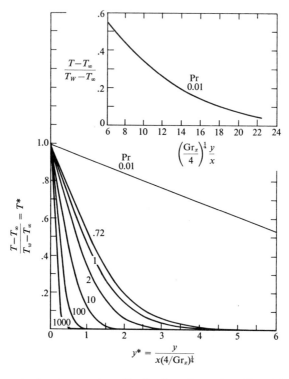

Fig. 10–5 Dimensionless temperature distributions for various Prandtl numbers [4]. [S. Ostrach, NACA Rept. 1111 (1953). Courtesy NASA.]

The Prandtl number appears as a parameter in the curves, since it is an arbitrary parameter in the energy equation, Eq. (10–6b). For a given Prandtl number, the dimensionless velocity is independent of position along the plate for any fluid and temperature difference. In Fig. 10–4, for example, only a single velocity profile is shown for a given Prandtl number; the position coordinate x appears in both the ordinate and abscissa parameters. The same can be said for the temperature profiles. Thus, the velocity profiles in laminar free convection are *similar* for a given Prandtl number, as are the temperature profiles.

The importance of the solution shown in Fig. 10–5 lies in the fact that heat transfer rates can be predicted from the temperature gradient at the wall by the use of the Fourier equation:

$$\left.\frac{\dot{q}}{A}\right|_w = -k\left(\frac{\partial T}{\partial y}\right)_w = -k(T_w - T_\infty)\left.\frac{\partial T^*}{\partial y}\right|_w$$

$$= -\frac{k}{x}(T_w - T_\infty)\left(\frac{\mathrm{Gr}_x}{4}\right)^{1/4}\left.\frac{dT^*}{dy^*}\right|_{y^*=0};$$

since

$$\left.\frac{\dot{q}}{A}\right|_w = h(T_w - T_\infty) \quad \text{and} \quad \mathrm{Nu}_x = \frac{hx}{k},$$

then

$$\mathrm{Nu}_x = \frac{hx}{k} = -\left(\frac{\mathrm{Gr}_x}{4}\right)^{1/4}\left.\frac{dT^*}{dy^*}\right|_{y^*=0}. \tag{10-7}$$

It can be seen from Fig. 10–5 that the value of $(dT^*/dy^*)_{y^*=0}$ is a function of the Prandtl number only. Therefore, Eq. (10–7) can be written as

$$\mathrm{Nu}_x = \frac{hx}{k} = \left(\frac{\mathrm{Gr}_x}{4}\right)^{1/4}\Phi_1(\mathrm{Pr}), \tag{10-8}$$

where $\Phi_1(\mathrm{Pr}) = (dT^*/dy^*)_{y^*=0}$ is the function of the Prandtl number displayed in Fig. 10–5. Ostrach obtained the following values for $\Phi_1(\mathrm{Pr})$:

Pr	0.01	0.72	0.733	1	2	10	100	1000
$\Phi_1(\mathrm{Pr})$	0.0812	0.5046	0.5080	0.5671	0.7165	1.1694	2.191	3.966

Equation (10–8) can be used in conjunction with this table to give local values of the heat transfer for specified fluids and temperature differences.

In many practical situations, the heat transfer rate at the surface (the heat flux) may be constant over the entire surface, and the surface temperature may be nonuniform. The equations developed by Ostrach (and other free convective solutions) appear to work well for the case of constant wall heat flux if the heat transfer coefficient and Grashof number are based on the temperature difference at the midpoint of the vertical wall [6]. Some writers have used a *modified Grashof number* for the case of constant heat flux $\dot{q}''$ at the surface:

$$\mathrm{Gr}_M = \left(\frac{g\beta\dot{q}''L^4}{k\nu^2}\right).$$

Similarity solutions have been obtained for several types of wall temperature variations with laminar free convection on vertical plates [7, 8, 9]. Eckert and Drake [10] presented an approximate method for solution of the laminar boundary layer equations described below.

280 FREE CONVECTION

The continuity and momentum equations, Eqs. (10–2a-c) can be combined and integrated with respect to y in a manner similar to that described in Section 8–4. The result is called the *integral momentum equation for free convection*:

$$\frac{d}{dx}\int_0^{\delta_F} u^2 \, dy = g\beta \int_0^{\delta_F} (T - T_\infty) \, dy - \frac{\tau_w}{\rho}. \tag{10–9}$$

In a similar fashion (see Section 8–4), Eqs. (10–2a) and (10–2c) can be combined to yield the *integral energy equation of free convection*:

$$\frac{d}{dx}\int_0^{\delta_F} u(T - T_\infty) \, dy = \frac{\dot{q}_w/A}{\rho c_p}. \tag{10–10}$$

Equations (10–9) and (10–10) must be solved simultaneously. This results from the dependence of the gravitational body force on the temperature. Also, both equations are integrated over an identical range of the y variable. This means that the velocity boundary layer thickness has been assumed equal to that of the thermal boundary layer. This would seem justified in free convection because the fluid motion is dependent upon temperature difference. For large Prandtl numbers, however, the solution of Ostrach [4] shows that the velocity variation extends much further into the fluid than the temperature variation.

These integral expressions allow the use of assumed, dimensionless velocity and temperature profiles to obtain expressions for δ_F as a function of x. Knowing the dependence of δ_F on x permits calculation of desired information from the assumed dimensionless profiles. This will now be illustrated.

To solve the integrated boundary layer equations for laminar flow, intuitive assumptions regarding the velocity and temperature profiles will be made. The temperature profile is assumed to be approximated by

$$\frac{T - T_\infty}{T_w - T_\infty} = \left(1 - \frac{y}{\delta_F}\right)^2. \tag{10–11}$$

The velocity profile is assumed to be

$$u = u_1 \left(\frac{y}{\delta_F}\right)\left(1 - \frac{y}{\delta_F}\right)^2. \tag{10–12}$$

where u_1 is a convenient velocity coefficient.

Equations (10–11) and (10–12) can be seen to give profiles similar in shape to those shown in Figs. 10–4 and 10–5.

It is further assumed, intuitively, that the free convection boundary layer thickness varies according to the relation

$$\delta_F = c_2 x^n, \tag{10–13}$$

and that the velocity coefficient u_1 varies as

$$u_1 = c_1 x^m. \tag{10-14}$$

By substituting Eqs. (10–13) and (10–14) into Eqs. (10–9) and (10–10), the unknowns, m, n, c_1, and c_2 can be evaluated. The resulting equations for the boundary layer thickness δ_F and u_1 are

$$\frac{\delta_F}{x} = 3.93 \, \text{Pr}^{-1/2}(0.952 + \text{Pr})^{1/4} \, \text{Gr}_x^{-1/4}, \tag{10-15a}$$

$$u_1 = 5.17 \frac{\nu}{x}(0.952 + \text{Pr})^{-1/2} \, \text{Gr}_x^{1/2}. \tag{10-15b}$$

Equation (10–11) can now be used to develop an expression for the Nusselt number by applying the Fourier-Biot law for heat conduction. Hence,

$$\text{Nu}_x = 0.508 \, \text{Pr}^{1/2}(0.952 + \text{Pr})^{-1/4} \, \text{Gr}_x^{1/4}. \tag{10-16}$$

Equation (10–16) can be used to calculate a local heat transfer coefficient for problems involving laminar free convection from a vertical flat surface. For air with a Prandtl number $\text{Pr} = 0.714$, Eq. (10–16) reduces to

$$\text{Nu}_x = 0.378 \, \text{Gr}_x^{1/4}. \tag{10-17}$$

It is of interest to note that for the same problem Ostrach [4], using the results shown in Fig. 10–5, obtained

$$\text{Nu}_x = 0.36 \, \text{Gr}_x^{1/4}, \tag{10-18}$$

which agrees very well with the approximate solution represented by Eq. (10–17).

Any of the expressions for the local heat transfer coefficient—Eqs. (10–8), (10–16), (10–17), or (10–18)—can be integrated with respect to x over the entire surface (assuming all fluid properties do not change with x) to give an average Nusselt number. In each case, or any case where $\text{Nu}_x = f(\text{Gr}_x)^{1/4}$, the following expression is valid:

$$\overline{\text{Nu}} = \tfrac{4}{3} \text{Nu}_{x=L}. \tag{10-19}$$

When Pr Gr exceeds approximately 10^9, the laminar boundary layer undergoes a transition to turbulence. The shape of the velocity profile changes and this affects the relationship between the Nusselt number and the Grashof and Prandtl numbers. Exact analytical methods are not available for predicting the relationship; however, the approximate integral method has been used to make such a prediction. In this method experimental relationships from forced convection relating τ_w and $\dot{q}_w$ to boundary layer thickness have been utilized.

It is also assumed that the turbulent boundary layer commences at the leading edge, whereas in reality it does not commence for some distance back.

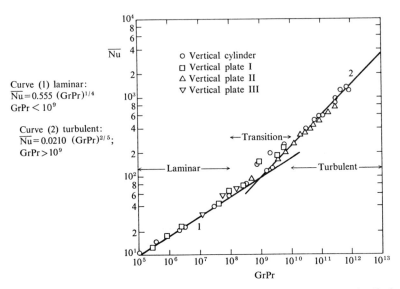

Fig. 10-6 Mean Nusselt number for free convection to vertical plates and cylinders [11]. [From E. R. G. Eckert and J. W. Jackson, "Analysis of Turbulent Free Convection Boundary Layer on a Flat Plate," NACA Report 1015 (1951). Courtesy NASA.]

This assumption does not create large error if most of the plate is covered by a turbulent boundary layer. The result of the integral analysis for the turbulent case is

$$\text{Nu}_x = 0.0295 \, \text{Gr}_x^{2/5} \, \text{Pr}^{7/15} (1 + 0.494 \, \text{Pr}^{2/3})^{-2/5}. \tag{10-20}$$

This rather complex equation can be approximated by a simplified equation and integrated to give an expression for the mean Nusselt number for vertical surfaces:

$$\overline{\text{Nu}} = 0.0210 (\text{Pr} \, \text{Gr})^{2/5}. \tag{10-21}$$

Equation (10-21) and a simplified, integrated form of Eq. (10-16) are shown in Fig. 10-6. Because of the assumption of equal velocity and temperature boundary layer thickness, Fig. 10-6 and Eqs. (10-16) and (10-21) should be used with caution for Prandtl numbers greatly different from 1.0.

The simplified equations such as are used in Fig. 10-6 show the Nusselt number as a function of the Prandtl-Grashof number product. The fact that both Pr and Gr have the same exponent can also be shown by dimensional reasoning if inertia terms are neglected.

Vertical Cylinders

Figure 10-6 as well as all equations developed in this chapter can be used for vertical cylinders if the diameter of the cylinder is large compared to the

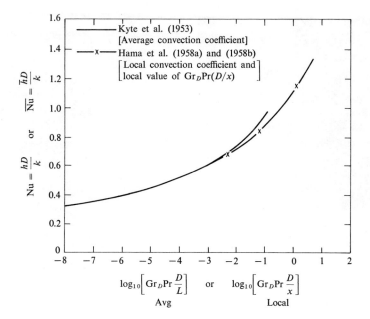

Fig. 10-7 Heat transfer coefficients for small vertical cylinders [26]. (From *Heat Transfer* by Gebhart. Copyright 1961 by McGraw-Hill Book Company, Inc. Used by permission of McGraw-Hill Book Company.)

boundary layer thickness δ_F. It should be emphasized that the correct length to use in the Grashof number is the height or vertical dimension. For a vertical cylinder this is the length, not the diameter, of the cylinder. All properties should be evaluated at the film temperature.

Gebhart [26] summarized the investigations that have been made for free convection to vertical cylinders (wires) where the diameter of the cylinder is small compared to the boundary layer thickness. Use of Fig. 10-7 is necessary when the calculated boundary layer thickness, given by Eq. (10-15), is of the same order as the wire diameter. It has been suggested that when

$$\frac{D}{L} \geq \frac{35}{\sqrt[4]{Gr_L}},$$

the error obtained in treating the thin wire as a vertical flat wall will be less than 5 percent. In the case of small vertical cylinders, a D/L or D/x term usually appears in the correlation. Figure 10-7 gives values of both the average and local coefficients (the Grashof number is based on diameter).

Example 10-1. Estimate the heat loss from a vertical pipe of 3-in. o.d., 4 ft long, if the pipe has a surface temperature of 140°F and is in air at atmospheric pressure and 60°F.

284 FREE CONVECTION

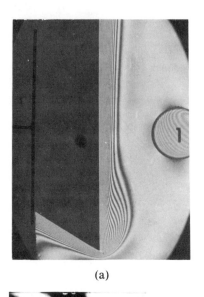

(a)

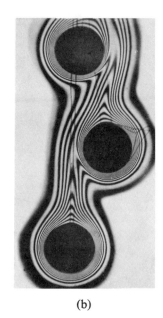

(b)

(c)

Fig. 10–8 Examples of interference photographs in free convection. (a) A sharp-edged vertical plate. (b) Staggered horizontal cylinders. (c) Horizontal flat plate. (Courtesy E. Soehngen.)

Solution. Evaluate properties of the air at $(140 + 60)/2 = 100°F$:

$$\text{Pr} = 0.72 \qquad \beta = 1.79 \times 10^{-3} \frac{1}{°F}$$

$$\nu = 0.18 \times 10^{-3} \frac{\text{ft}^2}{\text{sec}} \qquad k = 0.0154 \frac{\text{Btu}}{\text{hr ft }°F}.$$

Since the outside diameter of the pipe is 3 in., the curvature may be neglected and the pipe treated as a vertical wall. The characteristic length in this case is the vertical height or length, 4 ft. Evaluating the Grashof number, with $L = 4$ ft, $g = 32.2$ ft/sec², gives

$$\text{Gr} = \frac{(32.2) \text{ ft/sec}^2 \, (1.79 \times 10^{-3}) \, 1/°\text{F} \, (140 - 60)°\text{F} \, (4)^3 \, \text{ft}^3}{(0.18 \times 10^{-3})^2 \, \text{ft}^4/\text{sec}^2} = 9.1 \times 10^9.$$

$$\text{Pr Gr} = 6.55 \times 10^9.$$

This is greater than the transition value of 10^9. Therefore, using Eq. (10–21):

$$\overline{\text{Nu}} = \frac{\bar{h}L}{k} = 0.021(\text{Pr Gr})^{2/5},$$

$$\bar{h} = \frac{(0.021)(0.0154)}{4} (6.55 \times 10^9)^{2/5} = \frac{(0.021)(0.0154)}{4} (0.844 \times 10^4)$$

$$= 0.683 \, \frac{\text{Btu}}{\text{hr ft}^2 \, °\text{F}},$$

$$\dot{q} = \bar{h} A (T_w - T_\infty) = \frac{(0.683)(\pi)(3)(4)(80)}{12} = 172 \, \frac{\text{Btu}}{\text{hr}}.$$

Free convection may be conveniently studied by optical means, such as Schlieren photography or interferometry. Examples of interferometry are shown in Figs. 10–8(a) and 10–8(b). In such interferograms the local heat transfer coefficient is proportional to the reciprocal distance between fringes at the heater surface.

Interferometry is particularly helpful in visualizing the process of transition to turbulence in the boundary layer [15].

10–2 FREE CONVECTION FROM HORIZONTAL CYLINDERS

In the previous section equations were developed for the computation of temperatures and heat transfer rates in natural convection on plane and cylindrical vertical surfaces. With certain simplifying assumptions, the mathematical relationships were developed completely for the laminar case. The concept of similarity, used in developing the desired relationships in the previous chapter, is not always applicable in many common natural convection situations. In many of these situations empirical equations have been developed from dimensional analysis and experiment. These experiments have included direct measurement of temperature and velocity profiles and heat transfer rates and indirect measurements with interferometers, Schlieren photographs, and shadowgraphs. Discussions of these indirect methods are given by Hsu [16] and Schmidt [17]. This section will discuss some of the natural convection equations that have been found to be useful in engineering calculations for horizontal cylinders.

In the case of a horizontal cylinder, the entire surface is not parallel to the body force caused by gravity. The equations derived for the vertical cylinder cannot be expected to apply for this case. The variations of the angle between the surface and the gravity vector was considered by Hermann [18], who derived an equation for the local Nusselt number at various positions around a horizontal cylinder for air (Pr = 0.74). The equation is

$$\mathrm{Nu}_D = 0.604 \, \mathrm{Gr}_D^{1/4} \, \phi(\theta) \quad (\mathrm{Pr} = 0.74) \quad (10\text{--}22)$$

The function $\phi(\theta)$ depends upon the value θ, measured from the bottom of the cylinder, given below:

θ	0	30	60	90	120	150	165	180
$\phi(\theta)$	0.76	0.75	0.72	0.66	0.58	0.46	0.36	0

Notice that the Nusselt number and the Grashof number are based on the *diameter* of the cylinder, the significant length in all free convection problems with horizontal cylinders.

The values calculated by Eq. (10–22) seem to be in good general agreement with the values calculated by Eckert and Soehngen [19]. This is shown in Fig. 10–9. An integral approach to this problem by Levy [20] also agrees well with the theory of Hermann.

Equations for free convection from horizontal cylinders usually assume that the flow around the cylinder is laminar. Hyman, Bonilla, and Ehrlich

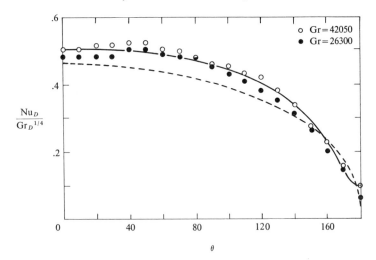

Fig. 10–9 Local nondimensionalized heat transfer coefficient for a horizontal cylinder in air. Dashed line is Eq. (10–22). [Solid line and data points are from E. R. G. Eckert and E. Soehngen, "Studies on Heat Transfer in Laminar Free Convection with the Zehnder-Mach Interferometer," USAF Tech. Rept. 5747 (1948). Courtesy USAF.]

10-2 FREE CONVECTION FROM HORIZONTAL CYLINDERS

[21] presented information to predict transition to turbulence for all types of fluids except liquid metals. Although the transition to turbulence is not common in free convection to horizontal cylinders, when such transition does occur, the heat transfer coefficient appears to increase rapidly from the transition value with very small change in temperature difference.

The average Nusselt number is usually of primary interest in free convection. For laminar flow around horizontal cylinders, Hyman et al. [21] found that the following equation for the average Nusselt number correlated the data well for water, toluene, silicones, mercury, lead, bismuth, lead-bismuth eutectic, sodium, and sodium-potassium alloy:

$$\overline{Nu}_D = 0.53 \left[\left(\frac{Pr}{Pr + 0.952} \right) (Gr_D\, Pr) \right]^{1/4}. \quad (10\text{-}23)$$

All properties in Eq. (10-23) should be evaluated at the film temperature.

Equation (10-23) is similar in form to Eq. (10-16), developed by Eckert and Drake [22] for free convection from a vertical wall. The similarity between Eqs. (10-23) and (10-16) can be seen by using a rule of thumb suggested by Hermann [18], who suggested that an equation for the average Nusselt number for a horizontal cylinder of diameter D could be obtained by treating the cylinder as a vertical wall of height 2.5 D. Using this approximation, Eq. (10-23) can be shown to be equivalent to Eckert's equation, Eq. (10-16), for the vertical wall.

A simplified form of Eq. (10-23) for the liquid metals (low Prandtl number) can be obtained by assuming that $Pr \ll 1$. The result is

$$\overline{Nu}_D = 0.53 (Gr_D\, Pr^2)^{1/4}. \quad (10\text{-}24)$$

For high Prandtl number fluids, assuming that $Pr \gg 1$, Eq. (10-23) reduces to

$$\overline{Nu}_D = 0.53 (Gr_D\, Pr)^{1/4}. \quad (10\text{-}25)$$

Equation (10-25) is identical to an equation suggested by McAdams [23] for a range $10^4 < Gr_D\, Pr < 10^8$ and, therefore, appears to have validity even at lower Prandtl numbers. A suggested curve (Fig. 10-10) was also given by McAdams [23]. Notice that the coordinates in the figure are in terms of $\log_{10}$. Thus, for example, a value of $Gr_D\, Pr = 10^4$ is shown as 4. A value of $\overline{Nu}_D = 2.52$ is shown as 0.4. It should be remembered that

$$\log_e N = 2.30 \log_{10} N.$$

Free convection problems of interest which occur in the range of $Gr_D\, Pr < 10^4$ usually are associated with cylinders of relatively small diameter such as wires. In such cases the boundary layer formed around the wire is comparable in thickness to the diameter of the wire. Boundary layer assumptions are no longer valid and a different approach is necessary.

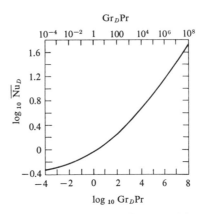

Fig. 10-10 Correlated data for determining coefficients of free convection between horizontal cylinders and gases or liquids [23]. (From *Heat Transmission* by McAdams. Copyright 1954 by William H. McAdams. Used by permission of McGraw-Hill Book Company.)

Langmuir [24] and Elenbaas [25] have suggested that the heat transfer from fine wires be considered as pure conduction. Their results appear to agree with experimental measurement.

10-3 FREE CONVECTION IN ENCLOSED SPACES

When fluid is placed between two horizontal, parallel plates with the upper plates at the higher temperature, a stable condition exists where the less dense fluid is above the more dense fluid. Heat transfer across the fluid layer is by pure conduction. If the lower plate is at a temperature above the upper plate, however, an unstable condition exists. At a value of Rayleigh number based on plate spacing ($Gr_L\, Pr$) of approximately 1700, a pattern of small cells develops in which the fluid circulates, rising in the center of each cell and descending at the outer edge. The cells often referred to as *Benard cells* appear to be almost perfectly hexagonal in shape. A sketch of the cells is shown in Fig. 10–11. Factors affecting the formation of these cells have been discussed by Palm et al. [34].

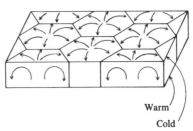

Fig. 10-11 Cellular flow pattern by free convection in horizontal air layers.

A correlation for several types of fluids over a wide range of the Rayleigh number based on spacing (Gr_L Pr) was given by Schmidt and Silveston [27] and later reported by Globe and Dropkin [28]. The correlation is shown in Fig. 10–12 where the ratio of apparent thermal conductivity to true thermal conductivity plotted vs. the Gr_L Pr product. The apparent thermal conductivity is defined as

$$k_{\text{app}} = \left(\frac{(\dot{q}/A)L}{T_1 - T_2}\right). \quad (10\text{–}26)$$

The characteristic length in Gr_L is the plate spacing. All fluid properties should be evaluated at the average of the two plate temperatures. Note that the apparent thermal conductivity and true thermal conductivity are equal up to a Gr_L Pr product of 1700, the point where the cellular motion commences. At a Gr_L Pr product of approximately 50,000, turbulence commences and the hexagonal pattern of Fig. 10–12 disappears. An equation given by Globe and Dropkin [28] for (0.02 < Pr < 8750) and ($3 \times 10^5 <$ Gr_L Pr $< 7 \times 10^9$) is

$$\overline{Nu}_L = 0.069 \sqrt[3]{Gr_L} \, Pr^{0.074}. \quad (10\text{–}27)$$

Vertical Layers

A limited amount of information is available in the literature on free convection in vertical fluid layers. Jakob [29] gives the following equations valid for air spaces:

$$\overline{Nu}_L = \frac{\bar{h}L}{k} = 0.18 \sqrt[4]{Gr_L} \left(\frac{H}{L}\right)^{-1/9} \quad 2 \times 10^4 < Gr_L < 2 \times 10^5, \quad (10\text{–}28)$$

$$\overline{Nu}_L = \frac{\bar{h}L}{k} = 0.065 \sqrt[3]{Gr_L} \left(\frac{H}{L}\right)^{-1/9} \quad 2 \times 10^5 < Gr_L < 11 \times 10^6, \quad (10\text{–}29)$$

in which H is the height of the vertical space and L is the distance between the plates (Fig. 10–13). For $Gr_L < 2000$, pure conduction exists. For values of $H/L < 3$, the equations for a single isolated vertical wall applies.

For liquids in the Prandtl number range of 3 to 30,000, Emery and Chu [30] suggested the following relations for a reasonable prediction of heat transfer across vertical layers:

$$\overline{Nu}_L = 1 \quad Gr_L \, Pr < 10^3 \quad (10\text{–}30)$$

$$\overline{Nu}_L = 0.280 \left(\frac{H}{L}\right)^{-1/4} (Gr_L \, Pr)^{1/4} \quad 10^3 < Gr_L \, Pr < 10^7 \quad (10\text{–}31)$$

The problem of free convection in inclined enclosures was studied by Dropkin and Somerscales [31].

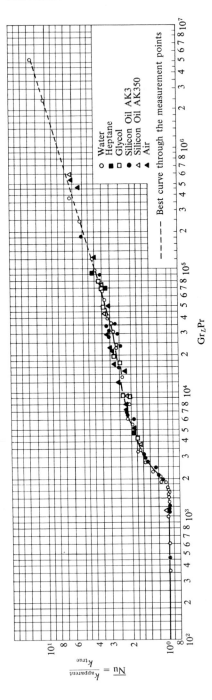

Fig. 10-12 Free convection in horizontal fluid spaces heated from below. [From E. Schmidt, "Heat Transfer by Natural Convection," paper presented at 1961 International Heat Transfer Conference, Boulder, Colo.]

10-3 FREE CONVECTION IN ENCLOSED SPACES

A convenient table for use in solving free convection problems has been compiled by the Society of Automotive Engineers [33] and is presented as Table 10–1. There is slight variation between some of these equations and equations for similar situations described elsewhere in the literature. The expression for laminar convection on a vertical wall in Table 10–1, for example, is $\overline{Nu}_L = 0.59(Gr_L\,Pr)^{1/4}$, whereas Jackson and Eckert suggest a coefficient of 0.555 (Fig. 10–6). Such variation is common in the literature and reflects the small variation obtained in experimental data. The engineer must often make a decision as to which equation is most suitable for a particular situation.

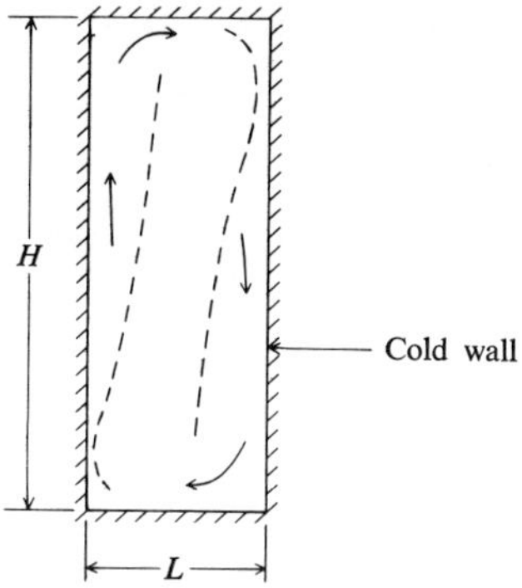

Fig. 10–13 Fluid circulation in an enclosure heated on one side and cooled on the other.

Since air is a common fluid in free convection, the general equations may be simplified for this case by assuming proper constant values for the properties and putting these into the equation. The results are shown in the last column of Table 10–1. It can be seen that the heat transfer coefficients in this case depend only on pressure, temperature difference, and, in most cases, a characteristic dimension.

The following symbols and units are utilized in Table 10–1:

$\overline{Nu}_L$—mean Nusselt number, $\dfrac{hL}{k}$

$\overline{Nu}_D$—mean Nusselt number, $\dfrac{hD}{k}$

Gr_L—Grashof number, $\dfrac{g\beta\,\Delta T L^3}{\nu^2}$

Gr_D—Grashof number, $\dfrac{g\beta\,\Delta T D^3}{\nu^2}$

Pr—Prandtl number, $\dfrac{\mu c_p}{k}$

ΔT—$(T_w - T_\infty)$, degrees F

L or D—in feet

h—mean heat transfer coefficient, $\dfrac{\text{Btu}}{\text{hr ft}^2\,{}^\circ\text{F}}$

p—pressure, psia

Table 10-1
SUMMARY OF EQUATIONS FOR FREE CONVECTION IN OPEN SPACES (RADIATION NEGLECTED)*

Situation	Range of validity	General equation	Simplified equation for air
Vertical plates and cylinders	$10^8 < Gr_L Pr < 10^9$	$\overline{Nu}_L = 0.59\,(Gr_L Pr)^{1/4}$	$\overline{h} = 0.29 \left(\dfrac{p}{14.7}\right)^{1/2} \left(\dfrac{\Delta T}{L}\right)^{1/4}$
	$10^9 < Gr_L Pr < 10^{12}$	$\overline{Nu}_L = 0.13\,(Gr_L Pr)^{1/3}$	$\overline{h} = 0.19 \left(\dfrac{p}{14.7}\right)^{2/3} (\Delta T)^{1/3}$
Horizontal cylinders	$10^8 < Gr_D Pr < 10^9$	$\overline{Nu}_D = 0.53(Gr_D Pr)^{1/4}$	$\overline{h} = 0.27 \left(\dfrac{p}{14.7}\right)^{1/2} \left(\dfrac{\Delta T}{D}\right)^{1/4}$
	$10^9 < Gr_D Pr < 10^{12}$	$\overline{Nu}_D = 0.126(Gr_D Pr)^{1/3}$	$\overline{h} = 0.18 \left(\dfrac{p}{14.7}\right)^{2/3} (\Delta T)^{1/3}$
Horizontal plates hot face up cold face down	$10^5 < Gr_L Pr < 2 \times 10^7$	$\overline{Nu}_L = 0.54(Gr_L Pr)^{1/4}$	$\overline{h} = 0.27 \left(\dfrac{p}{14.7}\right)^{1/2} \left(\dfrac{\Delta T}{L}\right)^{1/4}$
	$2 \times 10^7 < Gr_L Pr < 3 \times 10^{10}$	$\overline{Nu}_L = 0.14(Gr_L Pr)^{1/3}$	$\overline{h} = 0.22 \left(\dfrac{p}{14.7}\right)^{1/2} (\Delta T)^{1/3}$
Horizontal plates hot face down or cold face up	$3 \times 10^5 < Gr_L Pr < 3 \times 10^{10}$	$\overline{Nu}_L = 0.27(Gr_L Pr)^{1/4}$	$\overline{h} = 0.12 \left(\dfrac{p}{14.7}\right)^{1/2} \left(\dfrac{\Delta T}{L}\right)^{1/4}$
Sphere	$10^3 < Gr_D Pr < 10^7$	$\overline{Nu}_D = 0.51(Gr_D Pr)^{1/4}$	$\overline{h} = 0.487 \left(\dfrac{p}{14.7}\right)^{1/2} \left(\dfrac{\Delta T}{D}\right)^{1/4}$

* Adapted from "Aerospace Applied Thermodynamics Manual," SAE Committees A-9 (1960 rev. 1962).

10-3 FREE CONVECTION IN ENCLOSED SPACES

A numerical approach to three-dimensional laminar free convection problems has been described by Aziz and Hellums [35]. A recent summary of advances in free convection heat transfer was given by Ede [36].

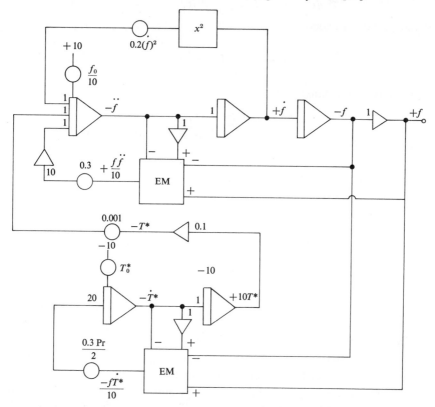

Fig. 10-14 Analog computer diagram for solution of Eqs. (10-33) and (10-34). [From Pace Application Note No. 2, Bulletin No. AN922-1. Courtesy Electronic Associates, Inc.]

Analog Computer Solution to Eqs. (10-6a-b)

The transformed energy and momentum equations, Eqs. (10-6a-b), are coupled ordinary nonlinear differential equations. They can be solved by use of an analog computer (Fig. 10-14). A change of variable is necessary to relate dimensionless distance y^* to time t. This is accomplished by letting t be proportional to y^*, with a the constant of proportionality:

$$t = ay^*$$

Then

$$\frac{d}{dy^*} = a\frac{d}{dt} \quad \text{and} \quad \frac{d^2}{dy^{*2}} = a^2\frac{d^2}{dy^{*2}}.$$

Equation (10-6a) then becomes, letting $f = \psi^*$,

$$a^3 \frac{d^3f}{dt^3} + 3a^2 f \frac{d^2f}{dt^2} - 2a^2 \left(\frac{df}{dt}\right)^2 + T^* = 0.$$

Equation (10-6b) becomes

$$a^2 \frac{d^2T^*}{dt^2} + 3a \operatorname{Pr} f \frac{dT^*}{dt} = 0. \tag{10-32}$$

These may be rearranged to give

$$\frac{d^3f}{dt^3} = -\frac{3f}{a}\frac{d^2f}{dt^2} + \frac{2}{a}\left(\frac{df}{dt}\right)^2 - \frac{T^*}{a^3},$$

$$\frac{d^2T^*}{dt^2} = -\frac{3 \operatorname{Pr} f}{a}\frac{dT^*}{dt}. \tag{10-33}$$

These two equations may now be solved by the circuit shown on the preceding page.

Since all of the initial conditions ($t = 0$) are not known, the problem is solved by trial and error. The initial conditions f_0 and T_0^* are assumed and the proper potentiometers set. The problem is then run, and a check is made on the boundary conditions for the free stream (large time). When the boundary conditions are satisfied the problem is solved.

REFERENCES

1. E. Schmidt, "Heat Transfer by Natural Convection," paper presented at 1961 International Heat Transfer Conference, Boulder, Colo.
2. E. Schmidt and W. Beckmann, "Das Temperatur—und Gerschwindigkeitsfeld von einer warme abgebenden senkrechten Platte bei naturlicher Konvektion," *Technische Mechanik und Thermodynamik,* **I** (1930).
3. H. Shuh, Grezschichten Monographien, Vol. B, Göttingen (1946).
4. S. Ostrach, "An Analysis of Laminar Free Convection Flow and Heat Transfer About a Flat Plate Parallel to the Direction of the Generating Body Force," NACA Report 1111 (1953).
5. E. M. Sparrow and J. L. Gregg, "The Variable Fluid Property Problems in Free Convection," *Trans. ASME* **80**, 879 (1958).
6. E. M. Sparrow and J. L. Gregg, "Laminar Free Convection from a Vertical Flat Plate," *Trans. ASME* **78**, 435 (1956).
7. E. M. Sparrow and J. L. Gregg, "Similar Solutions for Free Convection from Non-isothermal Vertical Plate," *Trans. ASME* **80**, 379 (1958).
8. K. Millsaps and K. Pohlhausen, *J. Aeronautical Sci.* **23**, 381 (1956).
9. K. Millsaps and K. Pohlhausen, *J. Aeronautical Sci.* **25**, 357 (1958).

10. E. R. G. Eckert and R. M. Drake, Jr., *Introduction to the Transfer of Heat and Mass*, McGraw-Hill, New York (1951).
11. E. R. G. Eckert and J. W. Jackson, "Analysis of Turbulent Free Convection Boundary Layer on a Flat Plate," NACA Report 1015 (1951).
12. A. A. Szewczyk, "Combined Forced and Free Laminar Flow," ASME Paper 63-WA-130 (1963).
13. A. Acrivos, "Combined Laminar Free and Forced Convection Heat Transfer in External Flows," *Trans. AIChE* **4**, 285 (1958).
14. E. M. Sparrow, R. Eichorn, and J. C. Gregg, "Combined Forced and Free Convection in a Boundary Layer Flow," *Phys. Fluids* **2**, 319 (1959).
15. E. R. G. Eckert and E. Soehngen, "Interferometric Studies on the Stability and Transition to Turbulence of a Free Convection Boundary Layer," *Proceedings General Discussion on Heat Transfer*, London (1951), p. 321.
16. H. J. Hsu, *Engineering Heat Transfer*, Van Nostrand, New York (1963), pp. 362–371.
17. E. Schmidt, "Heat Transfer by Natural Convection," paper presented at 1961 International Heat Transfer Conference, Boulder, Colo.
18. R. Hermann, "Warmeubergang bei freier Stromung am waagerechten Zylinder in zweiatomigen Gasen," VDI-Forschungsheft 379, Berlin (1936); translated in NACA TM 1366 (1954).
19. E. R. G. Eckert and E. Soehngen, "Studies on Heat Transfer in Laminar Free Convection with the Zehnder-Mach Interferometer," USAF., Tech. Rept. 5747 (1948).
20. S. Levy, "Integral Methods in Natural Convection Flow," *J. Appl. Mech.* **22**, 515 (1955).
21. S. C. Hyman, C. F. Bonilla, and S. W. Ehrlich, "Natural Convection Transfer Processes. I. Heat Transfer to Liquid Metals and Non-Metals at Horizontal Cylinders," *Chemical Engineering Progress Symposium Series* **49** (5), 21 (1953).
22. E. R. G. Eckert and R. M. Drake, Jr., *Heat Transfer*, 2d ed. McGraw-Hill, New York (1959), p. 315.
23. W. H. McAdams, *Heat Transmission*, 3d ed., McGraw-Hill, New York (1954), p. 177.
24. J. Langmuir, "Convection and Conduction of Heat in Gases," *Phys. Rev.* **34**, 401 (1912).
25. W. Elenbaas, "Dissipation of Heat by Free Convection from Vertical and Horizontal Cylinders," *J. Appl. Phys.* **19**, 1148 (1948).
26. B. Gebhart, *Heat Transfer*, McGraw-Hill, New York (1961), p. 270.
27. R. J. Schmidt, and P. L. Silveston, "Natural Convection in Horizontal Liquid Layers," Second National Heat Transfer Conference, Chicago (1958).
28. S. Globe and D. Dropkin, "Natural Convection Heat Transfer in Liquids Confined by Two Horizontal Plates and Heated from Below," *Trans. ASME, J. Heat Transfer* **81** (1), 24 (1959).

29. M. Jakob, *Heat Transfer*, Wiley, New York (1949), Vol. I.
30. A. Emery, and N. C. Chu, "Heat Transfer Across Vertical Layers," *Trans. ASME, J. Heat Transfer* **87** (1), 110 (1965).
31. D. Dropkin and E. Somerscales, "Heat Transfer by Natural Convection in Liquids Confined by Two Parallel Plates which are Inclined at Various Angles with Respect to the Horizontal," *Trans. ASME, J. Heat Transfer* **87**, (1), 71 (1965).
32. S. W. Churchill, "The Prediction of Natural Convection," Third International Heat Transfer Conference, Chicago, 1966.
33. Society of Automotive Engineers, SAE Committee A-9, "Aerospace Applied Thermodynamics Manual" (1960, rev. 1962).
34. E. Palm, et al., "On the Occurrence of Cellular Motion in Benard Convection," *J. Fluid Mech.* **30**, 651 (1967).
35. K. Aziz and J. D. Hellums, "Numerical Solution of the Three-Dimensional Equations of Motion for Laminar Natural Convection," *Phys. Fluids* **10**, 314 (1967).
36. A. J. Ede, "Advances in Free Convection," *Advances in Heat Transfer*, Academic Press, New York (1967), Vol. IV.
37. D. V. Julian and R. G. Akins, "Bibliography of Natural Convection Heat Transfer from a Vertical Flat Plate," Report 77, Kansas Engineering Exp. Station (1967).

PROBLEMS

1. Using the transformations suggested by Pohlhausen, Eq. (10–5), show that Eqs. (10–4a) and (10–4b) can be transformed into Eqs. (10–6a) and (10–6b).
2. Derive Eq. (10–19) by use of the definition for the average heat transfer coefficient:
$$\bar{h} = \frac{1}{L}\int_0^L h_x \, dx.$$
3. Make a plot of $(Nu_x/Gr_x)^{1/4}$ vs. Pr for a range of Pr from 0.01 to 1000 using both Eq. (10–8) and Eq. (10–16). Discuss any differences between the two curves.
4. Derive Eq. (10–15), using Eqs. (10–9), (10–10), (10–11), and (10–12), and making the suggested assumptions for variation of δ_F and u_1 with x. *Hint:* Equate coefficients of like powers of x in the expression obtained. For details, see Eckert and Drake [10].
5. Derive Eq. (10–16) from Eq. (10–11) and (10–15), using the Fourier-Biot law and the definition of the Nusselt number.
6. Estimate the heat loss from a vertical wall 4 ft high by 8 ft wide if the wall surface temperature is uniform and equal to 130°F and if it is exposed to oxygen at 1 atmosphere and 70°F.

7. Calculate the *maximum* velocity occurring in the boundary layer for Problem 6 at a point $x = 2$ ft from the bottom edge of the plate, using Eqs. (10–12) and (10–15). Compare your answer to that given by Fig. 10–4.

8. Calculate the mass rate of flow of oxygen past the point $x = 2$ ft for Problem 6 above.

9. An approximate equation for the average Nusselt number for free convection on a vertical plate in the turbulent range (Gr Pr > 10^9) is given as

$$\overline{\mathrm{Nu}}_L = 0.13(\mathrm{Gr}_L \, \mathrm{Pr})^{1/3}.$$

Show that the heat transfer coefficient is independent of the height L of the plate according to this equation. Discuss.

10. Use Eq. (10–20) to derive an expression for an average Nusselt number, assuming constant properties and assuming that the equation is valid over the entire plate of length L. Compare the equation with Eq. (10–21).

11. Estimate the rate of heat loss from a vertical cylinder 4 in. in diameter and 24 in. high to helium at 100°F. Assume that surface temperature of the cylinder is 300°F.

12. A vertical plane wall 8 ft wide by 4 ft high is maintained with a uniform surface temperature of 130°F. It is exposed to a still *air* environment at 70°F. Estimate the heat loss from the surface in watts.

13. Estimate the heat loss by free convection from a vertical wall 5 ft wide by 3 ft high. The wall surface temperature is 210°F and the undisturbed air is at 190°F and atmospheric pressure.

14. Estimate the rate at which heat would be lost from a vertical wall 4 ft high by 3 ft wide at a uniform temperature of 150°F if it is exposed to water at a temperature of 90°F.

15. Estimate the temperature required at the surface of a vertical wall, 0.3 ft tall by 1 ft wide if it is to lose 15 Btu/hr by free convection to air at 1 atmosphere and 70°F. Assume that the air properties can be approximated by the values for 100°F.

16. Use Hermann's approximation [18] to show the similarity between Eqs. (10–16) and (10–23). As a first step, Eq. (10–16) must be integrated to obtain an expression for the average heat transfer coefficient for the vertical plate. Refer to Problem 2.

17. Estimate the heat loss per foot by convection from a horizontal pipe, 3 in. o.d. if the pipe has a surface temperature of 140°F and is exposed to air at atmospheric pressure and 60°F.

18. Estimate the heat loss from a vertical wire of 0.01 in. o.d. and 6 in. long if the wire surface temperature is 250°F and the surrounding air is at atmospheric pressure and 150°F.

19. Estimate the maximum permissible current flow in a No. 14 AWG copper wire if the temperature in the wire is assumed uniform and may not exceed 300°F. Assume the electrical resistance to be 0.002525 ohms per foot. The wire is bare and exposed horizontally to air at 100°F and atmospheric pressure.

298 FREE CONVECTION

20. Saturated steam at 300 psia flows through a bare horizontal pipe of 4.0 in. o.d. The surrounding air is at 70°F and atmospheric pressure. Assuming that the outside surface temperature of the pipe is the same as the steam temperature and that the steam is flowing through the pipe at the rate of 5000 lb_m/hr, estimate the change in steam quality that would occur in 50 ft of pipe.

21. Determine the heat loss per foot of length by natural convection from a horizontal 1 in. o.d. cylinder if the surface temperature is 240°F and the surrounding fluid is at 160°F for sodium, water, air, and light oil.

    Assume for Sodium at 200°F:

    $\rho = 58.0 \ lb_m/ft^3$
    $c_p = 0.33 \ Btu/lb_m \ °F$
    $\nu = 8.1 \times 10^{-6} \ ft^2/sec$
    $k = 49.8 \ Btu/hr \ ft \ °F$
    $Pr = 0.011$
    $\beta = 0.15 \times 10^{-3} \ per \ °F$

    Assume for light oil at 200°F:

    $\rho = 54$
    $c_p = 0.51$
    $\nu = 4.6 \times 10^{-5}$
    $k = 0.074$
    $Pr = 62$
    $\beta = 0.42 \times 10^{-3}$

22. Estimate the temperature difference $(T_w - T_\infty)$ required to lose 300 Btu/hr of energy from a 3 in. o.d. cylinder 4 ft long by natural convection to air if the cylinder is placed horizontally and if the fluid properties are evaluated at 100°F as an approximation.

23. Estimate the surface temperature required to lose 200 Btu/hr by natural convection to surrounding air from a 2-in. o.d. horizontal cylinder 3 ft long. Assume that the air properties can be approximated by those at 1 atmosphere and 200°F, and that $T_\infty = 200°F$.

24. Estimate the temperature difference $(T_w - T_\infty)$ required to lose 10 Btu/hr of energy from a 4 in. o.d. sphere by natural convection to the surrounding air. Assume that film properties are evaluated at 200°F.

25. Estimate the surface temperature of a horizontal wire 0.6 in. o.d., cooled by free convection to air at 60°F and 1 atmosphere. The wire is dissipating 2 watts per foot of length. Assume air properties at 100°F. Check the Rayleigh number to assure that any assumptions made are valid.

26. Estimate the current necessary to maintain the surface temperature of an electrical heater at 200°F when in air at 50°F. Assume that the heater is a horizontal cylinder 0.5 in. in diameter and 15 in. long. Its electrical resistance is 5 ohms.

27. Estimate the temperature difference between two parallel horizontal plates separated by water if the two plates are 1 in. apart and exchange heat at the rate of 2400 Btu/hr ft². Assume that the average temperature of the water between the plates is 100°F and that the lower plate is at the highest temperature. Justify any assumptions made.

28. Wire up the analog computer circuit shown in Fig. 10–14 and obtain a solution for Eqs. (10–32) and (10–33).

CHAPTER 11

HIGH-SPEED FLOW AND HEAT TRANSFER

The development of jet propulsion, gas turbines, and high-speed flight has brought into importance a realm of fluid mechanics known as compressible flow. Prior to the 1940's most engineers had little contact with this subject either in the classroom or in practice. This is understandable when it is realized that it was not until 1947 that an airplane first "cracked the sound barrier." Since that time compressible flow has received considerable emphasis in the training of aeronautical and mechanical engineers. This chapter presents a general introductory treatment of compressible flow.

Initially compressible flow will be considered using a one-dimensional approach. While the one-dimensional approach may lack the precision of a two- or three-dimensional approach, it is much easier to handle mathematically and easier for a beginner to comprehend. The latter portions of the chapter deal with viscous compressible flow with heat transfer.

The principal difference between compressible and incompressible flow is in the behavior of the density of the fluid. The density varies significantly in compressible flow and, as will be shown later, this can result in the occurrence of strange phenomena (i.e., shock waves) not found in incompressible flow. Since most of the compressible problems in practice occur in gases, the bulk of this chapter will deal with the compressible flow of a perfect gas.

11-1 THE SPEED OF SOUND

It has long been known that sound is composed of pressure waves which travel in a medium at a speed which is nearly independent of loudness or pitch. The speed is, however, a function of the elasticity of the medium. For example, at normal atmospheric temperature the speed of sound in air is of the order 1100 ft/sec; in hydrogen 4200 ft/sec; in Freon refrigerating gases it is of the order of 300 ft/sec; and in water of the order of 4900 ft/sec.

Although a pressure disturbance in a fluid will propagate outward in all directions in the form of a spherical wave, it is easier to analyze by considering the wave to be flat or planar. Consider a long insulated tube of constant area A fitted with a piston at one end (Fig. 11-1). Let the piston move with a speed dV to the right. A pressure wave forms ahead of the piston and moves

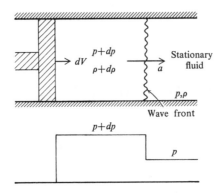

Fig. 11-1 Propagation of a small pressure wave.

with speed a. Ahead of the pressure wave the velocity of the fluid is zero, while behind the wave the velocity of the fluid is the same as that of the piston, dV. Assuming the pressure and density of the stationary fluid ahead of the wave to be p and ρ respectively, then the passage of the wave front will cause them to undergo small changes to become $p + dp$ and $\rho + d\rho$.

The development which follows applies the principles of conservation of mass and momentum to obtain an expression for the velocity of the pressure wave which is moving at the velocity of sound. This development then provides a way of calculating the velocity of sound.

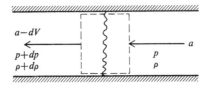

Fig. 11-2 Control volume for an observer moving with the wave front.

Consider an observer moving with the wave at speed a, the speed of sound. To this observer the relative speeds are those shown in Fig. 11-2. For steady flow (dV = constant), the rate of flow *out* of the control volume minus the rate of flow *in* will equal zero, and conservation of mass gives

$$(\rho + d\rho)(a - dV)A - \rho a A = 0.$$

By neglecting the higher-order differential term ($d\rho\, dV$), the above equation reduces to

$$dV = \frac{a}{\rho} d\rho. \qquad (11\text{-}1)$$

By neglecting any shear forces the conservation of momentum can be written as
$$pA - (p + dp)A = \rho a A[(a - dV) - a],$$
which reduces to
$$dp = \rho a \, dV. \tag{11-2}$$
By substituting Eq. (11-1) into Eq. (11-2),
$$a^2 = \left(\frac{\partial p}{\partial \rho}\right)_s. \tag{11-3}$$

The ratio $dp/d\rho$ is indicated as the partial derivative at constant entropy because the variations in pressure are vanishingly small, so that the flow is essentially isentropic. By taking the square root of Eq. (11-3), the speed of sound is
$$a = \left[\left(\frac{\partial p}{\partial \rho}\right)_s\right]^{1/2}. \tag{11-4}$$

For a liquid, Eq. (11-4) can be rewritten by utilizing Eq. (2-13):
$$a^2 = B_s/\rho, \tag{11-5}$$
where B_s is the isentropic bulk modulus of elasticity.

Example 11-1. Calculate the speed of sound in water at 15 psia and 68°F.

Solution. From Table 2-1, $B_T = 320{,}000$ psi. But for a liquid, $B_T \approx B_s$. Therefore,
$$a = \sqrt{\frac{320{,}000 \frac{\text{lb}_f}{\text{in}^2} \, 144 \frac{\text{in}^2}{\text{ft}^2}}{\left(62.4 \frac{\text{lb}_m}{\text{ft}^3}\right) \Big/ \left(32.2 \frac{\text{lb}_m \, \text{ft}}{\text{lb}_f \, \text{sec}^2}\right)}}$$
$$= 4870 \text{ ft/sec}.$$

The isentropic relation between pressure and density for a perfect gas can be expressed as
$$p = C_1 \rho^\gamma, \tag{11-6}$$
where γ is the ratio of specific heats c_p/c_v. Taking the logarithm of both sides and differentiating, the following is obtained:
$$\frac{dp}{p} = \gamma \frac{d\rho}{\rho} \tag{11-7}$$
Hence,
$$\left(\frac{\partial p}{\partial \rho}\right)_s = \gamma \left(\frac{p}{\rho}\right) = \gamma R_g T,$$
and, thus, the speed of sound for a perfect gas is
$$a = \sqrt{\gamma R_g T}. \tag{11-8}$$

In some cases it is convenient to write Eq. (11–8) in terms of the *universal gas constant* $\mathscr{R}$. The gas constant R_g is related to this universal gas constant through the molecular weight W:

$$R_g = \frac{\mathscr{R}}{W},$$

where

$$\mathscr{R} = 1545 \frac{\text{ft lb}_f}{(\text{lb mole})°\text{R}},$$

and W = molecular weight, in lb_m per lb-mole. Thus, Eq. (11–8) can be written

$$a = \sqrt{\gamma \mathscr{R} T / W}. \qquad (11\text{–}8a)$$

Example 11–2. Calculate the speed of sound in air at 60°F.

Solution. The molecular weight of air is 28.96. Therefore,

$$a = [(1.4)(1545)(32.2)(T)/(28.96)]^{1/2} = 49.01\sqrt{T°\text{R}} \text{ ft/sec}$$
$$= 49.01\sqrt{520} = 1118 \text{ ft/sec}.$$

11–2 THE MACH NUMBER AND FLOW REGIMES

The single most important parameter in the analysis of compressible fluids is the *Mach number* (M), named after the nineteenth century Austrian physicist Ernst Mach. It is defined as the ratio of the local velocity to the local speed of sound in the fluid:

$$\text{M} = \frac{V}{a}. \qquad (11\text{–}9)$$

The Mach number can be used to characterize flow regimes as follows:

Incompressible flow. The Mach number is very small compared to unity ($M < 0.3$). For practical purposes the flow is treated as incompressible.

Subsonic flow. The Mach number is less than unity but large enough so that compressible flow effects are discernible ($0.3 < M < 0.8$).

Transonic flow. The Mach number is very close to unity ($0.8 < \text{M} < 1.2$).

Supersonic flow. The Mach number is larger than unity but less than five ($1.2 < \text{M} < 5$).

Hypersonic flow. The Mach number is larger than five ($\text{M} > 5$).

A principal difference between high-speed and low-speed flow is demonstrated by studying the effect of the speed of a small particle on the pressure field in the region near the particle (Fig. 11–3). In the figure the location of a particle moving in a fluid is shown at three successive time periods (numbered

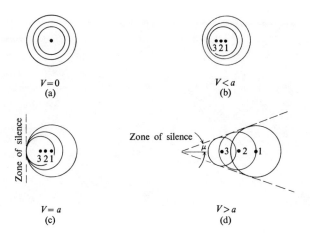

Fig. 11-3 Pressure patterns produced by a particle.

1, 2, and 3). Pressure waves are assumed to be propagated outward from the particle continuously. Three such waves are shown in the figure as spheres, the largest sphere corresponding to the wave emitted first. The radius of the sphere depends on the speed of the wave and the time since it was initially propagated.

At a speed $V < a$, the particle approaches the wave fronts but never quite "catches up," as in Fig. 11-3(b). At a speed $V = a$, the particle travels at the same speed as the wave fronts which merge at one point, as in Fig. 11-3(c). At a speed $V > a$, the particle travels faster than the wave fronts, which combine in an envelope to form a conical wave. This conical wave is called a *Mach wave*, or a *Mach cone*, and has a semivertex angle μ. The angle μ is easily obtained by inspection of Fig. 11-3(d):

$$\mu = \sin^{-1}\left(\frac{a}{V}\right) = \sin^{-1}\left(\frac{1}{M}\right). \tag{11-10}$$

Mach waves are infinitesimal pressure disturbances. The speed of the Mach wave normal to the wave is equal to the local speed of sound. In the region close to a finite-sized body a wave pattern will develop which is different from the Mach wave. These wave patterns are called shock waves and they represent regions where there is a finite increase in pressure as the fluid crosses the wave. At a very large distance from the body the shock wave degenerates into a Mach wave. Figure 11-4 shows a shadowgraph of supersonic flow around a missile model placed in a wind tunnel. The shock waves at the nose and coming off the rear flare are clearly visible. Shock waves will be discussed in more detail in Section 11-4.

Fig. 11-4 Shadowgraph of a missile model at M = 1.96. (Courtesy of E. A. Simon, George C. Marshall Space Flight Center, Huntsville, Ala.)

The effect of compressibility and shock waves on the flow behavior around an object is seen in Figure 11-5. The drag coefficient on a sphere is seen to depend strongly on the value of the Mach number, and, at sufficiently

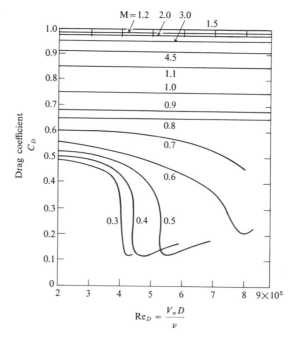

Fig. 11-5 Variation of drag coefficient on a sphere measured by A. Naumann. (From *Boundary Layer Theory* by Schlichting. Copyright 1968, 1960, 1955 by McGraw-Hill Book Company, Inc. Used by permission of McGraw-Hill Book Company.)

high Mach numbers, C_D is independent of the Reynolds number altogether. This is typical of many high-speed flow situations, where the Mach number is the most significant parameter. The curves in Fig. 11–5 are seen to approach the curve of Fig. 9–15 for low values of Mach number.

11–3 ISENTROPIC FLOW OF A PERFECT GAS

Many actual compressible flows (i.e., in nozzles) are almost isentropic (reversible adiabatic) when the frictional effects and heat transfer to the walls are negligible. The conservation of energy for a perfect gas between any two points in the flow can be written as

$$\tfrac{1}{2}V_1^2 + c_p T_1 = \tfrac{1}{2}V_2^2 + c_p T_2 = \text{constant.} \tag{11-11}$$

If one of the points is a stagnation point ($V = 0$) then,

$$T_T = T + \frac{V^2}{2c_p},$$

where T_T is the stagnation or total temperature. Since $V^2 = M^2 a^2 = M^2 \gamma R_g T$, and $c_p = R_g \gamma/(\gamma - 1)$, then

$$\left(\frac{T_T}{T}\right) = 1 + \frac{\gamma - 1}{2} M^2, \tag{11-12}$$

Equations (11–11) and (11–12) are valid for any adiabatic flow whether thermodynamically reversible or not. They are, therefore, valid across a shock wave which is irreversible.

The stagnation or total pressure can be obtained from the isentropic equation

$$\frac{p_T}{p} = \left(\frac{T_T}{T}\right)^{\gamma/(\gamma-1)} = \left(1 + \frac{\gamma - 1}{2} M^2\right)^{\gamma/(\gamma-1)}. \tag{11-13}$$

In an irreversible adiabatic flow the total pressure will not remain constant but will decrease. Since all real flows are irreversible to some extent, the total pressure always decreases in the absence of heat or work.

The stagnation (or total) density can be obtained by utilizing the isentropic equation again,

$$\frac{\rho_T}{\rho} = \left(\frac{T_T}{T}\right)^{1/(\gamma-1)} = \left(1 + \frac{\gamma - 1}{2} M^2\right)^{1/(\gamma-1)}. \tag{11-14}$$

These stagnation or total parameters p_T and ρ_T are the values of p and ρ at a point in the fluid if the flow were slowed to zero velocity isentropically. The equation for T_T, Eq. (11–12), is not restricted to isentropic flow. Equations (11–12), (11–13), and (11–14) are tabulated in Table 11–1 for Mach numbers between zero and 20 and are plotted in Fig. 11–6 for Mach numbers between zero and 5.

Table 11–1

ISENTROPIC FLOW PARAMETERS, $\gamma = 1.4$*

M	p/p_T	ρ/ρ_T	T/T_T	$\frac{1}{2}\rho V^2/p_T$	$A/A^\star$
0.0	1.0000	1.0000	1.0000	0.0000	∞
0.1	0.9930	0.9950	0.9980	0.6951 − 2	5.822
0.2	0.9725	0.9803	0.9921	0.2723 − 1	2.964
0.3	0.9395	0.9564	0.9823	0.5919 − 1	2.035
0.4	0.8956	0.9243	0.9690	0.1003	1.590
0.5	0.8430	0.8852	0.9524	0.1475	1.340
0.6	0.7840	0.8405	0.9328	0.1976	1.188
0.7	0.7209	0.7916	0.9107	0.2473	1.094
0.8	0.6560	0.7400	0.8865	0.2939	1.038
0.9	0.5913	0.6870	0.8606	0.3352	1.009
1.0	0.5283	0.6339	0.8333	0.3698	1.000
1.1	0.4684	0.5817	0.8052	0.3967	1.008
1.2	0.4124	0.5311	0.7764	0.4157	1.030
1.3	0.3609	0.4829	0.7474	0.4270	1.066
1.4	0.3142	0.4374	0.7184	0.4311	1.115
1.5	0.2724	0.3950	0.6897	0.4290	1.176
1.6	0.2353	0.3357	0.6614	0.4216	1.250
1.7	0.2026	0.3197	0.6337	0.4098	1.338
1.8	0.1740	0.2868	0.6068	0.3947	1.439
1.9	0.1492	0.2570	0.5807	0.3771	1.555
2.0	0.1278	0.2300	0.5556	0.3579	1.687
2.1	0.1094	0.2058	0.5313	0.3376	1.837
2.2	0.9352 − 1	0.1841	0.5081	0.3169	2.005
2.3	0.7997 − 1	0.1646	0.4859	0.2961	2.193
2.4	0.6840 − 1	0.1472	0.4647	0.2758	2.403
2.5	0.5853 − 1	0.1317	0.4444	0.2561	2.637
2.6	0.5012 − 1	0.1179	0.4252	0.2371	2.896
2.7	0.4295 − 1	0.1056	0.4068	0.2192	3.183
2.8	0.3685 − 1	0.9463 − 1	0.3894	0.2022	3.500
2.9	0.3165 − 1	0.8489 − 1	0.3729	0.1863	3.850
3.0	0.2722 − 1	0.7623 − 1	0.3571	0.1715	4.235
4.0	0.6586 − 2	0.2766 − 1	0.2381	0.7376 − 1	10.72
5.0	0.1890 − 2	0.1134 − 1	0.1667	0.3308 − 1	25.00
10.0	0.2356 − 4	0.4948 − 1	0.4762 − 1	0.1649 − 2	535.9
20.0	0.2091 − 6	0.1694 − 4	0.1235 − 1	0.5855 − 4	1538 + 1

* The plus and minus numbers indicate the number of spaces the decimal is to be moved: plus to the right, minus to the left. From "Equations, Tables and Charts for Compressible Flow," NACA Report 1135, 1953.

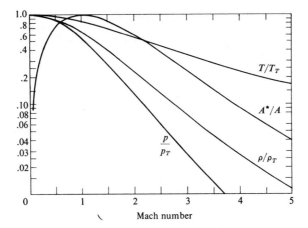

Fig 11-6 Isentropic flow variables.

It is important to note that the *local* value of a stagnation property depends only upon the local value of the static property and the local Mach number and is independent of the flow process. Equations (11–12), (11–13), and (11–14), Fig. 11–6, and Table 11–1 may be used to determine these local stagnation values, even for nonisentropic flow, assuming that the local static property and local Mach number are known.

Included in Table 11–1 are the values of the ratio of dynamic pressure to total pressure. The dynamic pressure can be expressed as a function of the Mach number and pressure:

$$\tfrac{1}{2}\rho V^2 = \tfrac{1}{2}\gamma p M^2. \tag{11-15}$$

Nozzles are flow passages which accelerate the fluid to higher speeds. *Diffusers* accomplish the opposite effect, that is, they are used to decelerate the flow. These devices are quite common in rockets, gas turbines, and flow-metering devices. In incompressible steady flow, the product VA is constant (conservation of mass when ρ is a constant), so that any passage which converges (causes A to decrease in the flow direction) is a nozzle, and any diverging passage is a diffuser. In fact, in any subsonic flow a converging channel accelerates and a diverging channel decelerates the flow. As will be shown in the following development, just the opposite is true in supersonic flow.

For the nozzle shown in Fig. 11–7, the conservation equation for mass can be written as $\rho A V =$ constant, which, by taking the logarithm and differentiating, can be expressed as

$$\frac{d\rho}{\rho} + \frac{dA}{A} + \frac{dV}{V} = 0. \tag{11-16}$$

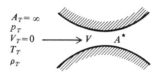

Fig. 11-7 Converging-diverging nozzle.

The conservation of energy in differential form can be obtained from Eq. (11-11):

$$c_p \, dT = -V \, dV. \tag{11-17}$$

A thermodynamic relation involving the differential of entropy is

$$T \, ds = c_p \, dT - dp/\rho. \tag{11-18}$$

From Eqs. (11-17) and (11-18) for an isentropic flow ($ds = 0$),

$$V \, dV = -dp/\rho. \tag{11-19}$$

This is also known as the *Euler equation* and could have been obtained from the Navier-Stokes equations (Eqs. 4-43a-c) for a one-dimensional inviscid flow. See Eqs. (6-1a-c), for example. Equation (11-19) states that the flow accelerates (positive dV) when the pressure decreases (negative dp) and vice versa.

By combining Eq. (11-19) with Eq. (11-16), one can obtain,

$$\frac{dA}{A} = -\frac{dV}{V}\left(1 - \frac{V^2}{dp/d\rho}\right) = -\frac{dV}{V}\left(1 - \frac{V^2}{a^2}\right),$$

or

$$\frac{dA}{dV} = \frac{A}{V}(M^2 - 1). \tag{11-20}$$

Referring to Fig. 11-8 and Eq. (11-20), one can examine how dV varies as a function of M and A.

1. *Subsonic flow* (M < 1)

    $\dfrac{dA}{dV} < 0$ (converging channel—velocity increases)
    (diverging channel—velocity decreases)

2. *Supersonic flow* (M > 1)

    $\dfrac{dA}{dV} > 0$ (converging channel—velocity decreases)
    (diverging channel—velocity increases)

3. *Sonic flow* (M = 1)

    $\dfrac{dA}{dV} = 0$

11-3 ISENTROPIC FLOW OF A PERFECT GAS

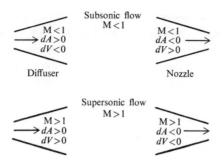

Fig. 11-8 Area and velocity changes for subsonic and supersonic gas flow.

The variation of velocity, pressure, temperature, and density in converging or diverging channels in both subsonic and supersonic flow is tabulated in Table 11-2. Thus we can see the surprising result that an area increase in supersonic flow results in an increase in velocity. This is just the opposite of the effect in subsonic flow.

Table 11-2

Type of flow passage	M	dA	dM	dV	dp	dT	$d\rho$
Subsonic converging nozzle	$M < 1$	−	+	+	−	−	−
Subsonic diverging diffuser	$M < 1$	+	−	−	+	+	+
Supersonic converging diffuser	$M > 1$	−	−	−	+	+	+
Supersonic diverging nozzle	$M > 1$	+	+	+	−	−	−

The Mach number-area variation in a nozzle can be determined by combining the conservation of mass with the definition of Mach number:

$$\frac{A_1}{A_2} = \frac{\rho_2 V_2}{\rho_1 V_1} = \frac{\rho_2 M_2 a_2}{\rho_1 M_1 a_1}, \quad \text{or} \quad \frac{A_1}{A_2} = \frac{M_2}{M_1}\frac{\rho_2}{\rho_1}\left(\frac{T_2}{T_1}\right)^{1/2}.$$

By substituting Eqs. (11-12) and (11-14) into the above,

$$\frac{A_1}{A_2} = \frac{M_2}{M_1}\left[\frac{1 + \left(\frac{\gamma - 1}{2}\right)M_1^2}{1 + \left(\frac{\gamma - 1}{2}\right)M_2^2}\right]^{(\gamma+1)/2(\gamma-1)} \qquad (11\text{-}21)$$

It is convenient to denote with a star those variables which occur at a section where the Mach number is unity (i.e., $T^\star$, $A^\star$, etc.). These conditions are called *critical conditions*. By setting the Mach number M_2 equal to unity in Eq. (11-21), a relation is obtained between the area A at any section in the passage and the critical area $A^\star$, where the areas are measured perpendicular to the direction of flow:

$$\frac{A}{A^\star} = \frac{1}{M}\left[\frac{1 + \left(\frac{\gamma - 1}{2}\right)M^2}{(\gamma + 1)/2}\right]^{(\gamma+1)/2(\gamma-1)}. \qquad (11\text{-}22)$$

This relation is tabulated in Table 11-1 along with the other isentropic parameters. When the area ratio $A/A^\star$ is plotted vs. Mach number M, as in Fig. 11-6, it can be seen that the minimum area occurs when the Mach number M = 1. Thus, if a converging-diverging nozzle is operating with small enough exit pressure, supersonic flow will occur in the diverging section, and M = 1 will occur in the *throat* (location of minimum area).

Mass Flow Calculation

The ideal mass flow per unit area can be determined by

$$\frac{\dot{m}}{A} = \rho V = \frac{p}{R_g T} V = \frac{p_T V}{\sqrt{\gamma R_g T}}\sqrt{\frac{\gamma}{R_g}}\sqrt{\frac{T_T}{T}}\frac{1}{\sqrt{T_T}}\frac{p}{p_T},$$

or by utilizing Eqs. (11-12) and (11-13),

$$\frac{\dot{m}}{A} = \sqrt{\frac{\gamma}{R_g}}\frac{p_T}{\sqrt{T_T}}M\left[1 + \left(\frac{\gamma - 1}{2}\right)M^2\right]^{-[(\gamma+1)/2(\gamma-1)]}. \qquad (11\text{-}23)$$

The value $(\dot{m}/A)$ has a maximum at M = 1:

$$\left(\frac{\dot{m}}{A}\right)_{max} = \frac{\dot{m}}{A^\star} = \left[\frac{\gamma}{R_g}\left(\frac{2}{\gamma + 1}\right)^{(\gamma+1)/(\gamma-1)}\right]^{1/2}\frac{p_T}{\sqrt{T_T}}. \qquad (11\text{-}24)$$

For air, with $\gamma = 1.4$ and $R_g = 53.3$ (ft-lb$_f$)/(lb$_m$ °R),

$$\frac{\dot{m}}{A^\star} = 0.53 \frac{p_T}{\sqrt{T_T}}, \qquad (11\text{-}25)$$

where $\dot{m}$ is in lb$_m$/sec, $A^\star$ is in in^2, T_T in °R, and p_T in psia.

Example 11-3. A tank is equipped with a converging-diverging nozzle. Air in the tank is at 100 psia and 80°F. Find the pressure, temperature, density, and nozzle area at a cross section of the nozzle where the Mach number is 2.0. The throat area is 1.0 in². The tank is assumed to be large enough so that the tank pressure does not change.

Solution.

$$T_T = 80°F = 540°R \qquad P_T = 100 \text{ psia}$$

$$\rho_T = \frac{p_T}{R_g T_T} = \frac{(100)(144)}{(53.3)(540)} = 0.5 \text{ lb}_m/\text{ft}^3$$

From Eq. (11–12) or Table 11–1:

$$\frac{T}{T_T} = \left[1 + \left(\frac{1.4-1}{2}\right)2^2\right]^{-1} = 0.5556,$$

$$T = (0.5556)(540) = 300°R = -160°F.$$

From Eq. (11–13) or Table 11–1:

$$\frac{p}{p_T} = \left[1 + \left(\frac{1.4-1}{2}\right)2^2\right]^{-(1.4/0.4)} = 0.1278,$$

$$p = (0.1278)(100) = 12.78 \text{ psia}.$$

From Eq. (11–14) or Table 11–1:

$$\frac{\rho}{\rho_T} = \left[1 + \left(\frac{1.4-1}{2}\right)2^2\right]^{-(1/0.4)} = 0.230,$$

$$\rho = (0.230)(0.5) = 0.115 \text{ lb}_m/\text{ft}^3$$

From Eq. (11–22) or Table 11–1:

$$\frac{A}{A^\star} = \frac{1}{2}\left[\frac{1 + \left(\frac{1.4-1}{2}\right)2^2}{(1.4+1)/2}\right]^{2.4/0.8} = 1.687,$$

$$A = (1.687)(1.0) = 1.687 \text{ in}^2.$$

Example 11–4. Determine the mass flow of air out of the tank described in Example 11–3.

Solution.

$$\dot{m} = A^\star(0.53)\frac{p_T}{\sqrt{T_T}}$$

$$= (1)(0.53)\frac{100}{\sqrt{540}} = 2.28 \text{ lb}_m/\text{sec}.$$

The *critical pressure ratio* is the pressure ratio (throat static pressure to stagnation pressure) necessary to accelerate the fluid to sonic velocity isentropically. It can be obtained for air by setting $M = 1$ in Eq. (11–13):

$$\left(\frac{p}{p_T}\right)_{cr} = \frac{p^\star}{p_T} = \left[1 + \left(\frac{1.4-1}{2}\right)1^2\right]^{-(1.4/0.4)} = 0.528. \qquad (11\text{–}26)$$

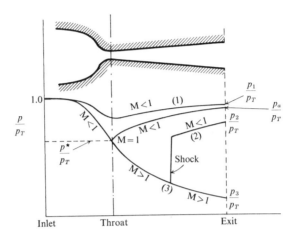

Fig. 11–9 Flow through a converging-diverging nozzle with various back pressures.

If the dimensionless pressure at the exit of a converging-diverging nozzle is greater than the value p_a/p_T, which would lead to a critical pressure $p^\star$ at the throat, then the flow will all be subsonic (curve 1 of Fig. 11–9). If the dimensionless exit pressure is greater than the isentropic pressure ratio associated with the exit area for supersonic flow, but less than the value which

Fig. 11–10 A flexible plate nozzle. (Courtesy USAF Arnold Engineering Development Center.)

would just lead to a sonic or critical value in the throat, then a portion of the flow downstream of the throat will be supersonic, followed by a rapid change at some position to subsonic flow (curve 2 of Fig. 11–9). This rapid change in pressure is a *shock wave* and is discussed in the next section.

If the dimensionless exit pressure is equal to or lower than the isentropic pressure associated with the exit area ratio for supersonic flow, then all of the flow in the diverging section of the nozzle will be supersonic, as shown in curve 3 of Fig. 11–9.

Because it may be desirable to use a supersonic wind tunnel for a variety of Mach numbers and Reynolds numbers, the nozzle in the tunnel may be constructed with flexible walls. This allows changing the shape of the nozzle to match the desired flow conditions. A flexible plate nozzle of this type is shown in Fig. 11–10.

11-4 NORMAL AND OBLIQUE SHOCK WAVES

One of the most interesting and unique phenomena that occur in supersonic flow is the shock wave. A shock wave can be considered as a discontinuity in the properties of the flow field. Fluid crossing a shock wave, normal to the flow path, experiences a sudden increase in pressure, temperature, and density, accompanied by a sudden decrease in speed from a supersonic to a subsonic value. The process is irreversible in that it is impossible for the reverse effect (an abrupt acceleration) to take place because this would violate the second law of thermodynamics. The thickness of the shock wave is of the order of only a few mean free paths for Mach numbers greater than 1.1. This would make the shock thickness less than one ten-thousandth of an inch at normal atmospheric air pressures and temperatures.

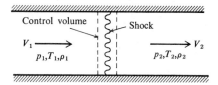

Fig. 11–11 Stationary normal shock wave.

For simplicity consider a perfect gas flowing in a duct as shown in Fig. 11–11.* For steady flow through a stationary normal shock, with no direction

---

* This section deals only with standing (stationary) shock waves in a moving stream. If the shock wave is moving, the steady flow conservation equations cannot be applied and the problem must be solved by use of a coordinate transformation.

change, area change, or work done, the mass, momentum, and energy equations are

Mass: $$\rho_1 V_1 = \rho_2 V_2. \tag{11-27}$$

Momentum: $$p_1 - p_2 = \rho_1 V_1 (V_2 - V_1). \tag{11-28}$$

Energy: $$\tfrac{1}{2} V_1^2 + c_p T_1 = \tfrac{1}{2} V_2^2 + c_p T_2 = c_p T_T. \tag{11-29a}$$

Since the flow is adiabatic, $T_{T_1} = T_{T_2}$, and therefore a modified form of the energy equation can be written as

$$T_{T_2} = T_2 \left[1 + \left(\frac{\gamma - 1}{2} \right) M_2^2 \right] = T_1 \left[1 + \left(\frac{\gamma - 1}{2} \right) M_1^2 \right] = T_{T_1}. \tag{11-29b}$$

These equations, together with the equation of state, the definition of Mach number, and $a = \sqrt{\gamma R_g T}$, will yield two solutions. One solution, which is trivial, states that there is no change and hence no shock wave. The other solution, which corresponds to the change across a normal shock wave, can be expressed in terms of the upstream Mach Number:

$$M_2^2 = \frac{M_1^2 + \dfrac{2}{\gamma - 1}}{\left(\dfrac{2\gamma}{\gamma - 1} \right) M_1^2 - 1}, \tag{11-30}$$

$$\frac{p_2}{p_1} = \left(\frac{2\gamma}{\gamma + 1} \right) M_1^2 - \frac{\gamma - 1}{\gamma + 1}, \tag{11-31}$$

$$\frac{p_{T2}}{p_{T1}} = \frac{\left\{ \left[\left(\dfrac{\gamma + 1}{2} \right) M_1^2 \right] \bigg/ \left[1 + \left(\dfrac{\gamma - 1}{2} \right) M_1^2 \right] \right\}^{\gamma/(\gamma-1)}}{\left[\left(\dfrac{2\gamma}{\gamma + 1} \right) M_1^2 - \dfrac{\gamma - 1}{\gamma + 1} \right]^{1/(\gamma-1)}}, \tag{11-32}$$

$$\frac{T_2}{T_1} = \frac{\left[1 + \left(\dfrac{\gamma - 1}{2} \right) M_1^2 \right] \left[\left(\dfrac{2\gamma}{\gamma - 1} \right) M_1^2 - 1 \right]}{\left[\dfrac{(\gamma + 1)^2}{2(\gamma - 1)} \right] M_1^2}. \tag{11-33}$$

In Table 11–3 the normal shock wave parameters are tabulated for the case of $\gamma = 1.4$.

Example 11–5. Air is flowing through a shock wave in a nozzle. The Mach number and temperature in front of the shock are 2.0 and 500°R respectively. What is the velocity immediately behind the shock?

Table 11-3
NORMAL SHOCK RELATIONS, $\gamma = 1.4$*

M	p_2/p_1	ρ_2/ρ_1	T_2/T_1	p_{T_2}/p_{T_1}	p_1/p_{T_2}	M_2
1.0	1.000	1.000	1.000	1.0000	0.5283	1.0000
1.1	1.245	1.169	1.065	0.9989	0.4689	0.9118
1.2	1.513	1.342	1.128	0.9928	0.4154	0.8422
1.3	1.805	1.516	1.191	0.9794	0.3685	0.7860
1.4	2.120	1.690	1.255	0.9582	0.3280	0.7397
1.5	2.458	1.862	1.320	0.9298	0.2930	0.7011
1.6	2.820	2.032	1.388	0.8952	0.2628	0.6684
1.7	3.205	2.198	1.458	0.8557	0.2368	0.6405
1.8	3.613	2.359	1.532	0.8127	0.2142	0.6165
1.9	4.045	2.516	1.608	0.7674	0.1945	0.5956
2.0	4.500	2.667	1.688	0.7209	0.1773	0.5774
2.1	4.978	2.812	1.770	0.6742	0.1622	0.5613
2.2	5.480	2.951	1.857	0.6281	0.1489	0.5471
2.3	6.005	3.085	1.947	0.5833	0.1371	0.5344
2.4	6.553	3.212	2.040	0.5401	0.1266	0.5231
2.5	7.125	3.333	2.137	0.4990	0.1173	0.5130
2.6	7.720	3.449	2.238	0.4601	0.1089	0.5039
2.7	8.338	3.559	2.343	0.4236	0.1014	0.4956
2.8	8.980	3.664	2.451	0.3895	0.9461 − 1	0.4882
2.9	9.645	3.763	2.563	0.3577	0.8848 − 1	0.4814
3.0	10.33	3.857	2.679	0.3283	0.8291 − 1	0.4752
3.1	11.04	3.947	2.799	0.3012	0.7785 − 1	0.4695
3.2	11.78	4.031	2.922	0.2762	0.7323 − 1	0.4643
3.3	12.53	4.112	3.049	0.2533	0.6900 − 1	0.4596
3.4	13.32	4.188	3.180	0.2322	0.6513 − 1	0.4552
3.5	14.12	4.261	3.315	0.2129	0.6157 − 1	0.4512
3.6	14.95	4.330	3.454	0.1953	0.5829 − 1	0.4474
3.7	15.80	4.395	3.596	0.1792	0.5526 − 1	0.4439
3.8	16.68	4.457	3.743	0.1645	0.5247 − 1	0.4407
3.9	17.57	4.516	3.893	0.1510	0.4987 − 1	0.4377
4.0	18.50	4.571	4.047	0.1388	0.4747 − 1	0.4350
5.0	29.00	5.000	5.800	0.6172 − 1	0.3062 − 1	0.4152
6.0	41.83	5.268	7.941	0.2965 − 1	0.2136 − 1	0.4042
7.0	57.00	5.444	10.47	0.1535 − 1	0.1574 − 1	0.3974
8.0	74.50	5.565	13.39	0.8488 − 2	0.1207 − 1	0.3929
9.0	94.33	5.651	16.69	0.4964 − 2	0.9546 − 2	0.3898
10.0	116.5	5.714	20.39	0.3045 − 2	0.7739 − 2	0.3876
15.0	262.3	5.870	44.69	0.4395 − 3	0.3446 − 2	0.3823
20.0	466.5	5.926	78.72	0.1078 − 3	0.1940 − 2	0.3804
30.0	1050	5.967	175.9	0.1453 − 4	0.8626 − 3	0.3790

* The minus numbers indicate the number of spaces the decimal is to be moved to the left.
From "Equations, Tables and Charts for Compressible Flow," NACA Report 1135, 1953.

Solution. From Table 11–3,

for $M_1 = 2.0$, $M_2 = 0.5774$, $T_2/T_1 = 1.688$.
$T_2 = (1.688)(500) = 844°R$.
$V_2 = M_2 a_2 = (0.5774)(49.01)\sqrt{844} = 822$ ft/sec.

Oblique Shocks

If a plane shock is not perpendicular to the flow but inclined at an angle, the shock will cause the fluid which flows through it to change direction, in addition to increasing its pressure, temperature, and density. An oblique shock is illustrated in Fig. 11–12, where the fluid flow is deflected through an angle δ. The conservation equations applied to the indicated control volume are

Mass:
$$\rho_1 V_{1n} = \rho_2 V_{2n}, \tag{11-34}$$

Energy:
$$\tfrac{1}{2}V_1^2 + c_p T_1 = \tfrac{1}{2}V_2^2 + c_p T_2, \tag{11-35}$$

Momentum normal to the shock:
$$p_1 - p_2 = \rho_2 V_{2n}^2 - \rho_1 V_{1n}^2, \tag{11-36}$$

Momentum tangent to the shock:
$$0 = \rho_1 V_{1n}(V_{2t} - V_{1t}), \tag{11-37}$$

where the subscripts t and n indicate directions tangent and normal to the shock respectively.

From Eq. (11–37),
$$V_{2t} = V_{1t} \tag{11-38}$$

Since the velocity components tangent to the shock are equal on opposite sides of the shock, it is obvious that an oblique shock is equivalent to a normal shock for an observer moving with velocity $V_{1t} = V_{2t}$. This situation permits the use of normal shock equations to calculate oblique shock

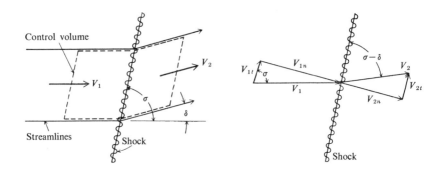

Fig. 11–12 Oblique shock wave.

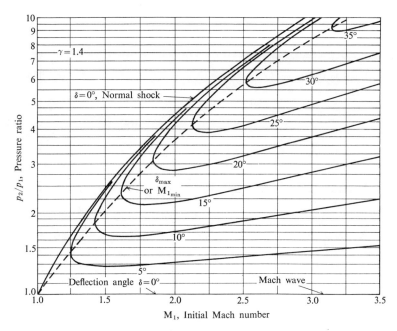

Fig. 11–13 Pressure ratio vs. inlet Mach number, with turning angle as parameter. (Ascher H. Shapiro—*The Dynamics and Thermodynamics of Compressible Fluid Flow*, Volume 1, Copyright 1953. The Ronald Press Company, New York.)

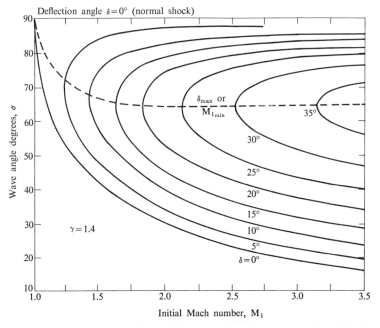

Fig. 11–14 Shock angle vs. inlet Mach number and turning angle. (Ascher H. Shapiro—*The Dynamics and Thermodynamics of Compressible Fluid Flow*, Volume 1, Copyright 1953. The Ronald Press Company, New York.)

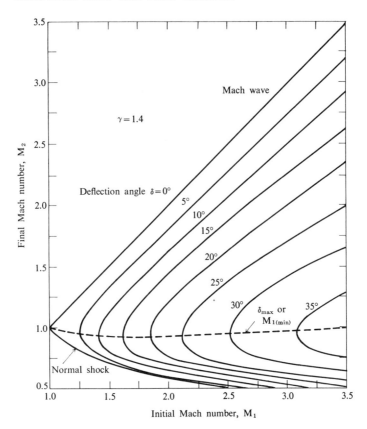

Fig. 11-15 Exit Mach number vs. inlet Mach number, with turning angle as parameter. (Ascher H. Shapiro—*The Dynamics and Thermodynamics of Compressible Fluid Flow*, Volume 1, Copyright 1953. The Ronald Press Company, New York.)

parameters. Equation (11-30) can be modified for oblique shock use by substituting M_{1n} and M_{2n} for M_1 and M_2 respectively. Similarly the ratios p_2/p_1 and T_2/T_1 can be obtained by replacing M_1 by M_{1n} in Eqs. (11-31) and (11-33) respectively, where

$$M_{1n} = M_1 \sin \sigma.$$

Graphical solutions of the oblique shock equations are shown in Figs. 11-13 through 11-16. These solutions are taken from Shapiro [1] for the case $\gamma = 1.4$ and for various values of δ. If any two variables are given, for example M_1 and the flow deflection angle δ, then all of the other downstream variables can be found in terms of upstream variables.

It is interesting to note in the oblique shock figures that, for a given initial Mach number M_1 and a given flow deflection angle δ, there are either two

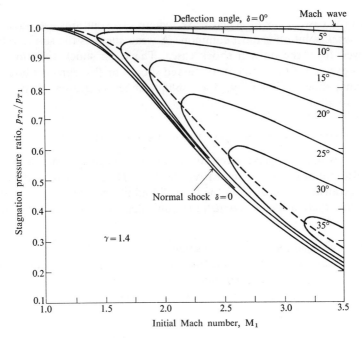

Fig. 11-16 Stagnation pressure ratio vs. inlet Mach number, with turning angle as parameter. (Ascher H. Shapiro—*The Dynamics and Thermodynamics of Compressible Fluid Flow*, Volume 1, Copyright 1953. The Ronald Press Company, New York.)

solutions or none at all. For a given M_1, a maximum deflection angle, δ_{max} can be found. This maximum varies from 0° at $M_1 = 1$ to about 45° as $M_1 \to \infty$. If δ_{max} is exceeded, the shock "detaches," that is, moves ahead of the turning surface and becomes curved (Fig. 11-17). For such a case no solution exists on the oblique shock figures (Figs. 11-13 to 11-16).

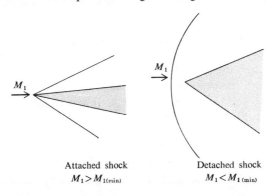

Fig. 11-17 Attached (*left*) and detached (*right*) shock waves.

In the case of the attached wave, where two solutions are possible, the *weak shock* solution is most common. A weak shock is the solution farthest removed from the normal shock case. The weak shock situation usually occurs except for (a) segments of curved shocks or (b) segments bounded by a normal shock. Generally, if an oblique shock is straight, it is the weak shock solution.

11–5 DIABATIC FLOWS AND EFFECTS OF FRICTION

Up to this point we have given consideration only to adiabatic flow, that is, flow without heat transfer to or from the fluid. Diabatic flow is flow in which heat is transferred to or from the fluid. In addition to heat transfer through the walls, there are effects which are equivalent to heat transfer to or from the fluid. They are:

1. Chemical reaction (combustion);
2. Phase change (i.e., condensation or evaporation of liquid droplets flowing with the fluid; and
3. Joule heating (electric current passing through a fluid of finite conductivity).

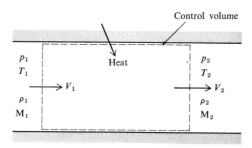

Fig. 11–18 Control volume for constant area flow with heating or cooling.

Frictional effects are always present in real flow, but they will be neglected in the following analysis. This diabatic flow process, without friction, at constant area is called *Rayleigh flow*.

In this section, the analysis is restricted to the steady flow of a perfect gas with no external work being done. The pertinent equations for the control volume shown in Fig. 11–18 are:

Mass: $\qquad\qquad\rho_1 V_1 = \rho_2 V_2,$ \hfill (11–39)

Momentum: $\qquad p_1 - p_2 = \rho_1 V_1 (V_2 - V_1),$ \hfill (11–40)

Energy: $\quad c_p(T_2 - T_1) + \tfrac{1}{2}(V_2^2 - V_1^2) = \dot{q} = c_p(T_{T2} - T_{T1}),$ \hfill (11–41)

where $\dot{q}$ is the rate of heat transfer per unit mass of the flowing fluid. When these equations are combined with the perfect gas equation ($p = \rho R_g T$), the Mach number equation ($M = V/\sqrt{\gamma R_g T}$), and the definitions of stagnation temperature and pressure, the following equations can be derived:

$$\frac{T}{T^{\star R}} = \left(\frac{1+\gamma}{1+\gamma M^2}\right)^2 M^2. \tag{11-42}$$

$$\frac{T_T}{T_T^{\star R}} = \frac{2(\gamma+1)M^2\left[1+\left(\frac{\gamma-1}{2}\right)M^2\right]}{(1+\gamma M^2)^2}. \tag{11-43}$$

$$\frac{p}{p^{\star R}} = \frac{1+\gamma}{1+\gamma M^2}. \tag{11-44}$$

$$\frac{p_T}{p_T^{\star R}} = \left(\frac{2}{\gamma+1}\right)^{\gamma/(\gamma-1)} \left(\frac{1+\gamma}{1+\gamma M^2}\right) \left[1+\left(\frac{\gamma-1}{2}\right)M^2\right]^{\gamma/(\gamma-1)}. \tag{11-45}$$

The superscript $\star R$ denotes those properties at $M = 1$ for the case of diabatic flow with negligible friction. The R is used since this type of flow is called *Rayleigh flow*. Equations (11-42) through (11-45) are plotted in Fig. 11-19 for the case of $\gamma = 1.4$.

Example 11-6. Air enters a constant area duct at a temperature of 40°F and a Mach number of 0.3. Heat is added to the air at the rate of 100 Btu/lb. Determine the temperature, Mach number, and stagnation temperature at the exit of the duct.

Solution.

$$T_{T1} = T_1\left[1 + \left(\frac{\gamma-1}{2}\right)M_1^2\right] = 500[1 + 0.2(0.3)^2] = 509°R.$$

$$T_{T2} - T_{T1} = \frac{\dot{q}}{c_p} = \frac{100}{0.24} = 417°R.$$

$$T_{T2} = 417 + 509 = 926°R.$$

From Fig. 11-19, at $M_1 = 0.3$,

$$\frac{T_1}{T^{\star R}} = 0.41, \qquad \frac{T_{T1}}{T_T^{\star R}} = 0.35.$$

$$\frac{T_{T2}}{T_{T1}} = \frac{926}{509} = 1.82, \qquad \frac{T_{T2}}{T_T^{\star R}} = \left(\frac{T_{T2}}{T_{T1}}\right)\left(\frac{T_{T1}}{T_T^{\star R}}\right) = (1.82)(0.35) = 0.64.$$

Therefore, from Fig. 11-19, $M_2 = 0.47$, $T_2/T^{\star R} = 0.73$, and

$$T_2 = \left(\frac{T_2}{T^{\star R}}\right)\left(\frac{T^{\star R}}{T_1}\right)T_1 = (0.73)\left(\frac{1}{0.41}\right)(500) = 890°R.$$

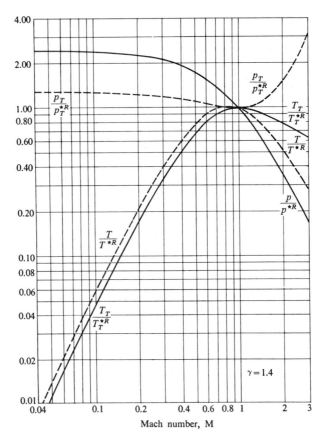

Fig. 11-19 Frictionless flow of a perfect gas in a constant-area duct with stagnation temperature change.

Referring to Fig. 11-19 and comparing the flow properties with the T_T/T_T^{*R} curve, several interesting facts are evident. First of all it is seen that for heating (increasing T_T/T_T^{*R}), the Mach number approaches unity for both subsonic and supersonic flow. From the curve of T/T^{*R}, it can be seen that the temperature reaches a maximum at $M = 1/\sqrt{\gamma}$. For air in the Mach number range of $0.85 < M < 1$, the temperature *decreases* during heat addition, and the temperature *increases* during heat removal.

Another interesting observation from Fig. 11-19 is that the stagnation pressure always decreases for heat addition to the stream. Hence, combustion will cause a loss in stagnation pressure.

A plot on the temperature-entropy (*T-S*) plane of the diabatic flow process without friction at constant area is shown in Fig. 11-20. It clearly pictures the effects of heating or cooling on the acceleration or deceleration of the fluid.

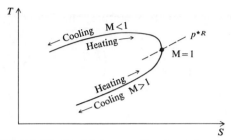

Fig. 11-20 Rayleigh curve for simple heating.

11-6 CONSTANT AREA ADIABATIC FLOW WITH FRICTION

Flow of a perfect gas through a constant-area adiabatic duct may be analyzed, assuming that the skin friction coefficient is constant over the entire length of the duct. Referring to Fig. 11-21, the conservation equations can be written as

Mass:
$$\frac{d\rho}{\rho} + \frac{dV}{V} = 0, \qquad (11\text{-}46)$$

Momentum:
$$\rho V \, dV = -dp - \tau_w \, dA_w, \qquad (11\text{-}47)$$

Energy:
$$c_p \, dT + V \, dV = 0, \qquad (11\text{-}48)$$

where dA_w is the wetted wall area.

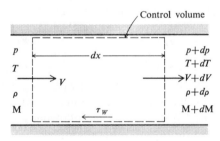

Fig. 11-21 Control volume for constant-area flow with friction.

From the definition of hydraulic diameter, one can write

$$D = \frac{4A}{dA_w/dx}. \qquad (11\text{-}49)$$

The wall skin friction stress τ_w can be expressed in terms of the skin friction coefficient C_f as

$$\tau_w = C_f \tfrac{1}{2} \rho V^2. \qquad (11\text{-}50)$$

When Eqs. (11-46) through (11-50) are combined with the perfect gas equation ($p = \rho R_g T$), the Mach number equation ($M = V/\sqrt{\gamma R_g T}$), and the definition of stagnation temperature, the following equations can be derived:

$$\frac{dM^2}{M^2} = \frac{\gamma M^2}{(1-M^2)}\left[1+\left(\frac{\gamma-1}{2}\right)M^2\right]4C_f\frac{dx}{D}, \quad (11\text{-}51)$$

$$\frac{dp}{p} = \frac{-\gamma M^2}{2(1-M^2)}[1+(\gamma-1)M^2]4C_f\frac{dx}{D}, \quad (11\text{-}52)$$

$$\frac{dT}{T} = -\frac{\gamma}{2(1-M^2)}[(\gamma-1)M^4]4C_f\frac{dx}{D}, \quad (11\text{-}53)$$

$$\frac{dp_T}{p_T} = -\frac{\gamma M^2}{2}4C_f\frac{dx}{D}. \quad (11\text{-}54)$$

Equation (11-51) shows that $dM^2 > 0$ for $M < 1$, and that $dM^2 < 0$ for $M > 1$. That is, friction always drives the Mach number toward unity. In this respect it acts just like heat addition to the gas. Equation (11-51) can be integrated between the limits of $M = 1$ and $M = M$ to yield

$$4C_f\frac{L^{\star F}}{D} = \frac{1-M^2}{\gamma M^2} + \frac{\gamma+1}{2\gamma}\ln\frac{(\gamma+1)M^2}{2\left[1+\left(\frac{\gamma-1}{2}\right)M^2\right]}, \quad (11\text{-}55)$$

where C_f is assumed constant and $L^{\star F}$ is length of duct required for the Mach number to change from M to unity. The length of duct required for Mach number M_1 to change to Mach number M_2 is

$$L = \frac{D}{4C_f}\left[\left(4C_f\frac{L^{\star F}}{D}\right)_1 - \left(4C_f\frac{L^{\star F}}{D}\right)_2\right]. \quad (11\text{-}56)$$

Adiabatic flow in a constant-area duct with friction is called *Fanno flow*. By denoting the pressure, temperature, and stagnation pressure at $M = 1$ by $p^{\star F}$, $T^{\star F}$, and $p_T^{\star F}$ respectively, the following relations may be derived [1] from Eqs. (11-51) through (11-54):

$$\frac{p}{p^{\star F}} = \frac{1}{M}\left(\frac{\gamma+1}{2\left[1+\left(\frac{\gamma-1}{2}\right)M^2\right]}\right)^{1/2}. \quad (11\text{-}57)$$

$$\frac{T}{T^{\star F}} = \frac{\gamma+1}{2\left[1+\left(\frac{\gamma-1}{2}\right)M^2\right]}. \quad (11\text{-}58)$$

$$\frac{p_T}{p_T^{\star F}} = \frac{1}{M}\left\{\frac{2}{\gamma+1}\left[1+\left(\frac{\gamma-1}{2}\right)M^2\right]\right\}^{(\gamma+1)/2(\gamma-1)}. \quad (11\text{-}59)$$

Equations (11-57) through (11-59) are plotted for $\gamma = 1.4$ in Fig. 11-22.

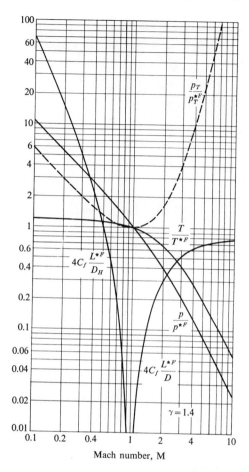

Fig. 11-22 Adiabatic flow of a perfect gas in a constant-area duct with friction.

Example 11-7. Air enters an insulated constant-area duct at a pressure of 10 psia and a Mach number of 0.3. If the hydraulic diameter is 0.2 ft and $C_f = 0.004$, what is the Mach number and pressure at a point 40 ft from the inlet?

Solution. From Fig. 11-22,

$$4C_f \frac{L^{\star F}}{D} = 5$$

and

$$\frac{p}{p^{\star F}} = 3.6.$$

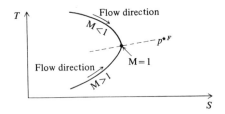

Fig. 11-23 Fanno line on T-S coordinates.

Therefore,
$$p^{\star F} = \frac{10}{3.6} = 2.78 \text{ psia,}$$

$$\left(4C_f \frac{L^{\star F}}{D}\right)_2 = \left(4C_f \frac{L^{\star F}}{D}\right)_1 - \left(4C_f \frac{L}{D}\right) = 5 - 4(0.004)\left(\frac{40}{0.2}\right) = 1.8.$$

From Fig. 11-22, $M_2 = 0.42$ for

$$4C_f \frac{L^{\star F}}{D} = 1.8 \quad \text{and} \quad \left(\frac{p}{p^\star}\right)_2 = 2.6.$$

Therefore,
$$p_2 = (2.6)(2.78) = 7.22 \text{ psia.}$$

A plot on the temperature-entropy (T-S) plane of the adiabatic, constant area friction flow is shown in Fig. 11-23. It clearly pictures the effect of friction on the acceleration or deceleration of the fluid.

11-7 FLOW ALONG A FLAT PLATE

In Section 5-1 it was shown that the dimensionless temperature and the Nusselt number were functions of four dimensionless groups, represented by the Reynolds, Prandtl, Eckert, and Grashof numbers. The dependence on Grashof number may be ignored in many problems where body forces due to buoyancy are small. The effects of compressibility and viscous heating were shown to be negligible when the Eckert number is small. There are many problems of interest, particularly in high-speed aerodynamics where the Eckert number,

$$\text{Ek} = \frac{V_\infty^2}{c_p(T_w - T_\infty)}, \tag{11-60}$$

is significant, and therefore the effects of compressibility and friction cannot be ignored. In such cases it must be expected that the correlations for dimensionless temperatures and for Nusselt numbers will contain the Eckert number as well as the Reynolds and Prandtl numbers.

11–7 FLOW ALONG A FLAT PLATE

The Eckert number is a function of the Mach number. This relationship can be easily derived by substituting $c_p = \gamma R_g/(\gamma - 1)$ and $a_\infty = \sqrt{\gamma R_g T_\infty}$ into Eq. (11–60). The equation for Eckert number is then

$$\mathrm{Ek} = \left[\frac{(\gamma - 1)T_\infty}{T_w - T_\infty}\right] M_\infty^2 \qquad (11\text{–}61)$$

Equation (11–61) indicates that $\mathrm{Ek} = f(M)$, and the Mach number can be used in place of the Eckert number in considering the effects of compressibility and viscous dissipation in correlations for high-speed flows.

The effect of compressibility on the temperature change in a flowing fluid can be shown by the use of the steady flow energy equation of a perfect gas. Assuming no work or heat flow and no changes in potential energy between points 1 and 2 in a fluid,

$$\frac{V_1^2}{2} + c_p T_1 = \frac{V_2^2}{2} + c_p T_2. \qquad (11\text{–}62)$$

When a fluid is slowed from the free stream velocity V_∞ and temperature T_∞ to a zero velocity under these conditions, then the temperature rise of the fluid will be, from Eq. (11–62),

$$\Delta T = (T_T - T_\infty) = \frac{V_\infty^2}{2c_p}. \qquad (11\text{–}63)$$

The temperature difference between the total and free stream temperature $(T_T - T_\infty)$ is called the *dynamic temperature*, which was previously defined in Section 6–1.

If the decrease in velocity to zero were accomplished reversibly (no friction), the rise in temperature would indicate a stored energy increase in the fluid equal to the work of reversible adiabatic compression on the fluid. Such an increase in temperature could also be accomplished by decelerating the fluid (either compressible or incompressible) to zero by viscous friction (in a boundary layer for example). The decrease in velocity to zero does not have to occur reversibly for Eq. (11–63) to apply. The assumptions restrict this equation to adiabatic but not necessarily to reversible flow.

In the boundary layer, friction is present and the decrease in the fluid velocity does not occur reversibly nor necessarily adiabatically. As the fluid is slowed to a lower velocity near the wall due to viscosity, a transfer of thermal energy may occur between adjacent layers of the fluid. Thus, the temperature of the fluid at the wall (i.e., the wall temperature) could be different from the stagnation temperature, even in the case where no net heat is being exchanged between the fluid and the wall.

The temperature that an insulated wall will attain in a fluid stream is referred to as the *adiabatic wall temperature*, T_{AW}. (Note that this means the *wall* is adiabatic and not necessarily the fluid in any particular flow channel or stream.) The adiabatic wall temperature is compared to the reversible adiabatic temperature rise by a factor r, called the *recovery factor*:

$$r = \frac{T_{AW} - T_\infty}{T_T - T_\infty}. \tag{11-64}$$

This definition of r, combined with Eq. (11-63), leads to the following:

$$T_{AW} - T_\infty = r\frac{V_\infty^2}{2c_p} = r\frac{\gamma - 1}{2}M_\infty^2. \tag{11-65}$$

Thus, the recovery factor relates the actual temperature rise at an insulated surface to the rise in temperature which would occur if the free stream fluid were decelerated to zero with no thermal energy exchange between adjacent layers.

The recovery factor and the adiabatic wall temperature can be determined by analysis for the flat plate in laminar flow with constant fluid properties. If the density is assumed constant, any temperature rise in the fluid will be due entirely to viscous dissipation. The results can be expected to be valid only where the effects of compression (i.e., changes in density, shock waves, etc.) are insignificant.

With constant fluid properties, the continuity and momentum equations are unchanged from the case with no viscous dissipation (Section 8–3) and are given by Eqs. (8–37) and (8–38). The thermal boundary layer equation for this case, including the effect of viscous dissipation, is

$$u\frac{\partial T}{\partial x} + v\frac{\partial T}{\partial y} = \alpha\frac{\partial^2 T}{\partial y^2} + \frac{\mu}{\rho c_p}\left(\frac{\partial u}{\partial y}\right)^2. \tag{11-66}$$

The boundary conditions for an adiabatic wall and constant free stream temperature are:

$$\frac{\partial T}{\partial y} = 0 \quad \text{at} \quad y = 0 \quad \text{and} \quad T = T_\infty \quad \text{as} \quad y \to \infty. \tag{11-67}$$

A solution to Eq. (11–66) and the boundary conditions, Eq. (11–67), was obtained by Pohlhausen [4], using a similarity technique. This will now be carried out in a slightly modified form.

Since the velocity boundary layer thickness is unchanged from the solution shown in Fig. 8–6,

$$\delta = 5\sqrt{(\nu x)/V_\infty}, \tag{11-68}$$

this characteristic length will be used to nondimensionalize length in Eq. (9–10). The temperature in this case will be nondimensionalized with the dynamic temperature, T_D. Thus,

$$y^* = \frac{y}{5\sqrt{(\nu x)/V_\infty}}, \quad T^* = \frac{T - T_\infty}{V_\infty^2/2c_p} = \frac{T - T_\infty}{T_D}. \tag{11-69}$$

The dimensionless stream function ψ^* is defined as before, in Eq. (8–41). As a result of these definitions, Eq. (11–66) becomes

$$\frac{d^2T^*}{dy^{*2}} + \frac{5}{2}\operatorname{Pr}\psi^*\left(\frac{dT^*}{dy^*}\right) + \frac{2\operatorname{Pr}}{25}\left(\frac{d^2\psi^*}{dy^{*2}}\right)^2 = 0. \quad (11\text{–}70)$$

As in Section 8–3, the Prandtl number Pr is a parameter in the resulting ordinary differential equation and will appear in the solution. The solutions to Eq. (11–70) for various values of the Prandtl number for the adiabatic wall are shown in Fig. 11–24.

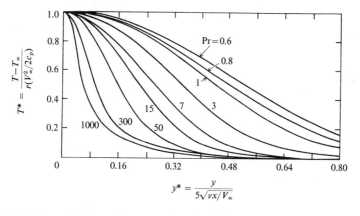

Fig. 11–24 Temperature profiles on an insulated (adiabatic) flat plate in laminar flow with constant fluid properties.

The solutions shown in Fig. 11–24 involve the recovery factor r, defined in the previous section.

Pohlhausen found the following to be approximately correct for laminar boundary layers:

$$r = \frac{T_{AW} - T_\infty}{T_T - T_\infty} = \sqrt{\operatorname{Pr}}. \quad (11\text{–}71)$$

Equation (11–71) was verified experimentally with air by Eckert [5] for subsonic velocities.

According to Eq. (11–71), an adiabatic surface exposed to a fluid stream will attain a temperature less than the stagnation temperature for $\operatorname{Pr} < 1$ and a temperature greater than stagnation for $\operatorname{Pr} > 1$. Since no energy would be added to the fluid stream by an adiabatic wall, the total energy content of the stream (described by the stagnation temperature) should remain unchanged. If the fluid at the wall is at a temperature less than the stagnation temperature ($T_{AW} < T_T$), then some portion of the boundary layer must have a local stagnation temperature greater than the free stream stagnation temperature. In other words, when $\operatorname{Pr} \neq 1$, the boundary layer

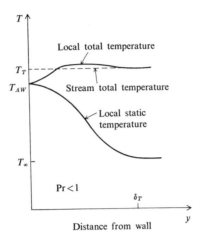

Fig. 11-25 Total and static temperatures in a boundary layer over an adiabatic wall

serves as an energy separator, resulting in different parts of the boundary having different local stagnation or total temperatures. This effect is shown in Fig. 11-25 for fluids having a value of Pr both less than and greater than 1.0 flowing over an adiabatic wall.

The special case of Pr = 1 is interesting because it implies that the local stagnation temperature is constant throughout the boundary layer for the case of an adiabatic wall. It can indeed be shown that T_T = constant is a solution to the energy equation, Eq. (11-66), for the case of Pr = 1.

Because Eqs. (11-70) and (8-58) are both linear in T, the solutions to these two equations can be combined to give the temperatures which would result when there is viscous heating with nonadiabatic walls.

This combination of solutions is carried out in the following manner. The local temperature T for the case of viscous heating with nonadiabatic (diabatic) walls may be expressed as

$$T - T_\infty = \left(\frac{T_1 - T_\infty}{T_D}\right) T_D + \left(\frac{T_2 - T_\infty}{T_W - T_\infty}\right)(T_W - T_{AW}) \quad (11\text{-}72)$$

where $(T_1 - T_\infty)/T_D$ is the solution to Eq. (11-66) for the case of viscous heating with adiabatic walls, and $(T_2 - T_\infty)/(T_w - T_\infty)$ is the solution to Eq. (8-58) for the case of diabatic walls with no viscous heating.

Equation (11-72) can be shown to satisfy the required boundary conditions at the wall and in the free stream. The result of combining the solutions is shown in Fig. 11-26 for a Prandtl number of 0.71. Each curve is the solution for a fixed value of $\frac{1}{2}$ of the product of the recovery factor and the Eckert number, $\frac{1}{2}(r \text{ Ek})$, which is given the symbol T^*_{AW}. This dimensionless

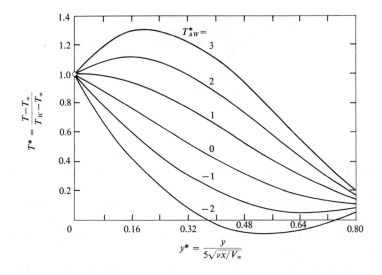

Fig. 11-26 Temperature profiles for laminar flow along a flat plate, with viscous dissipation and diabatic walls, Pr = 0.71.

quantity can also be expressed by the following relation:

$$T^*_{AW} = \frac{T_{AW} - T_\infty}{T_W - T_\infty}. \tag{11-73}$$

It can be seen that the derivative of temperature with respect to y at the wall is dependent upon the value of T^*_{AW}. When $T^*_{AW} = 1$, there is a zero temperature gradient and the wall is adiabatic, $(T_W = T_{AW})$. When $T^*_{AW} > 1$, heat flows from the fluid to the plate. When $T^*_{AW} < 1$, heat flows from the plate to the fluid. Obviously the direction of heat flow depends upon whether the wall temperature T_W is greater or less than the adiabatic wall temperature. This is shown in Fig. 11-27.

Solutions similar to that of Fig. 11-26, for a range of Prandtl numbers of approximately $(0.6 < \text{Pr} < 15)$, can be combined into a general correlation for heat transfer:

$$\text{Nu}_x = 0.332 \sqrt[3]{\text{Pr}} \sqrt{\text{Re}_x} \left(\frac{T_W - T_{AW}}{T_W - T_\infty} \right). \tag{11-74}$$

Note that the Nusselt number (and thus the heat transfer coefficient) could be negative for values of $T_{AW} > T_W$, where

$$h = \frac{\dot{q}}{A(T_W - T_\infty)}.$$

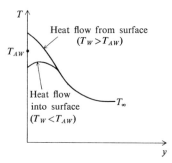

Fig. 11–27 Temperature profiles with nonadiabatic (diabatic) surfaces.

Negative values of h are foreign to our usual way of thinking. This problem is solved by use of a different temperature difference in this case.

It is logical for the heat transfer coefficient to be defined in terms of the temperature difference $(T_W - T_{AW})$ since this is the true driving potential and would always lead to positive coefficients. Therefore, defining h by

$$h = \frac{\dot{q}}{A(T_W - T_{AW})}, \tag{11-75}$$

and using this definition of h, the expression for the local Nusselt number is

$$\mathrm{Nu}_x = \frac{hx}{k} = 0.332 \sqrt[3]{\mathrm{Pr}} \sqrt{\mathrm{Re}_x}. \quad (0.6 < \mathrm{Pr} < 15) \tag{11-76}$$

The variation of fluid properties neglected in this solution can be taken into account by using Eq. (11–76) and properties (μ, c_p, ρ, k) evaluated at a reference temperature. Eckert [5] suggests the following reference temperature for flow over a flat plate:

$$T_{\mathrm{ref}} = T_\infty + 0.5(T_W - T_\infty) + 0.22(T_{AW} - T_\infty). \tag{11-77}$$

In recent years, it has become common to use a heat transfer coefficient defined in terms of enthalpy differences rather than temperature differences. This method will not be presented here.

Example 11–8. Determine the local rate of heat loss or gain from a flat plate at a point 1.5 in. from the leading edge when the plate is exposed to an airstream having a static temperature of 60°F and a velocity relative to the plate of 400 mph. The entire plate is maintained at a surface temperature of 140°F.

Solution. The recovery factor is evaluated from Eq. (11–71), using a mean film temperature of 100°F. (The reference temperature is unknown initially.)

$$r = \sqrt{\mathrm{Pr}} = \sqrt{0.72} = 0.848.$$

11-7 FLOW ALONG A FLAT PLATE

The adiabatic wall temperature is next determined.

$$T_{AW} = (T_T - T_\infty)r + T_\infty = \frac{V_\infty^2}{2c_p}r + T_\infty$$

$$= \frac{(586)^2}{(2)(32.2)(0.24)(778)}(0.848) + 60$$

$$= (28.6)(0.848) + 60 = 24.2 + 60 = 84.2°F.$$

The Reynolds number will be evaluated at the mean film temperature:

$$Re_x = \frac{V_\infty x}{\nu} = \frac{(586)(1.5)}{(0.18)(10^{-3})(12)} = 407,000.$$

The flow is assumed to be laminar at this point since $Re < 500,000$:

$$h = 0.332\frac{k}{x}\sqrt[3]{Pr}\sqrt{Re_x}$$

$$= \frac{(0.332)(0.0154)(12)}{1.5}(0.896)(638) = 23.4\frac{Btu}{hr\ ft^2\ °F}.$$

$$\frac{\dot{q}}{A} = h(T_W - T_{AW}) = (23.4)(140 - 84.2) = 1310\frac{Btu}{hr\ ft^2}.$$

The use of a reference temperature from Eq. (11–77) instead of an average film temperature could now be used to correct the above answer, but the change in the final answer will be fairly small.

The ability of engineers to predict the local temperatures that will occur in high-speed aircraft is illustrated in Fig. 11–28. For a 21-minute flight of

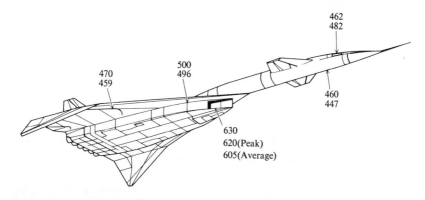

Fig. 11–28 Predicted and actual peak temperatures occurring on an XB-70A aircraft. (Courtesy *Aviation Week and Space Technology*. Used by permission.)

an XB-70A at Mach 3, the predicted peak temperatures, °F, at various locations are shown as the upper figures. The lower figures are those actually measured during such a flight. The predictions can be seen to be very good.

REFERENCES

1. A. H. Shapiro, "Dynamics and Thermodynamics of Compressible Flow," Wiley, New York (1953).
2. C. L. Dailey, and F. C. Wood, *Computation Curves for Compressible Fluid Problems*, Wiley, New York (1949).
3. "Equations, Tables and Charts for Compressible Flow," NACA Report 1135 (1953).
4. E. Z. Pohlhausen, *Angew. Math. Mech.* **78**, 1273 (1956).
5. E. R. G. Eckert, "Survey of Boundary Layer Heat Transfer at High Velocities and High Temperatures," WADC Tech. Rept. 59–624 (1960).
6. C. M. Plattner, "XB-70A Flight Research—Part 2," *Aviation Week* (June 13, 1966).
7. H. J. Schlichting, *Boundary Layer Theory*, 6ed., McGraw-Hill, New York (1968).

PROBLEMS

1. If the time between observation of a flash of lightning and the sound of thunder is 5 seconds, estimate the distance between the observer and the lightning. The air temperature is 80°F.
2. Calculate the velocity of sound at 60°F in (a) air, (b) neon, and (c) oxygen.
3. At a very large distance from a rifle bullet, the conical shock wave has a semivertex angle of 70°. Calculate the Mach number of the bullet.
4. Air flows from a tank through a converging-diverging nozzle. Stagnation conditions in the tank are: $p_T = 100$ psia and $T_T = 70°F$. Find the temperature, pressure and Mach number in the supersonic portion of the nozzle where $A/A^\star = 1.555$.
5. Determine the mass flow rate from the tank in Problem 4 if the throat area is 0.5 in.^{2}.
6. Air flows in a tube with a cross-sectional area of 1 ft^2 at the rate of 20 lb$_m$/sec. At one point in the tube the static pressure is 10 psia and the stagnation temperature is 40°F. (a) Calculate the Mach number at this section. (b) What is the smallest area the tube could be reduced without lowering the mass flow rate?
7. Air in front of a normal shock wave has a Mach number of 3.0 and a total temperature of 1200°R. What is the static temperature in front of and behind the shock wave?

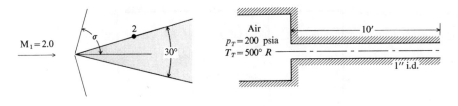

Problem 9   Problem 12

8. A Pitot tube designed for use in supersonic flow measures the total pressure behind the normal shock that forms at the tube entrance. If the supersonic Pitot tube measures a pressure of 20 psia in a supersonic wind tunnel where the static pressure is 1.38 psia, what is the Mach number of the undisturbed stream?

9. Find p_2/p_1, σ, M_2 and p_{T2}/p_{T1} for supersonic flow over the wedge shown in the figure.

10. Air enters a constant-area duct at a stagnation temperature of 700°R and a Mach number of 0.5. The air is cooled at the rate of 50 Btu/lb_m before it exits the tube. Determine the Mach number, static temperature, and stagnation temperature at the exit, assuming there is no friction.

11. The Mach number drops from 3.0 to 2.0 for air flowing in an insulated constant-area tube due to friction. If the intial pressure and temperature is 10 psia and 520°R respectively, find the final pressure and temperature.

12. A large stagnation tank is diagrammed in the figure. (a) Calculate the maximum flow rate. (b) What must the pressure be at the tube exit to achieve this maximum flow rate? (Assume $C_f = 0.004$)

13. A flat plate is maintained at a temperature of 100°F. The plate is exposed to an airstream whose free stream velocity and static temperature is 3300 ft/sec and 500°R respectively. Calculate the local rate of heat transfer $\frac{1}{4}$ in. from the leading edge.

14. For a free stream velocity of 650 mph in air and a stream temperature of 20°F, calculate (a) the Eckert number for a wall temperature of 180°F, and (b) the free stream Mach number.

15. Make a plot on log-log paper of dynamic temperature vs. velocity for air. Assume c_p is constant and equal to 0.24. Plot values for V_∞ from 100 to 1000 mph.

16. Calculate values of the Eckert number and Mach number for flow of air over a flat plate at 500 mph. The stream temperature is 100°F and the wall temperature is 300°F.

17. Air with a static temperature of 0°F flows around an object with a relative velocity of 600 mph. The object has a surface temperature of 100°F. Determine (a) the free stream Mach Number, (b) the stagnation temperature, and (c) the Eckert Number.

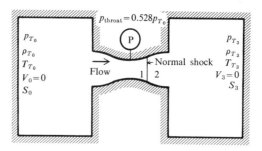

Problem 19

18. Solve Problem 6 by digital computer, assuming the mass flow rate is 30 lb_m/sec.
19. Air flows between two large reservoirs through a de Laval nozzle, as shown. Make the following terms into true statements by inserting the correct symbol ($<$, $=$, or $>$) between the terms. Subscripts 1 refer to properties just upstream of the shock, and subscripts 2 refer to properties just downstream of the shock.

dV_1	0	p_{T_1}	p_{T_2}	$d\rho_1$	0
dp_1	0	p_{T_0}	p_{T_3}	s_1	s_2
p_1	p_2	dV_2	0	ρ_1	ρ_2
T_1	T_2	dp_2	0	T_{T_0}	T_{T_3}
T_{T_1}	T_{T_2}	dT_1	0		

20. A blunt-nosed Pitot-static tube on a supersonic aircraft always lies in the subsonic flow field behind the normal shock produced by the tube itself. Two pressure taps on the tube are arranged to give the static to stagnation pressure ratio p/p_T. If p/p_T is 0.7209, at what Mach number is the aircraft flying?
21. A transport plane is in thermal equilibrium with its surroundings at 70°F while loading passengers. After takeoff it quickly climbs to altitude and attains cruising speed of Mach 0.9, where the air temperature is 10°F. Determine whether the wings will tend to heat or cool at this cruising condition.

CHAPTER 12

MULTIPHASE BEHAVIOR

This chapter provides an introduction to the behavior of fluid systems in which more than one phase is present. It will include a discussion of bubble dynamics, boiling, two-phase flow, cavitation, and condensation. Although these are widely varied topics, there are enough similarities in the basic phenomena to make it convenient to study them as a group.

The complexity of the phenomena associated with multiphase behavior has necessitated the use, in many cases, of empirical techniques for engineering design. Although most design has been essentially empirical in nature, a large amount of theoretical work has been done in nearly all aspects of multiphase behavior, and a good deal of insight has been gained from these studies. It would be impossible in this brief chapter to discuss all of the important and relevant material in the literature. New material is appearing at a very rapid rate.

The necessities for better design, prompted by economics and requirements such as those of the military and space programs, point out the need for even a better understanding of the basic phenomena.

12-1 BUBBLE AND DROPLET BEHAVIOR

In the following sections, boiling, cavitation, and the two-phase flow of a vapor or gas and liquid will be described. Bubbles are always present in boiling and cavitation. Bubbles are also present at very low vapor or gas flow rates in two-phase flow and change the character of the flow considerably from that of liquid flowing alone. In order to better understand the behavior of these bubbly flows as well as the fundamental process of boiling and cavitation, it is desirable to know something of the behavior of individual bubbles. The study of bubble growth and collapse is called *bubble dynamics*.

Bubbles may be either *gas bubbles* or *vapor bubbles*, depending upon whether the surrounding liquid is of different or of the same molecular structure, respectively, as the bubble. More precise definitions of the two types of bubbles are as follows:

Vapor bubble—An integral amount of gas submerged in a liquid phase of the same substance (e.g., a steam bubble in water). Its temperature is below the thermodynamic critical point.

Gas bubble—An integral amount of gas submerged in a liquid phase of unlike substance (e.g., an air bubble in water). The gas phase is usually at a temperature above its thermodynamic critical point. The gas is usually considered to be relatively insoluble in the liquid.

In reality both a gas and a vapor will usually exist in a bubble, since there is usually some gas (such as air) dissolved in the liquid which will diffuse into a vapor bubble and since some of the liquid will evaporate into a gas bubble. In analysis, these effects are often ignored. Both gas and vapor bubbles behave quite similarly in some respects and quite differently in others. Some of these similarities and differences will be pointed out in the following pages.

Surface Tension

Because a "free" surface of the liquid phase exists at the bubble interface, it will be necessary to consider a property of liquids that has not been discussed so far—surface tension. At a free surface of a liquid the molecules at the surface behave differently from those in the interior of the liquid. The absence of molecular forces across the interface causes tensile forces to exist between the molecules on the surface. These forces create a *surface tension*, which in some cases is very important in determining the shape of the free surface. Surface tension is particularly important in the behavior of liquids in space vehicles, where the absence of gravitational forces leaves liquid free to float about in its containing space.

A quantitative feeling for surface tension can be obtained by referring to Fig. 12–1, which shows a slider-wire arrangement with a liquid film present. The film pulls on the wire with a force per unit length equal to twice the surface tension. (The film has two surfaces.) A force F must be maintained on the slider wire to hold it in place against the pull of the surface tension σ, the force per unit length along the edge of the surface ($F = 2\sigma L$).

If the slider wire is moved to the right, the surface area of the film is increased and energy is required to bring additional molecules from the interior to the new surface. This energy is equal to the work required to move the slider wire. The film is not like a spring that is stretched, because the force F does not change with distance but depends only on the film molecular structure and the temperature. Equating the work done in moving the slider wire a distance dx to the increase in surface tension energy due to increasing area, gives

$$F\,dx = 2\sigma L\,dx$$

Surface tension is a liquid property which may be measured and tabulated in terms of other properties. It has dimensions of force per unit length. Common units are pounds force per foot and dynes per centimeter. For a pure substance, its value depends almost entirely on temperature and upon the type of substance across the interface. Impurities may affect the surface tension a great deal; for example, a small amount of detergent in water will

12-1 BUBBLE AND DROPLET BEHAVIOR 339

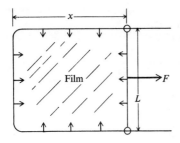

Fig. 12-1 Surface tension.

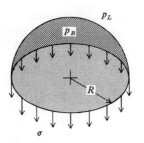

Fig. 12-2 Force balance on a static bubble.

cause a large decrease in surface tension. Electrical charge also affects surface tension. An excellent motion picture on surface tension is available in color [1].

The variation of surface tension with temperature for pure water in the presence of water vapor is shown in Table 12-1. It can be seen that the surface tension decreases with increase in temperature, having a zero value at the thermodynamic critical point. Surface tension of several important liquids is shown in Table 12-2.

Surface tension is the factor which tends to make a bubble spherical in shape. The tension causes the bubble to contract, so that it tends to assume the shape having the smallest surface area for a given volume—the sphere. In the absence of any other forces, such as gravity or viscous forces caused by liquid motion, the static bubble will be a perfect sphere.

Small drops of liquid in a gas or vapor are affected by surface tension in the same manner as bubbles. The surface tension, in the absence of other significant forces, causes a drop to contract to the shape of a sphere. Some of the relationships derived for bubbles can often be applied to predict drop behavior. The following derivation is such a case, being applicable to drops as well as bubbles.

Consider a small static bubble in the case where surface tension forces are significant enough to give the bubble an essentially spherical shape.

Surface tension creates a pressure within the bubble that is greater than the pressure existing in the surrounding liquid. This can be seen by a force balance on half of the bubble, as is shown in Fig. 12-2. The surface tension σ is the tensile force per unit length along a line on the interfacial surface. In this case the line is assumed to be a great circle around the bubble. The surface tension force along that line is

$$f_s = -2\pi R\sigma.$$

(A downward force is assumed negative.) The force downward on the upper half of the bubble due to liquid pressure is the product of the pressure p_L

Table 12-1
SURFACE TENSION OF SATURATED WATER

Temp. °F	$\sigma \times 10^4$ lb$_f$/ft
32	51.8
40	51.4
60	50.2
80	49.0
100	47.8
120	46.5
140	45.2
160	43.9
180	42.6
200	41.2
212	40.3
240	38.2
280	34.9
320	31.6
360	28.3
400	25.0
440	21.9
480	18.3
520	14.7
560	11.1
600	7.4
640	3.8
680	1.0
705.4	0

and the projected area:

$$f_L = -\pi R^2 p_L.$$

The force upward due to the vapor or gas pressure is

$$f_B = \pi R^2 p_B.$$

Summing the forces and setting them equal to zero,

$$-2\pi R\sigma - \pi R^2 p_L + \pi R^2 p_B = 0,$$

$$\frac{2\sigma}{R} = p_B - p_L. \tag{12-1}$$

It can be seen that the pressure of the vapor or gas in the bubble is greater than that in the liquid and that the difference is greater for higher surface tension and for smaller bubbles. This explains why very small bubbles

Table 12–2
SURFACE TENSION OF LIQUIDS IN CONTACT WITH THEIR VAPOR

Liquid	Temperature °C	Temperature °F	Surface tension $\frac{\text{dyne}}{\text{cm}}$	Surface tension $\frac{\text{lb}_f}{\text{ft}} \times 10^4$
Acetic acid	10	50	28.8	19.7
	20	68	27.8	19.0
	50	122	24.8	17.0
Acetone	0	32	26.21	17.96
	20	68	23.70	16.24
	40	104	21.16	14.50
Ammonia	11.1	51.8	23.4	16.0
	34.1	93.2	18.1	12.4
Argon	−188		13.2	9.04
Benzene	20	68	28.89	19.80
Bromine	20	68	41.5	28.4
Carbon dioxide	20	68	1.16	0.795
	−25	13	9.13	6.26
Carbon tetrachloride	20	68	26.95	18.47
	100	212	17.26	11.83
	200	392	6.53	4.47
Ethyl alcohol	10	50	23.61	16.18
	20	68	22.75	15.59
	30	86	21.89	15.00
Hydrogen	−255	−427	2.31	1.58
Hydrogen peroxide	18.2	64.6	76.1	52.1
Methyl alcohol	50	122	20.14	13.80
Methyl ether	−10	14	16.4	11.2
	−40	−40	21.0	14.4
Naphthalene	127	260.6	28.8	19.7
Nitrogen	−183	−297	6.6	4.52
	−193	−315	8.27	5.67
	−203	−333	10.53	7.22
Neon	−248	−414	5.5	3.77
N-octane	20	68	21.80	14.94
Oxygen	−183	−297	13.2	9.04
Tetrachlorethylene	20	68	31.74	21.75
Toluene	10	50	27.7	19.0
	20	68	28.5	19.5
	30	86	27.4	18.8

From *The Handbook of Chemistry and Physics*, 49th Edition, Robert C. Weast, ed., The Chemical Rubber Company, Cleveland, Ohio (1968–1969).

always tend to be spherical, and it is difficult to distort them. The above derivation would be applicable for a small static drop if the liquid pressure p_L were replaced by a gas pressure p_G and the bubble pressure p_B were replaced by the pressure in the drop p_D.

In order for a vapor bubble to grow in a liquid, the liquid around the bubble must be at some temperature greater than saturation temperature corresponding to the liquid pressure. Otherwise heat would flow from the bubble to the liquid and the vapor would condense. Although the concept of surface tension becomes meaningless at radii small enough for continuum concepts to be invalid, Eq. (12–1) does show a very large pressure difference exists for small bubbles, which would mean that a large amount of superheat would be required for very small bubbles to grow. Actually bubbles do not start from zero radii but must have finite-sized nuclei on which to begin. These nuclei usually consist of small gas or vapor pockets in the surface of solids in contact with the liquid. When great care is taken to remove all nucleation sites, very high degrees of superheat can be obtained in a liquid with no phase change [2]. In fact pressures can be reduced to negative values (tensile stress) in a pure liquid if care is taken to make the container scrupulously clean, or free of nucleation sites [3].

For saturated mixtures of liquid and vapor the Clausius-Clapeyron relationship of thermodynamics is valid. This relationship between saturation temperature and saturation pressure is useful to predict whether a vapor bubble will grow or collapse:

$$\frac{dp}{dT} = \frac{h_{fg}}{Tv_{fg}}, \qquad (12\text{--}2)$$

where h_{fg} is the specific enthalpy of vaporization and v_{fg} is the specific volume change during vaporization.

Assuming that the vapor can be approximated as an ideal gas and that the specific volume of the vapor v_v is much greater than that of the liquid, then $v_{fg} \approx v_v$, and Eq. (12–2) becomes

$$\frac{dp}{hT} = \frac{h_{fg}p}{R_g T^2}, \qquad (12\text{--}3)$$

where R_g is the gas constant. Equation (12–3) may be integrated between the limits of the vapor and the liquid saturation states and then combined with Eq. (12–1). Approximating the logarithmic term by the first term in a series expansion and solving for the bubble radius R, the following results:

$$R = \frac{(2\sigma T_{\text{sat}})}{\rho_v h_{fg}(T_v - T_{\text{sat}})} \qquad (12\text{--}4)$$

where R is the equilibrium radius of a vapor bubble.

Equation of Growth of a Spherical Bubble

The equation of motion of a growing or a collapsing bubble will now be derived following the work of Besant [4] and Rayleigh [5]. The following assumptions will be made:

1) The bubble is always spherical in shape, with radius R.
2) The bubble consists of an inert mixture of gas and vapor of uniform pressure.
3) The liquid surrounding the bubble is incompressible.
4) The motion of the liquid is radial and the viscous forces are not significant.

The spherical symmetry assumed above makes it desirable to use a spherical coordinate system with the origin at the center of the bubble. The continuity equation (for the liquid) reduces to

$$\frac{\partial v_r}{\partial r} + 2\frac{v_r}{r} = 0. \tag{12-5}$$

The momentum equation for the liquid becomes

$$\rho_L \left(\frac{\partial v_r}{\partial t} + v_r \frac{\partial v_r}{\partial r} \right) = -\frac{\partial p}{\partial r}. \tag{12-6}$$

The local velocity of the liquid at any radius r is obtained by integration of Eq. (12-5) with respect to r, remembering that v_r is still assumed to be a function of time t. The result is

$$v_r r^2 = f_1(t), \tag{12-7}$$

where $f_1(t)$ is an arbitrary function of time. Equation (12-7) is valid anywhere in the liquid. By letting the velocity of the bubble wall at $r = R$, that is, $(dr/dt)_{r=R}$, be designated as $\dot{R}$, Eq. (12-7) can be written as

$$\dot{R}R^2 = f_1(t). \tag{12-8}$$

Eliminating the $f_1(t)$ from Eqs. (12-7) and (12-8), an expression can be obtained for the local velocity of the liquid in terms of the bubble radius and bubble wall velocity:

$$v_r = \frac{\dot{R}R^2}{r^2} \qquad r \geq R. \tag{12-9}$$

The radial velocity of the liquid is a function of both radius r and time t, i.e., $v_r = v_r(r,t)$. Equation (12-9) may now be substituted into Eq. (12-6) and the resulting equation integrated with respect to r, holding t constant. The result is

$$p(r, t) = -\rho_L \left(-\frac{1}{r}(R^2\ddot{R} + 2\dot{R}^2 R) + \frac{1}{2r^4}(R^4\dot{R}^2) \right) + f_2(t), \tag{12-10}$$

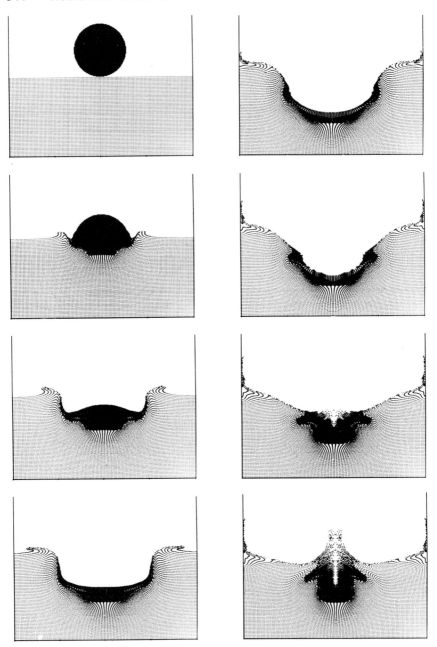

Fig. 12–3 The Marker-and-Cell Method, showing a cross-section of a splashing drop with $u_0 = 4.0$. Frames (read left, down; right, down) are at $t = 0, 5, 10, 15, 25, 28, 30,$ and 35. (Courtesy Francis H. Harlow and John P. Shannon, Los Alamos Scientific Laboratory.)

where $f_2(t)$ is an arbitrary function of time, and $\ddot{R} = (d^2r/dt^2)_{r=R}$. Substituting in the boundary conditions,

$$p(R, t) = p_g + p_v - \frac{2\sigma}{R}$$

and

$$p(\infty, t) = p_\infty.$$

Equation (12–10) becomes

$$R\ddot{R} + \tfrac{3}{2}(\dot{R})^2 = \frac{p_g + p_v - p_\infty}{\rho_L} - \frac{2\sigma}{\rho_L R}. \qquad (12\text{–}11)$$

Equation (12–11) is the equation of bubble motion for the conditions assumed. It is an ordinary nonlinear differential equation of second order. In order to attempt a solution, initial conditions on R, $\dot{R}$, and $\ddot{R}$, as well as the variation of the pressure in the liquid at relatively large distances from the bubble (p_∞), must be specified. In addition some information regarding the variation of the gas and vapor pressure within the bubble must be known or assumed. This involves a knowledge of thermodynamics and the use of energy or heat transfer relations. It is in this area that many recent investigations have been made and current research is being carried out [6–12].

An example of the theoretical work being carried out in the field of droplet behavior is shown in Fig. 12–3 [12]. Numerical techniques and a digital computer have been used to plot the motion of the liquid as a drop splashes into an undisturbed liquid surface. The predicted behavior is remarkably similar to actual behavior.

12–2 POOL BOILING

Phase changes are discussed extensively in the literature on thermodynamics. When a liquid changes phase and becomes a vapor at the free surface of a liquid, the process is referred to as *evaporation*. When the phase change occurs within the bulk of a liquid, usually in a vigorous manner, the process is referred to as *boiling*. The most common situation in which boiling occurs is where a solid surface in contact with a liquid is brought to a temperature above the saturation temperature of the liquid.

Boiling may occur in processes designed to accomplish such purposes as (1) the generation of large quantities of vapor, as in a steam power plant or in chemical processing, or (2) the production of large heat fluxes with moderate temperature differences.

These two objectives are not unrelated, since large heat fluxes are necessary to provide the latent heat of vaporization (h_{fg}) necessary when generating large quantities of vapor within equipment of reasonable size. The large heat fluxes in boiling are possible, since the energy being put into

346 MULTIPHASE BEHAVIOR

the process is being utilized in the phase change without raising the temperature of the liquid significantly.

The two general situations under which boiling occurs are called (1) pool boiling and (2) forced convection boiling. In pool boiling, the change of phase occurs at a heated surface immersed in a pool of liquid. Even in pool boiling there is liquid motion caused by free convection currents, as well as additional agitation because of the bubbles growing and collapsing. In the case of forced convection boiling, there is the added complexity of the forced flow of liquid through the tube or past the surface on which boiling occurs. Pool boiling will be considered first as it is the simpler of the two.

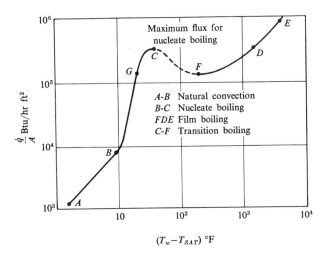

Fig. 12-4 The pool boiling curve.

Pool boiling occurs in a transition from free convection as local temperatures in the liquid exceed the saturation temperature. When the liquid is significantly below saturation temperature and the heater temperature is low, only free convection occurs. Free convection, transition to boiling, and changes in the boiling mechanism can best be seen from the results of a typical boiling experiment, shown in Fig. 12-4. The figure is a plot of heat flux vs. the temperature difference ($T_w - T_{sat}$) between a horizontal cylindrical heater surface and the saturation temperature of the surrounding liquid. This curve depicts the results of a common pool boiling experiment; in this case a small-diameter, electrically conducting wire, was submerged in a stationary body or pool of water at saturation temperature and atmospheric pressure.

Figure 12-4 shows that for values of ($T_w - T_{sat}$) less than 10°F (curve A–B) the heat flux is that value which is predicted for free convection with no

Fig. 12-5 Nucleate pool boiling from a cylindrical heater at low heat flux. (Courtesy J. W. Westwater and T. B. Dunskus.)

phase change. For water at this condition, when the heater surface is less than approximately 10°F above the saturation temperature, there is no significant effect due to phase change, and evaporation occurs only at the pool free surface. The heater surface in most cases exceeds the saturation temperature of the liquid by some perceptible amount before noticeable boiling commences. This required temperature excess, which is explained by Eq. (12-4), depends upon the type of fluid and the pressure. In addition, the type and condition of heater surface is significant since these factors determine the number and size of the small vapor and gas bubbles from which bubble growth must begin.

At a temperature difference of approximately 10°F, the curve shown in Fig. 12-4 begins to rise more sharply. This sharp increase in the slope of the curve between B and C, indicating a large increase in the heat transfer coefficient, is due to the inception of *nucleate boiling* at the heater surface. The liquid near the heater has attained sufficient superheat to cause significant numbers of bubbles to grow. The process of originating a bubble is referred to as *nucleation*. The term nucleate boiling is used since the vapor bubbles are formed at discrete nucleation sites. Nucleate boiling from a horizontal cylindrical heater is pictured in Fig. 12-5.

In a saturated liquid the vapor bubbles formed at the surface grow until the inertia and/or buoyancy forces overcome the surface forces holding the bubble to the heater, at which point the bubble breaks free and rises into the liquid. As the heat flux is increased, the rate at which bubbles are formed at a site increases and columns of vapor bubbles are soon apparent in the liquid. If the liquid above the heater is below the saturation temperature, sufficient heat may be transferred from the vapor bubble to the cooler liquid as the bubble rises so that it gets smaller and may completely condense and collapse before arriving at the free surface of the liquid. This situation is called *subcooled* or *local boiling*. With local boiling, little or no vapor reaches the free surface of the pool.

Again referring to Fig. 12–4, as the heat flux at the surface is increased, the number of nucleation sites increases until the surface becomes very crowded with sites. The discrete columns of vapor bubbles leaving the surface merge and form large globules of vapor and continuous vapor columns. Small discrete vapor bubbles are no longer apparent. Such a situation is shown in Fig. 12–6.

The crowding of the nucleation sites leads to a situation where much vapor and little liquid is present near the heater surface. Since the thermal conductivity of the vapor is much less than that of the liquid, the heat transfer coefficient will decrease with increasing quantities of vapor. If the rate of adding energy is maintained constant and the heat transfer rate is decreasing, energy will be stored in the heater, causing its temperature to rise until it is again losing energy at the same rate that it is being supplied. The boiling curve in Fig. 12–4 begins to level off sharply with increasing heat flux and decreasing heat transfer coefficient until finally, at point C, a limiting condition exists. A continuous blanket of vapor may at this point surround the heater, giving a very high resistance for heat transfer between the heater surface and the liquid. Even though the energy input to the heater surface is maintained constant, the temperature difference $(T_w - T_{\text{sat}})$ may rise sharply until an equilibrium condition is attained at D, where the heat loss through the vapor film to the liquid phase is equal to the heat flux at the heater surface. This film boiling condition involves a vary large temperature difference and, thus, for a fixed fluid temperature, a very high heater surface temperature. The heater surface temperature is usually so large that the heater material will melt or fail and the term *burnout* is commonly used to describe this condition. Since the heater may not actually fail or burn out under certain conditions, the terms *peak* or *critical heat flux* are more appropriate.

The case where the heater surface does not fail and the vapor film blankets the surface is commonly called *film boiling*. Film boiling is shown in Fig. 12–7. Note that there is a net production of vapor and globules of vapor are breaking away from the vapor blanketing the heater surface. Radiation may play an important role in film boiling, particularly at temperatures such as represented by point E in Fig. 12–4. Film boiling is usually undesirable, since the heater surface temperature is so high. Thus the heater flux is ordinarily maintained at some safe value well below that represented by point C, and nucleate boiling occurs. Nucleate boiling refers to the boiling mechanism occurring between points B and C, even though nucleation sites and bubbles may not be identifiable in the upper part of that range. Some authors use the term DNB or *departure from nucleate boiling* to describe the condition leading to point C.

Fig. 12-6 Formation of large vapor globules at higher heat fluxes: Methanol on a ⅜ in. o.d. copper tube. (Courtesy J. G. Santanglio and J. W. Westwater.)

Fig. 12-7 Film boiling: Methanol on a ⅜ in. o.d. copper tube. (Courtesy J. G. Santanglio and J. W. Westwater.)

If film boiling is attained without destroying the heater surface, the heat flux may be reduced and point F approached. Lower heat fluxes than that represented by F would usually lead to a sudden decrease in wall temperature back to values represented by point G on the curve, although, with care, points along a line between F and B may apparently be obtained.

The region shown between C and F is an unstable region most easily attained if the heater surface temperature rather than heat flux is the independently controlled variable. Such a situation might be obtained by condensing vapor at various pressures inside the heater tube instead of heating by

350 MULTIPHASE BEHAVIOR

electricity. In this region of the curve a combination of film boiling and nucleate boiling may occur locally along the heater surface.

Mechanism of Nucleate Boiling

The very high heat fluxes with relatively small temperature differences possible in nucleate boiling have intrigued investigators for decades.

One of the earlier mechanisms proposed to explain the high heat fluxes was the *bubble agitation mechanism*, Fig. 12–8(a). In this mechanism the rapidly growing, departing bubbles are assumed to mix or agitate the surrounding liquid. Schlieren and shadowgraph studies have indeed shown that there is appreciable agitation of the hot fluid (thermal layer) near the heater surface. Quantitative reasoning leads one to conclude, however, that agitation might be an important contribution to the high rates of heat transfer, but this does not appear to be the only cause. Apparently other impor-

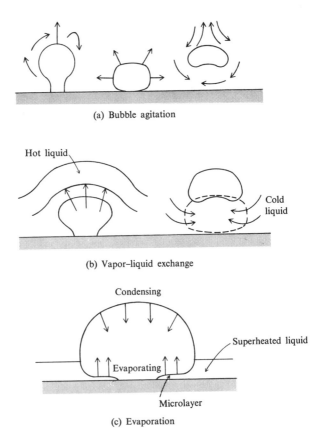

Fig. 12–8 Schematic representation of proposed nucleate boiling mechanisms.

tant factors must also be present. A mechanism proposed by Forster and Grief [13] is shown in Fig. 12–8(b).

When the liquid near the heater surface becomes sufficiently superheated, a small bubble forms at a nucleation site on the heater surface. This site is usually a small surface cavity which contains a pocket of a gas or vapor. The bubble grows rapidly in the superheated liquid, supposedly due to the transfer of heat from the heater surface through the liquid to the bubble [14]. The rapidly growing bubble displaces the adjacent superheated liquid, causing it to move away from the surface into the cooler bulk of the liquid. When the bubble grows sufficiently to break away from the surface, it leaves a space which must be immediately replaced by relatively cool liquid. In the case of subcooled boiling, the bubble may collapse before leaving the surface if it grows so large that much of it is in the subcooled liquid. In such a case, the collapsing bubble is likewise replaced by cooler liquid. The "pumping action" of the bubble thus is used to explain the high heat transfer rates found in nucleate boiling. This theory for the boiling mechanism seems to be verified by the findings of Rohsenow and Clark [15] who reported that, in subcooled boiling, only a small part of the total heat transferred was represented by the latent heat of the growing and collapsing bubbles. Photographic studies have confirmed that this proposed pumping action does in fact exist. The possibility of condensation occurring during bubble growth might invalidate the theory, however.

More recent studies [16–20] have indicated the possibility of the existence of a thin layer of liquid on the solid surface beneath the bubble during part of the growth period, as can be seen in Fig. 12–8(c). It can be shown that such a thin layer could lead to very high heat flux rates during its evaporation. This *microlayer* or *evaporation theory*, if validated by further studies, may require modification or even rejection of the older concepts, such as the pumping action theory described above. Grahams and Hendricks [20] have made a most excellent assessment of convection, conduction, and evaporation effects in nucleate boiling.

Several factors affect the rate at which bubbles are formed and leave a heater for a given liquid and temperature difference. The nature of the heater surface, its roughness, and its past history are quite important since they determine the number, size, and gas content of the nucleation sites. The boiling curve shown in Fig. 12–4 is only for saturated water at atmospheric pressure with a platinum heater. The curve would be quite different for other liquids or with other types of heater surfaces. Even small amounts of impurities in the liquid can affect the boiling curve.

Nucleate Boiling Correlation

Although several correlations of nucleate boiling data have been developed, no one correlation has been proven to be entirely reliable under all conditions.

Typical of these correlations is the one by Rohsenow [21] proposed to take into account the effect of pressure on pool boiling. Like most boiling correlations, it contains empirical constants. It is presented because it is one of the more popular correlations. It attributes the major heat transfer to conduction from the surface to the liquid. A bubble Reynolds number was used to account for the effect of increased liquid agitation because of bubble action. The result is

$$C_l \left(\frac{T_w - T_{sat}}{h_{fg}} \right) = C_{sf} \left(\frac{(\dot{q}/A)_b}{\mu_l h_{fg}} \sqrt{\frac{g_c \sigma}{g(\rho_l - \rho_v)}} \right)^{1/3} \Pr_l^{1.7}, \quad (12\text{-}12)$$

where $\Pr_l$ is the Prandtl number of the liquid, C_l is the specific heat of the liquid, C_{sf} is an empirical coefficient, and $(\dot{q}/A)_b$ is the heat flux due to boiling. Available experimental data gives the surface-fluid combination (C_{sf}) values listed in Table 12–3.

Table 12–3

Surface-fluid combination	C_{sf}
Water-nickel	0.006
Water-platinum	0.013
Water-copper	0.013
Water-brass	0.006
CCl_4-copper	0.013
Benzene-chromium	0.010
n-Pentane-chromium	0.015
Ethyl alcohol-chromium	0.0027
Isopropyl alcohol-copper	0.0025
35% K_2CO_3-copper	0.0054
50% K_2CO_3-copper	0.0027
n-Butyl alcohol-copper	0.0030

The correlation equation, Eq. (12–12), proposed by Rohsenow is shown plotted in Fig. 12–9 with the data of Addoms [22]. A reasonably good correlation is seen to exist between the theory and the experimental data used. The need for the empirical coefficient and the lack of good correlation for all conditions indicates a need for further work. The assumptions made in deriving Eq. (12–12) are certainly open to question in light of the more recent microlayer theory of boiling.

Reference to the literature will yield a large number of correlations on various aspects of nucleate boiling. The peak, or critical, heat flux (point C in Fig. 12–4) may be estimated for various fluids, pressure levels, and gravitational levels by an equation of Rohsenow and Griffith [23]:

$$\frac{(\dot{q}/A)_{max}}{\rho_v h_{fg}} = 143 \left(\frac{g}{g_0}\right)^{1/4} \left(\frac{\rho_l - \rho_v}{\rho_v}\right)^{0.6}. \tag{12-13}$$

The coefficient in Eq. (12–13) is proportional to the product of bubble diameter and the frequency of departure of bubbles from the heater surface and appears to be fairly constant over a wide range of conditions. The value of 143 is for units of ft/hr. Since the remaining terms on the right are dimensionless, the units selected for the quantities on the left-hand side of Eq. (12–13) must be consistent and also give units of ft/hr. This is accomplished by expressing the heat flux $(\dot{q}/A)$ in Btu/hr ft², the density ρ_v in lb$_m$/ft³, and the enthalpy of vaporization (h_{fg}) in Btu/lb$_m$. The ratio $(g/g_0)^{1/4}$ shows the effect of variation of gravity on peak heat flux.

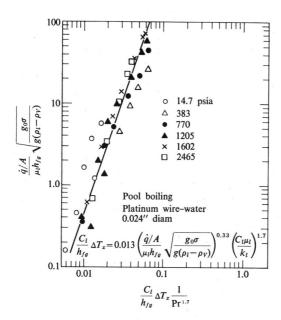

Fig. 12–9 Correlation of pool boiling data for water. [21] (Used with permission of the American Society of Mechanical Engineers.)

Another equation for predicting critical heat flux was given by Zuber [24]:

$$\frac{(\dot{q}/A)_{max}}{\rho_v h_{fg}} = 0.18 \left(\frac{g}{g_0}\right)^{1/4} \left(\frac{\sigma(\rho_l - \rho_v)}{\rho_v^2} gg_c\right)^{1/4} \left(\frac{\rho_l}{\rho_l + \rho_v}\right)^{1/2}. \tag{12-14}$$

The coefficient of Eq. (12-14) is dimensionless but both sides of the equation have dimensions of (length/time). Care must be taken to use consistent units.

Film Boiling Correlation

An equation has been suggested by Bromley [25] for the prediction of heat transfer by conduction through the vapor film in pool-film boiling around a horizontal cylinder of diameter D:

$$h_c = 0.62 \left[\frac{k_v^3 \rho_v (\rho_l - \rho_v) g [h_{fg} + 0.4 c_{p_v}(T_w - T_{\text{sat}})]}{D \mu_v (T_w - T_{\text{sat}})} \right]^{1/4}. \quad (12\text{-}15)$$

All properties in Eq. (12-15) should be evaluated at saturation temperature. The equation neglects radiation heat transfer between the heater surface and the liquid. Such an assumption is normally invalid due to the high heater surface temperature. The radiation can be taken into account by use of a *radiation heat transfer coefficient*, h_r, defined by

$$h_r = \frac{\sigma \epsilon (T_w^4 - T_{\text{sat}}^4)}{(T_w - T_{\text{sat}})}, \quad (12\text{-}16)$$

where σ is the Stefan-Boltzmann constant, and ϵ is the heater surface emissivity (see Chapter 15). The total heat loss in film boiling to a pool of liquid can then be calculated by the use of a total heat transfer coefficient h_T, defined by

$$\frac{\dot{q}}{A} = h_T (T_w - T_{\text{sat}}). \quad (12\text{-}17)$$

Bromley [25] showed that

$$h_T = h_c \left(\frac{h_c}{h_T} \right)^{1/3} + h_r. \quad (12\text{-}18)$$

Note that a trial-and-error solution is necessary since h_T is not expressed explicitly. Note also that the radiation loss decreases the conduction loss due to a thickening of the vapor layer. This effect is apparent in Eq. (12-18).

The film boiling behavior of liquid masses on hot surfaces, commonly called the *Leidenfrost phenomenon*, has been extensively studied in recent times. A summary of these studies has been made by Bell [51].

As may be surmised from the above discussion, the boiling process is extremely complicated, and relatively large amounts of experimentation have yielded insight into only a few of the perplexing problems. Empirical results have been the main product of the research and the chief basis for the design of new equipment. Although some theories have shown promise and an increasing amount of analytical work is being performed, it is still impossible to predict with certainty the boiling behavior of an untested liquid—and/or a new geometry.

Recent general references on boiling include a book by Tong [26] and review articles by Rohsenow [27], Leppert and Pitts [28], and Westwater [29].

12–3 LIQUID-GAS FLOWS

There are many engineering situations where two or more different substances flow simultaneously as a mixture. The substances may differ in phase, that is, the mixture may be some combination of solid, liquid, or gas. In such a case the flow would be referred to as *multiphase flow*. The most common case of multiphase flow is where only two phases exist. This is commonly referred to as *two-phase flow*. The possible types of two-phase flow are gas-liquid (or vapor-liquid), gas-solid (or vapor-solid), and liquid-solid. The case of gas-liquid flow will be discussed in this section.

The simultaneous flow of a liquid and a gas (or vapor) occurs in a number of types of situations, including pipeline flow of oil and gas, and the flow in evaporator tubes. These two examples illustrate the two basic types of flow to be considered in this section. The flow of oil and gas is a *two-component*, two-phase flow, since the oil and the gas are different in molecular structure as well as in phase. The flow of a liquid and its vapor in an evaporator tube is an example of a *single-component*, two-phase flow. The characteristics of these two types of two-phase flow can be quite different. For convenience, however, the term gas will be used to refer to either a vapor or to a gas that is different in molecular composition from the liquid phase.

The relative ratio of the gas phase flow rate and the liquid phase flow rate is the most important description of a two-phase flow. Nearly all of the other characteristics of importance depend upon the value of this ratio. The ratio determines the *flow pattern*, or distribution, of the two phases.

The types of flow patterns which might exist in horizontal two-phase gas-liquid flow are shown in Fig. 12–10, taken from Gouse [30]. The patterns are shown in the order of increasing ratio of gas to liquid flow. At gas flow rates that are low relative to liquid flow rates, only small bubbles of gas are present and these normally collect near the top of the tube. As the relative gas flow rate is increased, the bubbles coalesce into plugs of vapor which also tend to move along the top of the tube. With a further increase in gas flow, the plugs grow together and a continuous stream of gas flows above the liquid in a stratified manner.

If the relative gas flow rate is sufficiently high, waves will be formed on the liquid surface which may become large enough to cause the entire cross section to be filled with liquid slugs at distributed points along the flow. In annular flow, the gas rate is large enough to spread the liquid out into an annulus (not necessarily symmetrical) around the tube wall, with the gas occupying the core. Some liquid may be entrained in the core as droplets. At sufficiently large gas rates, the liquid annulus may be swept very thin, or

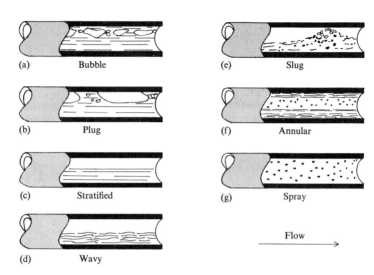

Fig. 12-10 Flow patterns in horizontal two-phase flow.*

possibly eliminated, and the liquid phase exists only in a dispersed or spray form. This is sometimes referred to as a *mist* or *fog flow*.

The flow patterns which exist in vertical upward flow of a gas and liquid, taken from Gouse [30], are shown in Fig. 12-11. In this case the gravitational force acts along the axis of the tube and the flow patterns are symmetrical so that neither the wavy nor stratified flow patterns found in horizontal flow exist. At very high gas flow rates, a mist flow exists, as in the horizontal case, although this is not shown in Fig. 12-11.

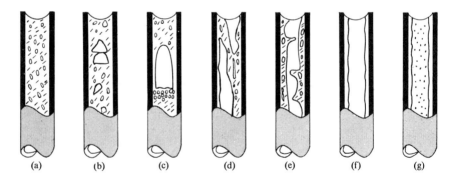

Fig. 12-11 Flow patterns in two-phase vertical upflow. a) homogeneous bubble or froth; b) nonhomogeneous bubble; c) slug; d) and e) semi-annular; f) annular; g) spray annular.*

* From S. W. Gouse, Jr., "An Introduction to Two-Phase Gas Liquid Flow," M.I.T. Report 8734 to Office of Naval Research.

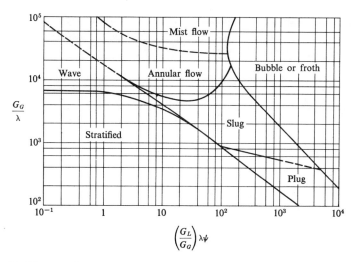

Fig. 12-12 Flow regime correlation for adiabatic, horizontal two-phase flow in which

$$\lambda = \left[\left(\frac{\rho_g}{0.075}\right)\left(\frac{\rho_L}{62.3}\right)\right]^{1/2} \qquad \psi = \left(\frac{73}{\sigma_L}\right)\left[\mu_L\left(\frac{62.3}{\rho_L}\right)^2\right]^{1/3}$$

G_L = mass velocity of liquid phase in $lb_m/hr\ ft^2$

G_G = mass velocity of gas phase in $lb_m/hr\ ft^2$

ρ_G = density of gas phase in lb_m/ft^3

ρ_L = density of liquid phase in lb_m per ft^3

μ_L = dynamic viscosity of liquid in centipoise

σ_L = surface tension of liquid in dynes/cm

(From O. Baker, "Design of Pipelines for Simultaneous Flow of Oil and Gas," *Oil Gas J.*, 185, July 26, 1954.)

Several investigators have attempted to predict which flow pattern would exist under certain specified conditions. There has not been general success or agreement in these attempts; however, the work of Baker [32] for adiabatic flow appears to be most widely accepted in the literature. His results, shown in Fig. 12-12, were obtained from a large number of horizontal flow measurements of other investigators.

The lines shown on Baker's charts do not represent sharp boundaries between flow regimes but instead show the general areas where transitions between patterns occur. The effect of increasing the relative flow rate of either phase is clearly indicated. Also, the liquid surface tension and viscosity and the density of the two phases are obviously important parameters.

In adiabatic flow, or where it is assumed that no liquid evaporates, the flow patterns usually remain fixed over a fairly large distance in the direction of flow. Large pressure drops may cause a sufficient change in gas density to cause a change in flow pattern along the tube, however.

Quantitative Relations for Two-Phase Flow

In two-phase, gas-liquid flow the two phases usually do not flow at the same velocity. This difference in velocity is usually referred to as *slip*. Because of slip the relative amounts of liquid and gas present in a given control volume of the conduit along the tube is not the same as the relative flow rates of the liquid and gas. The term *void fraction*, designated α, is used to specify the ratio of the volume occupied by the gas to the volume occupied by the mixture (or total volume):

$$\alpha = \frac{\text{volume of gas}}{\text{volume of gas} + \text{volume of liquid}}. \quad (12\text{-}19)$$

The relative quantity of gas and liquid may be defined in terms of either the *flowing mass quality* X_F or the *static mass quality* X_S, where

$$X_F = \frac{\text{mass rate of gas flow}}{\text{mass rate of mixture flow}}, \quad (12\text{-}20)$$

and

$$X_S = \frac{\text{mass of gas present in a control volume}}{\text{mass of mixture present in that control volume}}. \quad (12\text{-}21)$$

These two qualities become identical when there is no slip between phases.

The ratio of the average velocity of the gas to the average velocity of the liquid is the *phase velocity ratio* V_R:

$$V_R = \frac{V_G}{V_L}. \quad (12\text{-}22)$$

In a two-phase flow either phase might be assumed to flow along in the tube, that is, to occupy the entire cross section, but at the same mass flow rate, pressure, and temperature as in the two-phase flow. The resulting fictitious velocity of that phase is called the *superficial velocity:*

$$V_{LS} = \frac{\text{mass rate of liquid flow}}{(\text{density of liquid})(\text{area of tube})} = \frac{\dot{m}}{\rho_L A}, \quad (12\text{-}23)$$

$$V_{GS} = \frac{\text{mass rate of gas flow}}{(\text{density of gas})(\text{area of tube})} = \frac{\dot{m}}{\rho_G A}. \quad (12\text{-}24)$$

Several relations can now be developed, such as

$$\frac{V_G}{V_L} \equiv \frac{\rho_L}{\rho_G}\left[\frac{(\alpha-1)X_F}{\alpha(X_F-1)}\right], \quad (12\text{-}25)$$

$$\frac{\rho_G}{\rho_L} = \frac{(\alpha-1)X_S}{\alpha(X_S-1)}, \quad (12\text{-}26)$$

$$V_{LS} = V_{GS}\left(\frac{1-X_F}{X_F}\right)\frac{\rho_G}{\rho_L}. \quad (12\text{-}27)$$

Pressure Drop in Two-Phase Flow

The pressure drop which occurs with a gas-liquid flow is of primary interest. Experience has shown that pressure drops in two-phase flow are usually much higher than would occur for either phase flowing alone at the same mass rate. As in any flow, the total pressure drop along a tube depends upon three factors:

(1) friction, due to viscosity,
(2) change of elevation,
(3) acceleration of the fluid.

Friction is present in any flow situation, although in some cases it may contribute less than the other two factors. In horizontal flow the change in elevation is zero, and there would be no pressure drop due to this factor. Where there is a small change in gas density or little evaporation occurring, the pressure drop due to accleration would usually be small. In flow with large changes of density or where evaporation is present, however, the acceleration pressure drop may be very significant.

There has been a variety of schemes proposed for predicting the pressure drop in two-phase flow. Some investigators have attempted to develop theories and equations valid only for a particular flow pattern. These have the disadvantage of requiring that the type of flow pattern be known, and they lead to complexities in those situations where flow patterns change along the flow. Other investigators have attempted to develop single theories valid over several flow regimes. These attempts have not been completely successful.

A widely used method of predicting pressure drop in two-phase flow was developed by Lockhart and Martinelli [33]. Their method, which is essentially empirical, is based upon measurements of isothermal flow of air and several liquids under essentially incompressible conditions. *The pressure drop predicted is that due to friction only*. They arbitrarily divided all flows into one of four categories, depending upon whether the superficial Reynolds number of each phase indicated turbulent or laminar flow.

Turbulent flow was assumed to occur in a phase if the superficial Reynolds number of that phase was greater than 2000. The superficial Reynolds number of each phase was calculated, using the properties of that phase, the superficial velocity of that phase, and the pipe diameter. The two-phase frictional pressure drop was determined as a function of the pressure drops for each phase flowing alone, at the superficial phase velocity. A separate correlation was necessary for each category of flow, i.e., turbulent-turbulent, turbulent-laminar, laminar-turbulent, and laminar-laminar. The results are shown in Fig. 12–13, together with predicted values of void fraction for each phase.

The four types of flow defined by Lockhart and Martinelli do not necessarily distinguish the flow patterns that have been identified visually.

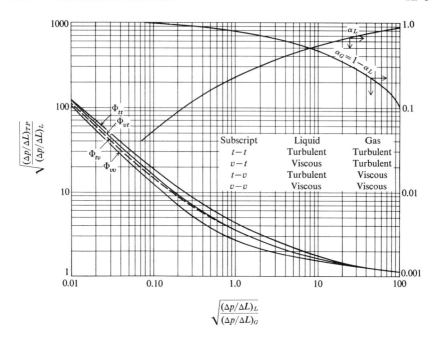

Fig. 12–13 Curves of Lockhart and Martinelli for horizontal two-phase flow without vaporization. (From *Chemical Engineering Progress*, Volume 45, 1949, with permission.)

In fact, experiments in which the liquid surface tension was changed by adding Nekal B-X indicated changes in the flow patterns observed at given flow rates but did not result in appreciable changes in pressure drop. Although this seems strange, it tends to support the idea of Martinelli and Lockhart that for pressure drop calculation, the type of flow should be defined by the value of the Reynolds numbers rather than by visually observed patterns. A theoretical basis for the Lockhart-Martinelli correlation has been presented by Chisolm [31].

Example 12–1. Calculate the pressure drop in 50 ft of 5-in. i.d. smooth pipe through which a mixture of air and water at 100°F and one atmosphere pressure is flowing. The mass flow rate of the water is 75 lb_m/sec and that of the air is 2.5 lb_m/sec.

Solution. The properties at 100°F and one atmosphere are:

$$v_L = 0.74 \times 10^{-5} \text{ ft}^2/\text{sec}$$
$$v_G = 0.18 \times 10^{-3} \text{ ft}^2/\text{sec}$$
$$\rho_L = 62.0 \text{ lb}_m/\text{ft}^3$$
$$\rho_G = 0.071 \text{ lb}_m/\text{ft}^3$$

The superficial velocities and Reynolds number can now be calculated from Eqs. (12–23) and (12–24):

$$V_{LS} = \frac{(75) \text{ lb}_m/\text{sec } (4)(144)}{(62.0) \text{ lb}_m/\text{ft}^3 (\pi)(5)^2 \text{ ft}^2} = 8.89 \text{ ft/sec},$$

$$\text{Re}_{LS} = \frac{V_{LS} D}{\nu} = \frac{(8.89) \text{ ft/sec } (5) \text{ ft}}{(0.74 \times 10^{-5}) \text{ ft}^2/\text{sec } (12)} = 5 \times 10^5,$$

$$V_{GS} = \frac{(2.5)(4)(144)}{(0.071)(\pi)(5)^2} = 258 \text{ ft/sec},$$

$$\text{Re}_{GS} = \frac{(258)(5)}{(0.18 \times 10^{-3})(12)} = 5.96 \times 10^5.$$

The friction factors for these respective Reynolds numbers can now be obtained from Fig. 9–2:

$$f_{LS} = 0.0130 \quad \text{and} \quad f_{GS} = 0.0127.$$

Using these friction factors and the superficial velocities, the superficial pressure drops can be calculated from Eq. 7–16:

$$\left(\frac{\Delta p}{\Delta L}\right)_L = \frac{-\rho_L V_{LS}^2 f_{LS}}{2 g_c D} = \frac{-(62.0)(8.89)^2(0.013)(12)}{(2)(32.2)(5)} = -2.38 \frac{\text{lb}_f}{\text{ft}^2\text{-ft}},$$

$$\left(\frac{\Delta p}{\Delta L}\right)_G = \frac{-(0.071)(258)^2(0.0127)(12)}{(2)(32.2)(5)} = -2.24 \frac{\text{lb}_f}{\text{ft}^2\text{-ft}},$$

$$\sqrt{\frac{(\Delta p/\Delta L)_L}{(\Delta p/\Delta L)_G}} = \sqrt{\frac{2.38}{2.24}} = \sqrt{1.063} = 1.03.$$

From Fig. 12–13 for turbulent-turbulent flow,

$$\sqrt{\frac{(\Delta p/\Delta L)_{TP}}{(\Delta p/\Delta L)_L}} = 4.1.$$

Therefore,

$$\left(\frac{\Delta p}{\Delta L}\right)_{TP} = (4.1)^2(-2.38) = -40 \frac{\text{lb}_f}{\text{ft}^2\text{-ft}},$$

$$\Delta p = -40 \frac{\text{lb}_f}{\text{ft}^2\text{-ft}} 50 \text{ ft} = 2000 \frac{\text{lb}_f}{\text{ft}^2},$$

or

$$\Delta p = \frac{2000 \frac{\text{lb}_f}{\text{ft}^2}}{144 \frac{\text{in}^2}{\text{ft}^2}} = 13.9 \text{ psi}.$$

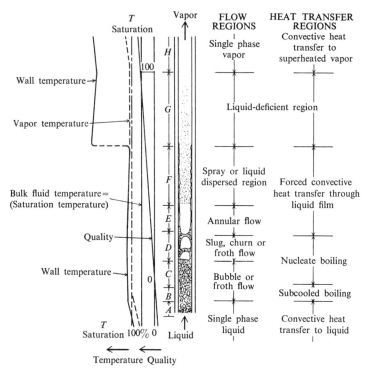

Fig. 12-14 Flow patterns in two-phase flow vertical upflow with heat transfer. (From *Two-Phase Flow and Heat Transfer*, with permission of John Collier and Graham B. Wallis.) [36]

The determination of the probable flow pattern and the void fraction is left as an exercise at the end of the chapter.

A discussion of two-phase flow and pressure drop has been presented by Parker [34]. An extensive collection of pressure drop data in two-phase flow has been collected by Dukler [35] and is available on IBM computer cards. Dukler used this data to develop a method for predicting pressure drop which appears to be quite useful.

Flow With Heat Addition

In flow with heat addition or pressure drop sufficient to cause a significant quantity of the liquid to evaporate, the flow patterns may change rapidly along a tube. This is shown at an exaggerated rate in Fig. 12-14 for vertical upward flow of an evaporating liquid.

In two-phase flow with heat addition, there is an interaction between the pressure drop and the heat transfer. Figure 12-15 shows the variation with length of the saturation temperature, mean (or bulk) coolant temperature,

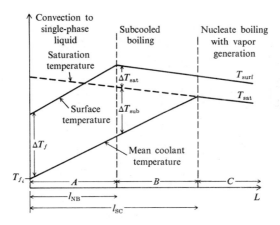

Fig. 12-15 Variation of temperatures in forced convection boiling. (From *Two-Phase Flow and Heat Transfer*, with permission of John Collier and Graham B. Wallis.) [36]

and the heater surface temperature for regions A, B, and C of Fig. 12-14. The entering liquid or coolant is assumed to be initially subcooled by an amount ΔT_{sub}. For constant heat addition along the tube the bulk coolant temperature increases linearly with distance until it reaches the saturation temperature. The pressure drop, due to friction, elevation, and momentum changes, causes the bulk fluid temperature to drop downstream from where nucleate boiling with net vapor generation begins, even though heat is being added. This is because the boiling mixture is essentially in thermodynamic equilibrium, and a decreasing pressure corresponds to a decreasing saturation temperature. This temperature drop and an increase in vapor formation causes the wall temperature to decrease in the direction of flow after having reached a maximum value at the start of subcooled boiling (in which bubbles form at the wall and then collapse in the cooler fluid in the stream).

The local heat transfer coefficient increases with increasing distance from inlet due to the higher fluid velocity caused by expansion in the evaporation process (Fig. 12-16). This velocity increase continues up to a point where there is insufficient moisture remaining in the tube (dryout) or where departure from nucleate boiling (DNB) occurs. For a given set of conditions, the magnitude of the heat flux determines which of these two possibilities will occur. The occurrence of the various boiling regions along the tube with variation in the magnitude of the wall heat flux is shown in Fig. 12-17. Dryout is seen to be most probable at low heat fluxes, whereas DNB occurs at the higher heat fluxes. In either case, the coefficient drops sharply to approximately the value for pure vapor flowing alone (Fig. 12-16). This sudden, large decrease in the local heat transfer coefficient leads to a large increase in the local wall temperatures, where heat input at the wall is constant

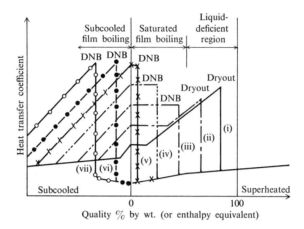

Fig. 12-16 Variation of heat transfer coefficient with quality, with increasing heat flux as parameter. (From *Two-Phase Flow and Heat Transfer*, with permission of John Collier and Graham B. Wallis.) [36]

along the tube length. This large wall temperature increase can lead to damage of the tube wall and possible failure due to melting or loss of strength. Such a condition is referred to as *burnout* and is a topic of considerable interest to research and design engineers working with evaporator tubes and boiling water nuclear reactors.

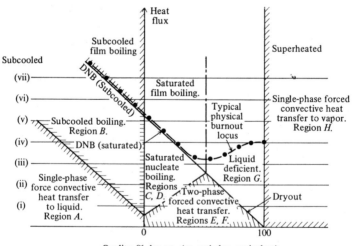

Fig. 12-17 Regions of two-phase forced convection heat transfer as a function of quality with increasing heat flux as ordinate. (From *Two-Phase Flow and Heat Transfer*, with permission of John Collier and Graham B. Wallis.) [36]

Most of the significant analytical studies of forced convection boiling and burnout have attempted to use in some way the knowledge gained from studies of pool boiling. In fact, since most commercial and practical boiling processes involve forced convection boiling rather than pool boiling, the real value of pool boiling studies are in the contribution that they have made to the field of forced convection boiling.

The prediction of heat transfer rates in forced convection boiling, for example, can be made by a method suggested by Chen [38], similar to a method previously proposed by Rohsenow [39]. Chen's method assumes that both boiling and forced convection affect the heat transfer rate and that superposition can be used to give an expression for the two-phase heat transfer coefficient h_{TP} in terms of the nucleate boiling coefficient h_{NB} and the forced convection coefficient h_c:

$$h_{TP} = h_{NB} + h_c. \tag{12-28}$$

The value of the convective coefficient h_c is obtained from a modified form of the Dittus-Boelter equation, Eq. (9–51), and the nucleate boiling coefficient is obtained from the equation of Forster and Zuber [41]. Wallis and Collier [36] give a step-by-step method for determining h_{TP} and suggest that this is the best correlation available at the present time for general use with a wide range of fluids in forced convection *saturated* boiling. Chen [40] has modified the method to make it suitable for liquid metals.

The general effect of velocity and of inlet subcooling on forced convection *subcooled* boiling can be seen in Fig. 12–18. The inception of nucleate boiling occurs approximately at the break in each curve. The effect of velocity can be seen to become insignificant after boiling commences. Subcooling has a significant effect after the inception of boiling.

Burnout caused by departure from nucleate boiling in forced convection boiling is difficult to predict for the general situation, although a large number of correlations have been proposed, each of which seem to be suitable for a certain limited range of parameters. Generally the heat flux at burnout increases with increased inlet subcooling and increased inlet velocity and decreases with the length-diameter ratio of the heater section. At lower heat fluxes, where dryout rather than DNB is likely to occur, a number of analyses have been presented in the literature. Most of these are based on an annular-mist model, which attempts to predict when there will no longer be a liquid film left at the heater surface.

In most practical evaporator designs, burnout is avoided by removing the two-phase fluid from the evaporator tubes at low quality and separating the liquid and vapor mechanically. The vapor may then be used in the saturated state or directed into a superheater if such is desired. The separated liquid is usually returned to the inlet of the evaporator tubes.

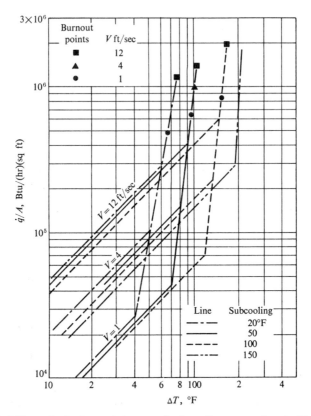

Fig. 12-18 Effect of velocity and degree of subcooling on surface boiling of water. (From *Heat Transmission*, by McAdams. Copyright 1954 by William H. McAdams. Used by permission of McGraw-Hill Book Company.)

Pressure drop in evaporating flows has been studied by Martinelli and Nelson [42], using a modification of the method of Lockhart and Martinelli [33], discussed earlier in this section.

12-4 CAVITATION

In the previous two sections the formation of vapor in a liquid due to heat addition was discussed. It is also possible to form vapor in a liquid without external heat addition by lowering the pressure to a value below the saturation pressure. Such a formation of vapor is referred to as *cavitation*, a name coined from the fact that the vapor formed appears as cavities or holes in the liquid.

Cavitation has been of interest for many years primarily because of the damage that can be done by the vapor cavities as they collapse upon encountering a higher pressure. The very small bubbles usually formed in cavitation

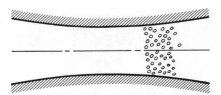

Fig. 12-19 Cavitation in a venturi.

collapse at a very rapid rate upon encountering pressure above saturation, and this creates extremely large local pressures in the liquid. When this occurs near a solid surface, as is often the case, the extremely large local pressure can dent the surface or knock off bits of material. Although the damage done by one bubble may be very slight, the cumulative effect over a period of time can be very significant. Cavitation damage to ship propellers and rudders was a serious problem that led to some of the earliest research work in this area. Recent studies have indicated that some of the damage in cavitation may be due to the very high temperatures created when the bubbles collapse.

Pressures below the vapor pressure of the liquid can be created in pump inlets, venturis, valves, and other flow restrictions. Cavitation in a venturi is shown in Fig. 12-19. Cavitation can also occur in stationary pools of liquid exposed to solid surfaces where severe levels of vibration occur. The water-cooled cylinder liner of a large diesel engine, for example, might vibrate and cause cavitation to occur on the water side with resulting damage. Pumps and bearings are commonly subjected to cavitation damage.

Cavitation can be analyzed by some of the same techniques used to study nucleate boiling, since the bubble formation and collapse are in many ways similar. Nucleation sites must be present before cavitation can occur, since, as shown in Section 12-1, bubbles cannot readily form from zero or near zero radii. The roughness of the solid surfaces in contact with the liquid, as well as the presence of dissolved gases in the liquid, is important and can cause significant variations in cavitating characteristics between otherwise similar systems.

Gaseous cavitation may occur as dissolved gases come out of solution with decreasing pressure (similar to the situation that occurs when a bottle of soda pop is opened). This gaseous cavitation may occur at pressures above the saturation pressure corresponding to the fluid temperature and thus may occur without vaporous cavitation taking place. Gas content of the liquid, therefore, may significantly affect the conditions at which cavitation occurs.

It is common to define a *cavitation number*, Cv, by a dimensionless ratio:

$$\text{Cv} = \frac{p - p_v}{\rho V_0^2/2}, \qquad (12\text{-}29)$$

Fig. 12-20 Supercavitation flow behind a $\frac{3}{4}$ in. diameter flat disk at a velocity of 50 ft/sec. (Courtesy Hydrodynamics Laboratories, California Institute of Technology.)

where V_0 is the liquid reference velocity, p is the static pressure at the point of interest, and p_v is the vapor pressure of the liquid corresponding to its temperature.

The *critical cavitation number* is the value of Cv at which cavitation inception occurs in a flowing system. The cavitation number is useful in modeling and in comparing experimental results. It is difficult to scale cavitation effects, however, due to the numerous factors that influence the nucleation process, such as surface roughness and gas content of the liquid. As in nucleate boiling there is a wide variation in experimental results and consequently great difficulty in predicting when cavitation will or will not occur.

In the flow of water around objects at high relative velocity—in hydrofoils, for example—trailing wakes may be formed that are practically void of liquid. This is known as *supercavitation*, a phenomenon that greatly affects the drag. Supercavitation is shown in Fig. 12-20.

12-5 CONDENSATION

Another important process involving two phases of the fluid is condensation. Here vapor is subjected to cooling sufficient to cause the formation of liquid. The liquid collects on the walls of the cool surface and is usually removed by the effect of gravity or the drag of the moving vapor.

As in the case of boiling, two distinct modes of condensation may occur. The more common case is called *film condensation*, in which the condensing vapor forms a thin liquid film over the entire surface. In order for this mode to occur, the surface must be wettable. Film condensation usually occurs when a noncontaminated vapor is in contact with a clean surface.

If a surface is contaminated or treated with a substance (such as a thin teflon coating) to create poor wettability or to make it nonwettable, the second mode of condensation, called *dropwise condensation*, may occur. Here the vapor condenses in minute drops which grow in place until they

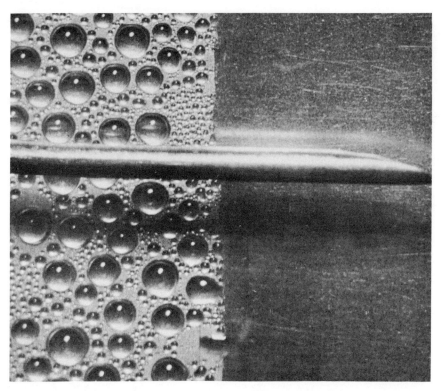

Fig. 12-21 Comparison of dropwise (*left*) and film condensation. (Courtesy J. F. Welch and J. W. Westwater.)

reach a certain size, when gravity or the motion of vapor causes them to move across the cooling surface, sweeping away all of the drops in their path (Fig. 12–21).

A surface in a practical case may have both types of condensation simultaneously. This is called *mixed condensation.*

Film Condensation

A simple analysis of the heat transfer in film condensation was first carried out by Nusselt [43] in 1916. Nusselt studied laminar film condensation on a vertical flat plate, using force, mass, and energy balances on a small element of the condensing film. His derivation has been frequently modified, and is, therefore, presented here in brief form. Nusselt's assumptions were:

1. The surface of the film is at saturation temperature. (Actually no condensation could occur for this case as a small temperature difference is necessary. This difference can be very small, however.)

2. The flow of liquid in the film is laminar. This is an important practical case, since many surfaces used for cooling are short, and transition to turbulence does not occur.

3. The momentum and shear forces at the interface can be neglected; the flow of liquid is entirely due to gravity. This assumes that the viscosity of the vapor is much less than that of the liquid and that the relative velocity between the vapor and liquid is not large.

4. The inertia forces in the liquid film are neglected.

5. A linear temperature profile is assumed to exist in the liquid film. More refined calculations have shown this assumption to be reasonable although not actually correct.

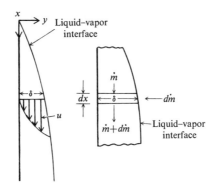

Fig. 12-22 Illustration for analysis of laminar film condensation.

The derivation is illustrated in Fig. 12-22. For this simple case the momentum equation applied to liquid, Eq. (4-44a), becomes

$$\nu \frac{d^2 u}{dy^2} = -g. \qquad (12\text{-}30)$$

The boundary conditions on Eq. (12-30) are: at $y = 0$, $u = 0$ (no slip at wall); at $y = \delta$, $du/dy = 0$ (no interface forces), where δ is the condensate layer thickness. The solution of Eq. (12-30) and the boundary conditions are easily obtained by integration:

$$u = \frac{\rho g}{\mu}\left(\delta y - \frac{y^2}{2}\right). \qquad (12\text{-}31)$$

The average velocity is found to be

$$u_{\text{avg}} = \frac{1}{\delta}\int_0^\delta u\, dy = \frac{1}{\delta}\int_0^\delta \frac{\rho g}{\mu}\left(\delta y - \frac{y^2}{2}\right) dy.$$

12-5 CONDENSATION

This yields

$$u_{\text{avg}} = \frac{\rho g \delta^2}{3\mu}. \tag{12-32}$$

At any section at elevation x, the mass flow downward is (for unit depth of film)

$$\dot{m} = \rho \delta u_{\text{avg}} = \frac{g\rho^2 \delta^3}{3\mu}, \tag{12-33}$$

and

$$d\dot{m} = \frac{g\rho^2}{\mu} \delta^2 \, d\delta. \tag{12-34}$$

This value $d\dot{m}$ is the excess mass flow rate at $x + dx$ over that at x (see Fig. 12-22). This excess mass flow rate must come from the condensation at the interface, which is given by

$$d\dot{m} = \frac{d\dot{q}}{h_{fg}}, \tag{12-35}$$

where h_{fg} is the enthalpy of vaporization.

The flow is laminar and parallel to the wall, and the energy given up by condensation must be transmitted to the cool wall by *conduction* through the liquid layer. This explains the fact that the film coefficient which will be derived depends only on this liquid thickness. Application of the Fourier-Biot equation gives

$$d\dot{q} = k \frac{T_{\text{sat}} - T_w}{\delta} (dx \cdot 1). \tag{12-36}$$

Combining Eqs. (12-34), (12-35), and (12-36) and solving for dx, gives

$$dx = \frac{g\rho^2 h_{fg}}{k\mu} \frac{1}{T_{\text{sat}} - T_w} \delta^3 \, d\delta. \tag{12-37}$$

Integrating with $\delta = 0$ at $x = 0$ gives an expression for the condensate layer thickness:

$$\delta = \left[\frac{4\mu k(T_{\text{sat}} - T_w)x}{g\rho^2 h_{fg}} \right]^{1/4}. \tag{12-38}$$

This equation for thickness of the liquid layer can now be used to obtain heat flow rates. Combining Eqs. (12-38) and (12-36) and integrating from $x = 0$ to $x = L$, the height of the plate, gives

$$\dot{q} = \frac{4}{3} \left[\frac{g\rho^2 h_{fg} k^3 (T_{\text{sat}} - T_w)^3 L^3}{4\mu} \right]^{1/4}. \tag{12-39}$$

Equation (12-39) is the historical result of Nusselt.

If a local coefficient of heat transfer is defined as $d\dot{q} = h\,dx(T_{\text{sat}} - T_w)$, then

$$h = \left[\frac{g\rho^2 h_{fg} k^3}{4\mu x(T_{\text{sat}} - T_w)}\right]^{1/4}. \tag{12-40}$$

A defined average heat transfer coefficient, $\dot{q} = \bar{h}A(T_{\text{sat}} - T_w)$, gives

$$\bar{h} = \frac{4}{3}\left[\frac{g\rho^2 h_{fg} k^3}{4\mu L(T_{\text{sat}} - T_w)}\right]^{1/4}. \tag{12-41}$$

Note that $\bar{h} = \frac{4}{3}h_{x=L}$ as in laminar free convection. In all the equations above, a mean condensate temperature, $\frac{1}{2}(T_w + T_{\text{sat}})$, should be used to evaluate liquid properties.

The condensate Reynolds number $\text{Re}_\delta = (u_{\text{avg}}\delta)/\nu$ is important in prediction of transition to turbulence in the liquid film. The expressions for u_{avg} and δ are obtained from Eqs. (12-32) and (12-38). The result is

$$\text{Re}_\delta = 0.943\left[\frac{k\rho^{2/3}g^{1/3}(T_{\text{sat}} - T_w)L}{\mu^{5/3}h_{fg}}\right]^{3/4}. \tag{12-42}$$

The best observations indicate that the critical Reynolds number for transition from laminar to turbulent flow is in the range from 300 to 500.

Several improvements and modifications of Nusselt's method have been suggested in the literature. Rohsenow [44] considered convection as well as conduction in the liquid layer but neglected inertia forces. The result was not significantly different from Nusselt's. Rohsenow et al. [45] suggested a scheme for accounting for the effect of vapor velocity. Correction for inertia of the liquid film has been correlated by Peck and Reddie [46]. Correction for particle subcooling gives a nonlinear temperature profile and has been correlated by Bromley [47]. Bromley suggested that the latent enthalpy h_{fg} should be multiplied by the term $[1 + 0.4(c_p\,\Delta T/h_{fg})]$.

Sparrow and Gregg [48] have shown the usefulness of boundary theory in the analysis of this problem, as briefly described below.

Neglecting viscous dissipation and assuming laminar film condensation with constant fluid properties on a vertical wall, the boundary layer equations for the condensate film are:

$$\frac{\partial u}{\partial x} + \frac{\partial v}{\partial y} = 0, \tag{12-43}$$

$$\rho\left(u\frac{\partial u}{\partial x} + v\frac{\partial u}{\partial y}\right) = g(\rho - \rho_v) + \mu\frac{\partial^2 u}{\partial y^2}, \tag{12-44}$$

$$\rho c_p\left(u\frac{\partial T}{\partial x} + v\frac{\partial T}{\partial y}\right) = k\frac{\partial^2 T}{\partial y^2}, \tag{12-45}$$

12-5 CONDENSATION

with the boundary conditions

$$\text{at} \quad y = 0, \quad u = v = 0 \quad T = T_w,$$
$$\text{at} \quad y = \delta, \quad \frac{\partial u}{\partial y} = 0 \quad T = T_{\text{sat}}. \tag{12-46}$$

Using the stream function defined in Eq. (4–30), the equations above become

$$\frac{\partial \psi}{\partial y} \frac{\partial^2 \psi}{\partial x \partial y} - \frac{\partial \psi}{\partial x} \frac{\partial^2 \psi}{\partial y^2} = g \frac{(\rho - \rho_v)}{\rho} + \nu \frac{\partial^3 \psi}{\partial y^3}, \tag{12-47}$$

$$\frac{\partial \psi}{\partial y} \frac{\partial T}{\partial x} - \frac{\partial \psi}{\partial x} \frac{\partial T}{\partial y} = \alpha \frac{\partial^2 T}{\partial y^2}. \tag{12-48}$$

The partial differential equations were reduced to ordinary differential equations by a transformation similar to the method of Blasius discussed in Section 8–3. The variables used in this transformation were

$$y^* = \frac{y}{(x/c)^{1/4}} \quad \text{where} \quad c = \left[\frac{gc_p(\rho - \rho_v)}{4\nu k}\right], \tag{12-49}$$

$$\psi^* = \frac{\psi}{4x^{3/4}\alpha c}, \quad \text{and} \quad \theta = \frac{T - T_{\text{sat}}}{T_w - T_{\text{sat}}}.$$

Both ψ^* and θ are assumed to be functions of y^* only.

Making the transformations of Eq. (12–49) on Eqs. (12–47) and (12–48) gives

$$\frac{d^3\psi^*}{dy^{*3}} + \frac{1}{\Pr}\left[3\frac{d^2\psi^*}{dy^{*2}}\psi^* - 2\left(\frac{d\psi^*}{dy^*}\right)^2\right] + 1 = 0, \tag{12-50}$$

$$\frac{d^2 T^*}{dy^{*2}} + 3\psi^* \frac{dT^*}{dy^*} = 0, \tag{12-51}$$

with the boundary conditions

$$\text{at} \quad y^* = 0, \quad \psi^* = 0 \quad \frac{d\psi^*}{dy^*} = 0 \quad \theta = 1,$$
$$\text{at} \quad y^* = y_\delta^*, \quad \frac{d^2\psi^*}{dy^{*2}} = 0 \quad \theta = 0. \tag{12-52}$$

An overall energy balance is employed to relate the dimensionless film thickness to known parameters. Equating the heat removed at the wall to the energy required for condensing and subcooling of the liquid flowing past some point x measured from the top of the vertical surface gives the following results:

$$\int_0^x k\left(\frac{\partial T}{\partial y}\right)_{y=0} dx = \int_0^\delta h_{fg} \rho u \, dy + \int_0^\delta \rho u c_p (T_{\text{sat}} - T) \, dy. \tag{12-53}$$

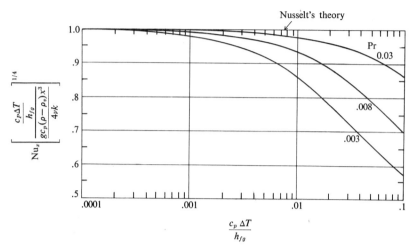

Fig. 12-23 Local heat-transfer results for low Prandtl number range, from solution of complete boundary layer equations ($h_{\text{avg}} = \frac{4}{3} h_x$). [48] (Used with permission of the American Society of Mechanical Engineers.)

Using the transformation variables, Eq. (12–53) becomes

$$\frac{c_p \Delta T}{h_{fg}} = -3 \frac{\psi^*_{y^*=\delta}}{(d\theta/dy^*)_{y^*=\delta}}. \tag{12-54}$$

The solution of Eqs. (12–50) and (12–51) can then be expressed in terms of the parameters Pr and ($c_p \Delta T/h_{fg}$). The temperature distribution in the condensate and the local heat transfer coefficient can be expressed in terms of these two parameters.

As a result of solving Eqs. (12–50) and (12–51), we find that there is little difference between this solution and those where acceleration is not considered for Prandtl numbers greater than 1.0. For lower Prandtl numbers, such as those for liquid metals, the variations between the solutions of Sparrow and Gregg and those of Nusselt are shown in Fig. 12–23.

Experience has shown that actual heat transfer coefficients on vertical surfaces are about 20 percent above those predicted by the theories of Nusselt and others mentioned above. Condensation rates for liquid metals cannot be predicted with any reliability using these theories.

Turbulent Flow of Liquid Film

It would be expected that for tall surfaces the liquid layer will become turbulent. Experience has shown that this is so, as indicated by the fact that the heat transfer coefficients in such cases do not continue to decrease with increasing plate length L, as expected for laminar flow in Eq. (12–41).

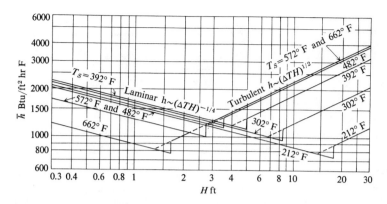

Fig. 12-24 Variation of average heat transfer coefficient with height for temperature difference $(T_{sat} - T_W) = 18°$ F. (From *Fundamentals of Heat Transfer* by Grober, Erk, and Grigull. Copyright 1961 by McGraw-Hill Book Company, Inc. Used with permission of McGraw-Hill Book Company.)

Grigull [49] suggested an empirical equation which can be used for the turbulent case:

$$\bar{h} = 0.30 \times 10^{-2}\left[\frac{k^3 g \rho^2 (T_{sat} - T_w) L}{\mu^3 h_{fg}}\right]^{1/2} \quad (12\text{-}55)$$

Equation (12-55) is a prediction of the average heat transfer coefficient over the entire surface, assuming that turbulence exists on the surface ($Re_\delta > 300$ to 500). No laminar flow calculation need be made in such cases.

An interesting representation of the condensation of steam on vertical surfaces is shown in Fig. 12-24, utilizing Eqs. (12-41) and (12-55) and assuming an arbitrary temperature difference: $(T_{sat} - T_w) = 18°F$.

The equations developed above may be used for condensation on vertical tubes as well as vertical plane surfaces provided the tube diameter is much larger than the film thickness δ.

Laminar Condensation on Horizontal Tubes

Condensation on a horizontal tube of diameter D can be analyzed in a manner similar to the Nusselt development. This analysis must account for the fact that the surface continually changes in its geometric orientation as the film flows from the top, around, and to the bottom of the tube. The result is

$$\bar{h} = 0.725\left[\frac{g\rho(\rho - \rho_v)k^3 h_{fg}}{D\mu(T_{sat} - T_w)}\right]^{1/4} \quad (12\text{-}56)$$

If the condenser consists of banks of tubes arranged one tube above the other, the condensate from higher tubes drops on the tubes below, increasing the condensate thickness and reducing the film coefficient. In such cases the coefficient may be predicted by

$$\bar{h} = 0.725 \left[\frac{g\rho(\rho - \rho_v)k^3 h_{fg}}{n D\mu(T_{sat} - T_w)} \right]^{1/4}. \tag{12-57}$$

Equation (12-57) is the average heat transfer coefficient for n tubes arranged in a vertical bank. The reduction predicted by Eq. (12-57) is offset by the splashing of condensate and condensation directly on the liquid layer between tubes. Empirical knowledge is often used in such cases.

Example 12-2. A horizontal tube with 1-in. o.d. is used to condense saturated steam on the outside surface. The saturation temperature of the steam is 120°F and the surface temperature of the tube is maintained at 80°F. Determine the number of pounds of vapor condensed per hour per foot of tube length. Neglect the effect of vapor velocity, the presence of noncondensable gases, liquid inertia, and subcooling.

Solution. For small-diameter horizontal tubes the flow is laminar and Eq. (12-56) is applicable. The density of the vapor will be neglected. At 120°F the value of h_{fg} is 1026 Btu/lb$_m$. The liquid properties are evaluated at the average liquid film temperature, 100°F:

$$\rho = 62 \text{ lb}_m/\text{ft}^3, \quad k = 0.364 \text{ Btu/hr ft °F},$$

$$\mu = 0.458 \times 10^{-3} \text{ lb}_m/\text{ft sec}.$$

$$\bar{h} = 0.725 \left[\frac{(32.2)\left(\frac{\text{ft}}{\text{sec}^2}\right)(62)^2 \left(\frac{\text{lb}_m}{\text{ft}^3}\right)^2 (0.364)^3 \left(\frac{\text{Btu}}{\text{hr ft °F}}\right)^3 \left(1026 \frac{\text{Btu}}{\text{lb}_m}\right)}{(\frac{1}{12})(\text{ft})(0.458 \times 10^{-3})\left(\frac{\text{lb}_m}{\text{ft sec}}\right)(40)(\text{°F})(\frac{1}{3600})\left(\frac{\text{hr}}{\text{sec}}\right)} \right]^{1/4}$$

$$= 0.725(14.4 \times 10^{12})^{1/4} = 1410 \text{ Btu/hr ft}^2 \text{ °F}.$$

$$\frac{\dot{q}}{L} = \frac{\bar{h}A}{L}(T_{sat} - T_w) = \frac{(1410)(\pi)}{12}(40) = 14{,}800 \text{ Btu/hr ft}.$$

$$\dot{m}_{cond} = \frac{\frac{\dot{q}}{L}}{h_{fg}} = \frac{14800 \text{ Btu/hr ft}}{1026 \text{ Btu/lb}_m} = 14.5 \text{ lb}_m/\text{hr ft}.$$

Dropwise Condensation

Very little data are available for the case of dropwise condensation shown in Fig. 12-21. Coefficients can be very large because the bare surface is exposed to vapor over a large percentage of its area. Film coefficients 10 times as large as for film condensation have been observed.

REFERENCES

1. L. M. Trefethen, "Surface Tension in Fluid Mechanics," Film #21610 Encyclopedia Britannica Educational Corp. Chicago.
2. J. F. Lee and F. W. Sears, *Thermodynamics*, 2d ed., Addison-Wesley (1963), pp. 168.
3. L. J. Briggs, *J. Appl. Phys.* **21,** 721 (1950).
4. Besant, *Hydrostatics and Hydrodynamics*, Cambridge, England (1859).
5. J. W. S. Rayleigh, *Phil. Mag.* **34,** 94 (1917).
6. S. A. Zwick and M. S. Plesset, "On the Dynamics of Small Vapor Bubbles in Liquids," *J. Math. Phys.* **33,** 308 (1955).
7. L. W. Florschuetz and B. T. Chao, "On the Mechanics of Vapor Bubble Collapse," ASME Paper No. 64-HT-23 (1964).
8. D. Hsieh, "Some Analytical Aspects of Bubble Dynamics," ASME Paper 65-FE-19 (1965).
9. H. K. Forster and N. Zuber, "Growth of a Vapor Bubble in a Superheated Liquid," *J. Appl. Phys.* **25,** 474 (1957).
10. N. Zuber, "The Dynamics of Vapor Bubbles in Non-uniform Temperature Fields," *International J. Heat Mass Transfer*, **2,** 83 (1961).
11. G. Birkhoff, R. S. Margulies, and W. A. Horning, "Spherical Bubble Growth," *Phys. Fluids*, **1** (3), 201 (1958).
12. F. H. Harlow and J. P. Shannon, "Distortion of a Splashing Liquid Drop," *Science* **57,** 549 (1967).
13. H. K. E. Forster and R. Grief, *Trans. ASME, J. Heat Transfer*, **81,** 43 (1959).
14. M. Jakob, *Heat Transfer*, Wiley, New York (1949), Vol. 1, Chapter 29.
15. W. M. Rohsenow and J. A. Clark, *Trans. ASME* **73,** 609 (1951).
16. F. D. Moore and R. B. Mesler, "The Measurement of Rapid Surface Temperature Fluctuations During Nucleate Boiling of Water," *AIChE J.* p. 620, (Dec. 1961).
17. T. F. Rogers and R. B. Mesler, "An Experimental Study of Surface Cooling by Bubbles during Nucleate Boiling of Water," *AIChE J.* p. 656, (Sept. 1964).
18. R. R. Sharp, "The Nature of Liquid Film Evaporation During Nucleate Boiling," NASA TN D-1997 (1964).
19. R. C. Hendricks and R. R. Sharp, "Initiation of Cooling Due to Bubble Growth on a Heating Surface," NASA TN D-2290 (1964).
20. R. W. Graham and R. C. Hendricks, "Assessment of Convection, Conduction and Evaporation in Nucleate Boiling," NASA TN D 3943 (1967).
21. W. M. Rohsenow, "A Method of Correlating Heat Transfer Data for Surface Boiling of Liquids," *Trans. ASME* **74,** 969 (1952). pp 969–975.
22. J. N. Addoms, ScD Thesis in Chemical Engineering, M.I.T. (1948).

23. W. M. Rohsenow and P. Griffith, "Correlation of Maximum Heat Flux Data Boiling of Saturated Liquids," Heat Transfer Symposium, AIChE, Louisville, 1955.
24. N. Zuber, *Trans. ASME* **80,** 711 (1958); also ASME Paper 57-HT-4 (1957).
25. L. A. Bromley, et al., *Indus. Eng. Chem.* **45,** 2639 (1953).
26. L. T. Tong, *Boiling Heat Transfer and Two-Phase Flow*, Wiley, New York (1965).
27. W. H. Rohsenow, "Heat Transfer with Boiling," *Modern Developments in Heat Transfer*, Academic Press, New York (1963).
28. G. Leppert and C. C. Pitts, "Boiling," *Advances in Heat Transfer*, Vol. 1, Academic Press, New York (1964).
29. J. W. Westwater, "Things We Don't Know About Boiling Heat Transfer," *Fundamental Research in Heat Transfer*, Macmillan, New York (1963).
30. S. W. Gouse, Jr., "An Introduction to Two-Phase Gas Liquid Flow," M.I.T. Report 8734 to Office of Naval Research.
31. D. Chisolm, "A Theoretical Basis for the Lockhart-Martinelli Correlation for Two-Phase Flow," *International J. Heat Mass Transfer* **10,** 1767 (1967).
32. O. Baker, "Design of Pipelines for Simultaneous Flow of Oil and Gas," *Oil Gas J.* 185 (July 26, 1954).
33. R. W. Lockhart and R. D. Martinelli, "Proposed Correlation of Data for Isothermal, Two-Phase, Two-Component Flow in Pipes," *Chem. Eng. Prog.* **45,** 39 (1949).
34. J. D. Parker, "Two-Phase Flow and Pressure Drop with Heat Transfer," Eleventh Annual Heat Transfer Conference, Oklahoma State University, School of Mechanical Engineering, 1965.
35. A. E. Dukler, et al., "Frictional Pressure Drop in Two-Phase Flow," *AIChE J.* 38 (Jan. 1964).
36. G. B. Wallis and John G. Collier, *Two-Phase Flow and Heat Transfer*, Short Course at Stanford University, Vol. III (1967).
37. W. H. McAdams, *Heat Transmission*, 3d ed, McGraw-Hill, New York (1954).
38. J. C. Chen, "A Correlation for Boiling Heat Transfer to Saturated Fluids in Convective Flow," ASME Paper 63-HT-34 (1963).
39. W. M. Rohsenow, "Heat Transfer with Evaporation," *Heat Transfer*, University of Michigan Press, Ann Arbor (1953).
40. J. C. Chen, "A Proposed Mechanism and Method of Correlation for Convective Boiling Heat Transfer with Liquid Metals," Third Annual Conference on High Temperature Liquid Metal Heat Transfer Technology, Oak Ridge National Laboratory, 1963; also BNL Report 7319.
41. H. K. Forster and N. Zuber, "Dynamics of Vapor Bubbles and Boiling Heat Transfer," *AIChE J.* **1** (4), 531 (1955).
42. R. C. Martinelli and D. B. Nelson, "Prediction of Pressure Drop during Forced Circulation Boiling of Water," *Trans. ASME* **70,** 695 (1948).

43. W. Nusselt, "Die Oberflachenkondensation des Wasserdampfes," *Z. Ver. Deutsch. Ing.* **60**, 541 (1916).
44. W. H. Rohsenow, "Heat Transfer and Temperature Distribution in Laminar Film Condensation," *Trans. ASME*, **78**, 1645 (1956).
45. W. M. Rohsenow, J. H. Weber, and A. T. Ling, *Trans. ASME*, **78**, 1637 (1956).
46. R. E. Peck and W. A. Reddie, "Heat Transfer Coefficients for Vapors Condensing on Horizontal Tubes," *Indus. Eng. Chem.* **43**, 2926 (1951).
47. L. A. Bromley, R. S. Brodkey, and N. Fishman, "Heat Transfer in Condensation," *Indus. Eng. Chem.* **44**, 2962 (1959).
48. E. M. Sparrow and J. L. Gregg, "A Boundary Layer Treatment of Laminar Film Condensation," *J. Heat Transfer* **81**, 13 (1959).
49. U. Grigull, *Forsch. Gebiete Ingenieurw.* **13**, 49 (1942).
50. H. Gröber and S. Erk, *Fundamentals of Heat Transfer*, 3d ed., rev. by U. Grigull, McGraw-Hill, New York (1961).
51. K. J. Bell, "The Leidenfrost Phenomenon—A Survey," *Chemical Engineering Progress, Symposium Series*, No. 79 (1967), p. 73.

PROBLEMS

1. Rohsenow and Griffith [23] correlated pool boiling peak heat flux data with

$$\frac{\dot{q}(/A)_{max}}{\rho_v h_{fg}} = 143 \left(\frac{g}{g_0}\right)^{1/4} \left(\frac{\rho_l - \rho_v}{\rho_v}\right)^{0.6} \text{ ft/hr.}$$

   a) Use this equation to determine the peak heat flux for water at atmospheric pressure and compare with Fig. 12–4. Assume $g = g_0$.
   b) Determine the pressure at which the largest peak heat flux will occur for water by plotting values of heat flux versus pressure at 500, 1000, 1500, 2000, and 2500 psia.

2. Zuber [24] presented an equation of the following form for the peak heat flux in pool boiling:

$$\frac{\dot{q}(/A)_{max}}{\rho_v h_{fg}} = 0.18 \left(\frac{g}{g_0}\right)^{1/4} \left(\frac{\sigma(\rho_l - \rho_v)}{\rho_v^2} gg_c\right)^{1/4} \left(\frac{\rho_l}{\rho_l + \rho_v}\right)^{1/2} \text{ ft/hr.}$$

   Use this equation for the case in Problem 1 and compare the answers.

3. Use Rohsenow's Eq. (12–12) to calculate the heat flux for pool boiling of water on a platinum wire at atmospheric pressure. Assume that the bulk of the water is saturated and the surface temperature of the wire is 250°F. Check the answer against Fig. 12–9.

4. Calculate the temperature of the surface of the heater (platinum) for pool boiling of water at 800 psia when the heat flux is 500,000 Btu/hr ft².

5. Estimate the heat flux that would occur in nucleate boiling of saturated water at 300°F with a brass heater at 315°F.

6. Estimate the heat flux in film boiling from a 1-in. o.d. tube at a surface temperature of 1000°F in saturated water at atmospheric pressure, neglecting radiation.

7. A mixture of air and water at 3 atmospheres pressure and 100°F flows through a 2-in. i.d. pipe. For a liquid flow rate of 3000 gal/hr and a gas flow rate of 200 ft^3/hr (measured at flow conditions), determine the flow pattern which would probably exist. Use Baker's chart (Fig. 12-12).

8. A saturated mixture of steam and water at 30 psia flows through a 1-in. i.d. tube at a mass flow rate of 2000 lb$_m$/hr. Determine the probable flow pattern if the flowing mass quality is (a) 0.001, (b) 0.01, and (c) 0.10.

9. Work Problem 8 for a pressure of 1500 psia and for flowing mass qualities of (a) 0.01, (b) 0.1, and (c) 0.2.

10. Work Problem 9 for a mass flow rate of 6000 lb$_m$/hr and for the same flowing mass qualities.

11. Water and steam at 50 psia flows through a 2-in. i.d. pipe at a rate of 4000 lb$_m$/hr with a mass flow quality of 5 percent. The phase velocity ratio of the two phases is 1.4. Calculate the void fraction and the liquid and vapor superficial velocities.

12. Derive Eq. (12-41).

13. Derive Eq. (12-42).

14. Check Eqs. (12-41) and (12-55) against several points in Fig. 12-24. Note that $(T_{sat} - T_w) = 18°F$ for the figure.

15. Saturated steam at 30 psia is being condensed on a vertical surface 8-in. tall and at a uniform temperature of 220°F. Estimate the film thickness at the bottom of the plate and the mass condensed per hour per foot of width of the surface.

16. Estimate the height of surface necessary for turbulent flow to exist in a condensing film of water. Assume that the vertical surface is at 180°F and the steam is saturated at 14.7 psia. Assume a critical Reynold's number $Re_\delta = 300$.

17. Calculate the heat transfer coefficient for condensing Freon 12 on a vertical wall 18 in. high. The wall temperature is $-5°F$ and the Freon vapor is at 0°F. Assume $h_{fg} = 68.8$ Btu/lb$_m$.

18. Steam at 14.7 psia condenses on a 1.0-in. o.d. horizontal tube 3.0 ft long. If the tube surface temperature is 195°F, calculate the condensing rate in lb$_m$/min.

19. Estimate the mean heat transfer coefficient h for the condensation of steam on a 0.1-ft o.d. horizontal tube whose surface temperature is 180°F. The steam is saturated and at a temperature of 220°F. Neglect the density of the vapor and assume that the enthalpy of vaporization for the process is 965 Btu/lb$_m$. Assume laminar film condensation.

20. A horizontal tube of 1-in. o.d. is used to condense saturated steam on the outside. The surface temperature of the tube is maintained at 80°F and the saturation temperature of the steam is 120°F. Neglecting the density of the

vapor, determine the number of pounds of vapor condensed per hour per foot of tube.

21. A vertical tube with 3.0 in. o.d. and a length of 2 ft has an outside wall temperature of 200°F. It is exposed to saturated steam at atmospheric pressure. Estimate the rate at which condensate will run off at the bottom of the tube, assuming laminar film condensation. Evaluate properties at saturated conditions.

22. Estimate the average heat transfer coefficient for saturated ammonia at 40°F condensing on a $\frac{1}{2}$-in. o.d. horizontal tube with a surface temperature of 24°F. Neglect the vapor density.

23. Saturated ammonia vapor at 40°F is condensing on a 0.3-in. o.d. horizontal tube, having a surface temperature of 24°F. Estimate the amount of vapor condensed on a 4-ft length of this tube per hour.

24. How many feet of $\frac{1}{2}$-in. o.d. horizontal tube is needed to condense saturated ammonia at 120°F at the rate of 3000 lb/hr. Assume that the outside surface of the tube is maintained at 80°F by water running inside the tube (condensation occurring on the outside) and that h_{fg} = 455 Btu/lb$_m$.

25. Use the rule of thumb—that a horizontal tube of diameter D can be treated as a vertical wall of height $2.5D$—to compute the mean heat transfer coefficient for laminar film condensation. Show the equivalence between Eqs. (12–41) and (12–56). (Neglect ρ_v compared to ρ_l).

26. Using the data in Example 12–2, estimate the number of pounds of vapor condensed per hour for four tubes arranged one above another. The tubes are 6 ft in length.

CHAPTER 13

ANALYSIS OF HEAT EXCHANGERS

In a great number of cases, heat transfer occurs from one fluid (liquid or gas) stream to a different fluid stream. The radiator of an automobile is an example—the energy transfer being from the coolant through the metal walls of the radiator to the air. Notice that the mechanism of convection must be combined with that of conduction for describing this energy transfer.

Many important problems involve both convection and conduction. An example is the wall of a building where heat exchange is usually a controlled but unavoidable factor and involves convective processes to and from the air on the inside and outside as well as conduction through the walls.

In these situations thermal radiation also occurs but may be neglected at moderate temperatures. Later a method will be given for accounting for thermal radiation.

13–1 THE OVERALL HEAT TRANSFER COEFFICIENT

For plane walls the transfer of heat by conduction was developed in Chapter 1 using the Fourier equation:

$$\frac{\dot{q}_n}{A} = -k_n \frac{\partial T}{\partial n},$$

which describes heat conduction in a direction n. The application of the equation to steady state conduction through a large plane slab with isothermal walls and constant thermal conductivity resulted in

$$\frac{\dot{q}}{A} = k\left(\frac{T_1 - T_2}{x_2 - x_1}\right) = \frac{k}{\Delta x}(T_1 - T_2),$$

where Δx is the thickness of the wall and $\dot{q}/A$ is the heat flux in the x direction. Rearranging the results gives

$$\dot{q} = -kA\frac{\Delta T}{\Delta x} \qquad \dot{q} = \frac{T_1 - T_2}{\left(\dfrac{\Delta x}{kA}\right)}. \tag{13–1}$$

This equation may be arranged in the form

$$\text{flow} = \text{potential} \div \text{resistance}$$

13-1 THE OVERALL HEAT TRANSFER COEFFICIENT

to give

$$\dot{q} = \frac{T_1 - T_2}{R_t}, \quad (13\text{-}2)$$

where R_t is the thermal resistance, $[R_t = \Delta x/(kA)$, for this simple case] and $(T_1 - T_2)$ is the potential.

It is useful to consider an equivalent electrical circuit for which the equation $I = \Delta E \div R$ holds. The thermal "circuit" may be drawn as an electrical circuit as an aid to understanding the situation. Figure 13-1 shows such a thermal circuit for a simple slab.

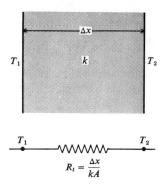 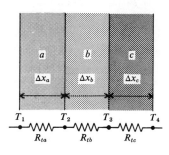

Fig. 13-1 Thermal circuit for simple slab.

Fig. 13-2 Thermal circuit for composite slab.

In case the plane slab is laminated or composed of several layers, like that shown in Fig. 13-2, the conduction can be calculated using the resistance concept. For simplification, the minus sign will be absorbed into the temperature differences. For example, ΔT_a will refer to the temperature difference $(T_1 - T_2)$ across the lamination designated a in Fig. 13-2. Other laminations are designated b, c, etc., and these designations will serve as subscripts to identify ΔT and Δx.

For the composite slab in Fig. 13-2, with steady state conduction, the following may be written:

$$\Delta T_a = \dot{q} R_{ta} = \dot{q} \frac{\Delta x_a}{k_a A},$$

$$\Delta T_b = \dot{q} R_{tb} = \dot{q} \frac{\Delta x_b}{k_b A}, \quad (13\text{-}3)$$

$$\Delta T_c = \dot{q} R_{tc} = \dot{q} \frac{\Delta x}{k_c A}.$$

Addition of Eqs. (13-3) gives the total temperature difference $(T_1 - T_4)$ or $\Delta T_{\text{overall}}$:

$$\Delta T_{\text{overall}} = \dot{q}(R_{ta} + R_{tb} + R_{tc}) = \dot{q} \sum R_t, \quad (13\text{-}4)$$

or
$$\Delta T_{\text{overall}} = \dot{q}\left(\frac{\Delta x_a}{k_a A} + \frac{\Delta x_b}{k_b A} + \frac{\Delta x_c}{k_c A}\right). \tag{13-5}$$

Rearranging,
$$\dot{q} = \frac{\Delta T_{\text{overall}}}{R_{ta} + R_{tb} + R_{tc}} = \frac{\Delta T_{\text{overall}}}{\sum R_t} \tag{13-6}$$

$$= \frac{\Delta T_{\text{overall}}}{\left(\frac{\Delta x_a}{k_a A}\right) + \left(\frac{\Delta x_b}{k_b A}\right) + \left(\frac{\Delta x_c}{k_c A}\right)}. \tag{13-7}$$

Obviously the number of terms in the denominator depends on the number of laminations. When two materials are placed together, imperfect contact usually occurs. A "contact" thermal resistance, therefore, should also be included. Such contact resistance is neglected in this chapter.

Newton's law of cooling, $\dot{q} = hA(T_1 - T_2)$, has been found useful for practical convection problems. Here the electrical analogy is expressed as

$$\dot{q} = \frac{\Delta T}{R_t} = \frac{\Delta T}{1/(hA)}, \tag{13-8}$$

so
$$R_t = 1/(hA). \tag{13-9}$$

These concepts will be applied to the case of steady one-directional heat transfer between fluids on each side of a large single slab (Fig. 13–3):

$$\dot{q} = \frac{\Delta T_{\text{overall}}}{R_i + R_a + R_o},$$

where i and o refer to the inside and outside fluids. $\Delta T_{\text{overall}}$ is $(T_i - T_o)$; the mean temperature difference between points outside the boundary layers in the two fluids. Assuming that h_i and h_o are constant over the surface gives

$$\dot{q} = \frac{T_i - T_o}{\frac{1}{h_i A} + \frac{\Delta x_a}{k_a A} + \frac{1}{h_o A}}. \tag{13-10}$$

Additional terms $(\Delta x/kA)$ result for the case of a laminated wall.

It is convenient to define an *overall coefficient of heat transfer*, U, by analogy to Newton's law of cooling:

$$\dot{q} = UA(T_i - T_0). \tag{13-11}$$

Combining Eq. (13–10) and (13–11), an equation for U is obtained for the case shown in Fig. 13–3:

$$U = \frac{1}{\left(\frac{1}{h_i}\right) + \left(\frac{\Delta x_a}{k_a}\right) + \left(\frac{1}{h_o}\right)}. \tag{13-12}$$

13–1   THE OVERALL HEAT TRANSFER COEFFICIENT   385

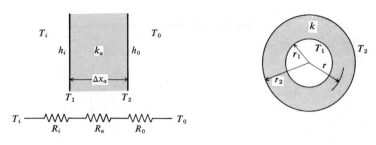

Fig. 13–3 Slab with fluid on each surface.   Fig. 13–4 Nomenclature for cylindrical shell.

Again, for the case of a composite slab, the additional values of $\Delta x/k$ are added to the denominator of Eq. (13–12).

Cylindrical Shapes

Before applying the concept of overall coefficient of heat transfer to cylinders such as pipes, tubes, or tanks, it is necessary to develop the equation for steady conduction through a cylinder. It will be assumed that the cylinder is long so that there are no end effects; that is, the heat flow is in a radial direction only. The outer and inner surfaces are at constant temperatures, T_2 and T_1 respectively, and the thermal conductivity of the material is constant. Applying the Fourier law to a cylindrical shell (Fig. 13–4),

$$\dot{q} = -kA\frac{\partial T}{\partial n} = -kA\frac{dT}{dr}.$$

For a cylinder of length L at some intermediate radius r, the area normal to the heat flow is $2\pi r L$, and

$$\dot{q} = -k(2\pi r L)\frac{dT}{dr}.$$

Rearrangement allows the following integration to be made:

$$\int_{T_1}^{T_2} dT = -\frac{\dot{q}}{2\pi kL}\int_{r_1}^{r_2}\frac{dr}{r},$$

$$T_2 - T_1 = \frac{-\dot{q}}{2\pi kL}\ln\left(\frac{r_2}{r_1}\right), \qquad (13\text{–}13)$$

$$\dot{q} = \frac{2\pi kL}{\ln(r_2/r_1)}(T_1 - T_2).$$

For this case,

$$R_t = \frac{\ln(r_2/r_1)}{2\pi kL}. \qquad (13\text{–}14)$$

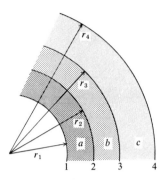

Fig. 13-5 Composite cylinder.

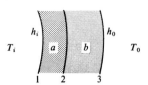

Fig. 13-6 Composite cylinder with convective surfaces.

For conduction through a cylinder made of three laminations (Fig. 13-5), the method used to obtain Eqs. (13-6) and (13-7) yields

$$T_1 - T_4 = \frac{\dot{q}}{2\pi L}\left(\frac{1}{k_a}\ln\frac{r_2}{r_1} + \frac{1}{k_b}\ln\frac{r_3}{r_2} + \frac{1}{k_c}\ln\frac{r_4}{r_3}\right). \quad (13\text{-}15)$$

For the case of convective heat transfer between fluids both inside and outside of a cylinder with two laminations (Fig. 13-6), the procedure above is used. Equating the heat flow by convection at the inside wall to that at the outside wall gives

$$\dot{q} = 2\pi r_1 L h_i(T_i - T_1) = 2\pi r_3 L h_o(T_3 - T_o). \quad (13\text{-}16)$$

This yields

$$\dot{q} = \frac{(T_i - T_o)}{\dfrac{1}{2\pi r_1 L h_i} + \dfrac{\ln(r_2/r_1)}{2\pi k_a L} + \dfrac{\ln(r_3/r_2)}{2\pi k_b L} + \dfrac{1}{2\pi r_3 L h_o}}. \quad (13\text{-}17)$$

If an overall coefficient is to be used, a selection must be made of the area to which it is to be referred—inside, outside, or some mean area. If the inside area is selected, then $\dot{q} = 2\pi r_1 L U_i(T_i - T_o)$. If Eq. (13-17) is multiplied and divided by $2\pi r_1 L$ (the inside area), then

$$U_i = \frac{1}{\dfrac{1}{h_i} + \dfrac{r_1}{k_a}\ln\dfrac{r_2}{r_1} + \dfrac{r_1}{k_b}\ln\dfrac{r_3}{r_2} + \dfrac{r_1}{r_3}\dfrac{1}{h_o}}, \quad (13\text{-}18)$$

where U_i means that the overall coefficient is based on the inside area. In a similar manner, an equation for U_o can be obtained.

Often in a practical situation the thermal conductivity of the layer (or layers) of solid material is very large and its thermal resistance may be neglected so that the value of U may be approximated by

$$U_i \approx \frac{1}{(1/h_i) + (r_i/r_o)(1/h_o)}. \quad (13\text{-}19)$$

13-1 THE OVERALL HEAT TRANSFER COEFFICIENT

The ratio of radii r_i/r_o may be very near to unity as in the case of large-diameter cylinders with thin walls, which further simplifies Eq. (13–19) to

$$U \approx \frac{1}{\left(\dfrac{1}{h_i} + \dfrac{1}{h_o}\right)}. \tag{13-20}$$

Example 13-1. It is desired to calculate the overall coefficient of heat transfer, based on the outside surface area, for the case where heat is passing through a tube wall, 1 in. i.d. and 1.2 in. o.d. Assume an average inside coefficient of heat transfer of 20 Btu/hr ft² °F, and an average outside coefficient of 10 Btu/hr ft² °F. The thermal conductivity of the tube wall is 10 Btu/hr ft °F.

Solution. Since the heat transferred is the same regardless of how the coefficient is defined, $U_o A_o = U_i A_i$, therefore

$$U_o = U_i \frac{A_i}{A_o} = U_i \frac{r_i}{r_o} = \frac{U_i}{\left(\dfrac{r_o}{r_i}\right)}.$$

Equation (13–18) can be modified to give an equation suitable for this case, where only one layer of tube material exists:

$$U_o = \frac{1}{\dfrac{r_o}{r_i}\dfrac{1}{h_i} + \dfrac{r_o}{k}\ln\dfrac{r_o}{r_i} + \dfrac{1}{h_o}}$$

$$= \frac{1}{\dfrac{1.2}{1.0}\dfrac{1}{20} + \dfrac{0.6}{(12)(10)}\ln\dfrac{1.2}{1.0} + \dfrac{1}{10}} = \frac{1}{0.06 + 0.0009 + 0.1}$$

$$= \frac{1}{0.1609} = 6.21 \,\frac{\text{Btu}}{\text{hr ft}^2\,°\text{F}}$$

Fouling Factors

All heat exchange surfaces are subject to fouling. The effectiveness of a surface in heat exchange may be reduced because of the deposition of a film (usually thin) of foreign matter which may have unusually high thermal resistance. In many cases a scale, sometimes quite thick, may form. It is usual to design for the amount of fouling that is anticipated between cleaning operations, which may be up to a year apart.

The fouling or "dirt" factors in the literature are usually given in terms of a thermal resistance or the reciprocal of an equivalent film coefficient. Values of normal fouling factors recommended by the Tubular Exchangers Manufacturers Association are given in Table 13–1. A method for determination of fouling factors has been given by Kern [2]. The "clean" overall

Table 13-1
NORMAL FOULING FACTORS* FOR HEAT TRANSFER EQUIPMENT†

Temperature of heating medium:	Up to 240°F		240°–400°F‡	
Temperature of water:	125°F or less		Over 125°F	
Types of water	Water velocity (ft/sec)		Water velocity (ft/sec)	
	3 ft/sec and less	Over 3 ft/sec	3 ft/sec and less	Over 3 ft/sec
Sea water	0.0005	0.0005	0.001	0.001
Distilled	0.0005	0.0005	0.0005	0.0005
Treated boiler feedwater	0.001	0.0005	0.001	0.001
Engine jacket	0.001	0.001	0.001	0.001
City or well water (such as Great Lakes)	0.001	0.001	0.002	0.002
Great Lakes	0.001	0.001	0.002	0.002
Cooling tower and artificial spray pond:				
Treated makeup	0.001	0.001	0.002	0.002
Untreated	0.003	0.003	0.005	0.004
Boiler blowdown	0.002	0.002	0.002	0.002
Brackish water	0.002	0.001	0.003	0.002
River water:				
Minimum	0.002	0.001	0.003	0.002
Mississippi	0.003	0.002	0.004	0.003
Delaware, Schuylkill	0.003	0.002	0.004	0.003
East River and New York Bay	0.003	0.002	0.004	0.003
Chicago sanitary canal	0.008	0.006	0.010	0.008
Muddy or silty	0.003	0.002	0.004	0.003
Hard (over 15 grains/gal)	0.003	0.003	0.005	0.005

* Fouling factor $= \dfrac{1}{h} \dfrac{\text{hr} \cdot \text{ft}^2 \cdot {}^\circ\text{F}}{\text{Btu}}$.

† From "Standards of the Tubular Exchanger Manufacturers Association," New York, 1959. Reproduced by permission of the publisher.

‡ Ratings in columns 3 and 4 are based on a temperature of the heating medium of 240°–400°F. If the heating medium temperature is over 400°F and the cooling medium is known to scale, these ratings should be modified accordingly.

coefficient U_C is related to the design or "dirty" overall coefficient, U_D, in the following way:

$$\frac{1}{U_D} = \frac{1}{U_C} + \frac{1}{h_{di}} + \frac{1}{h_{do}}, \qquad (13\text{-}21)$$

where $1/h_{di}$ is the inside fouling factor, and $1/h_{do}$ is the outside fouling factor.

13–1 THE OVERALL HEAT TRANSFER COEFFICIENT

Table 13–1 (cont.)

Fouling factors—industrial oils	
Clean recirculating oil	0.001
Machinery and transformer oils	0.001
Vegetable oils	0.003
Quenching oil	0.004
Fuel oil	0.005
Fouling factors—industrial gases and vapors	
Organic vapors	0.0005
Steam (nonoil-bearing)	0.0005
Alcohol vapors	0.0005
Steam, exhaust (oil-bearing from reciprocating engines)	0.001
Refrigerating vapors (condensing from reciprocating compressors)	0.002
Air	0.002
Coke oven gas and other manufactured gas	0.01
Diesel engine exhaust gas	0.01
Fouling factors—industrial liquids	
Organic	0.001
Refrigerating liquids, heating, cooling, or evaporating	0.001
Brine (cooling)	0.001

Here, as in the case of Eq. (13–20), the radius ratio is neglected in the last two terms. Inasmuch as fouling factor data is rather meager and inaccurate, this procedure is justified. Also the designer is seldom sure of the exact conditions under which the device will operate, especially over long periods of time. However, where thick scales are known to build up, the omission of the radius ratio could cause significant error. An extreme case of thick-scale buildup is shown in Fig. 13–7.

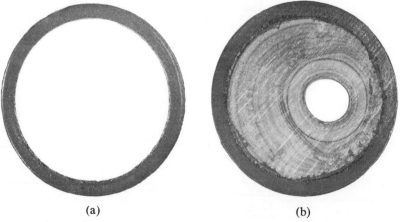

Fig. 13–7 An extreme case of scale buildup (a) clean (b) dirty. (Courtesy of Du Pont Company.)

Maximum Heat Flow Through Insulation

The engineer often must limit the heat losses from a pipe or tube by specifying the use of insulation. The heat loss can be calculated using Eq. (13–17).

The application of insulation would appear to reduce the heat loss. This is generally the case; however, for small wires or tubes, the addition of the extra thickness provides a larger outside area for the flow of heat which may overcome the insulating effect of added thickness.

For a given temperature of the *inside of the insulation*, the resistance to heat loss to a fluid surrounding the insulation per unit length is

$$R_t = \frac{1}{2\pi r_2 h_o} + \frac{1}{2\pi k_I} \ln \frac{r_2}{r_1},$$

where r_1 is the inside radius of the insulation and r_2 is the radius (a variable) of the outside of the insulation. This resistance can be minimized by solving $dR_t/dr_2 = 0$:

$$\frac{dR_t}{dr_2} = \frac{-1}{2\pi r_2^2 h_o} + \frac{1}{2\pi k_I r_2} = 0,$$

from which

$$(r_2)_{\max} = \frac{k_I}{h_o}.$$

This is the outside radius for which the maximum heat loss will occur. Heat losses increase with addition of insulation until $(r_2)_{\max}$ is reached. Increasing the radius beyond k_I/h_o will reduce the heat loss.

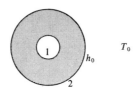

Fig. 13–8 Spherical shell.

The Sphere

The equation for conduction through a spherical shell (Fig. 13–8) can be shown to be

$$\dot{q} = \frac{4\pi k r_2 r_1 (T_1 - T_2)}{r_2 - r_1}. \qquad (13\text{–}22)$$

If convection occurs on the outside, then the overall equation is

$$\dot{q} = \frac{4\pi r_2^2 (T_1 - T_o)}{\left(\dfrac{1}{h_o}\right) + \dfrac{r_2}{r_1}\left(\dfrac{r_2 - r_1}{k}\right)} = U_o A_o (T_1 - T_o). \qquad (13\text{–}23)$$

13-2 THE HEAT EXCHANGER AND MEAN TEMPERATURE DIFFERENCE

In the above situation the inside surface might be maintained at a constant value by a nuclear reaction, by an electric heater, or by other means. The equation would, of course, hold any time that the inside surface temperature T_1 is a known, constant value.

13-2 THE HEAT EXCHANGER AND MEAN TEMPERATURE DIFFERENCE

The application of the equation $\dot{q} = UA(\Delta T)$ in practical engineering, such as its application in the design of heat exchangers, is seldom as easy as might be expected.

One reason for this is that the temperature difference may not be a constant but may vary throughout the heat exchanger. This is particularly true during the exchange of heat from one fluid stream to another. The temperature of one stream may be continuously increasing as it flows through the device and the temperature of the other may continuously decrease. Figure 13-9 shows a plot of the local temperatures for two streams in a cross-flow exchanger where the fluid in each stream does not mix with neighboring fluid in the same stream as it flows through the exchanger. In this case (unmixed fluids) fluid entering a particular channel, such as a tube, stays in that channel during its pass through the exchanger.

Another reason that the simple equation above is difficult to use is that the physical properties of the fluid streams may vary so as to cause the film coefficient for each fluid to change throughout the heat exchanger, giving rise to a continuously varying value of U. In addition, the prediction of a proper value of the overall coefficient may be difficult, due to the complexity of the geometry and to other factors, such as leakage and bypassing of fluid. The prediction of overall heat transfer coefficients in certain types of exchangers can be carried out by a method described by Bell [3].

If the equation $\dot{q} = UA(\Delta T)$ is to be utilized in connection with problems involving heat exchangers such as the one shown in Fig. 13-9, two facts seem apparent: (1) The overall heat transfer coefficient must be assumed to be some average value, and (2) a proper mean temperature difference must be used for ΔT. Such a mean temperature difference will be given the symbol ΔT_m.

One way to approach this problem for the case shown in Fig. 13-9 would be a step-by-step numerical procedure; at any location in the exchanger (indicated by the vertical line A-B), there is a discrete temperature for each stream, and, therefore, a local discrete temperature difference as well as a specific local overall heat transfer coefficient.

A large number of practical cases have been solved to determine a proper mean temperature difference, ΔT_m, for design purposes. Some simplifying assumptions are needed to solve most of these cases. In the following paragraphs this general problem will be treated.

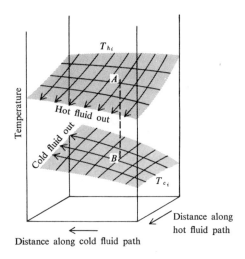

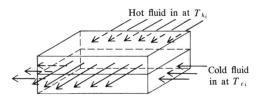

Fig. 13–9 Temperature variation in a flat plate cross-flow exchanger with neither fluid mixed.

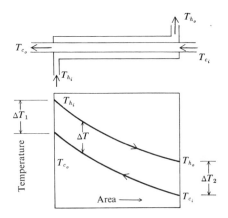

Fig. 13–10 Counterflow heat exchanger.

13–2 THE HEAT EXCHANGER AND MEAN TEMPERATURE DIFFERENCE

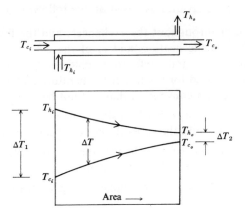

Fig. 13–11 Parallel-flow heat exchanger.

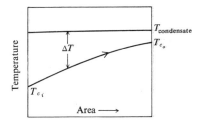

Fig. 13–12 Temperature variation in a condenser.

Figure 13–10 represents the situation which exists in a simple heat exchanger made of two concentric pipes arranged to allow heat flow between two streams. It is called a *counterflow* device, as the two streams generally flow in opposite directions. *Parallel-flow* (Fig. 13–11), where both fluids flow in the same direction, is obtained by switching the inlet and outlet of one of the streams. Figure 13–12 shows the plot of temperatures for the case where the hot stream is condensing (its temperature is constant). Evaporation (boiling) would result in the cold-stream temperatures remaining constant. In practice a heat exchanger might be a combination of these types. For example, the exchanger could have three sections: first, where the vapor is de-superheated; second, where condensation occurs; and, third, where subcooling of the condensate occurs. In each case, the area shown in Figs. 13–10, 13–11, and 13–12 is the net surface area for heat exchange between the two fluids for that part of the exchanger between the inlet and the specified point.

The derivation of mean temperature difference for steady state operation given below is for a counterflow exchanger, but the parallel-flow case leads to identical results. The derivation is based on the following assumptions:

1. U is constant throughout the exchanger, and the flow rates $\dot{m}_c$ and $\dot{m}_h$, as well as the specific heats c_c and c_h of the cold and hot fluids, are constant. The subscripts c and h refer to the cold and hot streams. (In many writings a capital T is used for the hotter temperature and a lower case t for the colder temperature, a convention that will not be used in this text.)

2. No heat is lost to any other heat sink or gained from any other source except the two streams, and the tubes do not conduct heat axially.

3. There is no mixed heat exchange, that is, sensible and latent type changes are not occurring for the same stream.

4. A single (bulk) temperature applies to each stream at a given cross section of the exchanger.

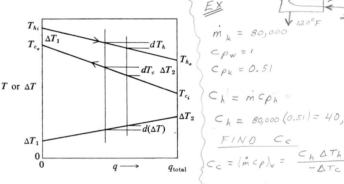

Fig. 13-13 Temperature variations vs. heat flow between fluids for a counterflow exchanger.

The subscripts i and o refer to inlet and outlet respectively. Figure 13-13 shows a plot of T vs. $\dot{q}$ for the two streams and also ΔT (the temperature difference between the two streams) plotted against q, the total heat exchanged between the hot and cold fluid from the inlet of one stream up to the specified section.

The rate of heat flow from the hot to the cold fluid at any surface area increment dA is

$$d\dot{q} = U \, \Delta T \, dA. \tag{13-25}$$

For the counterflow heat exchanger shown in Fig. 13-10, the heat transfer through an element of area dA may be written

$$d\dot{q}_h = \dot{m}_h c_h \, dT_h = C_h \, dT_h, \tag{13-26}$$

and

$$d\dot{q}_c = \dot{m}_c c_c \, dT_c = C_c \, dT_c, \tag{13-27}$$

$$\dot{q} = \dot{m} \, c_p \, \Delta T$$

13-2 THE HEAT EXCHANGER AND MEAN TEMPERATURE DIFFERENCE

where, for convenience, *fluid capacity rates*, C_h and C_c, are defined by

$$C_h = \dot{m}_h c_h, \tag{13-28}$$

and

$$C_c = \dot{m}_c c_c. \tag{13-29}$$

From Eqs. (13-26) and (13-27) it is apparent that for constant flow rates and heat capacities, the T vs. q curves are straight lines on this plot. Also the ΔT vs. q plot is a straight line and

$$\frac{d(\Delta T)}{d\dot{q}} = \frac{\Delta T_2 - \Delta T_1}{\dot{q}_t}, \tag{13-30}$$

where $\dot{q}_t$ is the total heat exchange rate, and ΔT_1 and ΔT_2 refer to the temperature difference between the streams at the left end and right end of the exchanger respectively. Substituting $d\dot{q} = U \Delta T \, dA$ into Eq. (13-30) and rearranging yields

$$\frac{d(\Delta T)}{U \Delta T} = \frac{\Delta T_2 - \Delta T_1}{\dot{q}_t} dA. \tag{13-31}$$

Integrating Eq. (13-31):

$$\frac{1}{U} \int_{\Delta T_1}^{\Delta T_2} \frac{d(\Delta T)}{\Delta T} = \frac{\Delta T_2 - \Delta T_1}{\dot{q}_t} \int_0^A dA \tag{13-32}$$

$$\frac{1}{U} \ln \frac{\Delta T_2}{\Delta T_1} = \frac{\Delta T_2 - \Delta T_1}{\dot{q}_t} A, \tag{13-33}$$

or

$$\dot{q}_t = UA \frac{\Delta T_2 - \Delta T_1}{\ln(\Delta T_2/\Delta T_1)}. \tag{13-34}$$

The same result is obtained for parallel-flow exchangers.

Comparison of Eq. (13-34) with $\dot{q} = UA \, \Delta T_m$ gives the expression for the proper mean temperature difference for a counterflow exchanger:

$$\Delta T_m = \frac{\Delta T_2 - \Delta T_1}{\ln(\Delta T_2/\Delta T_1)}. \tag{13-35}$$

The temperature difference defined by Eq. (13-35) is the *log mean temperature difference* and will be designated LMTD.

Figure 13-14 shows LMTD as a function of the terminal temperature differences between the two streams for a counterflow or parallel-flow exchanger. In using the figure, if a point falls off the chart, the values on the chart may all be multiplied by 10 and the chart will then accommodate the higher temperature differences.

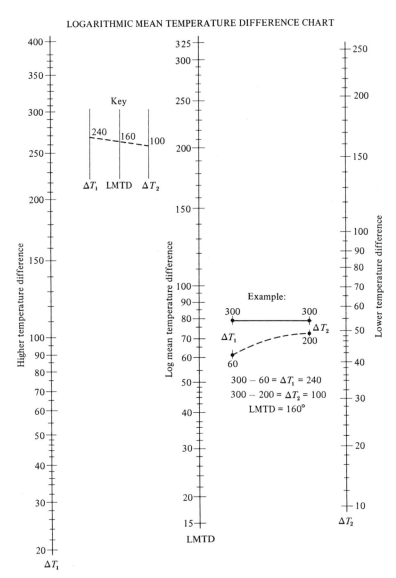

Fig. 13-14 Log mean temperature difference chart. (Courtesy Dean Products, Inc.).

13-2 THE HEAT EXCHANGER AND MEAN TEMPERATURE DIFFERENCE

Heat Exchange with Variable U

The development for LMTD in the last paragraph assumed constant properties and a constant value of U. Practical situations arise where U is a variable, however.

One case of variable U, where U may be assumed to vary linearly with temperature difference, is easily handled by analytical schemes. If the assumption is made that $U = a + b\,\Delta T$, the following result can be obtained, where the subscripts 1 and 2 refer to values at each end of the exchanger:

$$\dot{q}_t = A\,\frac{U_1\,\Delta T_2 - U_2\,\Delta T_1}{\ln\,(U_1\,\Delta T_2/U_2\,\Delta T_1)}\,. \tag{13-36}$$

If an exchanger has a nonlinear variation of U with ΔT, it may be handled by subdividing into a number of smaller exchangers each of which can be treated by the use of Eq. (13-36).

Fig. 13-15 Cutaway of a four tube-pass shell and tube heat exchanger. (Courtesy Perfex Corporation.)

Complex Flow Patterns

The path of flow in a practical heat exchanger is usually quite complex. In a few cases the true ΔT_m may be calculated analytically.

Many heat exchangers are of the shell and tube type. Figure 13-15 indicates such an exchanger. It may be thought of as an extension of the simple double-pipe exchanger of Figs. 13-10, 13-11, and 13-12, where the single inner pipe is replaced by many pipes or "tubes," with the ends of each tube fastened into a solid partition or "tube sheet" which provides for keeping the fluid streams separate. The outer pipe is called a "shell" and contains the hot fluid (usually); it provides the main structural support for the assembly.

Longitudinal baffles may be inserted to cause either shell-side or tube-side fluid to traverse the exchanger length more than one time (Fig. 13–15). Hairpin or U-shaped tubes are popularly used as they eliminate one tube sheet and many thermal expansion problems of the mechanical design. Here the tube fluid enters and leaves at the same end of the exchanger.

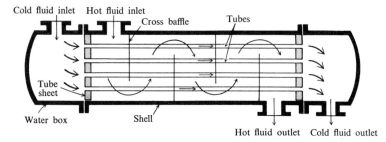

Fig. 13–16 Schematic drawing of a one shell-pass, one tube-pass heat exchanger.

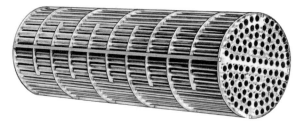

Fig. 13–17 Tube bundle with cross baffles. (Courtesy Perfex Corporation.)

Cross baffles cause the shell fluid to crisscross the exchanger, flowing mainly across the tubes but progressing generally in an axial direction from one end of the exchanger to the other as shown in Figs. 13–16 and 13–17.

Designations are given to the various flow arrangements, indicating the number of passes (traverses) of each fluid through the exchanger; for example:

1–2 *Exchanger*—One shell-pass and two tube-passes.

2–4 *Exchanger*—Two shell-passes and four tube-passes.

It is possible to develop equations for the mean temperature difference ΔT_m for some of the more complex flow arrangements found in practice.

Underwood [5] derived the equation for the 1–2 exchanger in 1934. He assumed steady flow and steady temperatures, constant properties

13-2 THE HEAT EXCHANGER AND MEAN TEMPERATURE DIFFERENCE

including U, single-phase fluids, constant temperature over each cross section of the flow (mixed flow), no heat losses, and constant conditions of the heating surfaces. His equation is

$$\Delta T_m = \frac{\sqrt{(T_{s,i} - T_{s,o})^2 + (T_{t,o} - T_{t,i})^2}}{\ln\left[\dfrac{T_{s,i} + T_{s,o} - T_{t,i} - T_{t,o} + \sqrt{(T_{s,i} - T_{s,o})^2 + (T_{t,o} - T_{t,i})^2}}{T_{s,i} + T_{s,o} - T_{t,i} - T_{t,o} - \sqrt{(T_{s,i} - T_{s,o})^2 + (T_{t,o} - T_{t,i})^2}}\right]},$$

(13–37)

where the subscripts s and t stand for shell fluid and tube fluid respectively, and i and o stand for in and out.

The complexity of this and other equations for different flow arrangements lead to the development of charts to replace the equations. Bowman, Mueller, and Nagle [6], published design charts that used the concept of a correction factor F_t, which was used as a multiplying factor for the equivalent counterflow LMTD to obtain an appropriate ΔT_m for these complex flow cases:

$$\dot{q} = UAF_t(\text{LMTD}),$$

(13–38)

where LMTD is the LMTD for an equivalent counterflow exchanger.

Figures 13–18, 13–19, and 13–20 show such design charts where

$$P = \frac{T_{c_o} - T_{c_i}}{T_{h_i} - T_{c_i}}, \quad \text{and} \quad R = \frac{T_{h_i} - T_{h_o}}{T_{c_o} - T_{c_i}}.$$

(13–39)

P may be thought of as an efficiency, or effectiveness, of the heat exchanger for the case $C_c < C_h$ (see also Eq. (13–40)). If the numerator and denominator are multiplied by C_c the denominator becomes

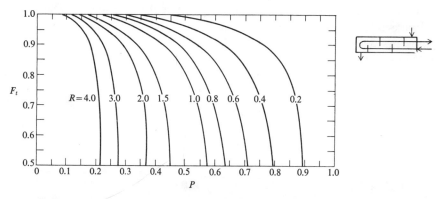

Fig. 13–18 Correction factors for a 1–2 parallel-counterflow exchanger.

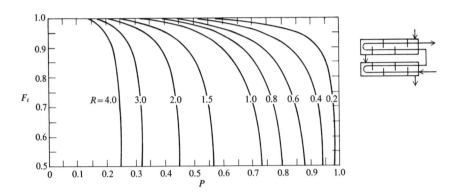

Fig. 13-19 Correction factors for a 2-4 multipass-counterflow exchanger.

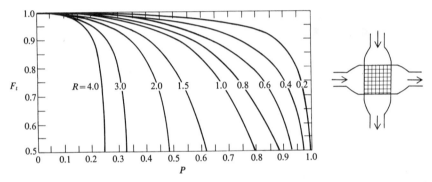

Fig. 13-20 Correction factor for a cross-flow exchanger with fluids unmixed.

$C_c(T_{h_i} - T_{c_i})$, which is the heat that could theoretically be added to the cold fluid if it could be heated to the temperature of the entering hot fluid. An infinite-area exchanger would be required for this maximum heating. P is thus the actual heat addition divided by the theoretical maximum heat addition for the cold fluid (the effectiveness). If $C_h < C_c$, then P may be shown to equal the heat exchanger effectiveness divided by R.

R is also subject to a useful interpretation. The numerator of R is $(T_{h_i} - T_{h_o})$, which equals $\dot{q}/C_h$, and the denominator is $(T_{c_o} - T_{c_i})$ which equals $\dot{q}/C_c$. So

$$R = \frac{\dot{q}/C_h}{\dot{q}/C_c} = \frac{C_c}{C_h}.$$

Thus, R is the ratio of the fluid capacity rates.

Normally heat exchangers are selected or designed so that values of F_t are greater than about 0.75.

13-3 NTU APPROACH TO THERMAL DESIGN OF HEAT EXCHANGERS

Example 13-2. Water at the rate of 125 lb_m/min is heated from 100° to 170°F by an oil having a specific heat of 0.5 Btu/lb_m °F. The fluids are used in a 2-4 multipass counterflow exchanger, and the oil enters the exchanger at 240°F and leaves at 140°F. The overall heat transfer coefficient is 65 Btu/hr ft² °F. Calculate the heat exchange area required.

Solution.

$$\dot{q} = \dot{m}_w c_{p_w} \Delta T_w = (125)(1)(170 - 100) = 8750 \text{ Btu/min}$$
$$= 525,000 \text{ Btu/hr.}$$

$$\text{LTMD} = \frac{(140 - 100) - (240 - 170)}{\ln[(140 - 100)/(240 - 170)]} = 53.6°.$$

$$P = \frac{170 - 100}{240 - 100} = 0.5, \quad \text{and} \quad R = \frac{240 - 140}{170 - 100} = 1.43.$$

From Fig. 13-19, $F_t = 0.89$. Then since $\dot{q} = UAF_t$ LMTD,

$$A = \frac{\dot{q}}{U \text{ LMTD } F_t} = \frac{525,000}{(65)(53.6)(0.89)} = 169 \text{ ft}^2.$$

13-3 THE NTU APPROACH TO THE THERMAL DESIGN OF HEAT EXCHANGERS

A procedure for heat exchanger design using a mean temperature difference approach has been discussed. If one knows the fluid inlet and exit temperatures and the correction factor F_t for a proposed exchanger, the design may be carried out. In some cases, however, it is more convenient to use the NTU (number of transfer units) method, which will now be described.

The NTU method has the advantage of ease in calculating a fluid's exit temperature if its inlet temperature and the inlet temperature of the other fluid are known for a given exchanger. Trial and error schemes are eliminated for most practical problems.

The total number of factors which enter into heat exchanger thermal design is large. Simple graphical representation is not easily accomplished. However, an appropriate grouping of some of the variables may be developed which simplifies the study of the design.

It is convenient to define an *effectiveness of heat transfer* (or heat exchanger effectiveness), ϵ, which provides a way to eliminate one of the terminal temperatures in the derived expressions:

$$\text{For} \quad C_c < C_h \quad \epsilon = \frac{T_{c_o} - T_{c_i}}{T_{h_i} - T_{c_i}}. \tag{13-40}$$

$$\text{For} \quad C_h < C_c \quad \epsilon = \frac{T_{h_i} - T_{h_o}}{T_{h_i} - T_{c_i}}. \tag{13-41}$$

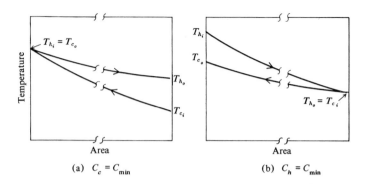

Fig. 13–21 Fluid temperature plot for infinite-area exchanger.

If the right side of Eq. (13–40) is multiplied by C_c/C_c, or if Eq. (13–41) is multiplied by C_h/C_h, the numerator of each equation becomes the actual heat transfer rate and the denominator becomes the maximum heat transfer rate that could be realized thermodynamically by an infinite-area counterflow exchanger. This defines the ratio ϵ.

Figures 13–21 (a and b) show the fluid temperatures for an infinite-area counterflow exchanger for the two cases ($C_c = C_{\min}$) and ($C_h = C_{\min}$). It can be seen why the effectiveness equations, Eqs. (13–40) and (13–41) were chosen in the form indicated. Since $\dot{q} = C_c \, \Delta T_c = C_h \, \Delta T_h$, where ΔT_c and ΔT_h are the changes in the cold and hot fluid respectively, then it follows that (1) if $C_c = C_{\min}$, then

$$C_c < C_h \quad \text{and} \quad \Delta T_c > \Delta T_h,$$

and (2) if $C_h = C_{\min}$, then

$$C_h < C_c \quad \text{and} \quad \Delta T_h > \Delta T_c.$$

The maximum temperature change which can occur in an exchanger for an infinite-area case is equal to the difference between inlet temperatures. Figure 13–21(a) shows—for the case of $C_c = C_{\min}$ ($C_c < C_h$)—that the *cold* fluid is heated through this maximum temperature difference and that $\dot{q}_{\max} = C_c(T_{h_i} - T_{c_i})$; and Fig. 13–21(b) shows—for the case $C_h = C_{\min}$ ($C_h < C_c$)—that the *hot* fluid is heated through this maximum temperature difference and that $\dot{q}_{\max} = C_h(T_{h_i} - T_{c_i})$.

Thus, two expressions for effectiveness are necessary: Eq. (13–40) and Eq. (13–41).

Considering the case $C_h < C_c$, Eq. (13–41) yields expressions for the outlet temperatures in terms of the inlet temperatures, the effectiveness, and the ratio of fluid capacity rates:

$$T_{h_o} = \epsilon(T_{c_i} - T_{h_i}) + T_{h_i}, \tag{13–42}$$

$C_h < C_c \qquad \dot{q} = C_h(T_{h_i} - T_{h_o})$

$C_c < C_h \qquad \dot{q} = C_c(T_{c_o} - T_{c_i})$

13-3 NTU APPROACH TO THERMAL DESIGN OF HEAT EXCHANGERS

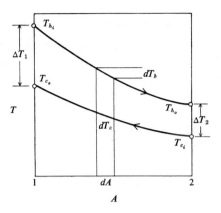

Fig. 13-22 Nomenclature for derivation of effectiveness in counterflow exchanger.

and in addition

$$T_{c_o} = \frac{\dot{q}}{C_c} + T_{c_i} = \frac{C_h}{C_c}(T_{h_i} - T_{h_o}) + T_{c_i} = \frac{C_h}{C_c}\epsilon(T_{h_i} - T_{c_i}) + T_{c_i}. \quad (13\text{-}43)$$

The following development relates temperature differences between hot and cold streams for a counterflow heat exchanger (Fig. 13-22). The development is for the case $C_h = C_{\min}$. For an infinitesimally small part of the heat exchanger surface, dA, the fluid temperature change is

$$d(T_h - T_c) = dT_h - dT_c, \quad (13\text{-}44)$$

but

$$d\dot{q} = -C_c \, dT_c = -C_h \, dT_h. \quad (13\text{-}45)$$

Substituting Eq. (13-45) into (13-44),

$$d(T_h - T_c) = -d\dot{q}\left(\frac{1}{C_h} - \frac{1}{C_c}\right), \quad (13\text{-}46)$$

and

$$d\dot{q} = U \, dA(T_h - T_c). \quad (13\text{-}47)$$

Combining Eqs. (13-46) and (13-47) and rearranging,

$$\frac{d(T_h - T_c)}{T_h - T_c} = -\left(1 - \frac{C_h}{C_c}\right)\frac{U}{C_h} \, dA. \quad (13\text{-}48)$$

Integrating Eq. (13-48) over the entire surface area yields

$$\ln\frac{T_{h_i} - T_{c_o}}{T_{h_o} - T_{c_i}} = \ln\frac{\Delta T_1}{\Delta T_2} = -\left(1 - \frac{C_h}{C_c}\right)\frac{1}{C_h}\int_A U \, dA. \quad (13\text{-}49)$$

The quantity $1/C_h \int_A U \, dA$ is now defined as the "number of heat transfer units" (NTU). For $U = $ constant, this becomes NTU $= (UA/C_h)$, and Eq. (13-49) becomes

$$\ln \frac{\Delta T_1}{\Delta T_2} = -\left(1 - \frac{C_h}{C_c}\right) \text{NTU}, \quad (13\text{-}50)$$

or

$$\frac{\Delta T_1}{\Delta T_2} = e^{-\text{NTU}[1-(C_h/C_c)]}. \quad (13\text{-}51)$$

Substituting Eqs. (13-42) and (13-43) into Eq. (13-51) and rearranging,

$$\epsilon = \frac{1 - e^{-\text{NTU}[1-(C_h C/c)]}}{1 - \frac{C_h}{C_c} e^{-\text{NTU}[1-(C_h/C_c)]}}, \quad (13\text{-}52)$$

which is an expression for effectiveness of a counterflow exchanger in terms of NTU and C_h/C_c.

Equation (13-52) can be generalized for more usefulness:

$$\epsilon = \frac{1 - e^{-\text{NTU}[1-(C_{\min}/C_{\max})]}}{1 - \frac{C_{\min}}{C_{\max}} e^{-\text{NTU}[1-(C_{\min}/C_{\max})]}}, \quad (13\text{-}53)$$

where now NTU $= AU/C_{\min}$. Note that $C_{\min}/C_{\max}$ has a range from 0 to 1.0. A value of zero would occur in practice when one fluid undergoes a phase change, since C for that fluid would be infinite. A plot of Eq. (13-53) is shown in Fig. 13-23(a).

The NTU can be thought of as a "heat transfer size" factor. Figure 13-23(a) shows that, for a constant value of $C_{\min}/C_{\max}$, large values of NTU give large values of the effectiveness and that small values of NTU give small values of the effectiveness. At large values of NTU, ϵ approaches a maximum value (not necessarily 1.0), which is limited by the flow arrangement and by thermodynamic considerations. The relationship NTU $= (UA)/C_{\min}$ indicates that a high NTU, and thus high effectiveness, is obtained by building exchangers with large area or by obtaining high values of U, usually at the cost of the increased velocities that are required for high values of the film coefficient.

The development above was for a counterflow heat exchanger. Given flow arrangement and surface geometry, the effectiveness of other heat exchangers is a function of NTU and the ratio $C_{\min}/C_{\max}$. Expressions for ϵ for a number of common types of exchangers are given in Table 13-2. Plots of ϵ vs. NTU for various values of $C_{\min}/C_{\max}$ are available in the literature for a variety of heat exchanger types [7]. Figures 13-23(a-d) show representative cases. Note that cases for which F_t was given in the preceding section are repeated here. A given problem can thus be handled using both methods of design.

13–3 NTU APPROACH TO THERMAL DESIGN OF HEAT EXCHANGERS

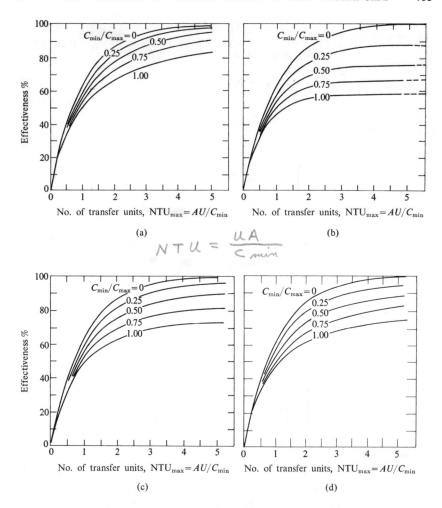

Fig. 13-23 Plots of heat exchanger effectiveness vs. number of transfer units (NTU) for several types of exchangers. (a) Counterflow exchanger. (b) 1–2 multipass exchanger, one shell-pass; 2,4,6,8, etc. tube-passes. (c) 2–4 multipass exchanger, two shell-passes; 4,8,12, etc. tube-passes. (d) Cross-flow exchanger, fluids unmixed. (From *Compact Heat Exchangers* by Kays and London. Copyright 1955 by W. M. Kays and A. L. London. Used with permission of McGraw-Hill Book Company.)

These charts show clearly the effect of the ratio of fluid capacity rates on heat exchanger effectiveness. The flow arrangement and geometry of the surface also strongly influence the value of effectiveness. Figure 13–24 shows this influence for a C_{min}/C_{max} ratio of 1.0 for a variety of cases. Note that for constant values of U and C_{min}, the counterflow arrangement requires a

Table 13-2
THERMAL EFFECTIVENESS OF HEAT EXCHANGERS WITH VARIOUS FLOW ARRANGEMENTS

Parallel-flow:
$$\epsilon = \frac{1 - \exp[-NTU(1 + C)]}{1 + C}$$

Counterflow:
$$\epsilon = \frac{1 - \exp[-NTU(1 - C)]}{1 - C\exp[-NTU(1 - C)]}$$

Cross-flow (both streams unmixed):
$$\epsilon = 1 - \exp\left\{\frac{C}{\eta}\left[\exp\left((-NTU)(C)(\eta)\right) - 1\right]\right\}^*$$
where $\eta = NTU^{-0.22}$

Cross-flow (both streams mixed):
$$\epsilon = NTU\left\{\frac{NTU}{1 - \exp(-NTU)} + \frac{(NTU)(C)}{1 - \exp[-(NTU)(C)]} - 1\right\}^{-1}$$

Cross-flow (stream C_{min} unmixed):
$$\epsilon = C\{1 - \exp[-C(1 - \exp(-NTU))]\}$$

Cross-flow (stream C_{max} unmixed):
$$\epsilon = 1 - \exp\{-C[1 - \exp(-(NTU)(C))]\}$$

1-2 Parallel-counterflow:
$$\epsilon = 2\left\{1 + C + \frac{1 + \exp[-NTU(1 + C^2)^{1/2}]}{1 - \exp[-NTU(1 + C^2)^{1/2}]}(1 + C^2)^{1/2}\right\}^{-1}$$
where $NTU = (UA/C_{min})$, and $C = C_{min}/C_{max}$.

* Approximate expression.

minimum area for a given effectiveness, as would be expected from thermodynamic considerations of reversibility. Some flow arrangements are subject to limiting values of the effectiveness that may be realized. For example, in the case of parallel flow, the curve of ϵ vs. NTU (for $C_{min}/C_{max} = 1.0$) is asymptotic to 0.5; a maximum effectiveness of only 50 percent is attainable, as contrasted to a theoretical maximum of 100 percent for counterflow.

Figure 13-25 allows for similar analysis of the effect of the number of shell-passes for the case of $C_{min}/C_{max} = 1.0$. Note that an infinite number of passes is required to equal counterflow performance.

Example 13-3. A cross-flow heat exchanger, fluids unmixed, is used to heat 10,000 ft³/min of air at 14.7 psia from 40°F to 100°F. Water enters the exchanger at 200°F. The total surface of the heat exchanger is 300 ft². Calculate the exit water temperature and the water flow rate. Assume that $U = 60$ Btu/hr ft² °F.

Solution. It is not known whether the air or water has the minimum value of C, so for the first trial it is assumed that air has the minimum C; the assump-

13-3   NTU APPROACH TO THERMAL DESIGN OF HEAT EXCHANGERS   407

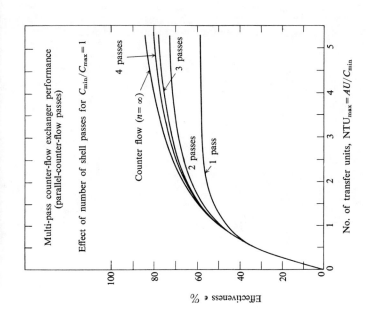

Fig. 13-25 Effect of number of shell-passes on effectiveness.*

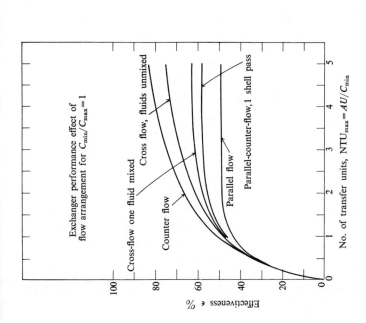

Fig. 13-24 Comparison of various flow arrangements.*

* (From *Compact Heat Exchangers* by Kays and London. Copyright 1955 by W. M. Kays and A. L. London. Used with permission of McGraw-Hill Book Company.)

Table 13-3

C_{min}	NTU	C_{min}/C_{max}	ΔT_w	ϵ_{calc}	ϵ_{chart}
5000	3.6	0.454	132	0.825	0.86
6000	3.0	0.545	110	0.688	0.80
4500	4.0	0.409	146.5	0.915	0.89
4700	3.83	0.427	140	0.877	0.88

tion will then be checked:

$$C_a = \dot{m}_a c_{p_a} = (10{,}000)(0.0764)(60)(0.24)$$
$$= 11{,}000 \text{ Btu/hr }°\text{F}.$$

$$\dot{q} = C_a \Delta T_a = (11{,}000)(60) = 660{,}000 \text{ Btu/hr}.$$

$$\text{NTU} = \frac{AU}{C_{min}} = \frac{(300)(60)}{11{,}000} = 1.635.$$

$$\epsilon = \frac{100 - 40}{200 - 40} = 0.375.$$

From Fig. 13-23(d) it can be seen that this is an impossible result. Therefore the water must be the fluid with the minimum value of C. One must now assume a flow rate of water and iterate until one can match a solution with Fig. 13-23(d). The assumptions and calculations are tabulated in Table 13-3. The calculations are based on the following equations:

$$\Delta T_w = \dot{q}/C_{min}, \qquad \epsilon_{calc} = \Delta T_w/(200{-}40).$$

From the table it is seen that the exit water temperature is approximately $200 - 140 = 60°\text{F}$. The water flow rate is

$$\dot{m}_w = C_w/c_{p_w} = 4700/(1.0) = 4700 \text{ lb}_m/\text{hr}$$
$$= 4700/(62.4) = 75.5 \text{ ft}^3/\text{hr} = 1.26 \text{ ft}^3/\text{min}.$$

The curves shown in Fig. 13-23 are in the form originally suggested by Kays and London [7]. A more recent publication by Mueller [8] discusses the use of these curves and points out that most practical exchangers have ϵ-NTU values in a rather limited range. Mueller suggested a method for plotting true mean temperature differences for exchangers that does not have the shortcomings of either the corrected LMTD or the ϵ-NTU method. A curve for the one shell-pass–two tube-pass exchanger is shown in Fig. 13-26. The true mean temperature divided by the difference in inlet temperatures (the temperature effectiveness) is plotted vs. P, with R as a parameter. P and R are defined by Eq. (13-39). Values of F_t are shown only as a guide for designers.

13-3 NTU APPROACH TO THERMAL DESIGN OF HEAT EXCHANGERS

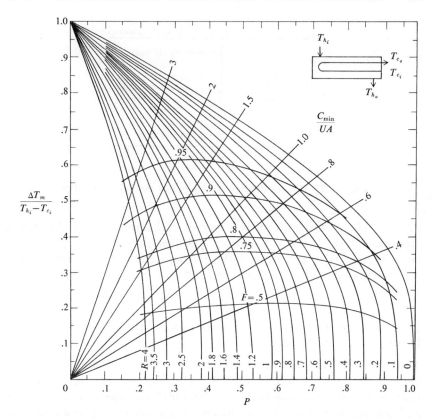

Fig. 13-26 Mueller's curve for 1-2 heat exchanger. (From *New Charts for True Mean Temperature Difference*, AIChE Preprint #10, August 1967, with permission.)

The reciprocal of NTU is also plotted on the graph, with particular values plotted as straight lines passing through the origin. If we know only two of the four parameters—P; R; $\Delta T_m/(T_{h_i} - T_{c_i})$; or $(\text{NTU})^{-1}$—the other two can be read from the graph.

Once ΔT_m is known the heat transfer can be calculated from

$$\dot{q} = UA\,\Delta T_m. \tag{13-54}$$

Storage-Type Exchangers

In some applications, such as in regenerators in gas turbine systems, the heat exchanger operates under transient conditions, most generally periodic. A rotating matrix in the exchanger may be exposed to a hot stream on one side and a cold stream on the other, as shown in Fig. 13-27. In this type exchanger energy is carried by the matrix material from the hot stream to the cold

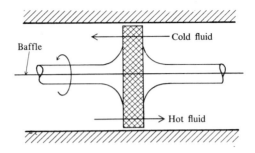

Fig. 13-27 A periodic-flow exchanger of the axial-flow type.

stream. The performance of this type of exchanger involves the thermal capacity of the matrix, and a transient type analysis is necessary [7]. Advantages of this type exchanger include compactness, the possible use of relatively inexpensive matrix material, and self-cleaning due to regular flow reversal.

13-4 HEAT TRANSFER RATES AND PRESSURE DROP IN HEAT EXCHANGERS

The heat exchanger must perform a prescribed job of effecting heat transfer between two (or more) fluid streams. The fluid streams are subjected to unavoidable pressure drop because of fluid friction, expansion, contraction, and changes in direction of the stream. The pressure drop is a loss which must be carefully accounted for in the design. In order to overcome the pressure drop, a pump or compressor must be inserted into the system. This means that a power charge for pumping must be made against the exchanger. No matter what scheme (such as a ram effect on an aircraft) is used to overcome pressure drop, a power requirement is always involved.

The ultimate design involves an optimization which balances pressure drop and heat transfer rates. That is, a higher heat transfer coefficient may be obtained by increasing the velocity of the stream; this results, however, in a higher pressure drop. The designer must decide how to "trade off" these two factors.

Often the amount of pressure drop which may be allowed is fixed by the overall design of the system. For example, other factors may fix the highest pressure allowed in the system. The sum of the pressure drops through all of the pieces of equipment and the piping of a system may, therefore, not exceed the total allowed for the entire plant or system. Other factors, such as the allowable pressure for a desired chemical reaction, may limit the allowable pressure drop.

Uncertain flow distribution in heat exchangers makes the calculation of mean values of the film coefficients subject to error. Attention will be called to certain problems in this area in the following pages.

The flow through a heat exchanger is nearly always so complex that the pressure drops for both hot fluid and cold fluid are difficult to estimate with accuracy.

Several schemes may be used in estimating pressure drop. The designer may use the fundamental relationships involving friction factor and attempt to calculate the pressure drop by adding the losses encountered in each portion of the exchanger as the flow proceeds. In industry it is common to develop "rules of thumb," or experience factors, which may be used to estimate the pressure drop. Some careful work has been done in correlating and generalizing existing information and in special research on both heat transfer and pressure drop for several types of heat exchangers. These generalizations are useful for design.

Although the complex geometry of heat exchangers, especially the compact types, makes accurate design difficult at best, some procedures are indicated in the following sections.

The Shell and Tube Exchanger

Tube-side heat transfer and pressure drop. A simple exchanger with two tube-passes is shown in Fig. 13–28 (only two tubes are shown). The pressure drop through the exchanger would be caused by any of the following: (1) sudden expansion at a; (2) sudden contraction at b; (3) friction in the tube from b to c; (4) sudden expansion at c; (5) loss from turning the fluid through 180° from c to d, (6) sudden contraction at d; (7) friction in the tube from d to e; (8) sudden expansion and turning loss at e; and (9) sudden contraction at f. Compared to the friction losses in long tubes, all of the other losses may usually be neglected. As the velocity increases in a design, however, consideration should be given to all of the above losses.

Calculation of the pressure drop and heat transfer coefficient inside the tube would follow the procedures developed in Chapters 6, 7, and 8. Careful note should be taken of the need to account for the nonisothermal nature of the flow in obtaining an appropriate friction factor.

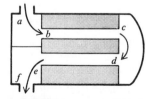

Fig. 13–28 Simple exchanger with two tube-passes.

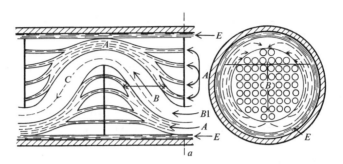

Fig. 13-29 Shell-side flow patterns. (From T. Tinker, "Shell Side Characteristics of Shell and Tube Heat Exchangers," *Proc. Gen. Disc. on Heat Transfer*, Institute of Mechanical Engineers, London, 1951.)

Shell-side heat transfer and pressure drop. The shell-side flow pattern is most complex, as was seen in Figs. 13-15, 13-16, and 13-17. The fluid enters the shell (sometimes impinging on a plate to deflect the flow and obtain better distribution). It then passes across a bank of tubes in a cross-flow arrangement. A baffle extends across a large portion of the shell to guide this flow. There is clearance between the baffle and the shell and between the baffle and the tubes which pass through it. Leakage occurs through these clearances. This means that only a portion of the shell-side fluid flows across the tube bank and through the "window" provided (area between the shell and the place where the baffle terminates). The fluid is subject to a variety of flow patterns from cross-flow through parallel-flow to counterflow in the various regions. Figure 13-29 shows Tinker's diagram [9], which illustrates this. The variation in velocities and in flow orientation to the tubes as well as the channeling effects cause great uncertainties in the pressure drop and heat transfer calculations.

A stepwise calculation similar to that suggested for the tube side would account for pressure drop due to the following: (1) entrance loss, including impingement plate; (2) flow across the tube bank; (3) turning the flow into the window; (4) flow through the window (orifice effect); and (5) flow through clearances (orifice effects). The flow through a clearance does not add pressure drop but its pressure drop must equal the pressure drop for the main stream's flow between the same two regions. Items (2) through (5) would have to be repeated for each traverse of the shell—between each pair of baffles.

Tinker [9], Bell [3], and others have correlated existing information for heat transfer and pressure drop. The heat transfer coefficient is usually based on the assumption that only a fraction of the flow is effective in passing over the tube bank. Similarly, a fractional part of the flow is used to determine the pressure drop.

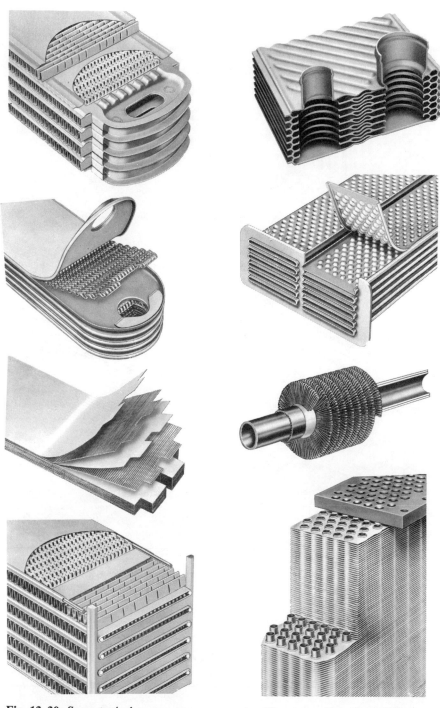

Fig. 13-30 Some typical compact arrangements. (Courtesy of Harrison Radiator Division, General Motors.)

Fig. 13-30 (cont.)

13–4 HEAT TRANSFER RATES AND PRESSURE DROP 415

Fraas and Ozisik [4] have arranged Tinker's results in a form which is easily understood. These details are lengthy and are therefore omitted from this text. The prediction methods described by Bell [3] are in common use.

The Compact Exchanger

A compact heat exchanger is one in which a large surface area is packed into a relatively small volume. (An arbitrary definition of a compact exchanger, given by Macklin [12], is an exchanger that has more than 200 square feet of surface area per cubic foot of volume.) Applications, in which the film

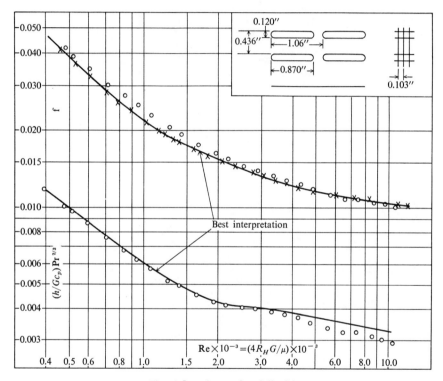

Finned flat tubes, surface 9.68—0.87
Fin pitch—9.68 per in.
Flow passage hydraulic diam—$4R_H = 0.01180$ ft.
Fin metal thickness—0.004 in.
Free-flow area/frontal area—$\sigma = 0.697$
Total heat transfer area/total volume—$\alpha = 229$ ft^2/ft^3.
Fin area/total area—0.795

Fig. 13–31 Heat transfer and friction data for a finned flat tube matrix. (From *Compact Heat Exchangers* by Kays and London. Copyright 1955 by W. M. Kays and A. L. London. Used with permission of McGraw-Hill Book Company.)

coefficient on one side (hot or cold) of the exchanger is very much larger than that obtainable on the other side of the exchanger, benefit from the utilization of fins or other geometric schemes to increase the area on the side of the low film coefficient. Typical cases are liquid-gas or condensing vapor-gas applications. The compact type exchanger is also used for gas-gas applica-

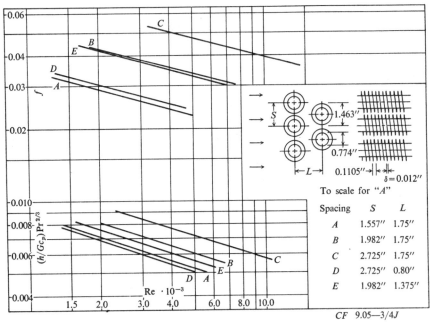

Finned circular tubes, surfaces CF—9.05-3/4J
(Data of Jameson)
Tube outside diam—0.774 in.
Fin pitch—9.05 per in.
Fin thickness—0.012 in.
Fin area/total area—0.835

	A	B	C	D	E	
Flow passage hydraulic diam—$4R_H$ =	0.01681	0.02685	0.0445	0.01587	0.02108	ft
Free-flow area/frontal area—σ =	0.455	0.572	0.688	0.537	0.572	
Heat transfer area/total volume—α =	108	85.1	61.9	135	108	ft^2/ft^3

Note: Minimum free-flow area in all cases occurs in the spaces transverse to the flow, except for D, in which the minimum area is in the diagonals. These data are included in this compilation because they include compact arrangements of interest not covered by Figs. 92–95.

Fig. 13-32 Heat transfer and pressure drop data. (From *Compact Heat Exchangers* by Kays and London. Copyright 1955 by W. M. Kays and A. L. London. Used with permission of McGraw-Hill Book Company.)

tions such as the gas turbine regenerator. Figure 13-30 shows some typical compact arrangements.

Analytical design techniques are of limited value in most compact surface design. Higher velocities and/or increased turbulence in the gas streams will increase the heat transfer coefficients. Turbulence increase is obtained by dimpling or crimping the surfaces and by use of strips and ribbons of complex geometry to induce turbulence. Increased turbulence is only partially effective in increasing the film coefficient, and the designer knows that a heavy penalty in pressure drop is incurred by increased velocity and turbulence. The usual design trade-off decision is inevitable.

Kays and London [7] tested a variety of compact and tube bank units and present extensive information regarding them. Figure 13-31 gives both heat transfer and friction factor for a finned flat tube matrix. This chart is for flow past the fin side surface (cross-flow around the tubes). The heat transfer and friction factor for the inside of the flattened tubes may be obtained using the hydraulic radius concept.

Typical of the data available is that of Jamison [11] for finned circular tubes, Fig. 13-32.

REFERENCES

1. "Standards of the Tubular Exchanger Manufacturer Association," New York (1959).
2. D. Q. Kern, "Heat Exchanger Design for Fouling Service," Third International Heat Transfer Conference, Chicago (1966).
3. K. J. Bell, "Exchanger Design—Based on the Delaware Research Program," *Petro. Chem. Eng.* No. 11, Vol. **32,** C-26 (Oct. 1960).
4. A. P. Fraas and M. N. Ozisik, *Heat Exchanger Design*, Wiley, New York (1965).
5. A. J. V. Underwood, "The Calculation of Mean Temperature Difference in Multipass Heat Exchangers," *J. Inst. Petro. Technologists* **20,** 145 (1934).
6. R. A. Bowman, A. C. Mueller, and W. M. Nagle, "Mean Temperature Difference in Design," *Trans. ASME* **62,** 283 (1940).
7. W. M. Kays and A. L. London, *Compact Heat Exchangers*, 2d ed., McGraw-Hill, New York (1967).
8. A. C. Mueller, "New Charts for True Mean Temperature Difference in Heat Exchangers," AIChE Preprint 10, Ninth Annual Heat Transfer Conference, Seattle (1967).
9. T. Tinker, "Shell Side Characteristics of Shell and Tube Heat Exchangers," *Proc. Gen. Disc. on Heat Transfer*, Institute of Mechanical Engineers, London (1951).

10. A. L. London, "Compact Heat Exchangers," Pts. I, II, III, *Mech. Eng.* 48, 31, 33 (May, June, July 1964).
11. S. L. Jamison, "Tube Spacing in Finned Tube Banks," *Trans. ASME* **67**, (1945).
12. M. Macklin, "A Guide to Optimum Design of Compact Heat Exchangers," *Machine Design*, 132 (Apr. 12, 1962).

PROBLEMS

1. A large plane wall is made up of three slabs of material, one 2 in. thick with $k = 0.09$ Btu/hr ft °F, one 1 in. thick with $k = 0.15$, and one 0.75 in. thick with $k = 0.025$. Estimate the heat flux through the wall if the outside surface temperature of the 2-in. slab is 350°F and the opposite surface temperature is 200°F. Calculate the temperatures at the two interfaces. Neglect any contact resistance between the slabs.

2. Calculate the heat flow through a composite wall made up of 8 in. of material A, 4 in. of material B, and 6 in. of material C. The mean values of k are 0.6, 0.15, and 0.24 Btu/hr ft °F for materials A, B, and C respectively. The temperature drop across the wall is 800°F.

3. Compute the temperatures at the two interfaces of a composite wall made up of 3 in. of material A, 5 in. of material B, and 6 in. of material C, if the temperature of the outside surface of A is 1000°F and the temperature of the outside surface of C is 200°F. The mean values of k are 0.06, 0.5, and 0.8 Btu/hr ft °F respectively for materials A, B, and C.

4. A large composite wall is made up of three layers. The outside layer is 3 in. thick and has a thermal conductivity of 0.5 Btu/hr ft °F. The middle layer is 6 in. thick and has a thermal conductivity of 0.4 Btu/hr ft °F. The inside layer is 4 in. thick and has k equal to 0.75. If 150 Btu/hr flows through each square foot of the wall when the inside temperature is 75°F, compute the temperature at the outside surface and at the two interfaces in the wall. The flow of heat is from the outside to the inside and is constant with time.

5. A wall 3 in. thick with thermal conductivity of 0.8 Btu/hr ft °F separates two fluids, one at 70°F and the other at 280°F. The convective heat transfer coefficients on the cold side and the hot side respectively are 2.0 and 60 Btu/hr ft² °F. Calculate the rate of heat transfer through the wall *and* the average temperature of the wall.

6. Heat is transferred from air at 100°F through a 4 in. thick wall to air at 40°F. Assume that the heat transfer coefficient on each side of the wall is 2 Btu/hr ft² °F and that the thermal conductivity of the wall is 0.1 Btu/hr ft °F. Calculate the temperature of the warmer surface of the wall.

7. The thermal conductivity of a certain material varies with temperature. According to this relation, $k = 0.0015T - 10^{-6}T^2$ where k is in Btu/hr ft °F and T is in degree F. What is the heat loss through one square foot of a plane wall made of this material if the wall is 6 in. thick and has surface temperatures of 800° and 400°F.

8. Estimate the overall heat transfer coefficient, based on outside area, for the case of heat flow from the fluid inside to the fluid outside a tube which has 2 in. i.d. and 3 in. o.d. The thermal conductivity of the tube material is 0.5 Btu/hr ft °F. Assume an inside heat transfer coefficient of 50 Btu/hr ft² °F and an outside coefficient of 10 Btu/hr ft² °F.

9. A hollow sphere of 6 in. i.d. and 9 in. o.d. has a thermal conductivity of 0.4 Btu/hr ft °F. A 5-kw heater is placed inside the sphere. Estimate the inside wall temperature of the sphere if the outside surface is maintained at 80°F.

10. Find the power in watts that must be supplied to a heater inside a composite hollow sphere to maintain the inside wall temperature at 1200°F if the outside wall temperature is at 100°F. The sphere is made up of two close-fitting concentric hollow spheres with the i.d. of the first sphere 4 in. and the o.d. 7 in. The o.d. of the outside sphere is 10 in. The mean k for the inside sphere is 0.5 and for the outside sphere 0.3 Btu/hr ft °F.

11. Derive Eq. (13-22) for one-dimensional conduction through a spherical shell.

12. Derive Eq. (13-23) a case in which the temperature of the inner surface of a spherical shell and of the ambient fluid surrounding the sphere are known.

13. An electric heater has the shape of a sphere of 2 in. o.d. It is covered with a 3-in. layer of insulation having $k = 0.2$ Btu/hr ft °F. If the outside of the insulation is assumed to be at a temperature of 100°F, estimate the inside temperature of the insulation when the heater is using energy at the rate of 1.0 watts.

14. Work Problem 13 assuming that the sphere is exposed to air at 70°F and 1 atmosphere and the outside surface temperature is unknown. Use the equation for free convective coefficients suggested in summary Table 10-1 (Chapter 10).

15. Fluid at a mean bulk temperature of 180°F flows through a 1-in. i.d. tube (o.d. = 1.4 in.). The tube material has a thermal conductivity k of 10 Btu/hr ft °F. The tube is covered with a 1-in. thickness of insulation ($k = 0.03$ Btu/hr ft °F) and exposed to air at 70°F. Assume an inside coefficient h_i of 25 Btu/hr ft² °F, and an outside coefficient h_o of 1.5 Btu/hr ft² °F. Estimate the heat loss per foot of pipe length.

16. Assuming an outside heat transfer coefficient h_o of 2 Btu/hr ft² °F, estimate the thickness of insulation that should be added to a $\frac{1}{16}$ in. diameter wire to maximize the heat flow. The insulation material has a thermal conductivity of 0.09 Btu/hr ft °F.

17. What temperature difference between the insulation surface and the surrounding air is necessary to make valid the assumption in Problem 16 that $h = 2$? Assume a film temperature of 100°F and a pressure of 1 atmosphere. Assume that the wire is horizontal.

18. A long cylinder, 3 in. in diameter is losing energy by heat transfer through the surrounding insulation at the rate of 40 Btu/hr per foot of cylinder length. The insulation surrounding the cylinder is 1.5 in. thick; $k = 0.08$ Btu/hr ft °F. The outer surface of the insulation is cooled by free convection to air (ambient temperature = 80°F). Estimate the temperature at the surface of the inner cylinder (the inside surface of the insulation).

19. A standard pipe with a nominal diameter of 4 in. is covered with 2 in. of insulation having a mean k of 0.06 Btu/hr ft °F, and this in turn is covered with 3 in. of insulation having a mean k of 0.075 Btu/hr ft °F. The temperature at the outer surface of the pipe is 600°F and the temperature at the outer surface of the second layer of insulation is 100°F. What is the temperature at the interface between the two layers of insulation?

20. A layer of insulating material with $k = 0.04$ Btu/hr ft °F is to be added to the outside of a pipe with a 6-in. o.d. If the temperature drop across the insulation is to be 180°F and the heat loss per foot of length is to be 100 Btu/hr, how thick should the layer be? Assume steady state and no end losses.

21. How much insulation ($k = 0.5$ Btu/hr ft °F) must be applied to a pipe with an o.d. of 4 in. if the surface of the insulation is not to exceed 100°F in a room where the air is 70°F. Assume the pipe temperature is constant at 320°F and the outside convective heat transfer coefficient at the insulation surface is 1.5 Btu/hr ft² °F.

22. How much heat will be lost per hour from a 100 ft length of pipe (i.d. = 3.5 in., o.d. = 4 in.) if it is carrying wet steam at 350°F through an environment at 50°F? The pipe is covered with 2 in. of insulation ($k = 0.05$ Btu/hr ft °F). Neglect the resistance of the steam and the pipe and assume that the heat transfer coefficient on the surface of the insulation is 2.0 Btu/hr ft² °F.

23. A 4 in. o.d. pipe is covered with a 1-in. thickness of insulation ($k = 0.022$ Btu/hr ft °F). The pipe's outside wall temperature is 200°F and the surrounding ambient air temperature is 40°F. An average value of the outside convective heat transfer coefficient $\bar{h}$ is assumed to be 3.0 Btu/hr ft² °F. For a 500-ft length of pipe, what reduction in heat transfer could be accomplished by the addition of an *extra* 1-in. thickness of the same type of insulation? In other words, how many Btu/hr will be saved by the use of the extra insulation? Neglect contact resistance and assume that the convective heat transfer coefficient remains unchanged.

Problem 24

24. For the composite cylinder shown in the figure, estimate the temperature at the interface of the two cylinders. Assume that T_i and T_o are uniform over the surfaces and that the flow is one-dimensional, that is, only in the radial direction. The dimensions given are in feet, and the values of thermal conductivity k in Btu/hr ft °F. Neglect contact resistance.

25. Let AMTD stand for the arithmetic mean temperature difference, $\frac{1}{2}(\Delta T_2 + \Delta T_1)$, and LMTD stand for the logarithmic mean temperature difference,

as in Eq. (13–35). Derive expressions for $\text{AMTD}/\Delta T_1$, $\text{LMTD}/\Delta T_1$, and the ratio AMTD/LMTD. Find the value of $\Delta T_2/\Delta T_1$ above which AMTD and LMTD differ by more than 5 percent.

26. What is the value of the average overall heat transfer coefficient for a cross-flow exchanger (fluids unmixed) operating under the following conditions: Water enters at 60°F and leaves at 92°F. Oil ($c_p = 0.46$ Btu/lb$_m$ °F, $\rho = 54.3$ lb$_m$/ft^3) enters at 100°F (50 gal/min) and leaves at 84°F. Assume that the exchanger has a surface area for heat transfer of 200 ft^2.

27. Estimate the surface area required in a cross-flow heat exchanger (both fluids unmixed) to cool 100,000 lbs$_m$/hr of helium from 140°F to 100°F with water at 60°F, flowing at the rate of 248,000 lbs$_m$/hr. Assume a good value of overall coefficient to be 40 Btu/hr ft^2 °F.

28. A counterflow heat exchanger has an area for heat transfer of 1000 square feet and the overall heat transfer coefficient for this case is 30 Btu/hr ft^2 °F. The cold fluid (specific heat = 0.6) passes through the exchanger at 25,000 lb/hr. The warm fluid (specific heat = 0.8) flows at the rate of 25,000 lb/hr. Determine the rate of heat transfer for the exchanger if the inlet temperature of the hot fluid is 150°F and the inlet temperature of the cold fluid is 70°F.

29. A heat exchanger heats a cold fluid from 150°F to 230°F while cooling a hot fluid from 300°F to 190°F. It operates under counterflow conditions. What inside area of tubes is required to transfer 1,500,000 Btu/hr if the overall heat transfer coefficient based on inside area is 300 Btu/hr ft^2 °F?

30. Water at 60°F is used to cool oil coming into a counterflow heat exchanger at 185°F. The exit temperature of the water is 72°F and the exit temperature of the oil is 100°F. Determine the LMTD and compare with the AMTD.

31. Work Problem 30 for a parallel-flow exchanger.

32. Determine the total heat transferred per hour in a heat exchanger containing 400 tubes (o.d. = 1 in., i.d. = 0.902 in.) each 12 ft long. The outside film coefficient is 1200 Btu/hr ft^2 °F and the inside coefficient is 600 Btu/hr ft^2 °F. The corrected mean temperature difference for the exchanger is 42°F. Neglect the thermal resistance of the tubes.

33. A counterflow heat exchanger is to be designed to cool tetrachlorethylene ($c_p = 0.216$) from 75° to 62°F at the rate of 200 lb/min. The cooling water enters at 50°F and leaves at 65°F. What surface area is required if the estimated overall heat transfer coefficient for the heat exchanger is 320 Btu/hr ft^2 °F?

34. Hot water at 100°F is used to heat 8700 lb/hr of fuel oil from 50°F to 65°F. Hot water flows at 2.5 ft/sec in a $\frac{1}{2}$-in. diameter copper pipe (i.d. = 0.731 in., o.d. = 0.840 in.). Oil is pumped through the annulus between this and a 1-in. steel pipe (i.d. = 1.182 in., o.d. = 1.315 in.) For "clean" use,
   a) What length of counterflow exchanger is needed?
   b) What length of parallel-flow exchanger is needed?
   c) If foul factors of 0.002 on water side are applied and 0.005 on oil side are applied, how are the above results altered?

422   ANALYSIS OF HEAT EXCHANGERS

35. A 1–2 parallel-counterflow heat exchanger is to be used to cool light oil from 150°F to 90°F with water at 70°F. The water flow rate is to be 200,000 lb_m/hr and the oil flow rate 100,000 lb_m/hr. Assume the specific heats of water and oil to be 1.0 and 0.5 Btu/lb_m °F respectively. If an overall coefficient of heat transfer for the exchanger is 40 Btu/hr ft² °F, estimate the necessary surface area required for the exchanger.

36. A heat exchanger is to be designed to heat 100,000 lb of water/hr from 60°F to 100°F by condensing steam at 230°F on the outside of the tubes. The tubes to be used have an i.d. of 0.1 ft and a wall thickness of 0.01 ft. The water-side heat transfer coefficient can be assumed to be 480 and the steam-side coefficient 2000 Btu/hr ft² °F. The thermal resistance of the tube wall is negligible. Compute the number of tubes required and the length of tubes for a one-pass exchanger if the inlet velocity of the water is 4 ft/sec.

37. Oil with a specific heat of 0.5 Btu/lb_m °F is being cooled from 180°F to 120°F by water which enters the tubes at 80°F and leaves at 100°F at the rate of 20,000 lb_m/hr. Assume a one shell-pass, two tube-pass exchanger and an overall coefficient of heat transfer of 40 Btu/hr ft² °F. Estimate the surface area required for the heat exchanger.

38. A one shell-pass, two tube-pass exchanger is used to cool oil ($c_p = 0.5$ Btu/lb_m °F) from 150°F to 120°F. Water ($c_p = 1.0$ Btu/lb_m °F) enters the exchanger shell at 60°F and leaves at 120°F. The exchanger has a surface area of 80 ft². Assume that a proper value of the overall heat transfer coefficient based on the above area is 60 Btu/hr ft² °F. Calculate (a) the total rate of heat exchange, and (b) the mass flow rates of the oil and water.

39. Work Problem 38, assuming that the cooling water enters the exchanger at 90°F.

40. Work Problem 38, assuming that the water flows through the tubes instead of the shell. Assume for comparative purposes that U remains unchanged.

41. Work Problem 38, assuming that the exchanger has *two* shell-passes and *four* tube-passes.

42. Determine the corrected mean temperature difference for an exchanger having two shell-passes and four tube-passes. Fluid enters the tubes at 60°F and leaves at 95°F. Another fluid enters the shell at 200°F and leaves at 130°F.

43. Determine the corrected log mean temperature difference for a heat exchanger having 2 shell-passes and twelve tube-passes. Fluid enters the tubes at 80°F and leaves at 110°F. The other fluid enters the shell at 180°F and leaves at 90°F.

44. Estimate the surface area required in a cross-flow heat exchanger (both fluids unmixed) to cool 100,000 lb_m/hr of helium from 140°F to 100°F with water at 60°F, flowing at the rate of 248,000 lb_m/hr. Assume a good value of overall coefficient to be 40 Btu/hr ft² °F.

45. A cross-flow exchanger with both fluids unmixed is used to cool air which is initially at 140°F and flowing at the rate of 10,000 lb_m/hr. The coolant is water at 60°F, flowing at a rate of 9600 lb_m/hr. Assume that a good value of the overall coefficient of heat transfer is 24 Btu/hr ft² °F. The surface area of the exchanger is 200 ft². Calculate the exit temperatures of the air and water.

46. Water at a temperature of 70°F is used to cool oil from a temperature of 160°F. The oil is to flow through the exchanger at the rate of 20,000 lb_m/hr and the water is to flow at the rate of 40,000 lb_m/hr. Using an assumed overall heat transfer coefficient of 60 Btu/hr ft² °F, estimate the exit temperature of the oil *and* of the water for a 1–2 parallel-counterflow exchanger with a surface area of 250 ft². Assume c_p of the oil and water to be 0.5 and 1.0 Btu/lb_m °F respectively.

47. A 1–2 parallel-counterflow heat exchanger is used to cool oil from 200°F by using water at 80°F. The oil has a specific heat of 0.5 Btu/lb_m °F and flows at the rate of 10,000 lb_m/hr. The water has a specific heat of 1.0 Btu/lb_m °F and flows at the rate of 20,000 lb_m/hr. The overall heat transfer coefficient U is assumed to be 75 Btu/hr ft² °F, and the surface area is 200 ft². Determine the rate of heat transfer between the two fluids.

48. The average overall heat transfer coefficient for a 1–2 shell- and tube-type heat exchanger is 300 Btu/hr ft² °F. The shell-side fluid (c_p = 1.0 Btu/lb_m °F) flow rate is 20,000 lb_m/hr, and the tube-side fluid (c_p = 0.85 Btu/lb_m °F) flows at the rate of 60,000 lb_m/hr. The fluid inlet temperatures are $T_{s\,i}$ = 500°F and $T_{t\,i}$ = 90°F. The total heat transfer area is 100 ft². Calculate the outlet temperatures $T_{s\,o}$ and $T_{t\,o}$.

49. Water at 80°F is used to cool light oil from 200°F in a 1–2 multipass exchanger. The water flow rate is to be 4000 gal/hr and the oil flow rate 2350 gal/hr. Assume that a good overall heat transfer coefficient U is 40 Btu/hr ft² °F, that the specific heat of water and oil is 1.0 and 0.5 Btu/lb_m °F respectively, that the average densities of oil and water are 53.0 and 62.1 lb_m/ft³ respectively, and that the surface area of the exchanger is 600 ft². Calculate the exit temperatures.

50. It is desired to cool 100,000 lb of air per hour from 250°F in a cross-flow exchanger with fluids unmixed. The cold fluid is water at 70°F, flowing at a rate of 80,000 lb/hr. The exchanger has a surface area for heat transfer of 3000 ft². Experience with this type of exchanger gives a predicted overall coefficient of heat transfer of 30 Btu/hr ft² °F. Calculate the exit temperature of the air.

51. A 1–2 parallel-counterflow heat exchanger is used to cool oil from 200°F by using water at 80°F. The oil has a specific heat of 0.5 Btu/lb_m °F and flows at the rate of 10,000 lb_m/hr. The water has a specific heat of 1.0 Btu/lb_m °F and flows at the rate of 20,000 lb_m/hr. The overall heat transfer coefficient U is assumed to be 75 Btu/hr ft² °F, and the surface area is 200 ft². Determine the rate of heat transfer between the two fluids.

52. Water at a temperature of 70°F is used to cool oil from a temperature of 160°F. The oil is to flow through the exchanger at the rate of 20,000 lb_m/hr, and the water is to flow at the rate of 40,000 lb_m/hr. Using an assumed overall heat transfer coefficient of 60 Btu/hr ft² °F, estimate the exit temperature of the oil *and* of the water for a 1–2 parallel-counterflow exchanger with a surface area of 250 ft². Assume c_p of the oil and water to be 0.5 and 1.0 Btu/lb_m °F respectively.

53. A 1–2 parallel-counterflow heat exchanger is to be used to cool light oil from 150°F to 90°F with water at 70°F. The water flow rate is to be 200,000 lb_m/hr

and the oil flow rate is 100,000 lb_m/hr. Assume the specific heats of water and oil to be 1.0 and 0.5 Btu/lb_m °F respectively. If an overall coefficient of heat transfer for the exchanger is 40 Btu/hr ft² °F, estimate the necessary surface area required for the exchanger.

54. Work Problem 38 by the NTU method.
55. Work Problem 27 by the NTU method.
56. Heat is to be exchanged in a 1–2 parallel-counterflow exchanger. Water at 70°F enters at the rate of 100,000 lb/hr in the shell. Light oil at 150°F at the rate of 80,000 lb_m/hr flows through the tubes. An overall coefficient U of 40 Btu/hr ft² °F is assumed and the surface area of the exchanger is 2000 ft². Determine the exit temperatures of two fluids.
57. A cross-flow exchanger with fluids unmixed is known to have a heat transfer surface area of 400 ft². It is used to cool air from 200°F by using air at 80°F. The hot air flow rate is 8000 lb_m/hr and the cold air flow rate is 12,000 lb_m/hr. Assume an overall heat transfer coefficient U for this situation to be 10 Btu/hr ft² °F and a value of c_p for air to be 0.24 Btu/lb_m °F. Calculate the exit temperatures of both airstreams.

CHAPTER 14

CONDUCTION

The transfer of heat in solids (using the Fourier-Biot equation) for the one-dimensional case has been discussed in Chapters 1 and 13. These simple problems—and more difficult ones involving temperature fields and heat transfer in solids—can be solved by use of the energy equation, Eq. (4-65), usually with some simplification in the form of the equation.

In the transfer of heat in solids, the heat exchange normally occurs at constant pressure and the specific heat at constant pressure is used. There is no viscous dissipation, no radiation, and normally no work of compression; therefore, these terms are eliminated. A term is needed, however, to allow for internal conversion of energy, such as might occur in a solid with electrical heating or nuclear reactions present.

14-1 STEADY CONDUCTION IN SOLIDS

The energy equation for a solid with variable thermal properties, Eq. (4-61), can be written in rectangular coordinates as

$$\rho c_p \left(\frac{\partial T}{\partial t} + u \frac{\partial T}{\partial x} + v \frac{\partial T}{\partial y} + w \frac{\partial T}{\partial z} \right)$$
$$= \frac{\partial}{\partial x}\left(k_x \frac{\partial T}{\partial x}\right) + \frac{\partial}{\partial y}\left(k_y \frac{\partial T}{\partial y}\right) + \frac{\partial}{\partial z}\left(k_z \frac{\partial T}{\partial z}\right) + \dot{q}''', \quad (14\text{-}1)$$

where $\dot{q}'''$ = energy conversion rate per unit volume.

Since there is no relative motion between elements in a solid, the energy equation can usually be solved independently of the momentum and mass conservation equations, provided appropriate boundary conditions are known.

In many solid conduction problems the coordinate system is fixed relative to the solid body under study, and the convective terms in Eq. (14-1) become zero. The equation for conduction in solids then becomes

$$\rho c_p \frac{\partial T}{\partial t} = \frac{\partial}{\partial x}\left(k_x \frac{\partial T}{\partial x}\right) + \frac{\partial}{\partial y}\left(k_y \frac{\partial T}{\partial y}\right) + \frac{\partial}{\partial z}\left(k_z \frac{\partial T}{\partial z}\right) + \dot{q}'''. \quad (14\text{-}2)$$

For constant thermal conductivity,

$$\rho c_p \frac{\partial T}{\partial t} = k\left(\frac{\partial^2 T}{\partial x^2} + \frac{\partial^2 T}{\partial y^2} + \frac{\partial^2 T}{\partial z^2}\right) + \dot{q}'''. \quad (14\text{-}3)$$

Dividing by ρc_p,

$$\frac{\partial T}{\partial t} = \alpha\left(\frac{\partial^2 T}{\partial x^2} + \frac{\partial^2 T}{\partial y^2} + \frac{\partial^2 T}{\partial z^2}\right) + \frac{\dot{q}'''}{\rho c_p}, \qquad (14\text{-}4)$$

where $\alpha = k/\rho c_p$ is the *thermal diffusivity*, a material property. Equation (14-4) can be written in cylindrical coordinates using the transformations in Section 3-1:

$$\frac{\partial T}{\partial t} = \alpha\left[\frac{\partial^2 T}{\partial r^2} + \frac{1}{r}\left(\frac{\partial T}{\partial r}\right) + \frac{1}{r^2}\left(\frac{\partial^2 T}{\partial \theta^2}\right) + \frac{\partial^2 T}{\partial z^2}\right] + \frac{\dot{q}'''}{\rho c_p}. \qquad (14\text{-}5)$$

Steady one-dimensional heat conduction with no internal conversion of energy could be solved by simplification of Eq. (14-4) or Eq. (14-5), leading to the same results obtained in Section 13-1. The solution of problems involving steady two- and three-dimensional heat conduction involves the use of partial differential equations. These types of problems have been intensively studied, and a large number of solutions are available. The texts of Carslaw and Jaeger [1], Schneider [2], Arpaci [3], and Churchill [4] present the mathematical solutions to many of the partial differential equations and boundary value problems of interest in conduction. In this section, however, the mathematics involved in obtaining solutions to partial differential equations will not be considered.

For steady conduction in a solid with no energy conversion and with constant thermal conductivity, the energy equation becomes

$$\frac{\partial^2 T}{\partial x^2} + \frac{\partial^2 T}{\partial y^2} + \frac{\partial^2 T}{\partial z^2} = 0. \qquad (14\text{-}6)$$

This type of equation, known as *Laplace's equation*, is a common equation in the fields of fluids and electricity and in physical situations where potential theory is applicable. It has already been discussed in Chapter 6. Any function that has continuous partial derivatives of the second order and which satisfies an equation of this type is called a *harmonic function*. An important consequence in heat transfer is that a family of curves ($T = $ constant) representing a solution to Eq. (14-6) is associated with a family of curves representing lines of constant heat flux. These two families of curves are everywhere mutually orthogonal. Thus if lines of constant temperature can be determined from Eq. (14-6), then lines representing the direction of heat flow can be easily obtained. This characteristic of orthogonality permits the drawing of these families of lines for two-dimensional cases, even where an analytical solution is not available.

A "flux plot" of this type is shown in Fig. 14-1. The inside surface is assumed to have a constant temperature of T_1 and the outside surface a constant temperature of T_2. If $T_1 > T_2$, then heat flows from the inside surface to the outside surface along the lines connecting the two surfaces. If $T_2 > T_1$, then heat flows along the same lines, but from the outside

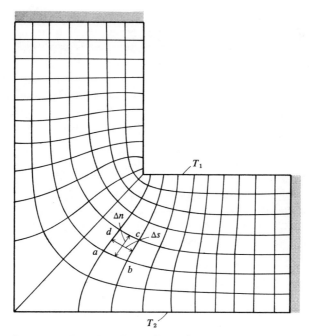

Fig. 14–1 Flux plot of two-dimensional heat conduction.

surface to the inside surface. Care has been taken in constructing the plot (by trial and error) to make the lines of constant temperature everywhere orthogonal to the lines of constant heat flow. Care has also been taken to make the local intervals between isotherms as nearly equal as possible to the local interval between adiabatic lines or lines of constant heat flow. This simplifies construction of the plot and calculations of the total heat flow between surface 1 and surface 2.

Consider the element $abcd$ shown in Fig. 14–1. The heat flow from ab to cd for a unit depth into the paper is calculated from the Fourier-Biot equation:

$$\dot{q}'_{ab-cd} = -k\,\Delta s\,\frac{\Delta T}{\Delta n}.$$

If Δs is made equal to Δn, then

$$\dot{q}'_{ab-cd} = -k\,\Delta T, \tag{14–7}$$

but $\Delta T = (T_1 - T_2)/N$ where N is the number of equal temperature intervals. There are several such channels conducting heat between the two surfaces, this number of channels is designated M.

Thus, the total heat flow per unit depth between surfaces 1 and 2 can be calculated by

$$\dot{q}'_{1-2} = kM\,\Delta T = k\,\frac{M}{N}\,(T_1 - T_2). \tag{14–8}$$

Obtaining heat flow rates by use of a flux plot and Eq. (14-8) is a crude technique but is often useful for obtaining approximate answers. Construction of a flux plot from an analytical solution is sometimes helpful and furnishes a check against possible error.

Example 14-1. Determine the heat flow across a piece of concrete, 4 ft high, with a cross section as shown in Fig. 14-1, if $T_1 = 100°F$ and $T_2 = 40°F$, and k is evaluated at the mean temperature $(100 + 40)/2 = 70°F$, so that $k = 0.54$ Btu/hr ft °F.

Solution. From Fig. 14-1, $M = 20$ and $N = 7$.

$$\dot{q} = \frac{(0.54)(4)(20)(100 - 40)}{(7)}$$

$$= 370 \text{ Btu/hr}.$$

The ratio M/N is called the *shape factor* per unit depth (S/L) for the particular geometry. Tables of shape factor for some common geometries have been published [5] and are useful for solution of some common problems in conduction. Such a table is shown as Table 14-1. The following equation is used to obtain the heat transfer rate between two isothermal faces of an object whose shape factor is known:

$$\dot{q} = (S) k\Delta T, \qquad (14-9)$$

where ΔT is the temperature difference between the two isothermal faces defining the shape of the cross section.

It can be seen from Eq. (14-9) that the shape factor S and the thermal resistance R_t of a given cross section are related by the thermal conductivity k of the material:

$$R_t = [(S)(k)]^{-1}. \qquad (14-10)$$

Shape factors may be determined by any one of several methods, including exact or numerical analysis and experimental methods. An easy way to determine shape factors experimentally is by cutting out a particular shape using electrically conducting paper and measuring the electrical resistance between two silvered edges of the pattern which represent lines of constant potential and which are analogs to isotherms of the heat flow problem.

Exact mathematical analysis of two-dimensional conduction problems such as are encountered in determining shape factors of some of the shapes shown in Table 14-1 is facilitated by the fact that the Laplace equation is linear in temperature, and *superposition* techniques may be employed. Some of the solutions shown are really the sum of solutions of more simple problems. This technique is discussed by Kutateladze [6].

Table 14-1 CONDUCTION SHAPE FACTORS

Shape	Diagram	$S = \dfrac{\dot{q}}{k\,\Delta T}$
Long hollow cylinder of length L.		$\dfrac{2\pi L}{\ln \dfrac{r_o}{r_i}}$
Hollow sphere.		$\dfrac{4\pi r_o r_i}{r_o - r_i}$
Cylinder with square insulation, length L.		$\dfrac{2\pi L}{\ln\left(1.08\,\dfrac{a}{D}\right)}$
Eccentric parallel cylinders of length L, with eccentricity e.		$\dfrac{2\pi L}{\ln\left(\dfrac{\sqrt{(r_o + r_i)^2 - e^2} + \sqrt{(r_o - r_i)^2 - e^2}}{\sqrt{(r_o + r_i)^2 - e^2} - \sqrt{(r_o - r_i)^2 - e^2}}\right)}$
Cylinder of diameter D and length L buried in a semi-infinite medium having a temperature at great distance T_∞. The cylinder is located a distance z from the surface.		$\dfrac{2\pi L}{\left(\ln \dfrac{2L}{D}\right)\left(1 + \dfrac{\ln(L/2z)}{\ln(2L/D)}\right)}$

Table 14-1 (continued)

Shape	Diagram	$S = \dfrac{\dot{q}}{k\,\Delta T}$
A vertical cylinder of length L and diameter D, placed in a semi-infinite medium having an adiabatic surface, and temperature T_∞ at large distance.		$\dfrac{2\pi L}{\ln\left(\dfrac{4L}{D}\right)}$
Sphere of diameter D, buried at distance z below adiabatic surface in a semi-infinite medium having temperature T_∞ at a large distance.		$\dfrac{2\pi D}{1 + \dfrac{D}{4z}}$
Half-sphere submerged into the surface of a semi-infinite medium with otherwise adiabatic surface and temperature T_∞ at a large distance.		πD
Disk placed on surface of semi-infinite medium with otherwise adiabatic surface and temperature T_∞ at a large distance.		$2D$

Table 14-1 (continued)

System	Shape factor
Thin disk in infinite medium having temperature T_∞ at a large distance.	$4D$
Rectangular plate on surface of semi-infinite medium with otherwise adiabatic surface and temperature T_∞ at large distance. Plate has dimensions $a \times b$.	$\dfrac{\pi a}{\ln \dfrac{4a}{b}}$ $a > b$
Thin rectangular plate of dimensions $a \times b$ buried in infinite medium having temperature T_∞ at large distance.	$\dfrac{2\pi a}{\ln \dfrac{4a}{b}}$ $a > b$
Conduction between cylinders of length L a distance Z apart, located in an infinite medium. No heat loss from cylinders to medium is considered.	$\dfrac{2\pi L}{\cosh^{-1}\left(\dfrac{4Z^2 - D_1^2 - D_2^2}{2D_1 D_2}\right)}$
Conduction between inside and outside surfaces of a rectangular box having uniform inside and outside surface temperatures. Wall thickness Δx is less than any inside dimensions.	$\dfrac{A}{\Delta x} + 0.54 \Sigma L + 1.2\, \Delta x$ ΣL = sum of all 12 inside lengths, Δx = thickness of walls, A = inside surface area.

14-2 NUMERICAL SOLUTIONS IN STEADY CONDUCTION

As shown in the previous section, problems involving steady conduction in three-dimensional isotropic solids with constant thermal conductivity lead to the Laplace equation. In rectangular coordinates this equation was given as Eq. (14–6):

$$\frac{\partial^2 T}{\partial x^2} + \frac{\partial^2 T}{\partial y^2} + \frac{\partial^2 T}{\partial z^2} = 0.$$

Equation (14–6) is a partial differential equation. For certain simple geometries and boundary conditions solutions have been obtained and may be used to determine local temperatures throughout the object [1–4]. In most of these cases the solutions are complex, and most often are expressed in the form of infinite series, making computed values of temperatures difficult to obtain. In many of these situations, and especially for more complicated geometries and boundary conditions, a solution to the problem based on approximate numerical methods is useful. In recent years, with widespread availability of digital computers, the numerical methods have come to be a very important way to solve practical heat transfer problems.

The numerical method utilizes approximations to derivatives such as those appearing in Eq. (14–6). The approximation is in terms of finite-sized differences between values of the particular variables involved. For this reason the method is sometimes referred to as a *finite-difference method*.

The approximation to a partial derivation may be obtained in various ways. A Taylor series expansion about a point will be used in this case. Suppose that the temperature in a substance depends only on position, that is, temperature is a function of x, y, and z. For a fixed or constant value of y and z, the temperature at points $-\Delta x$ and $+\Delta x$ away from a fixed point (x,y,z) can be expressed as:

$$T(x - \Delta x, y, z) = T(x,y,z) - \Delta x \left(\frac{\partial T}{\partial x}\right) + \frac{(\Delta x)^2}{2!}\left(\frac{\partial^2 T}{\partial x^2}\right) - \frac{(\Delta x)^3}{3!}\left(\frac{\partial^3 T}{\partial x^3}\right) + \cdots \quad (14\text{–}11)$$

and

$$T(x + \Delta x, y, z) = T(x,y,z) + \Delta x \left(\frac{\partial T}{\partial x}\right) + \frac{(\Delta x)^2}{2!}\left(\frac{\partial^2 T}{\partial x^2}\right) + \frac{(\Delta x)^3}{3!}\left(\frac{\partial^3 T}{\partial x^3}\right) + \cdots \quad (14\text{–}12)$$

If all terms involving Δx to the third power or higher are neglected, then Eqs. (14–11) and (14–12) may be added to give:

$$T(x - \Delta x, y, z) + T(x + \Delta x, y, z) - 2T(x,y,z) = (\Delta x)^2 \left(\frac{\partial^2 T}{\partial x^2}\right), \quad (14\text{–}13)$$

or

$$\frac{\partial^2 T}{\partial x^2} = \frac{T(x - \Delta x, y, z) + T(x + \Delta x, y, z) - 2T(x,y,z)}{(\Delta x)^2}. \quad (14\text{–}14)$$

14-2   NUMERICAL SOLUTIONS IN STEADY CONDUCTION

Equation (14–14) represents an approximation to the second derivative $\partial^2 T/\partial x^2$ for small values of Δx. The smaller the value of the difference Δx, the better the approximation. In the limit as $\Delta x \to 0$, Eq. (14–14) would represent the definition of the second derivative. It is apparent that the error involved in dropping the higher-order terms from Eq. (14–12) becomes small as Δx becomes small. In many practical problems the selection of a finite-sized Δx does not lead to significant error in the approximation. Thus, the second derivative of temperature with respect to x at a point can be approximated in terms of the temperatures existing at the point and at the points at distances $-\Delta x$ and $+\Delta x$ away. The derivative has, therefore, been expressed in terms of the finite differences in temperature between points that are a finite distance apart. Using the same reasoning for the two remaining derivatives in Eq. (14–6), that equation may be written:

$$\frac{T(x - \Delta x, y, z) + T(x + \Delta x, y, z) - 2T(x, y, z)}{(\Delta x)^2}$$

$$+ \frac{T(x, y - \Delta y, z) + T(x, y + \Delta y, z) - 2T(x, y, z)}{(\Delta y)^2}$$

$$+ \frac{T(x, y, z - \Delta z) + T(x, y, z + \Delta z) - 2T(x, y, z)}{(\Delta z)^2} = 0. \quad (14\text{–}15)$$

If $\Delta x = \Delta y = \Delta z$, then Eq. (14–15) may be simplified to:

$$T(x - \Delta x, y, z) + T(x + \Delta x, y, z) + T(x, y - \Delta y, z) + T(x, y + \Delta y, z)$$
$$+ T(x, y, z - \Delta z) + T(x, y, z + \Delta z) = 6T(x, y, z) \quad (14\text{–}16)$$

Equation (14–16) is an equation involving the temperature at seven points in the substance. Solving it for $T(x,y,z)$ gives an expression for the temperature at the point (x,y,z). It can be seen that this temperature is the average of the temperatures at six points surrounding the point of interest (Fig. 14–2). An expression identical to Eq. (14–16) could have been obtained by making an energy balance on a cube centered at (x,y,z) and with each surrounding temperature at a distance equal to the dimension of the cube. This derivation is asked for in one of the problems.

Equation (14–16) cannot be used to solve for the temperature at (x,y,z) directly, because generally speaking the temperatures at the surrounding points are also unknown. Instead the problem is approached by a mental subdivision of the entire object under study into finite-sized elements or *lumps*. In many problems it is convenient to use cubical-shaped lumps of dimension $\Delta x = \Delta y = \Delta z$. The center of each element or lump is referred to as the *node*. The temperature at the node is assumed to be the average temperature of the lump. Taking each lump in turn and letting (x,y,z) be the coordinates of the node, Eq. (14–16) can then be written for each particular

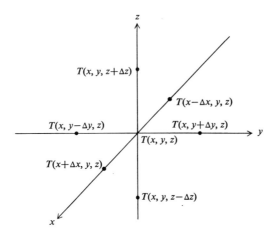

Fig. 14–2 Expressions for temperature surrounding a point (x,y,z).

lump. Note that in such a case the six temperatures on the left of Eq. (14–16) are the temperatures of the centers of the six lumps surrounding the one of interest. This leads to a series of equations, one equation for each node, with the equations expressed in terms of the temperatures of each node. With proper boundary conditions specified, the series of equations can be solved for all of the unknown temperatures. This set of temperatures at all of the nodal points is considered to be a solution to the problem.

The series of algebraic equations which might be written for a particular problem can be quite large and involve many unknown temperatures. These equations are usually solved by one of two methods, *relaxation* or *iteration*. Generally speaking the iteration method is most suitable for use where a digital computer is to be employed, whereas the relaxation method is most suitable for hand computation. Both methods can be programmed for a digital computer or worked out by hand calculation. The relaxation scheme will now be demonstrated by an example.

Example 14–2. In Fig. 14–3 is shown a cross section of a solid body in which heat transfer by conduction is taking place. This type of problem might occur when a rectangular duct is covered with flat insulation. The duct is considered to be long enough so that end effects can be ignored and the problem becomes one of two dimensions.

It will be assumed that a uniform temperature of 400°F exists over the inside surface and a uniform temperature of 100°F exists over the outside surface. The thermal conductivity of the material is 0.05 Btu/hr ft °F. It is desired to know the heat loss rate per foot of length. This will be done by numerical technique, using relaxation to obtain the temperatures at nodal points over the cross section.

14–2 NUMERICAL SOLUTIONS IN STEADY CONDUCTION

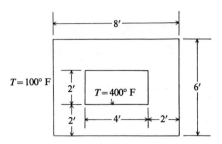

Fig. 14–3 Cross section of an insulated rectangular duct.

Solution. To reduce the number of computations, advantage is taken of the symmetry which exists over the cross section. Referring to Fig. 14–4, it can be seen that the temperature field existing over any quarter of the cross section is the mirror image of the adjacent quarter.

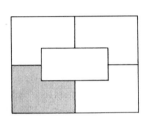
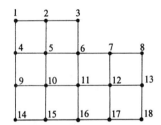

Fig. 14–4 The cross section divided into symmetrical quarters.

Fig. 14–5 Location of nodal points in the lower left quarter of cross section.

The lower left quarter will be selected for study in this example. Nodal points will be located on this quarter at a distance of one foot apart. Figure 14–5 gives 18 nodal points, representing the average temperature of the lump of material immediately surrounding the node. The selection and location of nodal points for any particular problem depends upon the geometry and the accuracy required. As a general rule, the more nodal points selected, the better the accuracy.

For this two-dimensional case, the steady conduction equation becomes

$$\frac{\partial^2 T}{\partial x^2} + \frac{\partial^2 T}{\partial y^2} = 0. \tag{14–17}$$

Writing Eq. (14–17) in finite-difference form, letting $\Delta x = \Delta y$, and simplifying, the following is obtained:

$$T(x - \Delta x, y) + T(x + \Delta x, y) + T(x, y - \Delta y) \\ + T(x, y + \Delta y) - 4T(x,y) = 0. \tag{14–18}$$

Equation (14–18) can now be applied to each of the interior nodes of Fig. 14–5. For node 5, for example, Eq. (14–18) becomes

$$T_4 + T_6 + T_{10} + T_2 - 4T_5 = 0; \qquad (14\text{–}19)$$

for node 10,

$$T_9 + T_{11} + T_{15} + T_5 - 4T_{10} = 0; \qquad (14\text{–}20)$$

for node 11,

$$T_{10} + T_{12} + T_{16} + T_6 - 4T_{11} = 0; \qquad (14\text{–}21)$$

and for node 12,

$$T_{11} + T_{13} + T_{17} + T_7 - 4T_{12} = 0. \qquad (14\text{–}22)$$

In each of the four equations above it can be seen that the temperature of a particular interior node is the average of the temperatures of the surrounding nodes. For example, from Eq. (14–19),

$$T_5 = \frac{T_4 + T_6 + T_{10} + T_2}{4}.$$

Figure 14–6 shows the arrangement of the nodes around node 5. The equations for the two nodes on the lines of symmetry T_2 and T_{13} can be obtained easily when it is noted that T_5 and T_{12} are matched by equal temperatures on the opposite sides of the lines of symmetry. Thus,

$$T_1 + T_3 + 2T_5 - 4T_2 = 0, \qquad (14\text{–}23)$$

and

$$2T_{12} + T_{18} + T_8 - 4T_{13} = 0. \qquad (14\text{–}24)$$

Equations (14–19) through (14–24) are six algebraic equations in terms of six unknown temperatures T_2, T_5, T_{10}, T_{11}, T_{12}, and T_{13}. Remember that the outside and inside surface temperatures are known. To solve these six equations by relaxation, the term *residual* will now be introduced. The residual of a particular equation, such as Eq. (14–19), is simply the value of the terms on the left-hand side of the equation for any arbitrary values of the

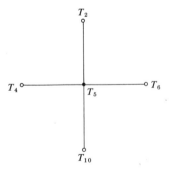

Fig. 14–6 Arrangement of nodes around node 5.

Table 14-2

Step	T_2	R_2	T_5	R_5	T_{10}	R_{10}	T_{11}	R_{11}	T_{12}	R_{12}	T_{13}	R_{13}
1	250		250		250		250		250		250	
2		0		0		−300		0		0		0
3	250		250		175		250		250		250	
4		0		−75		0		−75		0		0
5	250		231		175		250		250		250	
6		−38		1		−19		−75		0		0
7	250		231		175		231		250		250	
8		−38		1		−38		1		−19		0
9	241		231		175		231		250		250	
10		−2		−8		−38		1		−19		0
11	241		231		175		231		245		250	
12		−2		−8		−38		−4		1		−10
13	241		231		165		231		245		250	
14		−2		−18		2		−14		−9		−10
15	241		226		165		231		245		250	
16		−12		2		−3		−14		−9		−10
17	241		226		165		227		245		250	
18		−12		2		−7		2		−3		−10
19	241		226		165		227		245		247	
20		−12		2		−7		2		−6		2
21	238		226		165		227		245		247	
22		0		−1		−7		2		−6		2
23	238		226		165		227		243		247	
24		0		−1		−7		0		2		−2
25	238		226		163		227		243		247	
26		0		−3		1		−2		2		−2
27	238		225		163		227		243		247	
28		−2		1		0		−2		2		−2
29	238		225		163		226		243		247	
30		−2		1		−1		2		1		−2
31	238		225		163		226		243		246	
32		−2		1		−1		2		0		2
33	237		225		163		226		243		246	
34		2		0		−1		2		0		2
35	237		225		163		226.5		243		246	
36		2		0		0.5		0		0.5		2
37	237.5		225		163		226.5		243		246	
38		0		0.5		0.5		0		0.5		2
39	237.5		225		163		226.5		243		246.5	
40		0		0.5		0.5		0		1		0

temperatures appearing in that equation. Letting R_2 stand for the residual of Eq. (14–23), for example,

$$T_1 + T_3 + 2T_5 - 4T_2 = R_2. \qquad (14\text{–}25)$$

In a similar manner,

$$T_4 + T_6 + T_{10} + T_2 - 4T_5 = R_5, \qquad (14\text{--}26)$$

$$T_9 + T_{11} + T_{15} + T_5 - 4T_{10} = R_{10}, \qquad (14\text{--}27)$$

$$T_{10} + T_{12} + T_{16} + T_6 - 4T_{11} = R_{11}, \qquad (14\text{--}28)$$

$$T_{11} + T_{13} + T_{17} + T_7 - 4T_{12} = R_{12}, \qquad (14\text{--}29)$$

$$2T_{12} + T_{18} + T_8 - 4T_{13} = R_{13}. \qquad (14\text{--}30)$$

When the "correct" values of temperature are substituted into these equations, the residuals will all become zero—that is, the equations are satisfied. The term correct has been placed in quotes to emphasize that the solution to the set of equations is not the exact solution of the real problem, since finite approximations are involved. In other words, the values of temperature determined for the six nodes are not the actual temperatures occurring at those points in a real body. The relaxation scheme consists merely of assuming values of temperature, calculating values of the residuals, and correcting the temperatures to reduce the value of the residuals. This is best done in an orderly manner, as demonstrated in Table 14–2. For the first step, it will be assumed that each of the six unknown temperatures is equal to 250°F (the average of 100° and 400°F). The residuals for each node are then calculated using Eqs. (14–25) through (14–30).

The largest residual is seen to exist for node 10, where $R_{10} = -300$. The temperature is, therefore, corrected at that node to make the residual zero. This is called relaxation. From Eq. (14–27) it is seen that decreasing T_{10} from 250° to 175°F will make $R_{10} = 0$, assuming that the other temperatures remain unchanged. Changing T_{10}, however, will cause new values of residuals to exist at nodes 5 and 11, as is seen in step 4 of Table 14–2. In step 6, the residual at node 5 is reduced to 1 rather than zero to avoid introduction of fractional temperatures early in the computations. Such tricks are learned from experience, but the beginner may wish to relax to zero in each step until such experience is gained.

After the residuals at all nodes have been reduced to values very near zero, the temperatures at each node are then known to a sufficient degree of accuracy (step 39 in Table 14–2). (The temperatures are within one degree of the final value after step number 25.) Once the temperatures are known, then rates of heat transfer can be estimated. This is accomplished by assuming that heat flows along the lines connecting the nodes. Referring to Fig. 14–7, for example, the heat transfer from node 5 to node 4 is given by the Fourier law:

$$\dot{q}_{5-4} = kA\frac{\Delta T}{\Delta x} = (0.05)\,\frac{\text{Btu}}{\text{hr ft °F}}\,\frac{(1)(L)\text{ ft}^2}{1\text{ ft}}\,(225 - 100)\,°\text{F},$$

$$\frac{\dot{q}_{5-4}}{L} = (0.05)(125)\,\frac{\text{Btu}}{\text{hr ft}} = 6.25\,\frac{\text{Btu}}{\text{hr ft}}.$$

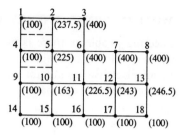

Fig. 14-7 Computed temperatures at nodal points using relaxation method.

The transfer between nodes 10 and 9, 10 and 15, 11 and 16, and 12 and 17 may be calculated in a similar manner. The transfer between nodes 2 and 1 and nodes 13 and 18 must involve division by 2, since only half of one lump is involved in the cross-sectional area in each case. Thus, the total heat transfer to the outer wall in the corner section is given by

$$\frac{\dot{q}}{L} = k\left[\frac{(T_2 - T_1)}{2} + (T_5 - T_4) + (T_{10} - T_9) + (T_{10} - T_{15}) + (T_{11} - T_{16})\right.$$
$$\left. + (T_{12} - T_{17}) + \frac{(T_{13} - T_{18})}{2}\right]$$
$$= (0.05)\left[\frac{(237.5 - 100)}{2} + (225 - 100) + (163 - 100) + (163 - 100)\right.$$
$$\left. + (226.5 - 100) + (243 - 100) + \frac{(246.5 - 100)}{2}\right]$$
$$= (0.05)(662.5) = 33.125 \frac{\text{Btu}}{\text{hr ft}}.$$

For all four corner sections, the heat loss would be

$$(33.125)(4) = 132.5 \text{ Btu/hr ft.}$$

As a check, the heat loss *from* the inner wall can be calculated using similar reasoning:

$$\frac{\dot{q}}{L} = 4k\left[\frac{(T_3 - T_2)}{2} + (T_6 - T_5) + (T_6 - T_{11}) + (T_7 - T_{12}) + \frac{(T_8 - T_{13})}{2}\right]$$
$$= 4(0.05)\left[\frac{(400 - 237.5)}{2} + (400 - 225) + (400 - 226.5)\right.$$
$$\left. + (400 - 243) + \frac{(400 - 246.5)}{2}\right]$$
$$= 132.9 \text{ Btu/hr ft,}$$

which checks within 0.4 Btu/hr ft with the heat loss to outer surface.

14-3 CONDUCTION WITH INTERNAL ENERGY CONVERSION

In many practical engineering problems there is an internal conversion or dissipation of energy within a solid. A common example is the dissipation of electrical energy within an electrical conductor due to the current. The equation describing this rate of energy dissipation is

$$P = I^2 R = \frac{V^2}{R}, \tag{14-31}$$

where P = power dissipated in watts, I = current in amperes, R = resistance in ohms, and V = potential in volts.

The rate of dissipation per unit volume within a solid is obtained by dividing the total power P by the total volume of the solid. This assumes that the dissipation within the solid is uniform.

The energy converted into the thermal form by this dissipation must be removed at the same rate, or additional thermal energy storage will occur. This increase in storage of thermal energy will cause a temperature rise within the substance. An equilibrium condition will be reached eventually when the local temperature gradient at each point is sufficient to cause the net heat per unit volume to be conducted away from the point at the same rate that it is being converted from the electrical to the thermal form.

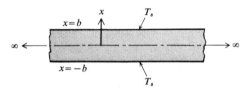

Fig. 14-8 One-dimensional conduction with energy conversion.

This equilibrium condition can be described quite easily for the one-dimensional case, assuming uniform conversion of heat and constant thermal conductivity. This condition might exist in a plate of finite thickness that is infinite in extent in the other two directions (Fig. 14-8). Assume that each surface of the plate is at a uniform temperature T_s. The thickness of the plate will be designated as $2b$ for convenience, and x will be measured from the center of the plate as shown. The energy conversion in the plate might be due to a nuclear reaction (fission) or a phase change or transformation.

The differential equation describing this situation is obtained from Eq. (14-3) by elimination of terms, which gives

$$\frac{d^2 T}{dx^2} = \frac{-\dot{q}'''}{k}. \tag{14-32}$$

14-3 CONDUCTION WITH INTERNAL ENERGY CONVERSION

The boundary conditions are:
$$T = T_s \quad \text{at} \quad x = b \quad \text{and} \quad x = -b.$$

Since both $\dot{q}'''$ and k are assumed to be constant, Eq. (14–32) can be solved by integration:

$$T = \frac{-\dot{q}'''}{2k} x^2 + C_1 x + C_2. \tag{14-33}$$

The constants of integration are evaluated by use of the boundary conditions to give

$$T - T_s = \frac{\dot{q}'''}{2k}(b^2 - x^2). \tag{14-34}$$

The temperature distribution is, therefore, a parabolic one. The maximum temperature $T_{\max}$ occurs at the center ($x = 0$). Therefore

$$T_{\max} - T_s = \frac{\dot{q}'''}{2k} b^2. \tag{14-35}$$

Example 14-3. A copper plate $\tfrac{1}{4}$-in. thick has an electrical potential between the surfaces of 0.01 volt. Determine the maximum temperature within the plate if the two plate surfaces are maintained at 80°F. The electrical resistivity of copper is 0.68 μohm inch.

Solution. The resistance across the narrow dimension of a piece of copper $\tfrac{1}{4}$-in. thick and 1 in.^{2} is obtained from the resistivity:

$$\frac{0.68 \ \mu\text{ohm in.} \times \tfrac{1}{4} \text{ in.}}{1 \text{ in.}^2} = 0.17 \ \mu\text{ohms}$$

The electrical power dissipation is, from Eq. (14–31),

$$P = \frac{V^2}{R} = \frac{0.0001}{0.17 \times 10^{-6} \text{ ohms}}$$

$$= 5.88 \times 10^2 \text{ watts in a } \tfrac{1}{4}\text{-in.}^3 \text{ lump } (\tfrac{1}{4} \times 1 \times 1 \text{ in.}).$$

Thus,
$$\dot{q}''' = \frac{5.88 \times 10^2}{\tfrac{1}{4} \text{ in.}^3 \times \frac{1}{1728} \frac{\text{ft}^3}{\text{in.}^3}} \text{ watts} = 4.07 \times 10^6 \ \frac{\text{watts}}{\text{ft}^3},$$

or
$$\dot{q}''' = 13.9 \times 10^6 \ \text{Btu/hr ft}^3.$$

Using $k = 212$ Btu/(hr ft °F),

$$T_{\max} - 80° = \frac{\dot{q}'''}{2k} L^2 = \frac{13.9 \times 10^6}{(2)(212)} \frac{1}{(8)^2(144)} = 3.55°\text{F},$$

$$T_{\max} = 83.55°\text{F}.$$

In case of large temperature variations within a solid, variations in both the thermal conductivity and the electrical resistivity might need to be considered.

A technique has been developed by Vidmar [7] to determine the maximum temperature in an electrical coil experimentally by observing the change in total electrical resistance of the coil as the voltage or current is increased. This technique utilizes the fact that the electrical resistivity of the copper varies linearly with temperature. The temperature rise in electrical coils was studied by Pemberton and Parker [8], who prepared charts for prediction of maximum temperatures. One problem in such studies is in defining the thermal properties of the heterogeneous matrix of wire and insulation making up the coil.

A common example of electrical energy conversion and heat conduction occurs in the case of electrical heating. Here the current is assumed to flow in an infinitely long cylinder. The maximum allowable current may be determined for a given wire size and composition. The appropriate equation for the steady state case with a uniform boundary condition along the surface is obtained from Eq. (14–5):

$$\frac{d^2T}{dr^2} + \frac{1}{r}\frac{dT}{dr} = \frac{-\dot{q}'''}{k}. \qquad (14\text{–}36)$$

This can be written

$$\frac{1}{r}\frac{d}{dr}\left(r\frac{dT}{dr}\right) = -\frac{\dot{q}'''}{k}. \qquad (14\text{–}37)$$

If the boundary conditions

$$T = T_s \text{ at } r = R \qquad \text{and} \qquad \frac{dT}{dr} = 0 \text{ at } r = 0$$

are applied to Eq. (14–37), the solution becomes

$$T - T_s = \frac{\dot{q}'''}{4k}(R^2 - r^2). \qquad (14\text{–}38)$$

Equation (14–38) is, therefore, the temperature distribution in a long cylinder or wire with uniform internal conversion of energy per unit volume and a uniform surface temperature. The temperature of the surface of the solid T_s in Eqs. (14–34) and (14–38) is *not* the temperature of the surrounding environment, since there is additional thermal resistance in the fluid surrounding the solid. In order to solve problems where the temperature of the environment is known instead of the surface temperature of the wire, it is convenient to use appropriate values of the convective heat transfer coefficient in applying boundary conditions to the differential equations.

An energy balance on a length L of wire with diameter D, cooled by a steady convective process described by a coefficient $\bar{h}$ would be

energy dissipated = heat removed by convection

$$\dot{q}''' \frac{\pi D^2}{4} L = \bar{h} \pi D L (T_s - T_\infty). \tag{14-39}$$

Therefore,

$$T_s - T_\infty = \frac{\dot{q}''' D}{4\bar{h}}. \tag{14-40}$$

Example 14-4. A 14-gage copper wire carries a current of 30 amperes. Assume that an average convective coefficient $\bar{h}$ for the bare wire is 2 Btu/hr ft² °F. Determine the maximum temperature in the wire if the environment is at 80°F. The resistance of the wire is 0.0025 ohms/ft and the diameter is 0.064 in.

Solution. The power dissipated per foot is

$$P = I^2 R = (30)(30)(0.0025) = 2.25 \text{ watts} = 7.65 \text{ Btu/hr}.$$

The heat loss is

$$\dot{q} = \bar{h} A (T_s - T_\infty).$$

Solving for the surface temperature and equating the power dissipated and the heat loss gives

$$T_s = \frac{\dot{q}}{\bar{h}A} + T_\infty = \frac{(7.65)(12)}{(2)(\pi)(0.064)(1)} + 80°F$$
$$= 228 + 80°F = 308°F.$$

The maximum temperature in the wire is obtained from Eq. (14-38):

$$T_{max} - T_s = \frac{\dot{q}''' D^2}{16k}.$$

A 14-gage wire has a cross-sectional area of 0.003225 in.², therefore,

$$\dot{q}''' = \left[\frac{7.65 \text{ Btu/hr ft}}{0.003225 \text{ in.}^2}\right]\left[\frac{144 \text{ in.}^2}{\text{ft}^2}\right],$$

and

$$T_{max} - T_s = \frac{(7.65)(144)(0.064)^2}{(0.003225)(16)(212)(144)} = 0.00286°F.$$

The temperature difference between the center and the surface of the wire is obviously negligible, and the maximum temperature is very close to 308°F.

The cooling of objects in which other forms of energy are being converted into thermal energy is particularly severe for objects having a large volume-to-surface ratio, high rates of energy conversion per unit volume, low thermal conductivity, and low convective heat transfer at the surface. Thermal contact resistance with adjacent solid surfaces also plays an important role in limiting heat removed by conduction.

In a nuclear reactor the conversion rate is so large that melting of the uranium metal may occur unless the heat removal rate is sufficient. In electronic devices, particularly solid state devices, the resulting high internal temperatures may damage the device and cause failure of the circuit [9].

The next section will point out ways in which heat transfer removal by convection at solid surfaces may be improved.

14-4 EXTENDED SURFACES WITH CONVECTIVE BOUNDARIES

In many structures the problem of calculating heat flow and temperature distribution is complicated by the unusual shapes which may exist. Instead of smooth walls from which convective effects can be calculated, "fins" or protruding structural parts are often found. The term *fin* will be used to refer to that portion of the structure or device which extends into a fluid and is cooled or heated by convection. A fin is also characterized by the conduction of heat to, or from, the parent wall or surface from which it extends. Such conducted heat passes through the *base* of area of attachment of the fin to the parent wall.

Fins are often used to decrease the overall thermal resistance in situations where a low convective heat transfer coefficient exists. The overall heat transfer coefficient U may be very low because of a low value of $\bar{h}$ and a correspondingly high value of thermal resistance, $R_t = 1/(\bar{h}A)$, on one surface. The overall coefficient can be improved by placing fins on the surface that has the low value of $\bar{h}$ in order to produce an effectively larger value of A. Thus, the product $\bar{h}A$ can be made larger in spite of the fact that $\bar{h}$ remains low. Some typical fins have already been shown in Fig. 13-30, and some typical pressure drop and heat transfer data were given in Figs. 13-31 and 13-32. To understand this behavior it is desirable to make some simple analyses of a few basic fin geometries.

An analysis of the straight fin of uniform but arbitrary cross section shown in Fig. 14-9 can be made from Eq. (14-3). It will be assumed that: (1) the base temperature T_b is greater than the fluid temperature T_∞ and that both are constant with time; (2) thermal conductivity of the fin is constant; (3) the heat transfer coefficient on the fin surface is constant; and (4) there are no temperature gradients except in the x direction. Of course, this would not be true in reality as there must be a gradient normal to the surface to cause the heat to be conducted to the surface. Analysis has shown that, for

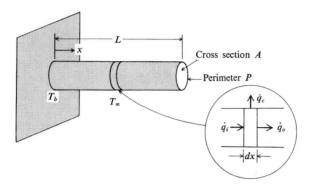

Fig. 14-9 A straight fin of uniform, arbitrary cross section and small control volume.

fins having a relatively long length compared to the smallest transverse dimension, only a small error is caused by this assumption [10].

Choosing the small control volume shown (thickness dx), heat ($\dot{q}_i$) is conducted in at the left face, and a smaller amount ($\dot{q}_o$) is conducted out at the right face of the element. The difference is equal to the heat removed by convection to the surrounding fluid at the surface. The convected heat can be treated as though it represented uniformly distributed heat sinks or negative energy conversion per unit volume. The value of $\dot{q}'''$ in Eq. (14-3) is made equal to the convected heat for the control element, $-(\bar{h})(P)(dx)(T - T_\infty)$, divided by the volume of the element $(dx)(A)$, where P is the "wetted" perimeter, or length exposed to the fluid.

Thus, with no storage (steady state), Eq. (14-3) for the one-dimensional case is

$$k\frac{d^2T}{dx^2} = -\dot{q}''',$$

and

$$\dot{q}''' = -\frac{(\bar{h})(P)(dx)(T - T_\infty)}{(dx)(A)} = -\frac{\bar{h}P}{A}(T - T_\infty).$$

Substituting,

$$k\frac{d^2T}{dx^2} = -\left[-\frac{\bar{h}P}{A}(T - T_\infty)\right],$$

and

$$\frac{d^2T}{dx^2} = \frac{\bar{h}P}{kA}(T - T_\infty). \quad (14\text{-}41)$$

The temperature in Eq. (14-41) may be nondimensionalized by defining

$$T^* = \frac{T - T_\infty}{T_b - T_\infty}. \quad (14\text{-}42)$$

Eq. (14-41) can now be written as

$$\frac{d^2 T^*}{dx^2} = \left(\frac{\bar{h} P}{kA}\right) T^*. \tag{14-43}$$

The general solution to Eq. (14-43) is well known:

$$T^* = c_1 e^{Nx} + c_2 e^{-Nx}, \tag{14-44}$$

where $N^2 = (\bar{h}P/kA)$. For boundary condition expressed in terms of the temperature T^*, the dimensionless temperature distribution in the fin will depend only on distance x from the fin base and the value of the parameter $(\bar{h}P/kA)$.

Fin with Adiabatic Tip

If the fin shown in Fig. 14-9 is relatively long compared to its thickness, the amount of heat which passes from the tip of the fin to the fluid is negligible. In such a case the assumption of an adiabatic tip is justified, and the temperature gradient in the fin at the tip is assumed to be zero. For this case the boundary conditions to use with the general solution, Eq. (14-44), are

$$T^* = 1 \text{ at } x = 0, \quad \text{and} \quad \frac{\partial T^*}{\partial x} = 0 \text{ at } x = L. \tag{14-45}$$

Substitution of these boundary conditions into Eq. (14-44) yields the temperature distribution in the fin:

$$T^* = \frac{\cosh N(L - x)}{\cosh NL}, \tag{14-46}$$

where $\cosh x = \frac{1}{2}(e^x + e^{-x})$.

A typical plot of Eq. (14-46) is shown in Fig. 14-10. It can be seen that the temperature decreases from the base temperature T_b to a value close to, but not exactly equal to, the fluid temperature T_∞, and that a zero temperature gradient exists at the tip, $x = L$.

The fin must have a temperature gradient along its length to provide for conduction. Therefore, it will be seen that the fin surface is less effective in

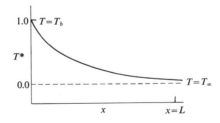

Fig. 14-10 Typical temperature variation in straight fin with adiabatic tip.

transferring heat per unit area than is the parent or base surface because there is a smaller average temperature difference between it and the surrounding fluid.

Since the tip is at a temperature very close to the fluid temperature, this part of the fin is less important in transferring heat than the fin surface near the base.

All of the heat lost by the fin over its entire surface must be transferred by conduction across the fin base area ($x = 0$). The Fourier law, Eq. (1–16), can be used to determine this heat flow, $\dot{q}_{x=0} = \dot{q}_0$, since the temperature distribution is known from Eq. (14–46):

$$\dot{q}_{x=0} = \dot{q}_0 = -kA\left(\frac{\partial T}{\partial x}\right)_{x=0} = \sqrt{\bar{h}PkA}\,(T_b - T_\infty)\tanh NL, \qquad (14\text{–}47)$$

where

$$\tanh x = \frac{\sinh x}{\cosh x} = \frac{e^x - e^{-x}}{e^x + e^{-x}}.$$

The assumption of an adiabatic tip leads to simple mathematical solutions, as seen above, but such a condition may not exist in the real case, particularly for short fins. To retain the usefulness of the simple equations derived above for the case where the assumption of an adiabatic tip is not valid, a *fictitious length* can be added to the fin to account for the tip loss. The fin is imagined to be increased in length by an amount that will give an increment in surface area equal to the tip area of the fin. If it is assumed that $\bar{h}$ over the tip surface is the same as $\bar{h}$ over the remaining fin surface, then the heat lost out of the incremented surface area would be approximately equal to that lost out of the tip. Thus, the corrected length

$$L_c = L + \Delta L, \qquad (14\text{–}48)$$

where $\Delta L = (A/P)$.

The corrected length can then be used in the simplified equations, such as Eqs. (14–46) and (14–47), giving some increase in accuracy over that obtained by using an uncorrected length.

General Solution for Fin with Nonadiabatic Tip

For the general case of the nonadiabatic tip, the boundary conditions are

$$T^* = 1 \text{ at } x = 0 \quad \text{and} \quad -k\frac{\partial T^*}{dx} = \bar{h}T^* \text{ at } x = L.$$

Substitution into the general solution, Eq. (14–44), gives

$$c_1 = \frac{1}{1 - Be^{2LN}},$$

$$c_2 = \frac{B1}{B - e^{-2LN}},$$

where $B = (\bar{h} + kN)/(\bar{h} - kN)$, and the dimensionless temperature distribution with length becomes

$$T^* = \frac{e^{LN(x/L)} - Be^{-LN[(x/L)-2]}}{1 - Be^{2LN}}. \tag{14-49}$$

The heat conducted across the fin base (and thus the total heat flow from the fin surface) is,

$$\dot{q}_0 = -kA\left(\frac{dT}{dx}\right)_{x=0} = AkN(T_b - T_\infty)\frac{Be^{2LN} + 1}{Be^{2LN} - 1}. \tag{14-50}$$

An alternative form which eliminates B is:

$$\dot{q}_0 = AkN(T_b - T_\infty)\frac{\bar{h}/Nk + \tanh LN}{1 + \bar{h}/kN \tanh LN}. \tag{14-51}$$

Schneider [2] discusses the usefulness of fins. For a value of $(\bar{h}A/kP) < 1$, the fin will have the useful effect of transferring *more* heat to or from the surface. For $(\bar{h}A/kP) > 1$, the fin merely serves as insulating material and is thus undesirable. In other words, if $\bar{h} < kP/A$, the fin is useful.

Fin Efficiency

In design practice, it is convenient to use a "fin efficiency" concept. The fin has a temperature gradient, which makes it less efficient as a heat-transfer surface than the base material, which possesses the maximum temperature difference for convective heat transfer.

The fin efficiency, e, is defined as the ratio of the actual heat flow from a finned surface to the amount of heat flow which would occur if the fin surface were all at the same temperature, T_b, as the base material:

$$e = \frac{\dot{q}_{\text{actual}}}{\dot{q}_{\text{ideal}}} = \frac{\dot{q}_{\text{actual}}}{\bar{h}A_f(T_b - T_\infty)}, \tag{14-52}$$

where A_f is the surface area of the fin.

Therefore,

$$\dot{q}_0 = \dot{q}_{\text{actual}} = e\dot{q}_{\text{ideal}},$$

and

$$\dot{q}_0 = e\bar{h}A_f(T_b - T_\infty). \tag{14-53}$$

For the simplified case of the adiabatic tip, $(dT/dx)_{x=L} = 0$, the efficiency can be shown to be

$$e = \frac{\tanh LN}{LN}. \tag{14-54}$$

Equation (14-54) may be used for the nonadiabatic tip case if the length, L, is the corrected length, L_c, from Eq. (14-48).

14-4 EXTENDED SURFACES WITH CONVECTIVE BOUNDARIES

The fin efficiency concept is particularly useful for the case of tapered fins or fins with a nonuniform cross section. Efficiencies of triangular and parabolic fins shown in Fig. 14–11 are compared with the efficiency of a straight fin of constant cross section. The area A_p shown in the figure is the *profile area* of the fin, the area formed by the intersection of the fin and a plane normal to both the fin base surface and the direction x.

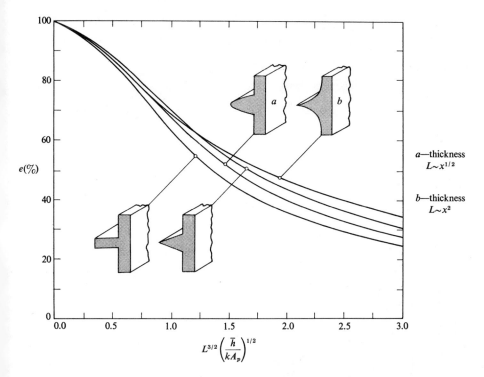

Fig. 14–11 Efficiency of four types of straight fin profiles having equal lengths, L, and equal profile areas A_p. (From *Conduction Heat Transfer*, by P. J. Schneider, Addison-Wesley, Reading, Mass., 1966.)

In practice many fins are "wrapped" around a tube and have a variable cross-sectional area normal to the radial heat flow, as in Fig. 14–12. In such cases, efficiency curves are again useful. This type of curve is shown in Fig. 14–13. The corrected length for nonadiabatic tips should be used for precise calculations.

Example 14–5. Determine the heat loss per foot from a straight triangular fin with base thickness of 0.5 in. and a length, L, of 2.5 in. assuming a base temperature of 200°F, a fluid bulk temperature of 80°F, and an average

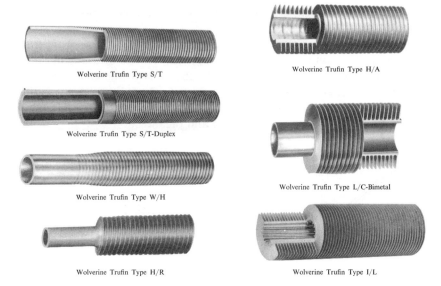

Fig. 14-12 Circular fins on round tubes. (Courtesy of Calumet and Hecla's, Wolverine Tube Division.)

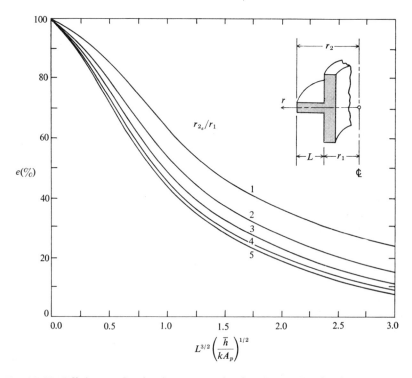

Fig. 14-13 Efficiency of a circular rectangular fin. (From Conduction Heat Transfer, by P. J. Schneider, Addison-Wesley, Reading, Mass., 1966.)

convective coefficient $\bar{h}$ of 4 Btu/hr ft² °F. The fin material has a thermal conductivity of 40 Btu/hr ft°F.

Solution. The profile area A_p is calculated first:

$$A_p = (\tfrac{1}{2})(0.5)(2.5) \text{ in.}^2 \frac{\text{ft}^2}{(144) \text{ in.}^2} = 4.34 \times 10^{-3} \text{ ft}^2.$$

$$L^{3/2}\left(\frac{\bar{h}}{kA_p}\right)^{1/2} = \left(\frac{2.5}{12}\right)^{3/2} \text{ft}^{3/2} \left[\frac{(4)\dfrac{\text{Btu}}{\text{hr ft}^2 \text{ °F}}}{(40)\dfrac{\text{Btu}}{\text{hr ft °F}}(4.34 \times 10^{-3}) \text{ ft}^2}\right]^{1/2} = 0.465.$$

From Fig. 14–11, $e = 0.90$. The heat loss is $\dot{q}_0 = e\bar{h}A_f(T_b - T_\infty)$.

$$A_f = \frac{(2)(1)\sqrt{\left(\dfrac{0.5}{2}\right)^2 + (2.5)^2}}{12} \frac{\text{ft}^2}{\text{ft}} = 0.425 \frac{\text{ft}^2}{\text{ft}}$$

$$\dot{q}_0 = (0.90)(4)(0.425)(200 - 80) = 184 \text{ Btu/hr ft}.$$

In the absence of the fin, the heat loss from the base material, over an area covered by the base of the fin, would have been only

$$\dot{q} = \bar{h}A(T_w - T_\infty) = (4)\frac{(0.5)(1)}{(12)}(200 - 80) = 20 \text{ Btu/hr ft}.$$

Fins are widely used in the electronics industry to cool rectifiers. transistors and other devices in which there is a large dissipation of energy (conversion of electrical energy to thermal energy). These fins may be used either in a forced convection or when the air motion is entirely due to heating. Examples of such fins are shown in Fig. 14–14.

14–5 TRANSIENT CONDUCTION

Some of the important problems in heat transfer include those in which the temperature at each point in a solid varies with time. Such conditions usually occur when the solid is exposed to a changing environment. This occurs, for example, in a quenching process when a hot ingot is suddenly placed in a cooler fluid. In these transient cases, energy storage within the substance must be considered, and the storage term must be retained in the differential equation describing the temperatures. In rectangular coordinates, the equation describing the local temperature for the unsteady case, with no heat generation and constant thermal conductivity, is obtained from Eq. (14–3).

$$\frac{\partial T}{\partial t} = \alpha\left(\frac{\partial^2 T}{\partial x^2} + \frac{\partial^2 T}{\partial y^2} + \frac{\partial^2 T}{\partial z^2}\right). \tag{14-55}$$

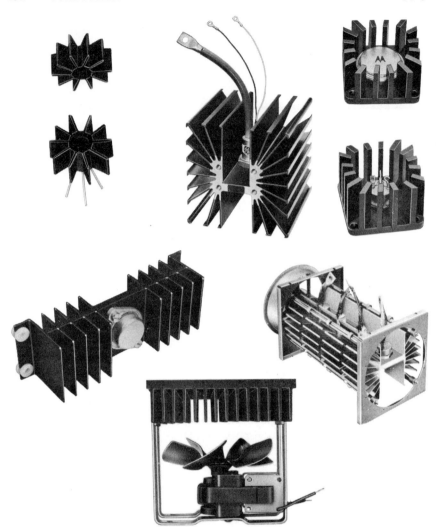

Fig. 14-14 Fins used in cooling of electronic devices. (Courtesy of Wakefield Engineering Company.)

The solution of Eq. (14–55) requires that suitable initial and boundary conditions be known. In the present discussion, only step changes in the boundary conditions at some initial time will be presented. The more complex boundary conditions are treated in advanced texts [1, 2, 3, 4].

In obtaining solutions it is convenient to nondimensionalize the equation by letting

$$T^* = \frac{T - T_s}{T_0 - T_s}, \quad x^* = \frac{x}{L}, \quad y^* = \frac{y}{L}, \quad z^* = \frac{z}{L}, \quad t^* = \frac{\alpha t}{L^2}. \quad (14\text{--}56)$$

In this case T_s is some reference temperature which does not change with time and T_0 is an initial temperature at some point, or points. The characteristic length, L, is any convenient length arbitrarily agreed upon. Time is nondimensionalized with respect to L^2/α. Equation (14–55) becomes

$$\frac{\partial T^*}{\partial t^*} = \frac{\partial^2 T^*}{\partial x^{*2}} + \frac{\partial^2 T^*}{\partial y^{*2}} + \frac{\partial^2 T^*}{\partial z^{*2}}. \tag{14–57}$$

The dimensionless time $\alpha t/L^2$ is referred to as the *Fourier modulus* (or *Fourier number*), Fo. If the boundary and initial conditions are described in terms of T^*, t^*, x^*, and z^* only, then the local value of T^* at a specified time depends only on the Fourier modulus and the dimensionless position:

$$T^* = f(\text{Fo}, x^*, y^*, z^*). \tag{14–58}$$

Assume, for example, that the temperature distribution in a semi-infinite slab (Fig. 14–15) is desired for any time t after the surface temperature is suddenly changed from T_0 to T_s and held at that value. It is assumed that the slab is initially at a uniform temperature, T_0.

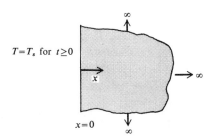

Fig. 14–15 Nomenclature for semi-infinite solid.

The differential equation, Eq. (14–57), becomes, for this one-dimensional case,

$$\frac{\partial T^*}{\partial t^*} = \frac{\partial^2 T^*}{\partial x^{*2}}. \tag{14–59}$$

The boundary and initial conditions are:

$T^* = 1$ for all x^* and for $t^* < 0$
$T^* = 0$ for $x^* = 0$ and for $t^* \geq 0$

The solution to Eq. (14–59) and these boundary and initial conditions is

$$T^* = \text{erf}\left(\frac{x^*}{2\sqrt{t^*}}\right) = \text{erf}\left(\frac{1}{2\sqrt{\text{Fo}_x}}\right), \tag{14–60}$$

where erf is the Gaussian error function and Fo_x is the Fourier modulus

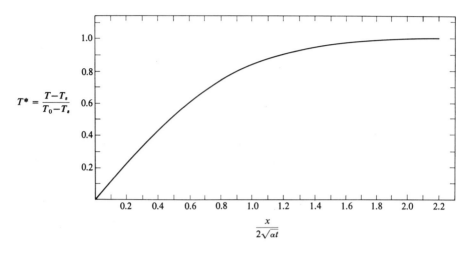

Fig. 14-16 Temperature distribution in a semi-infinite solid.

based on x:

$$\operatorname{erf} x = \frac{2}{\sqrt{\pi}} \int_0^x e^{-\tau^2} d\tau \tag{14-61}$$

Equation (14-60) is plotted in Fig. 14-16. Values of the error function are also tabulated in Table 14-3.

Example 14-6. A semi-infinite block of concrete is initially at a uniform temperature of 70°F. The surface temperature is suddenly changed to 120°F and held at that temperature. Calculate the temperature 1 inch from the surface after 30 minutes.

Solution.

$$\frac{1}{2\sqrt{\operatorname{Fo}_x}} = \frac{x}{2\sqrt{\alpha t}} = \frac{1/12 \text{ ft}}{2\sqrt{(0.021)\frac{\text{ft}^2}{\text{hr}}\left(\frac{30}{60}\right)\text{hr}}} = 0.428.$$

From Fig. 14-16,

$$\frac{T - T_s}{T_0 - T_s} = 0.46, \quad T = 120 - (50)(0.46) = 98°\text{F}.$$

In the solution above the surface temperature is given, but in practical cases the surface might be subjected to a fluid with a fixed, known bulk temperature, T_∞. The above solution would still be an excellent approximation (assuming $T_s = T_\infty$) if the resistance to heat transfer in the solid at the surface is high compared to the resistance to heat transfer in the fluid at the surface of the solid, that is, in the fluid above the concrete. In such cases,

Table 14-3
GAUSS'S ERROR FUNCTION (erf)

x	erf x	x	erf x	x	erf x
0.00	0.00000	0.88	0.78669	1.76	0.98719
0.02	0.02256	0.90	0.79691	1.78	0.98817
0.04	0.04511	0.92	0.80677	1.80	0.98909
0.06	0.06762	0.94	0.81627	1.82	0.98994
0.08	0.09008	0.96	0.82542	1.84	0.99074
0.10	0.11246	0.98	0.83423	1.86	0.99147
0.12	0.13476	1.00	0.84270	1.88	0.99216
0.14	0.15695	1.02	0.85084	1.90	0.99279
0.16	0.17901	1.04	0.85865	1.92	0.99338
0.18	0.20094	1.06	0.86614	1.94	0.99392
0.20	0.22270	1.08	0.87333	1.96	0.99443
0.22	0.24430	1.10	0.88021	1.98	0.99489
0.24	0.26570	1.12	0.88679	2.00	0.99532
0.26	0.28690	1.14	0.89308	2.02	0.99572
0.28	0.30788	1.16	0.89910	2.04	0.99609
0.30	0.32863	1.18	0.90484	2.06	0.99642
0.32	0.34913	1.20	0.91031	2.08	0.99673
0.34	0.36936	1.22	0.91553	2.10	0.99702
0.36	0.38933	1.24	0.92051	2.12	0.99728
0.38	0.40901	1.26	0.92524	2.14	0.99753
0.40	0.42839	1.28	0.92973	2.16	0.99775
0.42	0.44747	1.30	0.93401	2.18	0.99795
0.44	0.46623	1.32	0.93807	2.20	0.99814
0.46	0.48466	1.34	0.94191	2.22	0.99831
0.48	0.50275	1.36	0.94556	2.24	0.99846
0.50	0.52050	1.38	0.94902	2.26	0.99861
0.52	0.53790	1.40	0.95229	2.28	0.99874
0.54	0.55494	1.42	0.95538	2.30	0.99886
0.56	0.57162	1.44	0.95830	2.32	0.99897
0.58	0.58792	1.46	0.96105	2.34	0.99906
0.60	0.60386	1.48	0.96365	2.36	0.99915
0.62	0.61941	1.50	0.96611	2.38	0.99924
0.64	0.63459	1.52	0.96841	2.40	0.99931
0.66	0.64938	1.54	0.97059	2.42	0.99938
0.68	0.66378	1.56	0.97263	2.44	0.99944
0.70	0.67780	1.58	0.97455	2.46	0.99950
0.72	0.69143	1.60	0.97635	2.48	0.99955
0.74	0.70468	1.62	0.97804	2.50	0.99959
0.76	0.71754	1.64	0.97962	2.60	0.99976
0.78	0.73001	1.66	0.98110	2.70	0.99987
0.80	0.74210	1.68	0.98249	2.80	0.99992
0.82	0.75381	1.70	0.98379	2.90	0.99996
0.84	0.76514	1.72	0.98500	3.00	0.99998
0.86	0.77610	1.74	0.98613	∞	1.00000

the assumption that the surface temperature of the concrete is approximately equal to the fluid temperature may be valid. The ratio of internal to external thermal resistance can be approximated by looking at an energy balance on a surface element of a solid:

$$-k_s \frac{\partial T}{\partial n}\bigg|_{n=0} = h(T_w - T_\infty). \tag{14-62}$$

In this case n is considered the direction normal to the surface. Non-dimensionalizing Eq. (14-62) gives

$$-\frac{\partial T^*}{\partial n^*}\bigg|_{n=0} = \left(\frac{hL}{k_s}\right) T^*_{n^*=0}. \tag{14-63}$$

The dimensionless term $(hL)/k_s$ relates the dimensionless temperature gradient at the surface to the dimensionless surface temperature. This term, given the name *Biot modulus* (or *Biot number*), is usually written for the entire surface with an average heat transfer coefficient. For irregular shapes, it is common to select as a characteristic length the ratio of the volume of the solid to the surface area. With this arbitrary selection of L, a Biot number can be determined for any finite solid where both $\bar{h}$ and k are known:

$$\text{Bo} = \left(\frac{\bar{h}L}{k_s}\right). \tag{14-64}$$

The Biot number gives the relative magnitudes of internal and external resistance to heat transfer. For relatively high values of the Biot number, the surface temperature of a solid is very close to the surrounding fluid temperature.

In systems where the Biot number is small, $(\bar{h}L)/k_s < 0.1$, the solid can be considered to be at a nearly uniform temperature at all times. This would be the case where the internal resistance is negligible compared to the external resistance. Such a case would exist, for example, with a copper sphere ($k = 200$ Btu/hr ft °F) of 1.0-ft diameter cooled by free convection in air ($h \approx 2$ Btu/hr ft² °F). In this case,

$$\text{Bo} = \frac{(2)(1.0)}{(6)200} = 0.00167.$$

Since such a system is assumed to have a uniform temperature, it is quite easy to express an energy balance for the entire system:

rate of energy of decrease = rate of convective heat loss

$$-\text{Vol}\, \rho c_p \frac{dT}{dt} = \bar{h}A(T - T_\infty), \tag{14-65}$$

where Vol = volume of system and A = surface area of system. It is convenient to rewrite Eq. (14–65) as

$$\frac{d(T - T_\infty)}{T - T_\infty} = \frac{-\bar{h}A}{\text{Vol } \rho c_p} dt.$$

Integrating with $T = T_0$ at $t = 0$ and $T = T$ at $t = t$:

$$\ln \frac{T - T_\infty}{T_0 - T_\infty} = \frac{\bar{h}At}{\text{Vol } \rho c_p} \quad \text{or} \quad T^* = \exp\left(-\frac{\bar{h}At}{\text{Vol } \rho c_p}\right). \quad (14\text{–}66)$$

Letting Vol/A be the characteristic length L,

$$T^* = e^{-(\text{Bo})(\text{Fo})}. \quad (14\text{–}67)$$

The thermal conductivity k does not appear in either Eq. (14–66) or (14–67) since the internal resistance has been neglected. A typical plot of Eq. (14–66) or (14–67) is shown in Fig. 14–17 for the case of cooling.

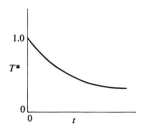

Fig. 14–17 Change of dimensionless temperature with time.

Notice that the group $(\bar{h}A)/(\text{Vol } \rho c_p)$ in the exponent of Eq. (14–66) has the units of 1 over time. The reciprocal $(\text{Vol } \rho c_p)/(\bar{h}A)$ is called the *time constant* for the system, the magnitude of which determines the time required for the system to undergo 63.2 percent of the total change that will occur in the temperature.

Equations (14–66) and (14–67) are identical to the equation describing the decay in electrical potential across a capacitor when it is discharging through a resistor, neglecting internal resistance in the capacitor. The electrical equation is

$$\frac{E - E_\infty}{E_0 - E_\infty} = e^{-[1/(C_e R_e)]t} \quad (14\text{-}68)$$

where E = electrical potential, C_e = electrical capacitance, and R_e = electrical resistance. This similarity between transient electrical and thermal systems provides another useful means of solving heat transfer problems.

458   CONDUCTION   14–5

By the proper choice of scaling factors, an electrical circuit can be designed and built to simulate a thermal system. The behavior of such an electrical system furnishes information on the behavior of the simulated thermal system. Such a simulation forms the basis for the thermal analyzer.

Example 14–7. It is desired to predict the temperature of a 1-ft diameter copper sphere 1 hr after being placed in air at 70F°. The copper sphere is initially at 300°F. Draw an equivalent electrical circuit. Assume a proper value of $\bar{h}$ to be 2.0 Btu/hr ft² °F.

Solution.

$$\frac{T - T_\infty}{T_0 - T_\infty} = \exp\left(-\frac{\bar{h}At}{\text{Vol } \rho c_p}\right)$$

$$\frac{T - 70}{300 - 70} = \exp\left(-\frac{2\pi D^2 (1)}{\frac{\pi D^3}{6}(558)(0.091)}\right) = e^{-0.236} = 0.79,$$

and

$$T = 70 + (230)(0.79) = 70 + 182 = 252°F.$$

The equivalent circuit is shown in Fig. 14–18. The switch, shown for $t < 0$, is grounded at $t = 0$, and the potential across the capacitor decreases exponentially with time according to Eq. (14–68).

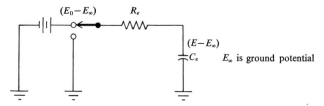

Fig. 14–18 Equivalent electrical circuit for Eq. (14–68).

In the case of common shapes, where the internal resistance is not negligible, solutions have been obtained by Heisler [11]. As would be expected, these solutions can be presented for any dimensionless location in the solid shape with dimensionless temperature a function only of the Fourier and Biot numbers. Curves for the flat plate, the cylinder, and the sphere are shown in Figs. 14–19, 14–20, and 14–21.

Example 14–8. Calculate the temperature 6 in. from the surface of an aluminum sphere of 12-in. radius (initially at 450°F) 10 minutes after it is exposed to a convective environment at 150°F. The heat transfer coefficient is 100 Btu/hr ft² °F.

Solution. The Heisler charts for spheres may be used for the solution of this problem:

$$T_0 - T_\infty = 450 - 150 = 300$$
$$\alpha = 3.66 \text{ ft}^2/\text{hr}$$
$$R = 12 \text{ in.} = 1 \text{ ft}$$
$$t = 10 \text{ min} = 0.167 \text{ hr}$$
$$k = 120 \text{ Btu/hr ft }^\circ\text{F}$$
$$\bar{h} = 100 \text{ Btu/hr ft}^2\text{ }^\circ\text{F}$$

$$\frac{k}{\bar{h}R} = \frac{120}{(100)(1)} = 1.2$$

$$\frac{\alpha t}{R^2} = \frac{(3.66)(0.167)}{(1)^2} = 0.611$$

$$\frac{r}{R} = \frac{6}{12} = 0.5$$

From Fig. 14–21:

$$\frac{T_M - T_\infty}{T_0 - T_\infty} = 0.35$$

$$T_M - T_\infty = 0.35(300) = 105^\circ\text{F}$$

$$\frac{T - T_\infty}{T_M - T_\infty} = 0.91$$

$$T - T_\infty = (0.91)(105) = 95.5^\circ\text{F}$$

$$T = 150 + 95.5^\circ = 245.5^\circ\text{F}$$

Two- and Three-Dimensional Bodies

The Heisler charts may also be used to determine the temperature time history of such finite size elements as rectangular blocks or cylinders of finite length. For example, the equation for conduction in a cylindrical rod with no conduction in the azimuthal (θ) direction is

$$\frac{\partial T}{\partial t} = \alpha\left(\frac{\partial^2 T}{\partial r^2} + \frac{1}{r}\frac{\partial T}{\partial r} + \frac{\partial^2 T}{\partial z^2}\right).$$

If the rod is infinitely long, the temperature is independent of the axial coordinate z, and its derivative with respect to z is zero. A solution is obtained by assuming that the solution can be expressed as $T(r,t) = R(r)\theta(t)$, which is the product of R (a function of r only) times θ (a function of t only). This leads to two ordinary differential equations which can be integrated and made to fit the boundary conditions of the problem. This solution has been presented graphically in the form of a Heisler chart for cylinders in Fig. 14–20.

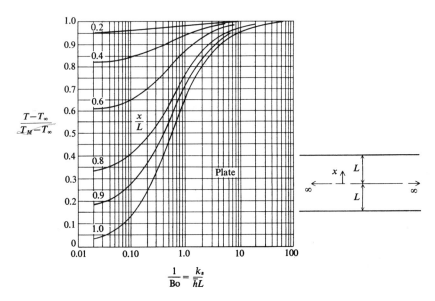

Fig. 14-19 Heisler chart for plates (L = half-thickness of plate). (a) Variation of local temperature T at various distances from midplane.

If the rod is not infinitely long compared to its diameter, then the solution involves z as well as r and t. The solution can be shown to have the following product form:

$$T(r,z,t) = T_r(r,t)T_z(z,t).$$

The solution for the short cylinder problem thus is equal to the product of the solutions $T(r,t)$ to the infinite cylinder (Fig. 14-20) and $T(z,t)$ for the infinite plate (Fig. 14-19). These solutions can both be obtained from the Heisler charts. The thickness of the infinite plate is the length of the cylinder.

Example 14-9. Calculate the temperature 3 in. from one end and 2 in. off the centerline of a solid cylinder, which was initially at 150°F, 6 minutes after it is exposed to a convective environment of 0°F. The heat transfer coefficient is 5.0 Btu/hr ft² °F. The cylinder is 12 in. long and 18 in. in diameter.

Solution.
$$T_0 - T_\infty = 150 - 0 = 150$$
$$\alpha = 2.72 \text{ ft}^2/\text{hr}$$
$$k = 0.41 \text{ Btu/hr ft }°\text{F}$$
$$\bar{h} = 5.0 \text{ Btu/hr ft}^2 \text{ °F}$$
$$t = 6 \text{ min} = 0.1 \text{ hr}$$

14–5 TRANSIENT CONDUCTION

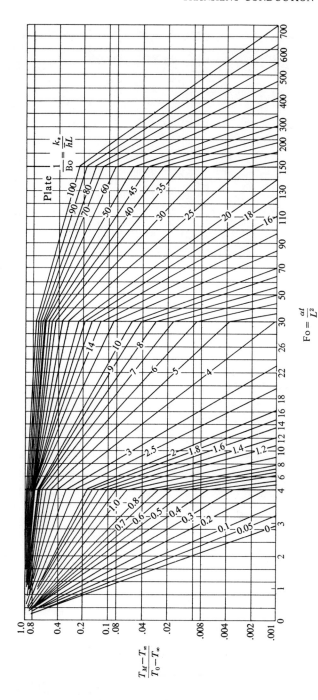

Fig. 14–19 (cont.) (b) Variation of midplane temperature T_M with dimensionless time. (Used with permission of the American Society of Mechanical Engineers.)

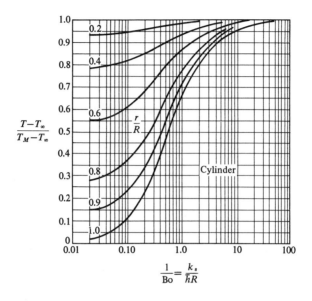

Fig. 14-20 Heisler chart for cylinders. (a) Variation of local temperature T at various distances from centerline.

For the *infinite plate solution* (Fig. 14–19), x is the distance from the center of the plate $= (12/2) - 3 = 3$ in. L is the half-thickness of the plate $= 6$ in. $= 0.5$ ft.

$$\frac{x}{L} = \frac{3}{6} = 0.5$$

$$\frac{k}{\bar{h}L} = \frac{0.41}{(5.0)(0.5)} = 0.164 \qquad \frac{\alpha t}{L^2} = \frac{(2.72)(0.1)}{(0.5)^2} = 1.08$$

From the plate chart (Fig. 14–19),

$$\left(\frac{T_M - T_\infty}{T_0 - T_\infty}\right)_p = 0.2 \quad \text{and} \quad \left(\frac{T - T_\infty}{T_M - T_\infty}\right)_p = 0.77.$$

Hence,

$$\left(\frac{T - T_\infty}{T_0 - T_\infty}\right)_p = (0.2)(0.77) = 0.154.$$

For the *infinite cylinder*, $r/R = 2/9 = 0.222$,

$$\frac{k}{\bar{h}R} = \frac{0.41}{(5.0)(9/12)} = 0.109 \quad \text{and} \quad \frac{\alpha t}{R^2} = \frac{(2.72)(0.1)}{(9/12)^2} = 0.483.$$

14–5 TRANSIENT CONDUCTION

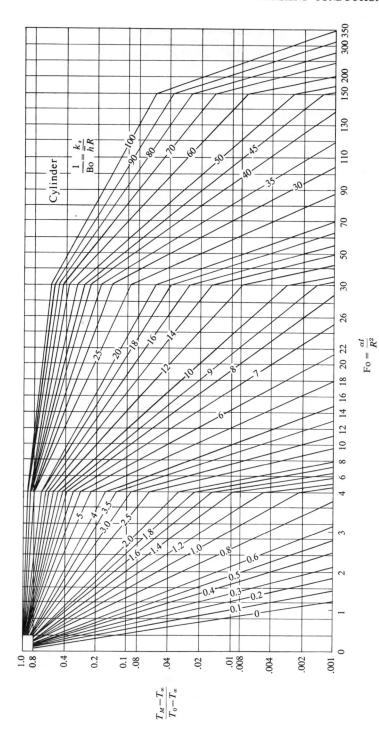

Fig. 14-20 (cont.) (b) Variation of centerline temperature T_M with dimensionless time. (Used with permission of the American Society of Mechanical Engineers.)

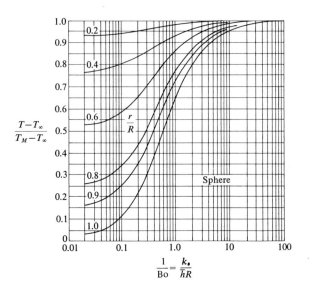

Fig. 14-21 Heisler chart for spheres. (a) Variation of local temperature T at various distances from center.

From the cylinder chart (Fig. 14–20),

$$\left(\frac{T_M - T_\infty}{T_0 - T_\infty}\right)_c = 0.2 \quad \text{and} \quad \left(\frac{T - T_\infty}{T_M - T_\infty}\right)_c = 0.94.$$

Hence,

$$\left(\frac{T - T_\infty}{T_0 - T_\infty}\right)_c = (0.2)(0.94) = 0.188.$$

Multiplying the solution of the infinite plate by the solution of the infinite cylinder,

$$\left(\frac{T - T_\infty}{T_0 - T_\infty}\right) = \left(\frac{T - T_\infty}{T_0 - T_\infty}\right)_p \left(\frac{T - T_\infty}{T_0 - T_\infty}\right)_c$$

$$= (0.154)(0.188) = 0.0290.$$

Hence,

$$T = T_\infty + (T_0 - T_\infty)(0.0290)$$
$$= 0 + 150(0.0290)$$
$$= 4.34°\text{F}$$

Similar problems can be solved by this method. For example, a long rectangular bar of dimensions $2a$ by $2b$ can be solved in terms of the product

14-5 TRANSIENT CONDUCTION 465

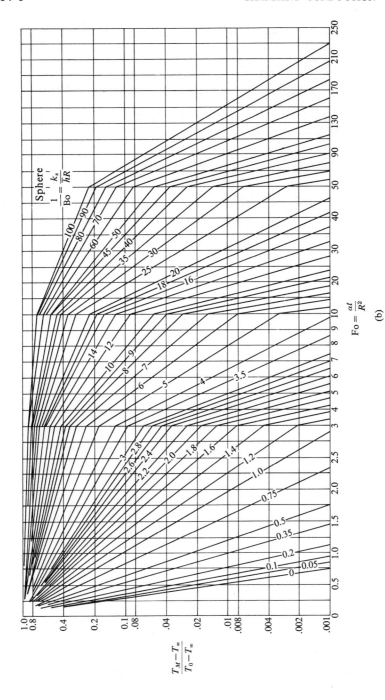

Fig. 14-21 (cont.) (b) Variation of temperature at center T_M with dimensionless time. (Used with permission of the American Society of Mechanical Engineers.)

of the solutions for the infinite plate of thickness $2a$ and the infinite plate of thickness $2b$:

$$\left(\frac{T-T_\infty}{T_0-T_\infty}\right)_{\substack{\text{bar}\\2a\times 2b}} = \left(\frac{T-T_\infty}{T_0-T_\infty}\right)_{\substack{\text{plate}\\2a\text{ thick}}} \left(\frac{T-T_\infty}{T_0-T_\infty}\right)_{\substack{\text{plate}\\2b\text{ thick}}}.$$

The rectangular parallelepiped or brick of dimension $2a$ by $2b$ by $2c$ can be determined by:

$$\left(\frac{T-T_\infty}{T_0-T_\infty}\right)_{\substack{\text{brick}\\2a\times 2b\times 2c}} = \left(\frac{T-T_\infty}{T_0-T_\infty}\right)_{\substack{\text{plate}\\2a\text{ thick}}} \left(\frac{T-T_\infty}{T_0-T_\infty}\right)_{\substack{\text{plate}\\2b\text{ thick}}} \left(\frac{T-T_\infty}{T_0-T_\infty}\right)_{\substack{\text{plate}\\2c\text{ thick}}}.$$

Solutions to conduction in semi-infinite bodies can likewise be used to determine heat conduction near corners and near edges of large solids.

14–6 NUMERICAL SOLUTIONS IN TRANSIENT CONDUCTION

The methods of numerical analysis are quite useful in those situations where the local temperatures in a substance vary with time as well as position. The method is particularly powerful in complex three-dimensional problems; however, an example for a one-dimensional case will be used initially to illustrate the derivation of finite difference relationships for the transient case.

In unsteady one-dimensional conduction in a solid which has constant thermal conductivity, the energy equation in rectangular coordinates is

$$\frac{\partial T}{\partial t} = \alpha \frac{\partial^2 T}{\partial x^2}. \tag{14–69}$$

In Eq. (14–69) the temperature is assumed to be a function of time t and position x:

$$T = T(x,t) \tag{14–70}$$

Therefore, referring to Eq. (14–14), one can write the second derivative of T with respect to x as

$$\frac{\partial^2 T}{\partial x^2} = \frac{T(x-\Delta x,t) - 2T(x,t) + T(x+\Delta x,t)}{(\Delta x)^2}. \tag{14–71}$$

For a fixed position x the change with time can be expressed in a similar manner:

$$T(x,t+\Delta t) = T(x,t) + \Delta t\left(\frac{\partial T}{\partial t}\right) + \frac{(\Delta t)^2}{2!}\left(\frac{\partial^2 T}{\partial t^2}\right) + \dots, \tag{14–72}$$

where the derivatives with respect to time are evaluated at the point x. Neglecting terms of order $(\Delta t)^2$ and higher, and letting the temperature at the new time $T(x, t + \Delta t)$ be designated by T^+,

$$T^+ = T + \Delta t \frac{\partial T}{\partial t},$$

or

$$\frac{T^+ - T}{\Delta t} = \frac{\partial T}{\partial t}. \tag{14-73}$$

Substituting Eqs. (14–71) and (14–73) into Eq. (14–69) and letting $M = (\Delta x)^2/\alpha \Delta t$,

$$T^+(x, t + \Delta t) = \frac{T(x - \Delta x, t) + (M - 2)T(x,t) + T(x + \Delta x, t)}{M}. \tag{14-74}$$

Equation (14–74) gives an approximation to the temperature at a fixed point x after some time period Δt has elapsed, in terms of the temperatures initially existing at the point and at points a distance $-\Delta x$ and $+\Delta x$ away from the point. The parameter M depends upon the size of the time steps (Δt), the size of the lumps (Δx) chosen, and the material property α. It is the reciprocal of a Fourier modulus.

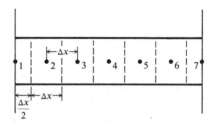

Fig. 14–22 One-dimensional transient heat conduction in a bar.

Suppose Eq. (14–74) were applied to the insulated bar shown in Fig. 14–22. The equation could be written for each interior node of the bar. It is assumed that the temperature of the exterior nodes T_1 and T_7 are known. For example, the "new" temperatures would be

$$T_2^+ = \frac{T_1 + T_3 + (M - 2)T_2}{M}, \tag{14-75}$$

$$T_3^+ = \frac{T_2 + T_4 + (M - 2)T_3}{M}, \tag{14-76}$$

and, for the nth node,

$$T_n^+ = \frac{T_{n-1} + T_{n+1} + (M - 2)T_n}{M}. \tag{14-77}$$

468 CONDUCTION 14-6

The value of M in the above equations is not entirely arbitrary. Values of M less than 2 cause the last term in the numerator of each equation to be negative. Such a situation can lead to mathematical instability, where it is possible for the value of the temperature at a node at successive new times to fluctuate and not converge to a definite value. Values of M greater than 2.0 assure convergence. For $M = 2$ the equations are convergent and simplify for the nth node to

$$T_n^+ = \frac{T_{n-1} + T_{n+1}}{2}. \tag{14-78}$$

Thus, the temperature of a node after a time period Δt is simply the average of the initial temperatures of the surrounding nodes. After the temperatures at $(t + \Delta t)$ are calculated for each node, the 'new' temperatures become the 'old' temperatures, and Eq. (14-78) is applied again to each node. The whole process is repeated as many times as needed to obtain the desired results. The technique for using Eq. (14-78) is best illustrated by use of an example.

Example 14-10. Assume that a round steel rod ($\alpha = 0.45$ ft^2/hr) has a uniform cross section and is 2 ft long. The rod is perfectly insulated except for the two ends, as shown in Fig. 14-22. The entire rod is initially at 200°F. The right end of the rod is maintained at that temperature while the left end is suddenly changed to 328°F and held at that temperature. We wish to know the temperature history of the rod.

Solution. The rod is assumed to be divided into seven lumps as shown in Fig. 14-22. The lumps at each end are one-half the size of the interior lumps so that the interior nodes are located at the center of the lumps, the surface nodes are at the surface, and every node is the same distance (Δx) apart. The choice of the number of lumps or nodes is arbitrary, although generally the greater the number of nodes (or the smaller the Δx), the higher the accuracy. In this case Δx is 4 in. or 0.333 ft. It is desired to have $M = 2$ so that the simplified equation, Eq. (14-78), can be applied and so that convergence is assured. Since

$$M = \frac{(\Delta x)^2}{\alpha(\Delta t)}, \quad \Delta t = \frac{(\Delta x)^2}{\alpha M} = \frac{(0.333)^2 \text{ ft}^2}{(0.45)(\text{ft}^2/\text{hr})(2)} = 0.12 \text{ hr.}$$

Thus, approximations to the temperature at all points will be obtained for successive time periods Δt apart. Table 14-4 shows the sequence of calculation. In the table the calculation is carried out for seven steps. However, it is easy to calculate what the temperature would be after an infinite time had elapsed, for steady state conditions would then be established, and the temperature distribution would be linear. The calculation for a linear temperature distribution gives the values tabulated in the last line of the table.

Table 14-4

Time	Node temperature (°F)						
	1	2	3	4	5	6	7
0 (initial time)	328	200	200	200	200	200	200
$\Delta t = 0.12$ hr	328	264	200	200	200	200	200
$2\Delta t = 0.24$ hr	328	264	232	200	200	200	200
$3\Delta t = 0.36$ hr	328	280	232	216	200	200	200
$4\Delta t = 0.48$ hr	328	280	248	216	208	200	200
$5\Delta t = 0.60$ hr	328	288	248	228	208	204	200
$6\Delta t = 0.72$ hr	328	288	258	228	216	204	200
$7\Delta t = 0.84$ hr	328	293	258	237	216	208	200
Infinite time	328	306.7	285.3	264	242.7	221.3	200

If the temperature distribution after a certain specified time period is desired in the above example, it may be obtained by linear interpolation.

If both Δx and α are specified, as is usually the case, then choice of M determines Δt or vice versa. The specification of Δt may make $M \neq 2$. In such cases, Eq. (14-77) must be used. Although not so simple in form as Eq. (14-78), it produces correct results, so long as $M \geq 2$ (for the one-dimensional case).

For the two-dimensional transient problem, the following equation is valid for interior nodes:

$$T^+_{i,j} = \frac{T_{i+1,j} + T_{i-1,j} + T_{i,j+1} + T_{i,j-1} + (M-4)T_{i,j}}{M}. \quad (14\text{-}79)$$

The nodes for a two-dimensional system are identified in Fig. 14-23, where a typical node i,j is shown with its surrounding nodes. For this case, $M \geq 4$ for assurance of convergence, and if $M = 4$, the last term becomes zero.

For the three-dimensional system, such as is shown in Fig. 14-24, the following equation describes the change of temperature at each node:

$$T^+_{i,j,k} = \frac{1}{M}(T_{i+1,j,k} + T_{i-1,j,k} + T_{i,j+1,k} + T_{i,j-1,k} + T_{i,j,k+1}$$
$$+ T_{i,j,k-1} + (M-6)T_{i,j,k}) \quad (14\text{-}80)$$

Example 14-9 specified the temperatures at the surfaces where heat transfer to the surroundings occurs; these temperatures were specified as constant. In many situations only the environment temperature history may

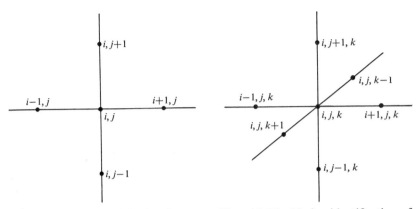

Fig. 14-23 Node identification for two-dimensional conduction.

Fig. 14-24. Node identification for three-dimensional conduction.

be known, and convective effects at the surface must be considered. If the proper heat transfer coefficients can be determined, the transient problem in these cases can be analyzed by modification of the above method. For a surface node, such as the one-dimensional node shown in Fig. 14-25, the new temperature, T^+, at the end of the first time interval, is obtained by using the boundary condition on the node and making an energy balance:

$$h(T_\infty - T_1) + k\frac{\partial T}{\partial x}\bigg|_{x=\Delta x/2} = \rho c_p \frac{\Delta x}{2} \frac{\partial T}{\partial t}. \qquad (14\text{-}81)$$

In finite difference form,

$$h(T_\infty - T_1) + k\frac{T_2 - T_1}{\Delta x} = \frac{\rho c_p \Delta x}{2} \frac{T_1^+ - T_1}{\Delta t}. \qquad (14\text{-}82)$$

Solving for the new temperature,

$$T_1^+ = \frac{1}{M}\left\{\frac{2h\,\Delta x}{k}T_\infty + 2T_2 + \left[M - \left(\frac{2h\,\Delta x}{k} + 2\right)\right]T_1\right\}. \qquad (14\text{-}83)$$

Similar expressions can be derived for the other one-dimensional exterior node and for exterior nodes for the two- and three-dimensional cases.

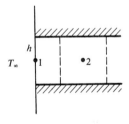

Fig. 14-25 One-dimensional surface node with convection.

These expressions for surface nodes may then be used with the appropriate expressions for interior nodes to obtain approximations to the local temperatures in a solid at various times after environmental conditions are changed.

Convergence is assured in Eq. (14–83) when the last term is zero, or positive, or when

$$M \geq \frac{2h\,\Delta x}{k} + 2. \qquad (14\text{–}84)$$

The parameter $(h\,\Delta x)/k$ is a type of Biot modulus which has been discussed previously.

The Implicit Method

The numerical procedure we have just discussed is referred to as an *explicit method*. The new temperatures to be calculated each time depend upon the old temperatures that are known or previously calculated. Each equation can be solved independently of the others in a particular time step. Another advantage of the method is that the computation is fairly fast for each time step. The disadvantage of being limited by stability requirements has already been discussed. This stability criteria often requires that an excessive number of steps be carried out for a given transient. The method also requires calculation through the complete transient to obtain the steady state solution.

The *implicit method* expresses the 'old' temperature at a node in terms of 'new' temperatures. If the temperatures in Eq. (14–71), for example, are evaluated at $t + \Delta t$ instead of t, then Eq. (14–74) would become

$$T(x,t) = -\frac{1}{M}[T(x - \Delta x, t + \Delta t) - (M + 2)T(x, t + \Delta t)$$
$$+ T(x + \Delta x, t + \Delta t)]. \qquad (14\text{–}85)$$

Applying Eq. (14–85) to the interior nodes of the bar in Fig. 14–22:

$$T_1^+ - (M+2)T_2^+ + T_3^+ \qquad\qquad\qquad\qquad = -MT_2$$
$$T_2^+ - (M+2)T_3^+ + T_4^+ \qquad\qquad\qquad = -MT_3$$
$$T_3^+ - (M+2)T_4^+ + T_5^+ \qquad\qquad = -MT_4$$
$$T_4^+ - (M+2)T_5^+ + T_6^+ \qquad = -MT_5$$
$$T_5^+ - (M+2)T_6^+ + T_7^+ = -MT_6.$$

We have five equations which can be solved for the five unknowns (T_2^+, T_3^+, T_4^+, T_5^+, and T_6^+). Remember that T_1^+ and T_7^+ are assumed to be known. Thus, it is obvious there is no convergence problem with the implicit method. The only problem is the simultaneous solution of the equations. Here the student may choose from a variety of schemes, such as Gaussian elimination, Gauss-Seidel iteration, or the matrix inversion method.

Numerical schemes similar to those presented here for conduction are also widely used in solving fluid mechanics and convective heat transfer problems. For very complex problems these numerical schemes are the only practical approaches to use. The advent of the modern digital computer has furnished a means to easily accomplish some solutions numerically.

REFERENCES

1. H. S. Carslaw and J. C. Jaeger, *Conduction of Heat in Solids*, Oxford, New York (1947).
2. P. J. Schneider, *Conduction Heat Transfer*, Addison-Wesley, Reading, Mass. (1955).
3. V. S. Arpaci, *Conduction Heat Transfer*, Addison-Wesley, Reading, Mass. (1966).
4. R. V. Churchill, *Fourier Series and Boundary Value Problems*, McGraw-Hill, New York (1941).
5. J. C. Smith, J. E. Lime, Jr., and D. S. Lermond, "Shape Factors for Conductive Heat Flow," *AIChE J.* **4**, 330 (1958).
6. S. S. Kutateladze, *Fundamentals of Heat Transfer*, ed. by R. D. Cess, Academic Press, New York (1963).
7. M. Vidmar, "Suggestion of an Addition to the Test Codes on Temperature Rise," *Elektrotechn. u. Maschinenbarr.* **26**, No. 49 (1918); cited in M. Jakob and G. A. Hawkins, *Elements of Heat Transfer*, Wiley, New York (1957).
8. T. Pemberton and J. D. Parker, "Prediction of Maximum Temperatures in Cylinders with Internal Heat Generation," *IEEE Trans. Parts, Materials and Packaging* **PMP-3**, 3 (1967).
9. A. D. Kraus, *Cooling Electronic Equipment*, Prentice-Hall, Englewood Cliffs, N.J. (1965).
10. W. B. Harper and D. R. Brown, "Mathematical Equations for Heat Conduction in the Fins of Air-Cooled Engines," NACA Report No. 158 (1922).
11. M. P. Heisler, "Temperature Charts for Induction and Constant Temperature Heating," *Trans. ASME*, (1947).

PROBLEMS

1. Determine the shape factor for a hollow sphere of homogeneous composition, having an inside radius of 3 in. and an outside radius of 9 in.
2. A conduit with square outside shape and a centered, circular inside shape is made of material having a thermal conductivity of 10^{-3} cal/sec cm °C. The diameter of the inside of the conduit is 8 in. and the outside dimension is 25 in.

If the temperature difference between the inside and the outside surfaces of the conduit is 220°F, what is the heat loss per foot of conduit length in Btu/hr? Assume steady state and no end losses.

3. A very long pipe of 4-in. o.d. is buried in soil ($k = 0.8$ Btu/hr ft °F) 2 ft below the surface. Assume that the outside temperature of the pipe is 200°F and the temperature of the soil is 70°F. Estimate the heat loss rate per 100 ft of pipe.

4. A 3-in. o.d. steam pipe and a 2-in. o.d. cold water pipe are to be buried in a thick concrete slab 1 ft apart between centers. The thermal conductivity of the concrete is 0.54 Btu/hr ft °F. The steam pipe temperature is 230°F and the water pipe average temperature is 60°F. Estimate the heat gained by the cold water per foot of pipe length. If the water is flowing at the rate of 600 gal/hr, estimate the change in water bulk temperature in 20 ft of pipe.

5. A nuclear reactor powers electronic equipment buried beneath the earth. The reactor has the shape of a sphere 2 ft in diameter and is buried 5 feet beneath the surface of the ground. If the unit must dissipate 1 kw continuously, estimate the temperature difference between the sphere surface and the ground at a large distance. Assume k for the soil to be 0.5 Btu/hr ft °F.

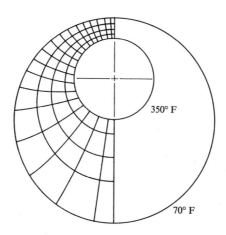

Problem 6

6. Compute the heat loss per hour for 1 ft of length of the shape shown in the figure. Assume $k = 0.06$ Btu/hr ft °F. The inside temperature is 350°F and the outside temperature 70°F.

7. A 2-in. o.d. tube is placed inside a 5-in. i.d. tube 1 in. off center. The space between the two tubes is then filled with insulation material. Draw the tubes to scale and sketch a flux plot to estimate the shape factor. Compare your answer with the value obtained from Table 14–1.

8. A box with outside dimensions of 4 × 5 × 6 ft has walls 6 in. thick, made of material with thermal conductivity $k = 0.025$ Btu/hr ft °F. The inside wall temperature is assumed constant and equal to −30°F. Estimate the rate of heat gain to the box when the outside wall is maintained at 100°F.

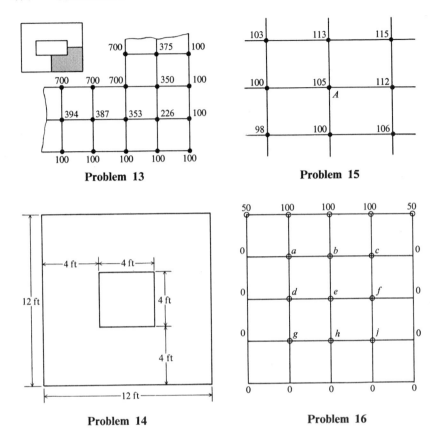

Problem 13

Problem 15

Problem 14

Problem 16

9. A box having inside dimensions of 3 × 4 × 5 ft and a wall thickness of 6 in. has a uniform inside surface temperature of 0°F and a uniform outside surface temperature of 80°F. The wall material has a thermal conductivity of 0.08 Btu/hr ft °F. Estimate the rate of heat transfer into the box.

10. A box with outside dimensions of 4 × 4 × 6 ft has walls 6 in. thick having a thermal conductivity of 0.48 Btu in./hr ft² °F. If the temperature difference between the inside and outside of the box is 80°F, estimate the rate of heat transfer into the box.

11. A cold storage box has outside dimensions of 42 × 30 × 78 in. and wall 1 ft thick. The wall material is insulation having a thermal conductivity of 0.018 Btu/hr ft °F. If the heat flow into the interior is to be less than 10 watts, what temperature difference can be maintained between the interior and exterior surfaces?

12. A rectangular box of outside dimensions 3 × 3 × 5 ft has walls 6 in. thick with thermal conductivity $k = 0.02$ Btu/hr ft °F. Inside the box is electrical equipment that dissipates 50 watts. What is the temperature difference across the box walls, assuming that the inside and outside surface temperatures are each uniform?

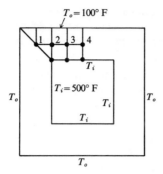

Problem 18

13. Determine the heat loss per foot of length for a duct having the cross section pictured in the diagram. Assume that the mean value of k for the material of the duct is 0.8 Btu/hr ft °F.

14. Write the energy balance equations for the two-dimensional steady state problem, shown in the figure. Write equations only for the nodes with unknown temperatures and reduce them to the form in which each term has the dimensions of temperature. $\Delta x = \Delta y = 2$ ft. $T_i = 600°F$ (surface), and $T_o = 100°F$ (surface).

15. Calculate the residual at point A in the figure and calculate the change in temperature at A necessary to reduce the residual to zero.

16. The figure represents the cross section of an infinitely long square solid member. Temperature excess values are given along the top and sides. Determine the steady state temperatures of the inner nodal points $a, b, c, d, e, f, g, h,$ and j by relaxation.

17. Write a computer program and solve the above problem by iteration.

18. The temperatures at nodal points in a relaxation problem are as shown in the figure. If k for the material is 0.5 Btu/hr ft °F, determine the heat loss from the center to the outside for the entire object per foot of length. $T_1 = 184°F$; $T_2 = 268°F$; $T_3 = 291°F$; $T_4 = 296°F$.

19. Work Example 14–3, assuming that the plate is made of iron with an electrical resistivity of 20μohm cm. Use $k = 36$ Btu/hr ft °F.

20. A long solid cylinder of 0.75-in. o.d. dissipates energy at the rate of 2 kw/ft of length. The heater is made of material with a thermal conductivity of 26 Btu/hr ft °F. What is the temperature difference between the axis of the heater and the surface, assuming that the surface temperature is uniform and constant with time?

21. How much power can be uniformly generated per foot of length in a long cylinder (radius = 3 in., thermal conductivity = 40 Btu/hr ft °F) if the cylinder is kept in a fluid stream having a temperature of 80°F, and if the unit surface conductance is 100 Btu/hr ft² °F? It is required that the temperature does not exceed 1000°F in the rod.

22. A 3-in. diameter solid cylinder, infinitely long, has a uniform internal heat generation of 5000 Btu/hr ft³. The cylinder material has a thermal conductivity

of 10 Btu/hr ft °F. The cylinder is exposed to an environment at 100°F. Assume a uniform heat transfer coefficient over the surface of the cylinder of 10 Btu/hr ft² °F and estimate the maximum temperature in the cylinder under steady state conditions.

23. Assume that energy is released at a steady uniform rate per unit volume ($\dot{q}'''$) in a sphere which is homogeneous and has a uniform surface temperature. Derive an expression for the temperature distribution $T(r)$ in the sphere.

24. A long cylindrical heater, 0.5 in. o.d., is used to heat an oil bath. The heater transfers heat to the oil at the rate of 400 Btu/hr for each foot of heater length. The surface temperature of the heater is measured and found to be 306°F. What is the maximum temperature in the heater if $k = 8$ Btu/hr ft °F?

25. How much heat will flow from the surface of a ripening orange if it is assumed that the ripening process releases 0.02 Btu/hr in.³ and the diameter of the orange is 3.5 in.?

26. What would the center temperature of the orange in Problem 25 be if the orange were assumed to be homogeneous with a thermal conductivity of 0.1 Btu/hr ft °F. Assume an air temperature of 70°F and an average heat transfer coefficient $\bar{h} = 2$ Btu/hr ft² °F.

27. Derive an expression for the temperature distribution in a large homogeneous plate with one face at temperature T_1 and the other at temperature T_2, assuming uniform energy generation per unit volume.

28. Electrical resistivity, and thus energy generation, per unit volume is dependent on temperature. If energy distribution per unit volume in a flat plate is

$$\dot{q}''' = \dot{q}_0'''[1 + \beta(T - T_0)],$$

derive an expression for the temperature distribution. T_0 is assumed to be the surface temperature of the plate.

29. For Example 14–4, determine the maximum permissible current if a temperature of 500°F is not to be exceeded in the wire.

30. Work Example 14–4 for a 12-gage and a 16-gage wire and compare the maximum temperatures.

31. What is the maximum electric current that can flow in a solid cylindrical conductor of 2-in. diameter if the centerline temperature of the conductor is not to exceed the surface temperature by more than 5°F? Assume that the thermal conductivity of the conductor is 24 Btu/hr ft °F and the electrical resistance is constant and is equal to 1 ohm/ft of conductor length.

32. A copper wire having a diameter of 0.1 in. carries a current of 100 amp. The resistance of the wire is 0.0008 ohm/ft. Assume that a convective heat transfer coefficient on the surface of the wire is 15.0 Btu/hr ft² °F. Estimate the temperature difference between the surface of the wire and the surroundings.

33. A cylindrical electric coil has an i.d. of 10 in. and an o.d. of 11 in. The coil is 15 in. long. When the coil is at a uniform temperature of 70°F, it has an electrical resistance of 90 ohms. When attached to a 120-volt d.c. source, the coil draws 1 amp of current, and the surface temperature of the coil rises to

120°F. Assume that the copper wire in the coil has a temperature coefficient of electric resistance equal to 0.0024 per °F. Assume no heat losses out the end of the coil and neglect the effect of curvature. Calculate the maximum temperature in the coil under load.

34. A straight fin which is 1 in. thick and 8 in. long is made of material with a thermal conductivity $k = 50$ Btu/hr ft °F. Assume that no heat is lost out of the tip of the fin and estimate the temperature at a point 5 in. from the base. Assume that the base temperature $T_b = 200°F$ and the fluid temperature $T_\infty = 80°F$. Assume $\bar{h} = 5$ Btu/hr ft² °F.

35. A rectangular fin 0.2 in. thick and 3 in. long is exposed to an environment at 80°F and has an average convective coefficient over the fin surface of 20 Btu/hr ft² °F. The base temperature of the fin is 200°F and the thermal conductivity of the fin material is 200 Btu/hr ft °F. Estimate the temperature of the fin 1 in. from the tip (outer end), assuming no heat loss out the tip.

36. Determine the total heat loss to air at 80°F from a rod 1 in. in diameter by 15 in. long. Assume the rod has a thermal conductivity $k = 8$ Btu/hr ft °F and an average heat transfer coefficient along the rod surface $\bar{h} = 1.5$ Btu/hr ft² °F. The base of the rod is maintained at 200°F. Assume that heat loss from the ends is negligible.

37. Make a plot of the temperature distribution along the rod in Problem 36.

38. Determine the heat loss in Problem 36 if the rod is shortened so that it is 7.5 in. long.

39. Derive an expression for the temperature distribution in a rod of uniform cross section A and length L if both ends of the rod are maintained at a fixed temperature T_b.

40. A square rod 0.5 × 0.5 × 10 in. long is heated at one end so that the end is held at a constant temperature of 300°F when the rod is cooled by air at 80°F. The average convective coefficient over the rod surface is approximately 2 Btu/hr ft² °F. Determine the average temperature at a cross section of the rod 3 in. from the heated end. Assume $k = 80$ Btu/hr ft °F.

41. A 2-ft rod of uniform cross section and 1 in. o.d. is exposed to still air at a temperature of 70°F. The average convective heat transfer coefficient along the rod is 1.2 Btu/hr ft² °F. One end of the rod is maintained at a temperature of 250°F. The thermal conductivity of the rod is 30 Btu/hr ft °F. What is the temperature of the rod at a distance 6 in. from the hot end? Assume that an adiabatic tip exists at the end.

42. A very long rod (1-in. diameter, $k = 50$ Btu/hr ft °F) extends out of a surface which is at 600°F. Assume a convective heat transfer coefficient of 2 Btu/hr ft² °F and air temperature of 70°F. At what distance from the furnace would the rod be at a temperature of approximately 150°F?

43. A hollow round rod (i.d. = $\frac{1}{4}$ in., o.d. = $\frac{3}{8}$ in., $k = 10$ Btu/hr ft °F) is 1 ft long. Each end of the rod is kept at 500°F while the rod is exposed to an environment of 1000°F. Assuming that the unit surface conductance has a value 1.0 Btu/hr ft² °F, calculate the temperature of the rod at the midpoint. Neglect conduction or cooling of the air inside the rod.

478   CONDUCTION

44. A round pin 0.1 in. in diameter and 2 in. long extends from a flat surface with a temperature of 300°F into a fluid stream which is at 60°F. The pin material has a thermal conductivity of 200 Btu/hr ft °F, and the average heat transfer coefficient over the fin surface is equal to 4 Btu/hr ft² °F. Estimate the rate of heat lost by the pin to the fluid stream.

45. A straight, rectangular aluminum fin is $\frac{1}{16}$ in. thick and 1.6 in. long. The base temperature of the fin is 500°F and the air surrounding the fin is at 70°F. (a) Assuming $\bar{h} = 2.00$ Btu/hr ft² °F, calculate the heat loss from the fin per foot of root length. (b) What is the fin efficiency?

46. Calculate the heat loss from the fin in Problem 45 if it is wrapped around a tube having an o.d. of 3.2 inch.

47. A 1-in. o.d. tube has rectangular aluminum fins $\frac{1}{32} \times \frac{1}{2}$ in., spaced at intervals of 128 fins/foot along the tube. For a tube temperature of 200°F, an air temperature of 50°F, and an $\bar{h}$ of 1.8 Btu/hr ft² °F, calculate the heat loss per foot of tube length.

48. A triangular fin has a base width of 0.5 in. and is 2 in. long. It is made of aluminum. The base temperature is 300°F and the surrounding air is 70°F. Assume that $\bar{h} = 3.0$ Btu/hr ft² °F and determine the heat loss per foot of base length. Compare with a rectangular fin of the same overall dimensions. Compare the two fins on the basis of heat loss per pound mass of fin material.

49. A long 10-in. diameter cylinder ($k = 10$ Btu/hr ft °F) ($\alpha = 0.20$ ft²/hr) is cooled from an initial uniform temperature of 800°F. Assume that the surrounding fluid temperature is 100°F and that an average value of the convective coefficient $\bar{h}$ is 40.0 Btu/hr ft² °F. Estimate the time for a point 3 in. from the center of the cylinder to come to a temperature of 150°F.

50. Assume the average thermal conductivity of a large metal plate to be 25 Btu/hr ft °F, the specific heat to be 0.125 Btu/lb$_m$ °F, and the density to be 500 lb$_m$/ft³. The plate is 2.4 in. thick and is initially at a uniform temperature of 1500°F. Estimate the time for the maximum temperature in the plate to be reduced to 500°F if the plate is suddenly exposed on both sides to a constant environmental temperature of 200°F. Assume that an average value of heat transfer coefficient over the plate surface is 50 Btu/hr ft² °F.

51. A large plate 2 in. thick, made of material with $k = 10$ Btu/hr ft °F and $\alpha = 0.15$ ft²/hr, is initially at a uniform temperature of 100°F. It is suddenly exposed on both faces to a fluid environment ($T_\infty = 200°F$, $\bar{h} = 50$ Btu/hr ft² °F). Estimate the temperature 0.4 in. below the surface after 5 minutes.

52. A large flat plate of steel 3 in. thick ($k = 26$ Btu/hr ft °F, $\alpha = 0.5$ ft²/hr) is perfectly insulated on one face. The plate is at an initially uniform temperature of 400°F. The uninsulated face is suddenly exposed to an environment at 100°F. Assuming that the average heat transfer coefficient $\bar{h}$ for the situation is 52 Btu/hr ft² °F estimate the time for the insulated face to cool to 130°F.

53. A large plate of SAE1010 steel, 3 in. thick, is at a uniform temperature of 500°F. It is suddenly placed in a large tank of water at 100°F. Assume that the heat transfer coefficient is constant and equal to 800 Btu/hr ft² °F. How long does it take for the center of the plate to come to 200°F?

54. A large plate 4 in. thick, made of material with $k = 40.0$ Btu/hr ft °F and $\alpha = 0.15$ ft^2/hr is initially at a uniform temperature of 400°F. It is suddenly exposed on each face to a fluid environment at 100°F. Assume that the heat transfer coefficient for the cooling process is constant and equal to 240 Btu/hr ft^2 °F. Estimate the time for the center of the plate to come to 160°F.

55. For Problem 54, estimate the temperature at the plane 1.2 in. from the surface of the plate when the centerline temperature reaches 250°F.

56. A semi-infinite block of steel ($\alpha = 0.5$ ft^2/hr) is initially at a uniform temperature of 300°F. The surface temperature is suddenly changed to 100°F. Estimate the time for the temperature to reach 200°F 1 ft below the surface.

57. A very long, 2.0-ft diameter cylinder (radial conduction only) has a uniform temperature of 100°F prior to time zero. At time zero the cylinder is subjected to an environment of 1100°F. The average heat transfer coefficient is 10 Btu/hr ft^2 °F. The conductivity of the cylinder is 60 Btu/hr ft °F and its thermal diffusivity is 10 ft^2/hr. Determine the temperature at a radial location halfway between the centerline and the outside surface after 1 hour.

58. A cylinder, 0.2-ft in diameter and 0.2 ft long, is initially at a uniform temperature of 100°F. It has a thermal diffusivity of 0.5 ft^2/hr and a thermal conductivity of 20 Btu/hr ft °F. Assume $\bar{h} = 100$ Btu/hr ft^2 °F and calculate the temperature at a point on the cylinder axis 0.04 ft from one end at 0.01 hr after the cylinder is placed in a fluid with a constant temperature of 300°F.

59. A long cylinder is 16 in. in diameter ($k = 8$ Btu/hr ft °F and $\alpha = 0.8$ ft^2/hr). It is initially at a uniform temperature of 300°F. Determine the time required for the points 7.2 in. from the axis to come to a temperature of 150 after the cylinder is suddenly exposed to an environment at 100°F. Assume $\bar{h} = 12$ Btu/hr ft^2 °F.

60. A semi-infinite cylinder initially at 100°F has a radius of 2 in., a thermal conductivity of 10 Btu/hr ft °F, and a thermal diffusivity of 0.2 ft^2/hr. What is the temperature in the cylinder at a point 0.8 in. from the axis and 2 in. from the exposed end 5 minutes after the cylinder is exposed to an environment of 200°F? Assume $\bar{h}$ is constant and the same for both the end and outside surfaces and has a value of 200 Btu/hr ft^2 °F.

61. A sphere having a radius of 1 ft is made of material with $k = 25$ Btu/hr ft °F and $\alpha = 0.5$ ft^2/hr. It is initially at a uniform temperature of 80°F and is suddenly exposed to a fluid environment at 200°F. Assuming that $\bar{h}$ for the process is 5 Btu/hr ft^2 °F, estimate the temperature (a) at the center and (b) at a point 6 in. from the center after 6 hours.

62. Calculate the temperature at the center of a sphere 9 in. in diameter 12 minutes after the surface of the sphere is changed suddenly to 80°F. The initial temperature of the sphere is uniform and equal to 1200°F. The sphere material has a density of 15 slugs/ft^3, a specific heat of 5 Btu/slug °F, and a thermal conductivity of 20 Btu/hr ft °F.

63. A sphere having a diameter of 1 ft is at a uniform initial temperature of 400°F. It is suddenly exposed to an environment at 100°F. A proper value of heat

transfer coefficient is 200 Btu/hr ft² °F, and the thermal conductivity of the sphere material is 20 Btu/hr ft °F. What is the temperature 1.2 in. from the sphere surface when the center temperature of the sphere has reached 300°F?

64. An aluminum alloy sphere, initially at a uniform temperature, with a radius of 1 ft experiences a sudden change in the surrounding temperature. Find the time required for $(T - T_\infty)$ to reach $0.38(T_0 - T_\infty)$ at a point 0.2 ft from the surface. For this alloy, $\alpha = 2.859$ ft²/hr and $k = 102$ Btu/hr ft °F. Assume $\bar{h} = 17$ Btu/hr ft² °F.

65. A bar of brass ($\alpha = 1.1$ ft²/hr) is 2 in. in diameter and 12 in. long. It is perfectly insulated except at each end and is at a uniform initial temperature of 70°F. One end face is held at 70°F and the other end face is suddenly changed to 200°F and held at that temperature. Divide the rod into nodes and, using $M = 2$, calculate the temperature distribution in the rod for each one of five successive time periods Δt, $2\Delta t$, $3\Delta t$, $4\Delta t$, and $5\Delta t$ after the initial temperature change occurs on the face. Express the time periods in terms minutes from the start. Describe the temperature profile at infinite time.

66. If both the time interval Δt and the nodal interval Δx in Problem 65 were specified, then M would obviously not be arbitrary. Using the bar above and the same nodal division ($\Delta x = 2$ in.) and a time interval of $\Delta t = 10$ seconds, calculate the temperature distribution in the bar after 30 seconds.

67. Work Problem 65 with an insulated face replacing the 70°F constant temperature face, using 10 time periods.

68. To illustrate divergence, work Problem 65 using $M = 1$. Note what happens to the temperature at certain points as time increases.

69. A bar 2 ft long is perfectly insulated except for the two bare ends and is at an initial uniform temperature of 300°F. The bar has a thermal conductivity k of 0.8 Btu/hr ft °F and an α of 0.05 ft²/hr. It is suddenly exposed on the ends to an environment at 80°F with a heat transfer coefficient on the ends of $\bar{h} = 1.5$ Btu/hr ft² °F. Estimate the temperature distribution in the rod after 3 hours, using 30-minute time periods.

70. A wall 6 in. thick and initially at a uniform temperature of 100°F is insulated on one face. The other face is exposed to an environment at 500°F. Assume that α for the wall is 0.05 ft²/hr, k is 5 Btu/hr ft °F, and that the heat transfer coefficient $\bar{h}$ is 30 Btu/hr ft² °F. Compute a proper value of time to allow use of the transient numerical explicit method for the case where the wall is divided into six equal increments. Show the temperature distribution in the wall after a lapse of 4 Δt's.

71. A rod 18 in. long is perfectly insulated on its circular face. It has a thermal conductivity $k = 8.0$ Btu/hr ft °F and a thermal diffusivity $\alpha = 0.25$ ft²/hr. Using two surface nodes and five interior nodes and letting $M = 2$, estimate the heat flux *into* the left end of the bar 1 hour after that end is suddenly brought to 272°F. Assume an initial temperature throughout the bar of 80°F. Assume that the right end is fixed at 80°F at all times.

72. An infinite wall, 0.6 ft thick, initially has a linear temperature gradient across the thickness with the surface temperatures at 700°F and 100°F. The two

surfaces are suddenly insulated perfectly. Use a transient numerical method and calculate the temperature distribution in the wall after 0.02 hours. Break the wall up into nodes 0.1 ft thick, except for surface nodes, and choose $M = 2$ in your equations. Assume $\alpha = 1$ ft^2/hr.

73. Assume that a 1×1 ft plate like that shown below has initial temperatures of $T_1 = 200°F$, $T_2 = 250°F$, $T_3 = 200°F$, and $T_4 = 300°F$, when the edges are suddenly fixed at a temperature $T_e = 100°F$. Plot the change of temperature with time for each point, assuming $\alpha = 0.5$ ft^2/hr. Let $M = 4$.

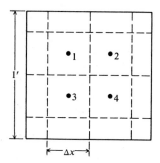

Problem 73

74. For the example described by Table 14–4 (in the solution to Example 14–10), supply the temperatures at nodes 2, 3, 4, 5, and 6 after 8 Δt's or 0.96 hr.

Problem 75

75. Initially the temperature through the insulated rod in the figure is linear, varying from 700°F on the left surface to 100°F on the right surface. The right surface is suddenly changed to 700°F while the left surface remains fixed at 700°F. Calculate the temperature distribution after a period of time equal to 5 Δt for points in the figure, assuming $\Delta x = 1.2$ in., $\alpha = 0.2$ ft^2/hr, and $M = 2$. A time 5 Δt corresponds to what length of time after change of right hand temperature?

76. Work Problem 75 by the implicit method.

77. A 1-in. diameter round steel bar, 15 in. long, is initially at 70°F. The left- and right-hand ends of the bar are suddenly changed to 100°F and 40°F respectively. Assuming the sides of the bar are insulated to prevent heat transfer (see Fig. 14–22), write a computer program and solve for the temperature-time history, using the implicit method.

CHAPTER 15

THERMAL RADIATION

Of increasing importance to the engineer is the mechanism of heat transfer called "thermal radiation." This importance has grown because of recent engineering developments. One of these is the trend toward the use of higher and higher temperatures for processes and operating systems. It will be seen that for such conditions thermal radiation heat transfer becomes more and more a significant or controlling factor.

The advent of the space age has caused man to place space vehicles, both manned and unmanned, in the outer reaches of the atmosphere where heat transfer by conduction and convection is small and often approaches zero. This leaves thermal radiation as the only possible means for transferring thermal energy to and from the device in order to maintain its operating condition. It appears that in all flights outside the atmosphere, both in the vicinity of the earth and in space, the proper management of the thermal energy aboard a vehicle or device will be a governing factor in its design and in the success of its operation.

Thermal radiation is an electromagnetic phenomenon. Its properties and characteristics are similar in many respects to other forms of electromagnetic radiations, such as radio waves, visible light, ultraviolet light, X rays, and gamma rays. In order to fully understand thermal radiation, an acquaintance with many of the laws of physics, including those of atomic physics, electromagnetic theory, optics, and spectroscopy, is necessary.

The entire electromagnetic spectrum (Fig. 15-1) is continuous, extending from very small wavelengths to those of many thousands of meters in length. All electromagnetic radiation, when absorbed by a system, produces thermal energy, that is, a heating effect. It has been discovered that this heating effect is strongly dependent upon wavelength and that practically all of the heating effect is concentrated in one particular wavelength band. To this type of radiation, which originates in the thermal agitation of particles of matter such as atoms, ions, or molecules, the term *thermal radiation* is applied. Thermal radiation is approximately that portion of the electromagnetic spectrum with wavelengths of from 0.1×10^{-6} meters up to approximately 100×10^{-6} meters. For the sake of convenience, the term *micron* is often used (1 micron = 10^{-6} meters); therefore, the approximate

THERMAL RADIATION 483

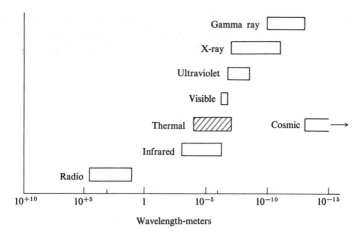

Fig. 15-1 Electromagnetic spectrum.

range of thermal radiation is from 0.1 to 100 microns. A portion of the shorter wavelengths in this range is visible to the human eye, as indicated in Fig. 15-1.

Thermal radiation travels at the speed of light, 186,000 miles per second, and is generated primarily by changes in the rotational and vibrational energy of molecules or of a system of molecules. The mechanism of this generation of energy changes with the wavelength of the radiation. For example, in the longer wavelengths (the "far infrared") this generation is associated with molecular rotation. At somewhat shorter wavelengths (the "near infrared"), the generation of energy is associated with vibrations within the molecule. Often the vibrational and rotational motions are interconnected and an exact separation of effects cannot be made. In general, thermal radiation beyond the visible spectrum is characterized by the motion of the molecule as a whole, or by the motions of atoms that can be considered as rigid bodies within the molecule. At the shorter wavelengths of thermal radiation, which include the visible portion, changes in the electron energies occur. In crystal structures, evidence indicates that thermal radiation arises from vibrations of the crystal lattice itself.

Like its companion forms of radiation, thermal radiation possesses both *wave* characteristics and *corpuscular* characteristics. Both wave theory and quantum theory are required to describe its behavior.

The total thermal radiation which impinges on a surface from all directions and from all sources is called the *total irradiation* (G). Its units are Btu/hr ft^2 in the English Engineering System.

The thermal radiation energy which falls on a surface is subject to absorption and reflection as well as transmission through transparent bodies (Fig. 15-2). *Absorption* is the transformation of the radiant energy into a different form of energy, such as thermal energy stored by the molecules. *Reflection* is the return of radiation by a surface without change of frequency. In effect the radiation is "bounced" off of the surface. *Transmission* is the passage of radiation through a medium without change of frequency.

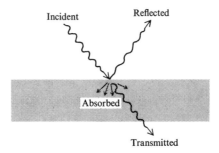

Fig. 15-2 The absorption, reflection, and transmission of incident radiation.

Energy falling on a surface must be subject to one of these three actions; therefore:

$$\alpha + \rho + \tau = 1, \qquad (15\text{-}1)$$

where

α is the absorptance—the fraction of the total incident thermal radiation absorbed;

ρ is the reflectance—the fraction of the total incident thermal radiation reflected; and

τ is the transmittance—the fraction of the total incident radiation transmitted through the body.

When the material is optically smooth and of sufficient thickness to show no change of reflectance or absorptance with increasing thickness, the terms *reflectivity* and *absorptivity* are used to describe the reflectance and absorptance respectively. In much of the literature there is no distinction between these terms.

Radiant energy may originate at a surface or from the interior of a medium due to the thermal excitation of the material. The rate of emission of energy is stated in terms of the *total emissive power* (E), where $E = E(T, \text{system})$—that is, its value depends only upon the temperature of the system and the characteristics of the material of the system. Some surfaces emit more energy than others at the same temperature. The units of E may be expressed in Btu/hr ft². E is the total energy emitted by the surface into the space and is a nondirectional total quantity.

It follows that radiant energy leaving an opaque surface ($\tau = 0$) comes from two sources: (1) the emitted energy and (2) the reflected irradiation.

This total energy leaving a surface is called the *radiosity* (J):

$$J = E + \rho G. \tag{15-2}$$

The radiosity will be important in problems of radiant exchange where total energy balances must be made on a surface.

A surface which reflects no radiation ($\rho = 0$) is said to be a *blackbody*, since in the absence of emitted or transmitted radiation it puts forth no radiation visible to the eye and thus appears black. A blackbody is a perfect absorber of radiation, and is a useful concept and standard for study of the subject of radiation heat transfer. The next section will discuss the characteristics of black and nonblack surfaces.

15-1 BLACK AND NONBLACK BODIES

The total emissive power E is a "total" value since it represents all of the radiant energy emitted by a surface. It is sometimes convenient to isolate a very narrow band of wavelengths of radiant energy for study. The band is narrow enough so that all of the radiant energy can be described in terms of one wavelength, λ. Such single-wavelength values are referred to as monochromatic values. The *monochromatic emissive power* E_λ is the rate of radiant energy emission per unit area and wavelength in the wavelength band from λ to $\lambda + d\lambda$. The total emissive power of a surface is obtained by integrating the monochromatic value over all wavelengths:

$$E = \int_0^\infty E_\lambda \, d\lambda. \tag{15-3}$$

A perfect emitter of radiant energy would be a surface that emits energy at the maximum rate in any wavelength band at a particular surface temperature. The maximum rate can be determined by thermodynamic reasoning. It is obvious from Eq. (15-3) that such a perfect emitter would also emit the maximum amount of total radiant energy at a given temperature. It can be shown that the perfect absorber of radiant energy is also a perfect emitter, thus the perfect radiant emitter is also given the name *blackbody*. The subscript B is applied to all symbols which refer to blackbodies, thus E_B and $E_{B\lambda}$ are the total and monochromatic emissive powers of a blackbody respectively.

The radiant emission of a perfect radiator or blackbody can be described by Planck's equation,

$$E_{B\lambda} = C_1 \frac{\lambda^{-5}}{\exp\left(\dfrac{C_2}{\lambda T}\right) - 1}, \tag{15-4}$$

where $C_1 = 1.1870 \times 10^8$ Btu μ^4/hr ft^2, $C_2 = 25{,}896 \, \mu \, °R$, and where μ is the symbol for microns and °R for degrees Rankine.

486   THERMAL RADIATION

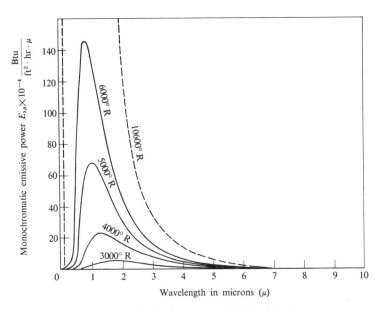

Fig. 15–3 Radiation curve of a blackbody.

This famous equation of physics was derived by Max Planck in 1900 [1] in an attempt to explain experimental results obtained in the late 19th century concerning the radiant emission of several types of surfaces. The laws of classical physics had failed to yield an analytical expression that agreed with the experimental results. Planck's solution of the problem assumed a quantization of energy. His idea was that the various oscillators which produced the radiation were permitted to have only certain values of energy or energy levels. His development of Eq. (15–4) was the beginning of the quantum theory, a concept that completely revolutionized physics and led to many important developments in our modern world.

Figure 15–3 shows the spectral distribution of a blackbody at several temperatures. These curves are a plot of Eq. (15–4) for the specified temperatures. A number of surfaces exist in nature (or can be fabricated by man) which approach the ideal behavior shown in Fig. 15–3. The sun is a common example of an emitter that approximates a blackbody—although it may be difficult to consider something as bright as the sun as a "blackbody." Many surfaces that approximate blackbody behavior do not look black to the human eye since they are emitting energy even if they reflect none. Equation (15–4) can be used to integrate Eq. (15–3) to give an expression for the total emissive power of a blackbody. The result is

$$E_B = \int_0^\infty E_{B\lambda}\, d\lambda = \sigma T^4, \tag{15-5}$$

where the Stefan-Boltzmann constant $\sigma = 0.1714 \times 10^{-8}$ Btu/hr ft^2 (°R)^4.

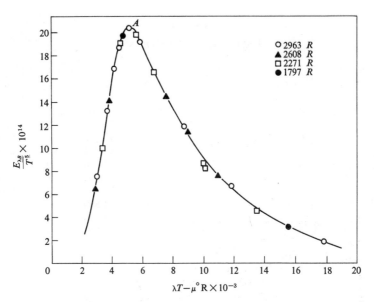

Fig. 15-4 Blackbody radiation as a function of λT. Maximum of $E_{B\lambda}/T^5$ occurs when $\lambda T = 5215.6 \, \mu \, R$.

Equation (15-5) is referred to as the *Stefan-Boltzmann equation* after the two men who derived the equation prior to the development of Planck's equation. Stefan [2], using the measurements of J. Tyndall, suggested in 1879 that the rate of emission was a function of the fourth power of the temperature. Boltzmann [3] used thermodynamic concepts to deduce this in 1884, and experimental confirmation was made in 1897 [4].

The radiation emitted by a blackbody, described by Eq. (15-4), is such that the following relationship can be shown to be true:

$$\frac{E_{B\lambda}}{T^5} = f(\lambda T). \tag{15-6}$$

Equation (15-6) is known as *Wien's distribution law* after the physicist who formulated it [5]. The function $f(\lambda T)$ is graphed in Fig. 15-4 and tabulated in Table 15-1. One important result of this law is obtained by observing the maximum which exists for the function at $(\lambda T) = 5215.6 \, \mu \, °R$. For a given temperature, a black emitter exhibits a maximum monochromatic emissive power at wavelength λ_{max}, given by

$$\lambda_{max} = \frac{5215.6}{T} \, \mu. \tag{15-7}$$

Equation (15-7) is known as *Wien's displacement law*. The maximum amount of radiation is emitted in the wavelengths around the value of λ_{max}

Table 15-1
RADIATION FUNCTIONS*

λT	$\dfrac{E_{B\lambda} \times 10^5}{\sigma T^5}$	$\dfrac{E_{B(0-\lambda T)}}{\sigma T^5}$	λT	$\dfrac{E_{B\lambda} \times 10^5}{\sigma T^5}$	$\dfrac{E_{B(0-\lambda T)}}{\sigma T^5}$	λT	$\dfrac{E_{B\lambda} \times 10^5}{\sigma T^5}$	$\dfrac{E_{B\lambda(0-\lambda T)}}{\sigma T^5}$
1000	0.0000394	0	7200	10.089	0.4809	13,400	2.714	0.8317
1200	0.001184	0	7400	9.723	0.5007	13,600	2.605	0.8370
1400	0.01194	0	7600	9.357	0.5199	13,800	2.502	0.8421
1600	0.0618	0.0001	7800	8.997	0.5381	14,000	2.416	0.8470
1800	0.2070	0.0003	8000	8.642	0.5558	14,200	2.309	0.8517
2000	0.5151	0.0009	8200	8.293	0.5727	14,400	2.219	0.8563
2200	1.0384	0.0025	8400	7.954	0.5890	14,600	2.134	0.8606
2400	1.791	0.0053	8600	7.624	0.6045	14,800	2.052	0.8648
2600	2.753	0.0098	8800	7.304	0.6195	15,000	1.972	0.8688
2800	3.872	0.0164	9000	6.995	0.6337	16,000	1.633	0.8868
3000	5.081	0.0254	9200	6.697	0.6474	17,000	1.360	0.9017
3200	6.312	0.0368	9400	6.411	0.6606	18,000	1.140	0.9142
3400	7.506	0.0506	9600	6.136	0.6731	19,000	0.962	0.9247
3600	8.613	0.0667	9800	5.872	0.6851	20,000	0.817	0.9335
3800	9.601	0.0850	10,000	5.619	0.6966	21,000	0.702	0.9411
4000	10.450	0.1051	10,200	5.378	0.7076	22,000	0.599	0.9475
4200	11.151	0.1267	10,400	5.146	0.7181	23,000	0.516	0.9531
4400	11.704	0.1496	10,600	4.925	0.7282	24,000	0.448	0.9589
4600	12.114	0.1734	10,800	4.714	0.7378	25,000	0.390	0.9621
4800	12.392	0.1979	11,000	4.512	0.7474	26,000	0.341	0.9657
5000	12.556	0.2229	11,200	4.320	0.7559	27,000	0.300	0.9689
5200	12.607	0.2481	11,400	4.137	0.7643	28,000	0.265	0.9718
5400	12.571	0.2733	11,600	3.962	0.7724	29,000	0.234	0.9742
5600	12.458	0.2983	11,800	3.795	0.7802	30,000	0.208	0.9765
5800	12.282	0.3230	12,000	3.637	0.7876	40,000	0.0741	0.9881
6000	12.053	0.3474	12,200	3.485	0.7947	50,000	0.0326	0.9941
6200	11.783	0.3712	12,400	3.341	0.8015	60,000	0.0165	0.9963
6400	11.480	0.3945	12,600	3.203	0.8081	70,000	0.0092	0.9981
6600	11.152	0.4171	12,800	3.071	0.8144	80,000	0.0055	0.9987
6800	10.808	0.4391	13,000	2.947	0.8204	90,000	0.0035	0.9990
7000	10.451	0.4604	13,200	2.827	0.8262	100,000	0.0023	0.9992
						∞	0	1.0000

* From R. V. Dunkle, "Thermal Radiation Tables and Applications," *Trans. ASME*, 1954. Reproduced by permission of the publisher.

(see Fig. 15-3). Thus, according to Eq. (15-7), as the temperature of a black emitter increases, the major part of the radiation that is being emitted shifts to shorter wavelengths. This is an important concept in engineering, since the concept may be applied to approximate behavior of many nonblack emitters. It implies that higher temperature surfaces are *primarily* emitters of short wavelength radiation and lower temperature surfaces are *primarily* emitters of long wavelength radiation. The sun, which has a surface temperature of approximately 10,000°F, emits radiation with a maximum in the visible range. The earth, which is at a much lower temperature, emits radiation with a maximum at a much longer wavelength.

The Nonblack Body

The Stefan-Boltzmann law applies to the emittance of a blackbody. Practical engineering situations involve bodies which do not exhibit this behavior. In order to handle such surfaces or systems the *monochromatic emittance* is defined as the ratio of the monochromatic emissive power of the nonblack body to that of a blackbody at the same temperature and wavelength of radiation. The symbol ϵ_λ is used for this ratio.

$$\epsilon_\lambda = \frac{E_\lambda}{E_{B\lambda}} \tag{15-8}$$

or

$$E_\lambda = \epsilon_\lambda E_{B\lambda} \tag{15-9}$$

Figure 15-5 shows typical data for ϵ_λ vs. λ.

The emittance of a layer of material having an optically smooth surface and of such thickness that there is no change in emittance with thickness is sometimes defined as the emissivity. The words emittance and emissivity have identical meaning in much of the literature.

The concept of monochromatic emittance is of great importance in the calculation of difficult radiation problems. The tendency to accept an average value of ϵ as a constant value should be guarded against, however, as it may result in large errors in individual situations.

The Gray Body

Many bodies emitting radiation may be treated in a simple way. By examining data such as shown in Fig. 15-5, it can be seen that, even for concrete, the curve of ϵ_λ vs. λ is rather smooth and that ϵ_λ is nearly constant over a limited range (approximately 2.8 to 6.3 microns). This line of reasoning leads to the concept of the *gray body*, in which the monochromatic emittance is a constant. The modified Stefan-Boltzmann equation becomes, for a gray body,

$$E_g = \epsilon \sigma T^4, \tag{15-10}$$

where E_g is the gray body total emissive power and ϵ is a constant, average value.

This constant emittance of a gray body indicates that the monochromatic emissive power of a gray body is always the same fixed percentage of the blackbody value at the same wavelength.

Two important relations will now be presented without proof, although such proof is not difficult to show [7]. For a body exchanging radiant energy only with other bodies at the same temperature, i.e., for thermal equilibrium,

$$\alpha = \epsilon. \tag{15-11}$$

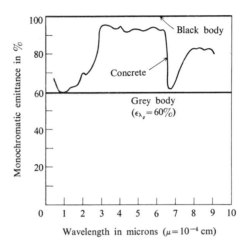

Fig. 15-5 Monochromatic emittance of concrete.

Equation (15-11) is a statement of one of Kirchhoff's laws and applies only in the case of thermal equilibrium.

For *any* surface, even with nonequilibrium conditions existing, the monochromatic absorptance and monochromatic emittance are equal:

$$\alpha_\lambda = \epsilon_\lambda. \tag{15-12}$$

Equation (15-12) verifies the previous statements concerning the fact that a blackbody is a perfect emitter.

Intensity of Radiation

Directional effects must be considered when calculating radiation heat transfer. Some surfaces reflect energy in a *specular* manner, that is, the reflections come from the surface at the same angle as the incident radiation. Most engineering surfaces possess roughness qualities which result in reflections taking place in all directions. In the special type of surface called a *diffuse surface*, the intensity of the thermal radiation, *I*, is not a function of direction but is constant. *Intensity* is defined as the thermal radiation flux from a surface included in a unit solid angle per unit area of the surface projected normal to the line connecting the area and the observer (or detector). An alternative definition states that intensity is the quotient of the radiant flux emitted by an element of surface at a point (and propagated in directions defined by an elementary cone containing the given direction) divided by the product of the solid angle of the cone and the area of the orthogonal projection of the element of surface on a plane perpendicular to the given direction. The term *radiance* is sometimes used for this quantity.

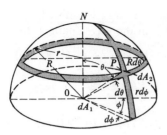

Fig. 15–6 Radiation from a small area into a hemispherical space.

In Fig. 15–6, the rate at which energy leaves an emitting surface dA_1 with intensity I and falls on an area dA_2 is $I\, dA_1 \cos\theta\, dA_2/r^2$. This can be thought of as an infinitesimal part of the total emitted energy leaving dA_1:

$$d(E\, dA_1) = I\, dA_1 \cos\theta\, \frac{dA_2}{r^2},$$

$$E\, dA_1 = \int d(E\, dA_1).$$

Now the solid angle dA_2/r^2 can be written in terms of Fig. 15–6 as

$$\frac{dA_2}{R^2} = \sin\theta\, d\theta\, d\varphi.$$

So

$$d(E\, dA_1) = I\, dA_1 \cos\theta \sin\theta\, d\theta\, d\varphi,$$

and

$$E\, dA_1 = I\, dA_1 \int_0^{2\pi} d\varphi \int_0^{\pi/2} \cos\theta \sin\theta\, d\theta = I\, dA_1 \pi,$$

or

$$E = \pi I. \tag{15-13}$$

This development assumes that the radiation is diffuse and that no reflected energy comes from dA_1. However, a similar development would show that for the case of reflected energy with diffuse emission *and* reflection, the equation would become

$$J = \pi I. \tag{15-14}$$

Example 15–1. Determine the intensity of radiation of a black surface at 440°F.

Solution.

$$I = \frac{E}{\pi} = \frac{\sigma T^4}{\pi} = \frac{(0.1714 \times 10^{-8})(900)^4}{\pi} = 358\, \frac{\text{Btu}}{\text{hr ft}^2\, \text{steradian}}$$

The solid angle is expressed in a unit called the *steradian*.

Example 15–2. How much energy is emitted by 1 square foot of a black radiating surface at 740°F in the wavelengths below 10 microns.

Solution. From Table 15–1, with $\lambda T = (10)(1200) = 12{,}000$,

$$\frac{E_{B(0-12,000)}}{\sigma T^4} = 0.7876,$$

$E_{B(0-12,000)} = (0.7876)(0.1714 \times 10^{-8})(1200)^4 = 2.79 \times 10^3$ Btu/hr ft².

15–2 CONFIGURATION FACTORS AND RADIANT EXCHANGE

In many radiation problems it is convenient to use a term called a configuration factor. The *configuration factor* is the fraction of the diffuse radiation leaving one surface which would fall directly on another surface. This factor is sometimes referred to in the literature as the *angle factor*—or the *view, shape, interception,* or *geometrical factor*. For diffuse radiation this factor is a function only of the geometry of the surface or surfaces to which it is related. It is important to note that the configuration factor is useful for any type of diffuse radiation. For this reason information obtained in illumination, radio, or nuclear engineering studies is often useful to engineers interested in thermal radiation.

The symbol for configuration factor always has two subscripts describing the surface or surfaces which it describes. For example, the configuration factor, F_{12}, applies to the two surfaces numbered 1 and 2. F_{12} would be the fraction of the diffuse radiation leaving surface 1 which would fall directly on surface 2. F_{11} would be the fraction of the diffuse radiation leaving surface 1 which would fall on itself and obviously applies only to nonplane surfaces.

A mathematical description of the configuration factor for the general case can be derived by the use of Fig. 15–7, showing two arbitrary surfaces in space.

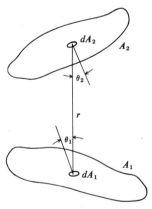

Fig. 15–7 Determination of configuration factor.

Infinitesimal areas dA_1 and dA_2 are chosen on surfaces 1 and 2 respectively. Radiation leaving dA_1 and striking dA_2 would travel along the straight line connecting the two small areas. The amount of radiant energy leaving dA_1 in the direction of the connecting line, per unit time and per unit of solid angle, would be

$$d\dot{q}_r = I_1 \, dA_1 \cos \theta_1. \tag{15-15}$$

The solid angle $d\Omega$ defined by the area dA_2 is

$$d\Omega = \frac{dA_2 \cos \theta_2}{r^2}.$$

Therefore, the total radiation leaving dA_1 and striking dA_2 would be

$$d^2\dot{q}_r = I_1 \frac{\cos \theta_1 \cos \theta_2 \, dA_1 \, dA_2}{r^2}. \tag{15-16}$$

The energy which would leave the surface 1 and strike surface 2 is obtained by integrating Eq. (15–16) over both A_1 and A_2, or

$$\int_{A_1} \int_{A_2} I_1 \frac{\cos \theta_1 \cos \theta_2}{r^2} \, dA_1 \, dA_2. \tag{15-17}$$

The total diffuse radiant energy which would leave surface A_1 is $A_1 \pi I_1$. Therefore, the fraction of the diffuse radiation which would leave dA_1 and strike dA_2, would be

$$F_{12} = \frac{1}{A_1} \int_{A_1} \int_{A_2} \frac{\cos \theta_1 \cos \theta_2}{\pi r^2} \, dA_1 \, dA_2. \tag{15-18}$$

If one of the two surfaces were very small compared to the other, or if the configuration factor for radiation from a point to a finite surface is desired, then only one integration of Eq. (15–17) is necessary.

Equation (15–18) is the general expression for the configuration factor F_{12}. Notice that the same Eq. (15–18) would have resulted from the determination of F_{21}, except that A_2 would appear outside the integral instead of A_1. This shows a very important and useful characteristic of configuration factors:

$$A_1 F_{12} = A_2 F_{21}. \tag{15-19}$$

Equation (15–19) is called the *reciprocity relationship*. Its usefulness is in determining configuration factors when the reciprocal factor is known or when the reciprocal factor is more easily obtained than the desired factor. An example of the usefulness of the reciprocity relationship is shown in the case of net heat transfer by radiation between two black surfaces.

Using the definition of configuration factor for the blackbody (which is a diffuse emitter), the transport of energy by radiation from a surface 1 to a surface 2 is

$$\dot{q}_{1 \to 2} = \sigma T_1^4 A_1 F_{12}. \tag{15-20}$$

The transport of energy by radiation from surface 2 to surface 1 is

$$\dot{q}_{2\to 1} = \sigma T_2^4 A_2 F_{21}. \tag{15-20a}$$

$$\dot{q}_{1-2\,\text{net}} = \sigma T_1^4 A_1 F_{12} - \sigma T_2^4 A_2 F_{21}.$$

Since

$$A_1 F_{12} = A_2 F_{21},$$

$$\dot{q}_{1-2\,\text{net}} = \sigma A_1 F_{12}(T_1^4 - T_2^4). \tag{15-21}$$

The definition of the configuration factor leads directly to another important relationship between factors:

$$F_{11} + F_{12} + F_{13} + F_{14} + \cdots F_{1n} = 1. \tag{15-22}$$

where n represents the total number of surfaces "seen" by surface 1. For example, the floor of a closed, square room can "see" the ceiling and four walls. The sum of the configuration factors from the floor to each of the five surfaces would add up to 1.0.

The integration of Eq. (15–18) can be accomplished in closed form for several simple configurations. For more complex configurations a number of techniques have been developed. These include (1) the unit sphere method; (2) mechanical integrators, based on the unit sphere method; (3) numerical integration, by subdividing areas into smaller areas; and (4) contour integration.

Several graphs of configuration factors are given in Figs. 15–8, 15–9, 15–10, and 15–11. Figure 15–8 shows the configuration factor for a small element dA_1 and a *parallel* rectangular element A_2 with one corner directly above dA_1. In the figure,

$$F_{12} = \frac{1}{2\pi}\left[\frac{a}{\sqrt{a^2 + l^2}}\tan^{-1}\frac{b}{\sqrt{a^2 + l^2}} + \frac{b}{\sqrt{b^2 + l^2}}\tan^{-1}\frac{a}{\sqrt{b^2 + l^2}}\right];$$

$$F_{1-(2+3+4+5)} = F_{12} + F_{13} + F_{14} + F_{15}.$$

Figure 15–8 is used to calculate F_{12}, F_{13}, F_{14}, and F_{15}. Figure 15–9 shows the configuration factors for a small element dA_1 and a *perpendicular* rectangular element A_2 with one corner directly above dA_1. Here,

$$F_{12} = \frac{1}{2\pi}\left[\tan^{-1}\frac{a}{l} - \frac{b}{\sqrt{b^2 + l^2}}\tan^{-1}\frac{a}{\sqrt{b^2 + l^2}}\right];$$

$$F_{1-(2+3+4+5)} = F_{1-(4+3)} + F_{1-(5+2)},$$

$$F_{1-(3+2)} = F_{1-(2+3+4+5)} - F_{1-(4+5)},$$

$$F_{13} = F_{1-(3+4)} - F_{14}.$$

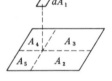

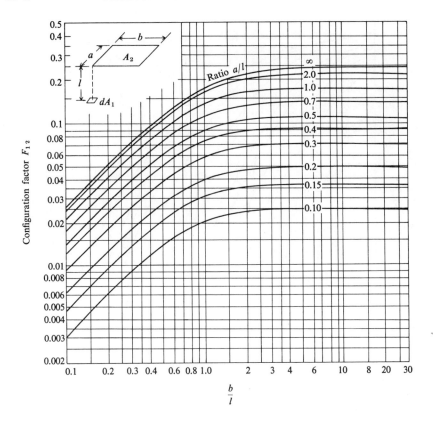

Fig. 15-8 Configuration factor for a small element dA_1 and a *parallel* rectangular element A_2 with one corner directly above dA_1 [8].*

The configuration factor for adjacent perpendicular rectangles is shown in Figure 15-10, where

$$F_{12} = \frac{1}{\pi}\left\{ \tfrac{1}{4}\log_e\left[\frac{(1+Y^2+X^2)^{[Y-(1/Y)+(X^2/Y)]}(Y^2)^Y(X^2)^{(X^2/Y)}}{(1+Y^2)^{[Y-(1/Y)]}(1+X^2)^{[(X^2/Y)-(1/Y)]} \times (Y^2+X^2)^{[Y+(X^2/Y)]}}\right] \right.$$

$$\left. + \tan^{-1}\frac{1}{Y} + \frac{X}{Y}\tan^{-1}\frac{1}{X} - \sqrt{1+\frac{X^2}{Y^2}}\tan^{-1}\frac{1}{\sqrt{X^2+Y^2}} \right\};$$

$$X = \frac{x}{l}, \quad Y = \frac{y}{l}.$$

---

* Figures 15-8 through 15-11 are reproduced from "Radiant Heating and cooling" by C. O. Mackey et al., Bulletin 32, Cornell University Engineering Experiment Station, 1943.

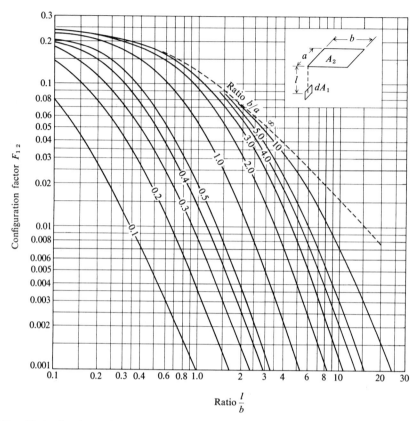

Fig. 15–9 Configuration factors for a small element dA_1 and a *perpendicular* rectangular element A_2 with one corner directly above dA_1 [8].

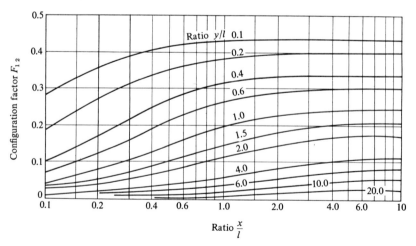

Fig. 15–10 Configuration factor for adjacent perpendicular rectangles [8].

$$F_{13} = F_{1-(2+3)} - F_{12}$$
$$F_{1-(2+3)}.$$

Figure 15-10 is used to obtain $F_{31} = \dfrac{A_1}{A_3} F_{13}$ from reciprocity relation.

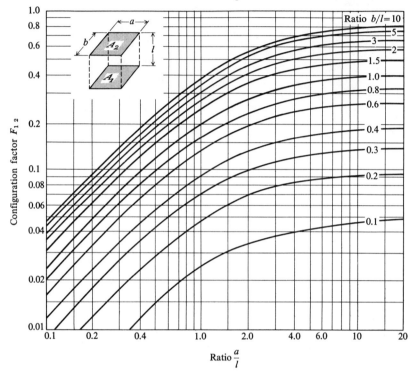

Fig. 15-11 Configuration factor for opposed parallel rectangles [8].

In Figure 15-11, we see the configuration factor for opposed parallel rectangles. In this figure,

$$F_{12} = \dfrac{2}{\pi AB} \left\{ \log_e \left[\dfrac{(1+B^2)(1+A^2)}{1+B^2+A^2} \right]^{1/2} + A\sqrt{1+B^2}\, \tan^{-1}\!\left(\dfrac{A}{\sqrt{1+B^2}}\right) \right.$$
$$\left. + B\sqrt{1+A^2}\, \tan^{-1}\!\left(\dfrac{B}{\sqrt{1+A^2}}\right) + A \tan^{-1} A - B \tan^{-1} B \right\},$$

where $A = a/l$ and $B = b/l$;

$$\text{for } a \gg l, \quad F_{12} = \sqrt{1 + \dfrac{1}{B^2}} - \dfrac{1}{B}.$$

$$\text{for } a \gg l \quad \text{and} \quad b \gg l \quad F_{12} = 1.$$

A rather complete collection of configuration factors will be found in References [9], [10], and [11]. The usefulness of the tabulated values of configuration factors is increased tremendously by the use of a technique called *flux algebra*, which involves the algebraic manipulation of Eqs. (15–19) and (15–22) and similar relationships between configuration factors. A complete description of this method is given by Chapman [12].

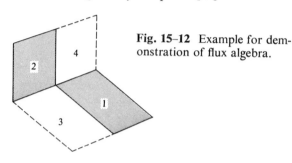

Fig. 15–12 Example for demonstration of flux algebra.

An example of the use of flux algebra is given in Fig. 15–12. Suppose it is desired to find F_{12} for the surfaces shown. The following relationship is used as a starting point:

$$(A_1 + A_3)F_{(1+3)(2+4)} = A_1 F_{12} + A_3 F_{32} + A_1 F_{14} + A_3 F_{34}.$$

Solving for F_{12},

$$F_{12} = \frac{1}{A_1}[(A_1 + A_3)F_{(1+3)(2+4)} - A_1 F_{14} - A_3 F_{32} - A_3 F_{34}]. \quad (15\text{–}23)$$

Obviously $F_{(1+3)(2+4)}$, F_{14}, and F_{32} can be obtained from Fig. 15–10. The factor F_{34} is as difficult to obtain as F_{12}, the desired factor. The use of Eq. (15–18) for determining both F_{12} and F_{34} would lead to equations which are identical except for the order of integration. Since the order of integration is not significant, a special type of reciprocity relationship is obtained for this particular case:

$$A_1 F_{12} = A_3 F_{34}. \quad (15\text{–}24)$$

The relationship of Eq. (15–24) does not depend upon the areas A_1 and A_3 being equal. Substituting Eq. (15–24) into Eq. (15–23) gives

$$F_{12} = \frac{1}{2A_1}[(A_1 + A_3)F_{(1+3)(2+4)} - A_1 F_{14} - A_3 F_{32}]. \quad (15\text{–}25)$$

All the configuration factors on the right may be determined from Fig. 15–10.

Example 15–3. Determine the net heat transfer between surfaces 1 and 2, shown in Fig. 15–12, assuming both surfaces are black, with $T_1 = 1000°F$ and $T_2 = 800°F$. Area A_2 and A_4 are 4×4 ft, and area A_1 and A_3 are 4×8 ft.

15-3 RADIANT EXCHANGE BETWEEN NONBLACK SURFACES

Solution. From Eq. (15–21),

$$\dot{q}_{1-2 \text{ net}} = \sigma A_1 F_{12}(T_1^4 - T_2^4).$$

To determine F_{12} we use Eq. (15–25) and Fig. 15–10. For

$$\frac{y}{l} = 1.0 \quad \text{and} \quad \frac{x}{l} = 0.5, \quad F_{(1+3)(2+4)} = 0.15,$$

$$\frac{y}{l} = 2.0 \quad \text{and} \quad \frac{x}{l} = 1.0, \quad F_{14} = 0.12,$$

$$\frac{y}{l} = 2.0 \quad \text{and} \quad \frac{x}{l} = 1.0 \quad F_{32} = 0.12.$$

MUST BE RANKINE

$$F_{12} = \frac{1}{(2)(4)(8)} [(8)(8)(0.15) - (4)(8)(0.12) - (4)(8)(0.12)]$$
$$= \tfrac{1}{2}[2(0.15) - 0.12 - 0.12] = \tfrac{1}{2}(0.06) = 0.03.$$

$$\dot{q}_{1-2 \text{ net}} = 0.1714 \times 10^{-8} \frac{\text{Btu}}{\text{hr ft}^2 \, {}^\circ\text{R}^4} (4)(8) \text{ ft}^2 (0.03)(1460^4 - 1260^4){}^\circ\text{R}$$

$$= 3330 \text{ Btu/hr}.$$

Notice that accuracy in reading F from the graphs is very critical, since values read are subtracted from each other. The answer obtained in this example is very dependent upon the accuracy of the values read from Fig. 15–10.

15-3 RADIANT EXCHANGE BETWEEN NONBLACK SURFACES

The previous section discussed the exchange of radiant energy in systems with black surfaces. In nonblack systems the characteristics of the surfaces must be considered and energy reflected from surfaces must be accounted for. The radiosity, Eq. (15–2), of a surface is significant since it includes both emitted and reflected energy. For a nonblack surface the rate at which energy leaves the surface is given by

$$J = \epsilon E_B + \rho G, \tag{15–26}$$

where
  J = radiosity,
  E_B = total emissive power of a black body,
  ϵ = emittance of surface,
  G = irradiation on surface, and
  ρ = reflectance of surface.

For an opaque surface $\rho = (1 - \alpha)$, and for a gray surface $\alpha = \epsilon$; thus, radiosity of a gray surface is

$$J = \epsilon E_B + (1 - \epsilon)G. \tag{15–27}$$

Letting the symbol $\dot{q}/A$ stand for the rate at which energy is brought to the surface per unit area by any means except radiation, then for steady state conditions $\dot{q}/A$ must equal the net rate at which radiant energy leaves the surface:*

$$\frac{\dot{q}}{A} = J - G = \epsilon E_B + (1 - \epsilon)G - G. \tag{15-28}$$

Elimination of G from Eqs. (15-27) and (15-28) gives

$$\frac{\dot{q}}{A} = \frac{\epsilon}{1 - \epsilon}(E_B - J). \tag{15-29}$$

The net radiant heat flow from the surface is

$$\dot{q} = \frac{E_B - J}{(1 - \epsilon)/\epsilon A}. \tag{15-30}$$

The electrical analogy can be utilized by considering that the surface potential for flow is $E_B - J$, and the surface resistance to flow is $(1 - \epsilon)/\epsilon A$, as shown in Fig. 15-13.

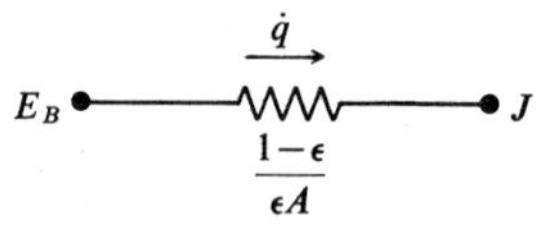

Fig. 15-13 Representation of surface resistance in radiant heat transfer.

When both the emitted and reflected energy J_1 leaving a surface A_1 do so in a diffuse manner, then the total rate of energy leaving A_1 and reaching another surface A_2 is $J_1 A_1 F_{12}$.

Likewise, the total rate of energy leaving surface A_2 and reaching A_1 is $J_2 A_2 F_{21}$. The net rate of exchange of energy between surfaces A_1 and A_2 is, therefore,

$$\dot{q}_{1-2} = J_1 A_1 F_{12} - J_2 A_2 F_{21} = A_1 F_{12}(J_1 - J_2), \tag{15-31}$$

or

$$\dot{q}_{1-2} = \frac{J_1 - J_2}{\dfrac{1}{A_1 F_{12}}} = \frac{J_1 - J_2}{\dfrac{1}{A_2 F_{21}}}. \tag{15-32}$$

Applying the electrical analogy to Eq. (15-32), the difference in radiosity between two surfaces is the *space potential* and $1/(A_1 F_{12})$ is the *space resist-*

---

* This sign convention is opposite to the usual thermodynamic convention, in which the surface is assumed to be the system and the radiation is the heat leaving the system. The authors are following the convention most common in the radiation literature.

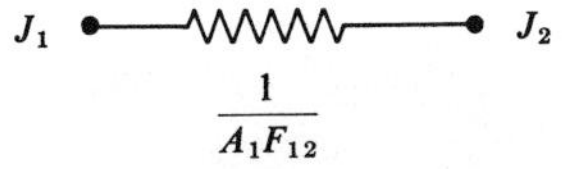

$J_1 \bullet\!\!-\!\!\mathrm{WWW}\!\!-\!\!\bullet J_2$

$\dfrac{1}{A_1 F_{12}}$

Fig. 15-14 Equivalent circuit for radiant exchange between two surfaces.

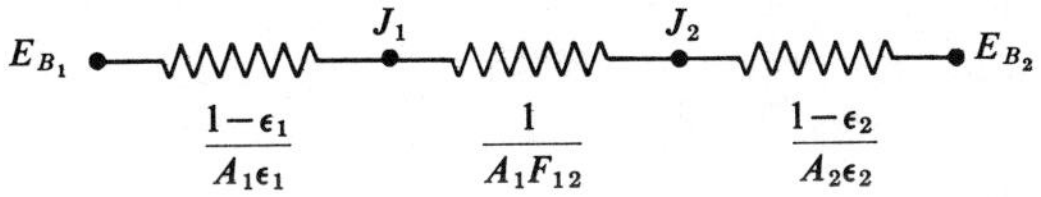

$E_{B_1} \bullet\!\!-\!\!\mathrm{WWW}\!\!-\!\!\overset{J_1}{\bullet}\!\!-\!\!\mathrm{WWW}\!\!-\!\!\overset{J_2}{\bullet}\!\!-\!\!\mathrm{WWW}\!\!-\!\!\bullet E_{B_2}$

$\dfrac{1-\epsilon_1}{A_1\epsilon_1} \qquad \dfrac{1}{A_1 F_{12}} \qquad \dfrac{1-\epsilon_2}{A_2\epsilon_2}$

Fig. 15-15 Equivalent circuit for radiant exchange between two gray surfaces that see only each other.

ance for flow between the two surfaces, assuming that no other surfaces are present. The equivalent electrical circuit is shown in Fig. 15-14. Equations (15-30) and (15-32) can be combined by adding the two surface potentials and space potentials and by adding the two surface resistances and space resistances to produce a circuit for the radiant exchange between two gray surfaces which "see" only each other. The circuit is shown in Fig. 15-15. The equation for net exchange is derived from the circuit

$$\dot{q}_{\substack{1-2\\ \text{net}}} = \dfrac{E_{B_1} - E_{B_2}}{\dfrac{1-\epsilon_1}{A_1\epsilon_1} + \dfrac{1}{A_1 F_{12}} + \dfrac{1-\epsilon_2}{A_2\epsilon_2}}$$

$$= \dfrac{\sigma(T_1^4 - T_2^4)}{\dfrac{1-\epsilon_1}{A_1\epsilon_1} + \dfrac{1}{A_1 F_{12}} + \dfrac{1-\epsilon_2}{A_2\epsilon_2}}.$$

A network can be devised for the net radiant heat exchange between any number of diffuse gray surfaces with $\epsilon = \alpha$ for each surface. For example, a three-surface network would appear as in Fig. 15-16. No other surfaces are

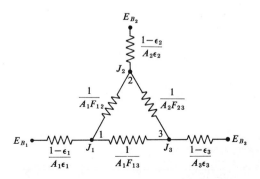

Fig. 15-16 Network for three nonblack surfaces.

assumed to exist in view of any of these three surfaces. The potentials E_{B_1}, E_{B_2}, and E_{B_3} are related to the surface temperatures through the Stefan-Boltzmann relation, Eq. (15-5). Flow balances can be written for each of the three nodes to determine the respective radiosities and, thus, the heat flux at each surface. If one of the surfaces is adiabatic (all energy striking it is reflected or re-radiated), then that surface can be represented by a "floating" node whose potential is proportional to the equilibrium temperature of that surface. Such a case is shown in Fig. 15-17, where surface 2 is assumed to be such an adiabatic surface.

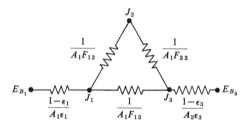

Fig. 15-17 Network for a three-surface system with one adiabatic surface.

Example 15-4. Use the network method to determine the heat exchange between two very large parallel equal size planes A_1 and A_2 with emissivities ϵ_1 and ϵ_2 and at different temperatures.

Solution. Since the planes are large, $F_{12} \approx 1$ and A_1 is assumed equal to A_2, the network becomes

$$\dot{q}_{1-2 \atop \text{net}} = \frac{E_{B_1} - E_{B_2}}{\dfrac{1}{A_1}\left(\dfrac{1-\epsilon_1}{\epsilon_1} + 1 + \dfrac{1-\epsilon_2}{\epsilon_2}\right)},$$

and

$$\left(\frac{\dot{q}_{1-2}}{A}\right)_{\text{net}} = \frac{E_{B_1} - E_{B_2}}{\left(\dfrac{1}{\epsilon_1} + \dfrac{1}{\epsilon_2} - 1\right)}.$$

For a set of n surfaces which see only each other and which are assumed to have a uniform radiosity J over each surface, Eq. (15-32) can be generalized for each surface (the jth surface) to give

$$\dot{q}_{\text{net }j} = J_j A_j - \sum_{k=1}^{n} J_k F_{kj} A_k. \qquad (15\text{--}33)$$

Equation (15–33) can be combined with Eq. (15–29), written for each node, to give a set of n equations for the radiosities of each of the n surfaces. Once the radiosities are known, the heat flow from each individual surface can be completed.

15-4 THERMAL CONTROL OF SPACECRAFT

Objects traveling in free space, for example the Apollo spacecraft, represent a unique type of thermal problem because of the absence of heat transfer by conduction or convection to any surrounding substance. In high vacuum the molecules are so few and far apart they can transport only a negligible amount of energy, and therefore radiation is the only significant means of energy exchange. A space vehicle can exchange radiant energy with the sun, with any nearby planet such as the earth, and with the vast expanse of outer space. The engineer is often concerned with achieving a proper balance in this radiant exchange so that the spacecraft operates in a favorable range of temperature.

To illustrate a simple problem in spacecraft temperature control, a spherical space vehicle, assumed to be at a uniform temperature and operating far from any planet, will be considered. Irradiation from the sun, G_s, is striking the space vehicle from one side. The amount of energy absorbed by the sphere is equal to the product of the solar irradiation, the projected area of the sphere $\pi D^2 / 4$, and the absorptance of the sphere for sunlight, α_s:

$$\dot{q}_A = G_s \frac{\pi D^2}{4} \alpha_s. \qquad (15\text{--}34)$$

FOR CYLINDERS
$G_s L D \alpha_s$

The irradiation from the remainder of space "seen" by the vehicle is assumed to be negligible due to the assumption that no planets are nearby. Outer space has an apparent temperature very near to absolute zero and thus emits very little energy. It also acts like a blackbody, reflecting none of the sphere's emitted energy.

The energy emitted by the sphere is given by

FOR CYLINDERS
$\epsilon \sigma \left(\pi D L + \frac{\pi D^2}{2} \right) T^4$

$$\dot{q}_E = \epsilon \sigma \pi D^2 T^4, \qquad (15\text{--}35)$$

where ϵ is the total emittance of the sphere surface at the absolute temperature T. It should be noted that the sphere is radiating energy in all directions, whereas it is receiving energy from the sun only on one side. This would normally cause the sphere to have a hot and cold side just as the moon does.

CYLINDER END

$$T = \left[\frac{\alpha_s}{\epsilon} \frac{G_s}{4\sigma} \frac{D}{(L + \frac{D}{2})} \right]^{1/4}$$

Table 15-2
RATIOS OF (α_s/ϵ) FOR VARIOUS SURFACES AT TEMPERATURE T

	$\left(\dfrac{\alpha_s}{\epsilon}\right)$	T in °R
White epoxy resin paint	0.28	450
Anodized titanium	0.59	450
Flat black epoxy resin paint	1.07	450
Titanium heated to 800°F in air for 300 hours	3.88	555
410 Stainless steel heated to 1300°F in air	5.88	555
Ebanol C on copper	8.25	555
Tabor solar-collector chemical treatment 110-30 on nickel-plated copper	17.4	555

Rotation of the sphere at a sufficient rate, however, could help to maintain the isothermal temperature which was initially assumed.

Since there are no other forms of energy exchange the rate at which radiant energy is absorbed by the sphere must be equal to the rate at which radiant energy is emitted, and, therefore,

$$G_s \frac{\pi D^2}{4} \alpha_s = \epsilon \sigma \pi D^2 T^4. \qquad (15\text{-}36)$$

Solving for the temperature of the spherical spacecraft, _CYLINDER_ ¼

$$T = \left[\frac{\alpha_s}{\epsilon}\frac{G_s}{4\sigma}\right]^{1/4} \cdot \left[\frac{\alpha_s G_s L}{\epsilon (L + \frac{D}{2})\pi\sigma}\right] \quad (15\text{-}37)$$

It can be seen that the spacecraft temperature depends on the magnitude of the solar irradiation G_s, and on the ratio (α_s/ϵ). The solar irradiation is dependent on distance from the sun, decreasing with the square of the distance from the sun's center. At the earth's distance, 93 million miles, the irradiation of the sun is equal to approximately 442 Btu/hr ft² or 2.0 cal/min cm².

The other factor, (α_s/ϵ), the ratio of the spacecraft surface absorptance to sunlight to its emittance at its own surface temperature, determines the surface temperature at any fixed distance from the sun. The temperature of the spacecraft can be changed by changing values of this ratio. Special paints and surface treatments have been developed to give a wide range of equilibrium surface temperatures. Some typical values of (α_s/ϵ) are shown in Table 15-2.

The variation of surface temperature with (α_s/ϵ) for the spherical spacecraft described above, at earth's distance from the sun, is given in Fig. 15-18. It can be seen that it is possible to vary the spherical spacecraft's tempera-

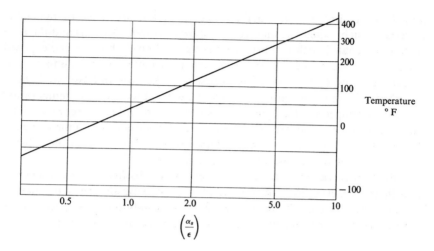

Fig. 15-18 Temperature of a spherical spacecraft vs. the ratio α_s/ϵ.

ture between approximately $-90°F$ and $420°F$ by varying (α_s/ϵ) from about 0.3 to about 10.

The use of a fixed, painted, or specially treated surface, referred to as *passive temperature control*, would cause the spacecraft to have a temperature that varies with time if the spacecraft were near a planet and moving in and out of the planet's shadow. The rate of temperature variation would depend on the thermal capacity of the spacecraft as well as its radiative characteristics. To overcome some of the problems created by sudden changes in a spacecraft's irradiation, *active systems* of control are used. The active system utilizes moving parts such as shutters to change the spacecraft's apparent (α_s/ϵ) as the irradiation, and thus the surface temperature of the vehicle, changes. One such type of active control is that suggested by Wiebelt [13]

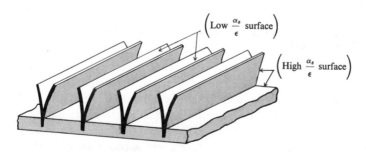

Fig. 15-19 Active thermal control system of Wiebelt with fins open. (From "Design Considerations for Thermostatic Fin Spacecraft Temperature Control" by J. A. Wiebelt et al., NASA CR-500, 1966.)

and shown in Fig. 15–19. In this system, the fins are made of bimetallic material, which "open" as their temperature increases. The fins and the base surface are coated with a substance having a relatively high (α_s/ϵ), exposed when the fins are closed. As the fin temperature increases and the fins open, a relatively low (α_s/ϵ) surface is exposed to the sun and less sunlight is absorbed relative to the radiant energy emitted. Thus, a further temperature increase is hampered. The system allows operation of the spacecraft within a narrower range of temperature variation, which reduces many of the engineering problems concerned with performance of equipment within the vehicle, or with comfort in the case of manned spacecraft.

Most space vehicles utilize equipment which dissipates thermal energy. This thermal energy must be considered when making an energy balance on the spacecraft. Generally then, in equilibrium conditions, the space vehicle must be losing more radiant energy than it receives from the sun and nearby planets.

Fig. 15–20 Typical finned radiator surfaces.

In situations where space vehicles must lose or gain large quantities of energy, finned surfaces are sometimes utilized to give increased surface area. Typical examples of such finned radiator surfaces are shown in Fig. 15–20. The finned tube and tube sheet are used with a fluid which carries the thermal energy to or from the radiating surface.

15–5 COMBINED CONDUCTION, CONVECTION, AND RADIATION

In the preceding sections of this book, heat transfer by conduction or convection and radiation have been treated as separate phenomena. Radiation has been neglected in all convection and conduction problems, and likewise conduction and convection have been neglected in all radiation problems. An exception was considered in the discussion of film boiling in Section 12–2, where the effect of radiation was taken into account along with conduction through the vapor film.

Radiation is known to have a significant effect on the transfer of heat through porous or loose material such as plastic forms, powders, glass wool,

15-5 COMBINED CONDUCTION, CONVECTION, AND RADIATION

and other common insulating materials. This effect is particularly significant where the material is placed in a vacuum, and conduction through any gas in the pores is drastically reduced. The only conduction path in such cases is through the solid material. In the case of a powder or loose material such as glass wool, the conduction path is poor since heat must flow across numerous contacts, each with fairly high thermal resistance. The heat transfer by radiation through such a material, although small, may be significant compared to the small amount of heat by conduction. To simplify computation, the heat transfer through such materials is often considered to be simply conduction, using an "apparent" thermal conductivity with an equation such as Eq. (1–17). Since the heat transfer by radiation is approximately proportional to the fourth power of temperature difference instead of the first power as in the case of conduction, the apparent thermal conductivity of such materials must be adjusted with temperature, the adjustment being more than can be explained by any change in the true thermal conductivity of the solid material. Rock wool, for example, has a k of 0.017 Btu/hr ft °F at 20°F, which increases to 0.030 Btu/hr ft °F at 200°F.

Combined Convection and Radiation

Radiation effects may be very significant in some convection problems, particularly in the case of gases and where high temperatures exist in either the gas or the solid surface. In free convection with gases particularly, where the convective coefficients may be small (on the order of 1.0 Btu/hr ft² F), radiation heat transfer may be significant compared to the heat transfer by convection.

Example 15–5. Compare the heat transfer by radiation to that by free convection for the case of a 1-ft o.d. horizontal heating duct, having a surface temperature of 200°F, exposed to a room with air and wall temperatures at 0°F. Assume the effective emittance of the duct to be 0.8.

Solution. For the free convection, the simplified equation for air from Chapter 10 will be used:

$$\bar{h} = 0.27 \left(\frac{\Delta T}{D}\right)^{1/4} = (0.27)\left(\frac{200}{1}\right)^{1/4}$$

$$= 1.02 \frac{\text{Btu}}{\text{hr ft}^2 \text{ °F}}.$$

The heat loss per square foot of duct surface would be

$$\left(\frac{\dot{q}}{A}\right)_{\text{conv}} = \bar{h}(\Delta T) = (1.02)(200) = 204 \frac{\text{Btu}}{\text{hr ft}^2}.$$

For the radiation, the duct can be assumed to be small compared to the surrounding walls, therefore, it receives little radiation back by reflection

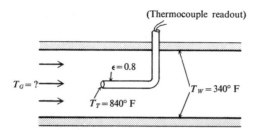

Fig. 15-21 Determination of gas temperature.

from the surrounding, and it is effectively in a black environment. The radiant heat transfer is

$$\left(\frac{\dot{q}}{A}\right)_{\text{rad}} = \sigma\epsilon(T_D^4 - T_s^4) = (0.1714)(0.8)\left[\left(\frac{660}{100}\right)^4 - \left(\frac{460}{100}\right)^4\right]$$
$$= 199 \text{ Btu/hr ft}^2.$$

In this particular example the heat loss by free convection and radiation are approximately the same.

Thermometer Error Due to Radiation

A large error in temperature measurement is possible when attempting to measure the temperature of a gas flowing through a duct which has an inside wall temperature significantly different from the gas temperature. This might occur if a metal duct is not insulated and carries a gas which is at a temperature greatly different from the ambient temperature. Such an example is shown in Fig. 15-21. A thermocouple at the center of a pipe through which gas at an unknown temperature is flowing indicates a temperature of 840°F. The wall temperature of the pipe is measured and found to be 340°F. The thermocouple will evidently radiate energy to the pipe wall and must receive energy at a similar rate by convection from the gas. Therefore, the thermocouple must be at a temperature *less* than the gas stream, and its readout cannot be considered to be the true gas temperature. The error may be further increased if significant energy is lost by conduction through the thermocouple stem to the surroundings. Proper design of the thermocouple support and wire arrangement can usually reduce the conduction error to a negligible amount.

Example 15-6. Compute the true gas temperature for the example illustrated in Fig. 15-21. Neglect any conduction loss through the thermocouple stem. Assume that a suitable value of convective heat transfer coefficient is 30 Btu/hr ft² F.

Solution. The rate of heat gain by convection to the thermocouple is equated to the rate of heat loss by radiation. The thermocouple is assumed small compared to the pipe diameter and, thus, is effectively in a black environ-

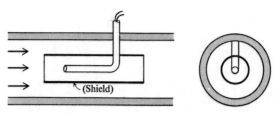

Fig. 15-22 A thermocouple radiation shield.

ment. Gas radiation and absorption are neglected.

$$\frac{\dot{q}}{A} = \bar{h}(T_G - T_T) = \sigma\epsilon(T_T^4 - T_W^4),$$

or

$$T_G = \frac{\sigma\epsilon}{\bar{h}}(T_T^4 - T_W^4) + T_T$$

$$= \left\{ \frac{(0.8)(0.1714)}{(30)} \left[\left(\frac{1300}{100}\right)^4 - \left(\frac{800}{100}\right)^4 \right] + 840 \right\}°\text{F}$$

$$= (112 + 840)°\text{F} = 952°\text{F}.$$

The error of 112°F is quite significant.

Thermocouple errors like the above can be reduced by means of a radiation shield such as the ones shown in Fig. 15-22, which prevents the thermocouple from seeing the adjacent areas of the pipe wall.

The radiation shield thus offers additional thermal resistance between the thermocouple and pipe wall and reduces the rate of heat transfer to and from the thermocouple. This reduces the magnitude of the temperature difference between the gas and the thermocouple, and, therefore, reduces the error in measurement. This effect is demonstrated in Problem 46 at the end of the chapter.

15-6 GAS RADIATION

Radiation exchange between gases and surfaces is complex if the gas absorbs and emits radiation. Fortunately the molecules of many of the common gases (such as N_2, O_2, Na) and other nonpolar symmetrical molecules are practically transparent to thermal radiation. Gases like CO_2, H_2O, CO, O_3, and SO_2 and various hydrocarbon gases emit and absorb appreciable amounts of radiant energy. At ordinary temperatures dry air is transparent to radiation, but the temperatures generated in the stagnation regions of reentry vehicles are high enough to cause dissociation and ionization. At these conditions, the air is not transparent to radiation. Unlike most solids which radiate at all wavelengths over the entire spectrum, most gases absorb and radiate only in narrow bands of wavelength.

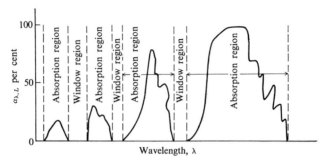

Fig. 15-23 Typical monochromatic gas absorptance.

Another difference between solid-body radiation and gas radiation is that while solid-body radiation is essentially a function of the surface, gas radiation is dependent upon the gas pressure and shape of the gas volume as well as the surface area. In general, the absorption of radiant energy in a gas depends on the number of molecules in the path of a beam. Consider monochromatic radiation of intensity $I_{\lambda 0}$ passing through a gas layer of thickness ds. The change in intensity along the path can be expressed as

$$dI_\lambda = -\rho K_\lambda I_\lambda \, ds, \tag{15-38}$$

where ρ = mass density of the gas, and K_λ = monochromatic mass absorption coefficient.

If the gas mass density ρ is uniform along the path, then Eq. (15-38) can be integrated along the path length L to obtain

$$I_\lambda = I_{\lambda 0} \exp(-\rho K_\lambda L). \tag{15-39}$$

It is seen from the above equation that if the gas is infinitely thick, then all of the radiation will be absorbed. Equation (15-39) is known as *Beer's law*. The monochromatic absorptance can be expressed as:

$$\alpha_{\lambda, L} = \frac{I_{\lambda 0} - I_\lambda}{I_{\lambda 0}} = 1 - \exp(-\rho K_\lambda L). \tag{15-40}$$

Figure 15-23 illustrates the typical variation of monochromatic absorptance vs. wavelength. To obtain the effective value of the absorptance (or emmitance, assuming Kirchoff's law is valid), one must sum over all radiation bands

$$\alpha_g = \epsilon_g = 1 - \left[\sum_\lambda I_{\lambda 0} \exp(-\rho K_\lambda L)\right] \bigg/ \sum_\lambda I_{\lambda 0}. \tag{15-41}$$

Equation (15-38) is in reality an approximate equation because no account was made of the radiation emitted by the gas itself. For those conditions where the incidence intensity is large and the gas temperature is relatively low, then Eq. (15-38) and Eq. (15-39) are good approximations.

When these conditions are not met, it is necessary then to account for the radiant emission of the gas in the following manner:

$$dI_\lambda = -\rho K_\lambda (I_\lambda - I_{B,\lambda})\, ds. \tag{15-42}$$

$I_{B,\lambda}$ is the monochromatic blackbody intensity and it can be obtained from Eqs. (15-4), (15-5), and (15-13):

$$I_{B,\lambda} = \frac{C_1}{\pi} \frac{\lambda^{-5}}{\exp\left(\dfrac{C_2}{\lambda T}\right) - 1}. \tag{15-43}$$

The expression $\rho K_\lambda\, ds$ is called the *differential optical depth*, $d\tau$. The optical distance in the direction s can be expressed as

$$\tau = \int_{s_1}^{s_2} \rho K_\lambda\, ds. \tag{15-44}$$

Equation (15-42) can be rewritten in terms of the optical depth to form a first order linear ordinary differential equation:

$$\frac{dI_\lambda}{d\tau} + I_\lambda = I_{B,\lambda}. \tag{15-45}$$

This equation is valid only for intensity in a single direction and it must be solved with a known boundary condition. As an example, suppose I_λ has the boundary value

$$I_\lambda = I_{\lambda 0} \quad \text{at} \quad s = 0, \tag{15-46}$$

which could be the location of a solid wall or any other point where I_λ is known. If the surface at $s = 0$ radiates as a blackbody at a temperature T_w, then

$$I_{\lambda 0} = I_{B,\lambda}(T_w). \tag{15-47}$$

If Eq. (15-45) is solved by standard means with the boundary conditions of Eq. (15-46), then

$$I_\lambda = e^{-\tau}\left[I_{\lambda 0} + \int_0^\tau I_{B,\lambda} e^\tau\, d\tau\right]. \tag{15-48}$$

It is obvious that Eq. (15-48) is a solution in the sense that I_λ is expressed in the form of an integral that can be evaluated only if the state of the gas is known. Solutions to Eq. (15-45) can be very complicated for even the simplest boundary conditions, and the reader should refer to References [7, 10, 14, and 15] for more detailed information on methods of solution.

The *radiation heat flux vector* $\dot{\mathbf{q}}_r$, is the radiant heat flow per unit area per unit time. It is in reality the vector sum of all of the intensities at a given point over all solid angles and all wavelengths, and it can be expressed as

$$\dot{\mathbf{q}}_r = \int_0^\infty \int_\Omega \mathbf{s} I_\lambda(\mathbf{s})\, d\Omega\, d\lambda, \tag{15-49}$$

where $d\Omega$ is a differential solid angle, and $\mathbf{s}$ is a unit vector in the direction of $I_\lambda(\mathbf{s})$. The amount of radiant energy transmitted to a unit volume per unit time (see Eq. 4–59) is

$$-\text{div}\,\dot{\mathbf{q}}_r = -\left(\frac{\partial q_{rx}}{\partial x} + \frac{\partial q_{ry}}{\partial y} + \frac{\partial q_{rz}}{\partial z}\right). \tag{15-50}$$

By substituting Eqs. (15–42) and (15–49) into Eq. (15–50), the radiant energy term can be written as

$$-\text{div}\,\dot{\mathbf{q}}_r = \int_0^\infty \rho K_\lambda \left[\int_0^{4\pi} I_\lambda\, d\Omega - 4\pi I_{B,\lambda}\right] d\lambda. \tag{15-51}$$

The inclusion of Eq. (15–51) in the general energy equation, Eq. (4–61), results in an unwieldy integro-differential equation that must be solved by numerical means to yield an approximate solution. However, in a number of cases of practical interest one can deal with certain limiting states of the gas, which simplifies Eq. (15–51). Consider the following limiting cases.

Optically Thin Gases

If a gas is optically thin,

$$\tau = \int_{s_1}^{s_2} \rho K_\lambda\, ds \ll 1, \tag{15-52}$$

then the gas will be emission dominated with $I_\lambda \ll I_{B,\lambda}$, and Eq. (15–51) will reduce to

$$-\text{div}\,\dot{\mathbf{q}}_r = -4\alpha_p \sigma T^4, \tag{15-53}$$

where the Planck or emission mean absorption coefficient α_p is defined by

$$\begin{aligned}\alpha_p &= \frac{\int_0^\infty \rho K_\lambda I_{B,\lambda}\, d\lambda}{\int_0^\infty I_{B,\lambda}\, d\lambda} \\ &= \frac{\pi}{\sigma T^4}\int_0^\infty \rho K_\lambda I_{B,\lambda}\, d\lambda.\end{aligned} \tag{15-54}$$

Optically Thick Gases

If the gas is optically thick ($\tau \gg 1$), then it has been shown [16] that

$$\dot{\mathbf{q}}_r = -\frac{16\sigma T^3}{3\alpha_R}\frac{\partial T}{\partial s}, \tag{15-55}$$

where α_R is called the Rosseland mean absorption coefficient:

$$\alpha_R^{-1} = \frac{\int_0^\infty \frac{1}{\rho K_\lambda}\frac{dI_{B,\lambda}}{dT}\, d\lambda}{\int_0^\infty \frac{dI_{B,\lambda}}{dT}\, d\lambda} = \frac{\pi}{4\sigma T^3}\int_0^\infty \frac{1}{\rho K_\lambda}\frac{dI_{B,\lambda}}{dT}\, d\lambda. \tag{15-56}$$

Gray Gas Approximation

If ρK_λ is assumed frequency-independent and if we let this value equal α, then Eq. (15–52) will reduce to

$$-\text{div } \dot{\mathbf{q}}_r = \alpha \left(\int_0^\infty I \, d\Omega - 4\sigma T^4 \right). \tag{15-57}$$

This equation, which has been widely used, simplifies the many complexities associated with the coupling of radiation and gas dynamics.

There are many heat transfer problems which involve the radiant energy passing through a medium which contains local inhomogeneities which absorb, emit, reflect, and transmit varying percentages of the radiant energy. The combination of reflection and transmission is called radiation scattering. Some examples of radiation scattering are dust particles in air, microscopic carbon particles in a luminous flame, particles in fluid bed reactors, fibers in fibrous thermal insulation, pigment particles in paint, and luminous particles in solid propellant rocket exhaust.

Energy which is scattered from a ray of radiation will contribute to the intensity of radiation in another direction. This characteristic of scattering complicates the equation of transfer, (Eq. 15–42), because it now must contain an integral term to account for the radiation scattered in the direction of I_λ.

This field of radiation-scattering heat transfer has been largely neglected. However, recently there has been increased interest and activity in this area. For detailed information on methods of solution to radiant heat transfer in an absorbing, emitting, and scattering media, the reader should consult the work of Love [14].

REFERENCES

1. M. Planck, *The Theory of Heat Radiation*, Dover, New York (1959).
2. J. Stefan, *Wien. Berichte* **79**, 391 (1879).
3. L. Boltzmann, *Wied. Annalen* **22**, 291 (1884).
4. O. Lummer and E. Pringsheim, *Wied. Annalen* **63**, 395 (1897); and *Annalen Physik*, **3**, 159 (1900).
5. W. Wien, *Sitzungsberichte Akad. Wissensch*, Berlin (Feb. 9, 1893), p. 55.
6. R. V. Dunkle, "Thermal Radiation Tables and Applications," *Trans. ASME* **76**, 549 (1954).
7. J. A. Wiebelt, *Engineering Radiation Heat Transfer*, Holt, New York (1966).
8. C. O. Mackey et al., "Radiant Heating and Cooling," Bulletin 32, Cornell University Engineering Experiment Station (1943).
9. F. Krieth, *Radiation Heat Transfer*, International Textbook Co., Scranton (1962).

10. E. M. Sparrow and R. D. Cess, *Radiation Heat Transfer*, Brooks/Cole, Belmont, Calif. (1966).
11. D. C. Hamilton and W. R. Morgan, "Radiant Interchange Configuration Factors," NACA TN 2836 (1952).
12. A. J. Chapman, *Heat Transfer*, 2d ed., Macmillan, New York (1967).
13. J. A. Wiebelt et al., "Design Considerations for Thermostatic Fin Spacecraft Temperature Control," NASA CR-500 (1966).
14. T. J. Love, *Radiative Heat Transfer*, Charles E. Merill, Columbus, Ohio (1968).
15. H. C. Hottel and A. F. Sarofim, *Radiative Transfer*, McGraw-Hill, New York (1967).
16. W. G. Vincenti and C. H. J. Kruger, "Introduction to Physical Gas Dynamics," Wiley, New York (1965).
17. J. C. Richmond, *Measurement of Thermal Radiation Properties of Solids*, NASA SP-31 (1963).
18. A. K. Oppenheim, "Radiation Analysis by Network Method," *Trans. ASME* **78**, 725 (1956).

PROBLEMS

1. Calculate the rate energy emitted from a 1-ft² black surface at 100°F, 1000°F, and 10,000°F.
2. Using Eq. (15–3) and (15–4), derive Eq. (15–5).
3. Use Eq. (15–4) to calculate the rate of energy emission per micron wavelength from a 1-ft² black surface (500°F) at 1, 5, and 10 microns.
4. Calculate the wavelength at which maximum radiation is emitted from a black body at 100°R, 1000°R, and 10,000°R.
5. Estimate the amount of radiation emitted per unit time and area by a black surface at 740°F in the wavelength range between 3.9 and 4.1 microns. At what wavelength would the maximim amount of radiation occur for that temperature?
6. Estimate the total amount of energy emitted per square foot of surface area by a black surface at 4000°R in the wavelength range between 1.5 and 2.0 microns.
7. Work Problem 1 for a gray body with $\epsilon = 0.8$.
8. A gray body ($\epsilon = 0.8$) is at a temperature of 800°F. How much energy does it emit per square foot of surface in the wavelength range from 5 to 20 microns?
9. Show that the double integral used in deriving Eq. (15–13) is equal to π.
10. What is the solid angle subtended by the sun to an observer on earth? It is 93 million miles to the sun from the earth and the sun's radius is 433,000 miles.
11. What fraction of the energy emitted by the sun strikes the earth?

PROBLEMS 515

12. The thermal radiation from the sun closely approximates that of a black body at 10,000 °R. If the sun's diameter is 865,000 miles and its mean distance from the earth is 93 million miles, what is the total radiation energy flux rate in Btu/hr ft² on a surface which faces the sun just outside our atmosphere?

13. Consider a sphere with area A_1 which is completely surrounded by a second sphere with area A_2. What is a general expression for the configuration factor F_{21} in terms of areas A_1 and A_2?

14. Using Eq. (15–18), show that the configuration factor F_{12} for a *small* surface dA_1 centered directly above a parallel, circular disk A_2 of radius R, and located a distance l away, is equal to $R^2/(R^2 + l^2)$.

15. Use Fig. 15–8 to determine the configuration factor F_{12} for a small surface dA_1, located directly above a parallel surface A_2, which is 2 ft away and which has dimensions of 4 × 6 ft. The normal drawn from the small surface strikes the large surface at a point 2 ft from one short edge and 2 ft from one long edge.

16. For Problem 15, estimate the rate of heat transfer per square foot of dA_1 if both surfaces are black and $T_1 = 1000°R$ and $T_2 = 600°R$.

17. In the small figure below Fig. 15–10, assume that A_1 is 4 × 3 ft, A_2 is 4 × 2 ft and A_3 is 4 × 1 ft. Estimate the configuration factor F_{31}.

18. Two black rectangular parallel planes, 3 × 5 ft, are spaced directly opposed and 1 ft apart and are transferring heat by radiation. Estimate the percent reduction in radiant heat exchange when the plates are moved to (a) 2 ft apart and (b) 4 ft apart.

19. A room 12 ft long by 18 ft wide is 8 ft high. Assuming that all surfaces are perfectly black, estimate the fraction of the radiant energy leaving the floor and going to each of the walls and to the ceiling.

20. For the room in Problem 19, an opening of 4 × 4 ft is made in the exact center of one of the 12 × 8-ft walls. What is the fraction of the radiant energy leaving the floor which escapes through the opening?

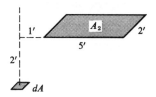

Problem 21

21. The small plane surface dA in the figure emits energy diffusely. What fraction of this energy strikes a parallel plane A_2, 2 × 5 ft, with one corner 1.0 ft from the normal to dA, and 2 ft above it?

22. Estimate the net radiant exchange between two black, parallel, 6 × 9-ft rectangles located 3 ft apart. One surface is at 2000°F and the other is at 600°F. Assume that the walls connecting the two surfaces are also black.

23. Estimate the net heat transfer by radiation from the floor of a room 10 × 20 ft to the four walls, which are 8 ft high. Assume that the walls and floors are black surfaces at 40°F and 80°F respectively.

24. A box 3 × 4 × 5 ft has black inside surfaces which are maintained at 1040°F. One wall of the box (4 × 5 ft) is missing. What is the amount of radiant energy, Btu/hr, which *leaves* through the opening formed by the missing wall?

25. A rectangular furnace has interior dimensions of 4 × 2 × 3 ft high. A radiant heater is located over the entire ceiling of the furnace and emits thermal radiation diffusely. Estimate the fraction of the emitted energy which directly strikes (a) the lower half of all four furnace walls, and (b) the floor of the furnace.

26. A sphere with $\epsilon = 0.6$ and surface area of 2 ft² is placed in a large environment at 200°R. Estimate the rate of heat loss from the sphere by radiation when the surface temperature of the sphere is 400°R.

27. Two infinite, parallel gray planes are at 140°F and 840°F. A thin gray plane is between the two planes and parallel to them. Assume $\epsilon = 0.5$ for all surfaces and calculate the temperature of the center plane.

28. Two large parallel gray planes have emittances $\epsilon_1 = 0.11$ and $\epsilon_2 = 0.91$. The temperatures of the two planes are $T_1 = 75°F$ and $T_2 = 190°F$. What is the heat transferred per hour per square foot of surface?

29. Two very large parallel planes have emittances of 0.4 and 0.8 and temperatures of 40°F and 540°F respectively. A very thin sheet of foil, the same size as the planes, with emittance of 0.1, is placed between the planes. Estimate the equilibrium temperature of the foil.

30. Two 5-ft² parallel flat plates are 2 ft apart. Plate 1 ($\epsilon = 0.4$) is at a temperature of 1540°F and plate 2 ($\epsilon = 0.6$) is at a temperature of 440°F. The surroundings are black and at 0 degrees absolute. Draw the electrical equivalent network for the system and determine the radiosity of plate 1.

31. Four arbitrary surfaces form an enclosure. The characteristics of the surfaces and the product of the area and the configuration factor are as follows:

Surface No.	Area	Product of A × F	Emittance	Total Emissive Power
1	5	$A_1F_{12} = 1.5$, $A_1F_{13} = 3.0$	0.2	100
2	10	$A_2F_{23} = 2.5$	0.4	200
3	15		0.5	re-radiating
4	20		0.25	300

a) Complete the A-F product table, i.e., obtain A_1F_{14}, A_2F_{24}, and A_3A_{34}.
b) Draw the gray body network with all resistance values indicated as fractions.

32. An enclosure is composed of three surfaces. Two of the surfaces (1 and 2) are flat walls with sides 5 × 5 ft, placed parallel to each other 1.0 ft apart. Surface 3 is composed of four flat walls 5 × 1 ft, place to form a rectangular box. Surface 1 is a gray surface ($\alpha_1 = \epsilon_1 = 0.5$) at 1000°R. Surface 2 is a gray

surface ($\alpha_2 = \epsilon_2 = 0.8$) at 600°R. Surface 3 is a blackbody ($\alpha = \epsilon = 1.0$) at 500°R.

a) Draw the radiation network for this problem labeling all potential points and resistances.

b) Calculate the rate that energy must be supplied to surface 1 (from beneath the surface) to maintain its temperature at 1000°R.

33. Calculate the temperature of the spherical spacecraft discussed in Section 15–4 at a distance of (a) 50 million and (b) 20 million miles from the center of the sun. Assume (α_s/ϵ) equal to 1.0.

34. Work Problem 33 for an (α_s/ϵ) equal to (a) 0.5 and (b) 5.0.

35. Estimate the equilibrium surface temperature of a thin flat plate, located in space at earth's distance from the sun, and oriented normal to the sun's rays. Assume (α_s/ϵ) to be (a) 1.0, (b) 0.4, or (c) 6. Neglect planet radiation.

36. Work Problem 35, assuming that the plate is oriented so that the sun's rays strike the plate at an angle from the normal of (a) 45°, (b) 75°.

37. Estimate the time required for a 1-ft diameter black sphere ($c_p = 0.1$ Btu/lb_m °R, $\rho = 550$ lb_m/ft³), initially at a uniform temperature of -160°F, to come to a uniform temperature of 40°F, after it is placed in sunlight at earth's distance from the sun, with no nearby planets. Assume that the sphere rotates and that it has such a high thermal conductivity that the assumption of uniform temperature can be justified.

38. Estimate the equilibrium temperature of a long rotating cylinder, 1 ft in diameter, oriented in space with its axis normal to the sun's rays. Assume that it is 93 million miles from the sun and a great distance from any planet and that (α_s/ϵ) is equal to 1.0.

39. Write the differential equation describing the temperature distribution in a long, solid cylindrical fin, held at a base temperature T_B on one end and seeing only a black environment at 0°K. Assume a uniform temperature at any distance x measured from the fin base.

40. Estimate the equilibrium temperature of a sphere 2 ft in diameter if it must dissipate an additional 10 watts of energy due to electrical equipment in the interior. Assume the sphere is in free space at earth's distance from the sun, that $\alpha_s = 0.3$, $\epsilon = 0.6$, and no nearby planets affect the temperature.

41. Work Problem 40, assuming the sphere is a blackbody.

42. Work the Example 15–5 for the heating duct, assuming that the emittance is reduced to 0.2.

43. Work the problem in Example 15–5, assuming that the duct surface temperature is (a) 100°F and (b) 300°F but that the surrounding temperature and emittance are unchanged.

44. Work the problem in Example 15–6 for the gas temperature measurement, assuming that emittance of the thermocouple is reduced to 0.1.

45. Work the problem in Example 15–6 assuming that the convective coefficient is (a) 40 and (b) 20 Btu/hr ft² °F, with the emittance equal to 0.8.

46. Two coaxial pipes are at temperature T_o (outside) and T_i (inside). The area of the outer pipe is 3 ft² per linear foot and the area of the inner pipe is 1 ft² per linear foot. Derive an expression for the percent reduction of energy flow due to radiation if a thin tube with area 2 ft² per foot of length is inserted between the two pipes. Consider all surfaces black.

47. A black sphere (1-in. diameter) is placed in a large oven where all the walls are maintained at 700°F. The temperature of the air in the oven is 200°F, and the heat transfer coefficient for convection between the surface of the sphere and the air is 4 Btu/hr ft² °F. Estimate the net rate of heat flow to the sphere when its surface temperature is 400°F.

48. What temperature will a 6-in. diameter sphere attain 1 hour after it is placed in a large environment at 0°R if it is originally at a temperature of 540°F? Assume that the sphere has infinite thermal conductivity, a specific heat of 0.14 Btu/lb$_m$ °F, and a density of 500 lbs/ft³. The emittance equals 0.8.

49. How many steradians are described by a 5-ft diameter disk with its center located 10 ft away from the observer, assuming that the disk surface is a plane normal to the line connecting the observer and the disk center?

50. The sun has a diameter of 752,000 nautical miles. The earth, located 80.7 × 10⁶ nautical miles from the sun has an average irradiation of 442 Btu/hr ft². What is the irradiation on Mars, located 123 × 10⁶ nautical miles from the sun?

51. If the earth has a surface temperature of 288°K and an effective emittance of of about 0.7, how much energy would it lose by radiation to outer space in a 24-hr day?

52. What is the value of the configuration factor F_{12} for the case of a 3-in. o.d. sphere (surface 2) located concentrically inside a 9-in. i.d. sphere (surface 1), assuming both spheres have diffusely emitting surfaces?

53. Write an expression for the configuration factor F_{13} in the accompanying figure, in terms of configuration factors for adjacent perpendicular planes and for some or all of the areas A_1, A_2, and A_3.

CHAPTER 16

MASS TRANSFER

Mass can be transferred by diffusion of one component of a fluid mixture from a region of high concentration to one of low concentration. This is analogous to the transfer of heat from a high-temperature region to a low-temperature region. Important applications of mass transfer occur in many branches of engineering, but it has traditionally been the chemical engineers who have distinguished themselves as designers of mass transfer equipment and processes.

For the sake of simplicity we shall restrict our attention in this chapter to two-component systems that have a diffusive mass transfer due to density gradients of the components. Mass transfer due to thermal diffusion (Soret effect), pressure diffusion, and force diffusion will not be considered here.

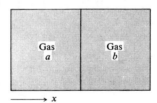

Fig. 16-1 Vessel containing two gases.

16-1 FICK'S LAW

Consider a vessel containing two gases a and b, separated by a partition as shown in Fig. 16-1. If the partition is removed, the two gases will diffuse into one another by the action of the random motion of the molecules. After a period of time the concentration of the molecules in the vessel will be uniform. At a given cross section at any instant of time the rate of mass transfer per unit area (mass flux) of component a can be expressed as

$$\dot{m}_a = -\rho D_c \frac{d\omega_a}{dx}, \qquad (16\text{-}1)$$

where

D_c = diffusion coefficient or diffusivity,
$\dot{m}_a$ = mass transfer of species a per unit time per unit area,
ω_a = mass fraction of species $a = \rho_a/\rho$.

D_c has units of (length)^2/(time) which are the same units as thermal diffusivity or kinematic viscosity.

Equation (16–1) is the one-dimensional scalar form of *Fick's law* (sometimes called Fick's first law), proposed in 1855 by Adolph Fick, a German doctor of medicine [1]. The more general form of Fick's law for the mass current $\dot{\mathbf{m}}$ can be expressed as a three-dimensional vector:

$$\dot{\mathbf{m}}_a = -\rho D_c \left[\frac{\partial \omega_a}{\partial x}\mathbf{i} + \frac{\partial \omega_a}{\partial y}\mathbf{j} + \frac{\partial \omega_a}{\partial z}\mathbf{k} \right]. \tag{16-2}$$

Notice the similarity between Eq. (16–1) and the equation for shear stress in a fluid:

$$\tau = \mu \frac{du}{dy}. \tag{16-3}$$

Notice also the similarity between Eq. (16–1) and Fourier's heat conduction equation:

$$\dot{q} = -k \frac{dT}{dx}. \tag{16-4}$$

Equations (16–1), (16–3), and (16–4) express the rate of transfer of mass, momentum, and energy respectively. The minus sign in Eqs. (16–1) and (16–2) is necessary because the mass flux is in the direction of the decreasing density gradient.

Fick's law can also be expressed in terms of molal concentrations and fluxes, in addition to mass fractions and mass fluxes. The choice is one of convenience.

When the fluid is a gas, which is closely approximated by the equation of state $p = \rho R_g T$, then it is possible to express Fick's law in terms of the partial pressures of the gases. From the equation of state of an ideal gas and from Dalton's law of partial pressure, $p = p_a + p_b$, we can write

$$\rho_a = \frac{p_a W_a}{\mathcal{R} T} \qquad \rho_b = \frac{p_b W_b}{\mathcal{R} T}, \tag{16-5}$$

where $R_g = \mathcal{R}/W$, and where W_a and W_b are the molecular weights of gas a and b respectively. If the diffusion is isothermal and at constant pressure and density, then Eq. (16–1) for component a can be expressed as

$$\dot{m}_a = -\frac{D_c W_a}{\mathcal{R} T} \frac{dp_a}{dx}. \tag{16-6}$$

Fick's law has been found to apply to gases, liquids, and solids. The diffusivities of ideal gases and dilute liquid solutions are functions of temperature and pressure and independent of concentrations. However, for solids, concentrated liquid solutions, and nonideal gases, the diffusivity may be a function of the relative concentrations.

Fick's law, Eq. (16–2), is valid for determining the mass flux only if the fluid is stagnant (bulk velocity is zero). When the fluid is stagnant and if all conditions are steady state, then Fick's law may be integrated directly to find the rate of diffusion in terms of boundary conditions.

Example 16–1. A pipe of length L contains two gases a and b diffusing isothermally at constant pressure and density. The partial pressures at $x = 0$ are fixed at p_{a0} and p_{b0}, while at $x = L$ they are fixed at p_{aL} and p_{bL}. If the bulk velocity is zero, what is the molal rate of diffusion of gas a and gas b, assuming the total pressure $p = p_a + p_b =$ constant?

Solution. From Eq. (16–6),

$$\dot{m}_a = \frac{-D_c W_a}{\Re T} \frac{dp_a}{dx}$$

or

$$\frac{\dot{m}_a}{W_a} = N_a = -D_c \frac{1}{\Re T} \frac{dp_a}{dx},$$

where N_a is the molal diffusion rate. Integrating,

$$N_a \int_0^L dx = \frac{-D_c}{\Re T} \int_{p_{a0}}^{p_{aL}} dp_a.$$

Therefore,

$$N_a = \frac{D_c}{\Re T} \frac{(p_{a0} - p_{aL})}{L}$$

$$= \frac{D_c}{L W_a} (\rho_{a0} - \rho_{aL}).$$

Similarly,

$$N_b = \frac{D_c}{\Re T} \frac{(p_{b0} - p_{bL})}{L}$$

$$= \frac{D_c}{L W_b} (\rho_{b0} - \rho_{bL}).$$

Since $p_a + p_b = p =$ constant,

$$\frac{dp_a}{dx} + \frac{dp_b}{dx} = 0,$$

and hence $N_a = -N_b$.

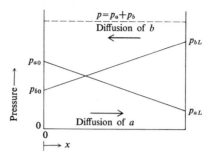

Fig. 16-2 Partial pressure gradients for equimolal counterdiffusion.

The molal flux of *a* in one direction is countered by an equal molal flux of *b* in the other direction. This process is called *equimolal counterdiffusion* and the partial pressure curves are seen to be straight lines in Fig. 16-2.

16-2 DIFFUSIVITIES

It is possible to derive expressions for diffusivity for monatomic gases from the kinetic theory of gases. For polyatomic gases, Fuller et al. [2] used the same equational form as that for monatomic diffusivity and obtained the following semiempirical equation for diffusivities in gases:

$$D_c = 0.00138 \frac{(T)^{1.75}}{p(\text{Vol}_a^{1/3} + \text{Vol}_b^{1/3})^2} \left(\frac{1}{W_a} + \frac{1}{W_b}\right)^{1/2} \quad (16\text{-}7)$$

where D_c is in ft²/hr, T in °R, and p in atmospheres; W_a and W_b are molecular weights, and Vol_a and Vol_b are the atomic diffusion volumes of components *a* and *b* given in Table 16-1.

Since Eq. (16-7) is semiempirical, it should be used only if experimental values are not available. For more information on diffusivities, Reid and Sherwood [3], Jost [4], and Bird et al. [5] should be consulted.

Example 16-2. Calculate the diffusivity of CH_4 in air at atmospheric pressure and 60°F.

Solution. Utilizing Table 16-1,

for air, $\quad \text{Vol}_a = 20.1, \quad W_a = 28.9,$

for CH_4, $\quad \text{Vol}_b = 4(1.98) + 16.5 = 24.4, \quad W_b = 16.$

$$D_c = 0.00138 \frac{520^{1.75}(1/28.9 + 1/16)^{1/2}}{1(20.1^{1/3} + 24.4^{1/3})^2} = 0.775 \text{ ft}^2/\text{hr}.$$

The experimental value [2] for CH_4 in air was 0.85 ft²/hr.

Table 16-1
SPECIAL ATOMIC DIFFUSION VOLUMES

Atomic and structural diffusion volume increments			
C	16.5	(Cl)	19.5
H	1.98	(S)	17.0
O	5.48	Aromatic or hetero-	
(N)	5.69	cyclic rings	20.2

Diffusion volumes of simple molecules			
H_2	7.07	CO_2	26.9
D_2	6.70	N_2O	35.9
He	2.88	NH_2	14.9
N_2	17.9	H_2O	12.7
O_2	16.6	(CCl_2F_2)	114.8
Air	20.1	(SF_6)	69.7
Ne	5.59		
Ar	16.1	(Cl_2)	37.7
Kr	22.8	(Br_2)	67.2
(Xe)	37.9	(SO_2)	41.1
CO	18.9		

Parentheses ( ) indicate that listed values are based on only a few data points. From E. N. Fuller, P. D. Schettler, and J. C. Giddings, "A New Method for Prediction of Binary Gas-Phase Diffusion Coefficients," *Indust. Eng. Chem.*, 1966. Reproduced by permission of the publisher.

It is much more difficult to estimate the diffusivities of liquids than gases. The kinetic theory of liquids is not well developed, and hence diffusivities for liquid solutions are usually obtained by experimental or empirical methods. Since liquids usually diffuse much slower than gases, the diffusivities for liquids are usually much smaller. Liquid diffusivities range from 10^{-6} to 10^{-5} ft²/hr.

Diffusion in solids is of interest to metallurgists. Solids diffuse into solids if both the temperature and concentration gradient are large. For example if iron is heated in a bed of coke, the carbon will diffuse into the iron with the highest concentration occurring near the surface. Theories are of little use in predicting diffusivities of solids. One has to rely almost exclusively on experimental data.

16-3 BULK MOTION AND DIFFUSION

When a fluid system is flowing, mass will be transferred by bulk motion as well as molecular diffusion. Consider a flowing mixture containing components a and b such that the mass density may be written as

$$\rho_a + \rho_b = \rho, \tag{16-8}$$

or, in terms of mass fraction,

$$\omega_a + \omega_b = 1. \tag{16-9}$$

The total mass flux is then given by

$$\dot{m} = \dot{m}_a + \dot{m}_b = \rho V, \tag{16-10}$$

where V is the net bulk velocity of the fluid. The flux of a is due to both diffusion (if there is a mass fraction gradient of component a) and bulk motion:

$$\dot{m}_a = -\rho D_c \,\text{grad}\, \omega_a + \dot{m}\omega_a. \tag{16-11}$$

Similarly for b,

$$\dot{m}_b = -\rho D_c \,\text{grad}\, \omega_b + \dot{m}\omega_b. \tag{16-12}$$

Since $(\omega_a + \omega_b) = 1$, it is obvious that Eqs. (16–11) and (16–12) satisfy Eq. (16–10). Note that $\dot{m}_a$ or $\dot{m}_b$ in Eq. (16–11) or (16–12), respectively, expresses the total mass transfer rate of a or b, not just the diffusion rate as was the case in Section 16–2.

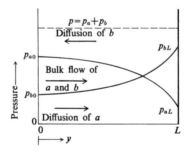

Fig. 16-3 Partial pressure gradients for a semipermeable wall.

Example 16–3. Consider the case where a gas mixture of a and b is adjacent to a semipermeable wall which absorbs a but not b. Assume that a is continually supplied to the system so that steady state is maintained. The species partial pressures are shown in Fig. 16–3. Derive an expression for the mass flux of species a, assuming the diffusion takes place at constant temperature and pressure.

Solution. At first glance one might guess that b must flow away from the surface since a is diffusing toward it. But it must be remembered that b is not being removed from or added to the system. This means that

$$\dot{m}_b = 0,$$

and

$$\dot{m} = \dot{m}_a + \dot{m}_b = \dot{m}_a.$$

Since T and p are constant and since $\dot{m} = \dot{m}_a$, Eq. (16–11) can be rewritten as

$$\dot{m}_a = -\rho D_c \frac{d\omega_a}{dy} + \dot{m}_a \omega_a.$$

But since $\omega_a + \omega_b = 1$, then

$$\dot{m}_a(1 - \omega_a) - \dot{m}_a \omega_b = -\rho D_c \frac{d\omega_a}{dy}. \tag{16–13}$$

Since $p_a + p_b = p = $ constant, and $p = \rho R_g T = \rho \frac{\Re T}{W}$

$$\frac{dp_a}{dy} = -\frac{dp_b}{dy} \quad \text{and} \quad \omega_b = \frac{\rho_b}{\rho} = \frac{p_b}{p}.$$

Therefore Eq. (16–13) can be written as

$$\dot{m}_a = \frac{p}{p_b} \frac{D_c W_a}{\Re T} \frac{dp_b}{dy}. \tag{16–14}$$

Since the total mass flux of b is zero, Eq. (16–12) can be written as

$$\dot{m}_b = -\frac{D_c W_b}{\Re T} \frac{dp_b}{dy} + \dot{m} \omega_b = 0,$$

or

$$\dot{m} \omega_b = \frac{D_c W_b}{\Re T} \frac{dp_b}{dy}. \tag{16–15}$$

The diffusion caused by the gradient of p_b is exactly counterbalanced by the bulk movement of b in the other direction. Thus b is actually a stagnant gas. This type of process is known as diffusion (of a) through a stagnant fluid (b).

In most cases, the gradients of diffusion through a stagnant fluid occur in a layer adjacent to the surface. A reference plane can be chosen outside this layer, where the pressure gradients are essentially zero—such as plane 0 in Fig. 16–3. Since $\dot{m}_a$ is constant, Eq. (16–14) can be integrated by:

$$\dot{m}_a \int_0^L dy = \frac{p D_c W_a}{\Re T} \int_{p_{b0}}^{p_{bL}} \frac{dp_b}{p_b}, \tag{16–16}$$

or

$$\dot{m}_a = \frac{D_c}{L} \frac{pW_a}{\Re T} \ln \frac{p_{bL}}{p_{b0}}, \tag{16–17}$$

or

$$\dot{m}_a = \frac{D_c}{L} \frac{pW_a}{\Re T} \ln \frac{p - p_{aL}}{p - p_{a0}}. \tag{16–18}$$

If the partial pressure $p_a \ll p$ then a series expansion of Eq. (16–18) yields the following equation:

$$\dot{m}_a = \frac{D_c W_a}{L \Re T} (p_{a0} - p_{aL}). \tag{16–19}$$

16-4 MASS TRANSFER COEFFICIENT

A mass transfer coefficient is defined by the following equation:

$$\dot{m}_a = h_D(\rho_{a1} - \rho_{a2}), \quad (16\text{--}20)$$

where h_D = mass transfer coefficient, and ρ_{a1} is the density of the diffusing component at a phase change boundary (i.e., interface between a liquid and its vapor), and ρ_{a2} is the density of the component at some reference point in the fluid. This definition for mass transfer coefficient is very similar to the definition of heat transfer coefficient. For the case of steady state diffusion with linear species density gradients (see Example 16-1), we have

$$\dot{m}_a = \frac{D_c}{L}(\rho_{a0} - \rho_{aL}) = h_D(\rho_{a0} - \rho_{aL}).$$

Thus

$$h_D = \frac{D_c}{L}. \quad (16\text{--}21)$$

For the semipermeable wall in Example 16-3,

$$h_D = \frac{D_c p W_a}{L\mathcal{R}T(\rho_{a0} - \rho_{aL})} \ln \frac{p - p_{aL}}{p - p_{a0}}. \quad (16\text{--}22)$$

16-5 SPECIES CONSERVATION EQUATION

Consider the control volume fixed in a flow field of a two-component mixture, a and b, as shown in Fig. 16-4. For the sake of clarity, only the flow in the x direction is shown. For the x direction, the net rate of flow of component a out of the control volume is

$$\left[\dot{m}_{ax} + \frac{\partial}{\partial x}(\dot{m}_{ax})\, dx\right] dy\, dz - \dot{m}_{ax}\, dy\, dz = \frac{\partial}{\partial x}(\dot{m}_{ax})\, dx\, dy\, dz. \quad (16\text{--}23)$$

Similar expressions could be derived for the y and z directions. The species conservation equation for component a can be expressed as

net rate of outflow of a + rate of accumulation of a
   = rate of production of a.

The rate of accumulation of a can be expressed as

$$\frac{\partial}{\partial t}(\rho_a)\, dx\, dy\, dz. \quad (16\text{--}24)$$

The rate of production of a (due to chemical reaction) can be expressed as

$$\dot{r}_a\, dx\, dy\, dz. \quad (16\text{--}25)$$

where $\dot{r}_a$ is the rate of production a per unit volume.

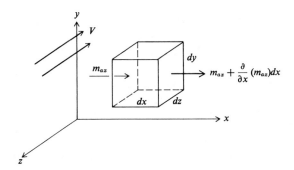

Fig. 16-4 Control volume for deriving species conservation equation.

The species conservation equation for a, after dividing through by $dx\,dy\,dz$, is

$$\frac{\partial \dot{m}_{ax}}{\partial x} + \frac{\partial \dot{m}_{ay}}{\partial y} + \frac{\partial \dot{m}_{az}}{\partial z} + \frac{\partial \rho_a}{\partial t} = \dot{r}_a, \qquad (16\text{--}26\text{a})$$

or

$$\frac{\partial \rho_a}{\partial t} + \text{div}\,\dot{\mathbf{m}}_a = \dot{r}_a. \qquad (16\text{--}26\text{b})$$

For species b, a similar equation can be derived:

$$\frac{\partial \rho_b}{\partial t} + \text{div}\,\dot{\mathbf{m}}_b = \dot{r}_b. \qquad (16\text{--}27)$$

By adding Eqs. (16-26a-b) and (16-27) and noting that $\dot{\mathbf{m}}_a + \dot{\mathbf{m}}_b = \rho \mathbf{V}$, we obtain

$$\frac{\partial \rho}{\partial t} + \text{div}\,\rho\mathbf{V} = 0, \qquad (16\text{--}28)$$

which is the overall, or global, conservation of mass equation for the mixture. This is the same conservation of mass equation as derived previously (Eq. 4-14) for a pure fluid. Equation (16-28) is valid only if ρ is the mass density. It is not a valid equation if one assumes ρ to be the molal density. In obtaining Eq. (16-28), use was made of the relation $\dot{\mathbf{m}}_a + \dot{\mathbf{m}}_b = \dot{\mathbf{m}} = \rho\mathbf{V}$ and the conservation of mass $\dot{r}_a + \dot{r}_b = 0$.

When Eq. (16-11) is substituted into Eq. (16-26b), we get

$$\frac{\partial \rho_a}{\partial t} + \text{div}\,(\rho_a \mathbf{V}) = \text{div}\,(\rho D_c \,\mathbf{grad}\,\omega_a) + \dot{r}_a. \qquad (16\text{--}29)$$

The divergence of $\rho_a \mathbf{V}$ can be written as

$$\text{div}\,(\rho_a \mathbf{V}) = \rho_a \,\text{div}\,\mathbf{V} + \mathbf{V} \cdot \mathbf{grad}\,\rho_a. \qquad (16\text{--}30)$$

If the density ρ is assumed constant, then, from Eq. (16-28), div $\mathbf{V} = 0$, and, if D_c is constant, Eq. (16–29) can be written as

$$\frac{\partial \rho_a}{\partial t} + \mathbf{V} \cdot \mathbf{grad}\ \rho_a = D_c \nabla^2 \rho_a + \dot{r}_a. \qquad (16\text{–}31\text{a})$$

The expanded form of Eq. (16–31a) is

$$\frac{\partial \rho_a}{\partial t} + u\frac{\partial \rho_a}{\partial x} + v\frac{\partial \rho_a}{\partial y} + w\frac{\partial \rho_a}{\partial z} = D_c\left(\frac{\partial^2 \rho_a}{\partial x^2} + \frac{\partial^2 \rho_a}{\partial y^2} + \frac{\partial^2 \rho_a}{\partial z^2}\right) + \dot{r}_a. \qquad (16\text{–}31\text{b})$$

A similar equation can be written for component b. If there are no chemical changes, $\dot{r}_a = 0$, and if the bulk velocity $\mathbf{V} = 0$, then

$$\frac{\partial \rho_a}{\partial t} = D_c\left(\frac{\partial^2 \rho_a}{\partial x^2} + \frac{\partial^2 \rho_a}{\partial y^2} + \frac{\partial^2 \rho_a}{\partial z^2}\right). \qquad (16\text{–}32)$$

This equation is called *Fick's second law of diffusion*. This equation can be applied to solids or stationary liquids. Notice the similarity between this equation and the heat conduction equation, Eq. (15–5). A large number of solutions to Eq. (16–32) have been obtained by Carslaw and Jaeger [6].

16–6 MASS TRANSFER COOLING

Many problems in engineering are concerned with the simultaneous transfer of energy and mass (by both bulk motion and diffusion). The analysis, for example, of a steady flow laminar boundary layer of a two-component, nonreacting, compressible fluid with combined energy and mass transfer requires a solution of the following two-dimensional conservation equations:

Mass:

$$\frac{\partial(\rho u)}{\partial x} + \frac{\partial(\rho v)}{\partial y} = 0 \qquad (16\text{–}33)$$

Species:

$$\frac{\partial}{\partial x}(\rho_a u) + \frac{\partial}{\partial y}(\rho_a v) = \frac{\partial}{\partial y}\left(\rho D_c \frac{\partial \omega_a}{\partial y}\right) \qquad (16\text{–}34)$$

Momentum:

$$\rho u \frac{\partial u}{\partial x} + \rho v \frac{\partial u}{\partial y} = -\frac{\partial p}{\partial x} + \frac{\partial}{\partial y}\left(\mu \frac{\partial u}{\partial y}\right) \qquad (16\text{–}35)$$

Energy:

$$\rho u c_p \frac{\partial T}{\partial x} + \rho v c_p \frac{\partial T}{\partial y}$$

$$= \frac{\partial}{\partial y}\left(K \frac{\partial T}{\partial y}\right) + \mu\left(\frac{\partial u}{\partial y}\right)^2 + u \frac{\partial p}{\partial x} + D_c(c_{pa} - c_{pb})\frac{\partial T}{\partial y}\frac{\partial \rho_a}{\partial y}. \qquad (16\text{–}36)$$

The solution to the above equation would be very difficult, as numerical schemes would probably have to be used. If, however, the fluid is assumed incompressible, with constant properties and a zero pressure gradient, then the above equations reduce to

$$\frac{\partial u}{\partial x} + \frac{\partial v}{\partial y} = 0 \tag{16-37}$$

$$u\frac{\partial \rho_a}{\partial x} + v\frac{\partial \rho_a}{\partial y} = D_c \frac{\partial^2 \rho_a}{\partial y^2} \tag{16-38}$$

$$u\frac{\partial u}{\partial x} + v\frac{\partial u}{\partial y} = \nu \frac{\partial^2 u}{\partial y^2} \tag{16-39}$$

$$u\frac{\partial T}{\partial x} + v\frac{\partial T}{\partial y} = \alpha \frac{\partial^2 T}{\partial y^2}. \tag{16-40}$$

The boundary conditions for these differential equations are:

for $y = 0$ $u = 0$ $v = v_w$ $T = T_w$ $\rho_a = \rho_{aw}$
for $y = \infty$ $u = u_\infty$ $T = T_\infty$ $\rho_a = \rho_{a\infty}$

In Chapter 8 it was shown that ν/α (the Prandtl number) was the connecting link between the solutions of Eqs. (16–39) and (16–40) for the dimensionless velocity and temperature profiles. For the case where $\Pr = 1$, the velocity and temperature profile are similar. In like manner, when one compares Eqs. (16–38) and (16–40) and the boundary conditions for ρ_a and T, it is possible to show that the dimensionless temperature and species density profile are similar for any value of wall velocity v_w, provided the *Lewis number* (Le) $= 1$, where

$$\text{Le} = \frac{\alpha}{D_c}. \tag{16-41}$$

Figure 16–5 presents the dimensionless temperature and species density profiles for $\Pr = 0.7$ and $\text{Le} = 1.0$, according to Hartnett and Eckert [8].

If the wall velocity $v_w = 0$, then the dimensionless velocity profile and species density profile are similar when the *Schmidt number* (Sc) $= 1$, where

$$\text{Sc} = \frac{\nu}{D_c}. \tag{16-42}$$

From the definitions of Pr, Le, and Sc, it is easy to show that if $\text{Le} = 1.0$, then $\Pr = \text{Sc}$. It was shown in Chapter 8 that the dimensionless heat transfer coefficient, the Nusselt number (Nu), was a function of the Reynolds number and Prandtl number,

$$\text{Nu} = \frac{hL}{k} = f(\text{Re}, \Pr). \tag{16-43}$$

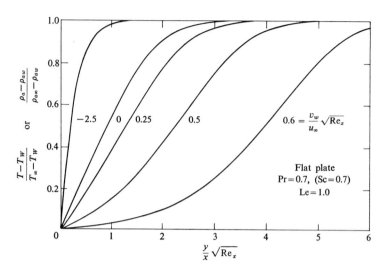

Fig. 16-5 Temperature and species density profiles for a laminar boundary layer over a flat plate. (From J. P. Hartnett and E. R. G. Eckert, *Trans. ASME*, **79**, 247, 1957.)

But for mass transfer, a mass transfer Nusselt number, called the *Sherwood number* (Sh), is used and can be shown to be a function of the Reynolds number and Schmidt number:

$$\text{Sh} = \frac{h_D L}{D_c} = f(\text{Re}, \text{Sc}). \tag{16-44}$$

For example, Gilliland [9] found that

$$\text{Sh} = 0.023(\text{Re})^{0.83}(\text{Sc})^{0.44} \tag{16-45a}$$

or

$$\frac{h_D D}{D_c} = 0.023 \left(\frac{\rho \bar{V} D}{\mu}\right)^{0.83} \left(\frac{\nu}{D_c}\right)^{0.44} \tag{16-45b}$$

represented the vaporization of liquids into air, flowing inside circular columns with the liquid wetting the wall. ($\bar{V}$ is the mean velocity of the air relative to the wall, not the liquid surface.) Equations (16-45a-b) are applicable for Reynolds number and Schmidt number ranges of

$$2{,}000 < \text{Re} < 35{,}000 \quad \text{and} \quad 0.6 < \text{Sc} < 2.5.$$

Compare Eqs. (16-45a-b) with the Dittus-Boelter equation, Eq. (9-51).

The Colburn j factor, introduced in Chapter 9, related the skin friction coefficient to the forced convection heat transfer coefficient by a simple equation. It has been determined that the Colburn j factor correlations hold for convective-mass transfer as well. Defining,

$$j_H = \frac{h}{\rho c_p V} (\text{Pr})^{2/3}, \qquad (16\text{-}46)$$

and

$$j_D = \frac{h_D}{V} (\text{Sc})^{2/3}. \qquad (16\text{-}47)$$

Then for fully developed flow in circular tubes,

$$j_H = j_D = \frac{C_f}{2}. \qquad (16\text{-}48)$$

Smooth pipes. The skin friction coefficient is determined from $C_f = 4f$, where f is the Moody friction factor (see Fig. 7–13).

Smooth flat plate. For flow over a smooth flat plate the skin friction coefficients for laminar and turbulent incompressible flow can be obtained from Eqs. (8–53) and (9–31) respectively. Therefore,

Laminar:
$$j_H = j_D = 0.332 \, \text{Re}_x^{-0.5} \qquad (16\text{-}49)$$

Turbulent:
$$j_H = j_D = 0.0288 \, \text{Re}_x^{-0.2}. \qquad (16\text{-}50)$$

Single spheres. Froessling [10] found the following equation could be used to determine the mass transfer to or from a liquid (or solid) nonrotating sphere in forced convection:

$$\text{Sh} = \frac{h_D D}{D_c} = 2.0(1 + 0.276 \, \text{Re}_D^{1/2} \, \text{Sc}^{1/3}). \qquad (16\text{-}51)$$

Cylinder in transverse flow. Bedingfield and Drew [11] found the following equation for cylinders in forced convection to be generally accurate to less than 1 percent error for a Reynolds number range of 1000 to 50,000:

$$\text{Sh} = \frac{h_D D}{D_c} = 0.281(\text{Re}_D)^{0.6}(\text{Sc})^{0.44}. \qquad (16\text{-}52)$$

If heat and mass transfer occur simultaneously, the mass transfer coefficient can be obtained from the first equality in Eq. (16–48), $j_H = j_D$, or

$$\frac{h}{h_D} = \rho c_p \left(\frac{\text{Sc}}{\text{Pr}}\right)^{2/3} = \rho c_p \, \text{Le}^{2/3}. \qquad (16\text{-}53)$$

Example 16–4. Calculate the mass transfer coefficient h_D for a spherical drop of water $\frac{1}{8}$ in. in diameter falling in air at 47 ft/sec. Assume $D_c = 22.5 \times 10^{-5}$ ft²/sec and Sc = 0.6 for water. Assume standard conditions for air.

Solution.

$$\text{Re}_D = \frac{V}{\nu} = \frac{(47)(\frac{1}{8})}{(1.55 \times 10^{-4})(12)} = 3150,$$

$$\text{Sh} = 2.0[1 + 0.276(3150)^{1/2}(0.6)^{1/3}] = 32.44.$$

Therefore,

$$h_D = \frac{(\text{Sh})(D_c)}{D} = \frac{(32.44)(22.5 \times 10^{-5})(12)}{(\frac{1}{8})} = 0.701 \text{ ft/sec}.$$

REFERENCES

1. A. Fick, *Annalen Physik* **94**, 59 (1855).
2. E. N. Fuller, P. D. Schettler, and J. C. Giddings, "A New Method for Prediction of Binary Gas-Phase Diffusion Coefficients," *Indus. Eng. Chem.* **58** (5), 19 (1966).
3. R. C. Reid and T. K. Sherwood, *The Properties of Gases and Liquids*, 2d. ed., McGraw-Hill, New York (1966).
4. W. Jost, *Diffusion*, Academic Press, New York, (1952).
5. R. B. Bird, W. E. Stewart, and E. N. Lightfoot, *Transport Phenomena*, Wiley, New York (1960).
6. H. S. Carlslaw and J. C. Jaeger, *Conduction of Heat in Solids*, 2d ed., Oxford University Press, New York (1959).
7. S. Chapman and T. G. Cowling, *The Mathematical Theory of Non-Uniform Gases*, Cambridge University Press, New York (1958).
8. J. P. Hartnett and E. R. G. Eckert, *Trans. ASME* **79**, 247 (1957).
9. E. R. Gilliland, "Diffusion Coefficients in Gaseous Systems," *Indus. Eng. Chem.* **26**, 681 (1934).
10. N. Froessling, *Gerlands Beitr. Geophys.* **32**, 170 (1938).
11. C. H. Bedingfield, Jr., and T. B. Drew, "Analogy Between Heat Transfer and Mass Transfer," *Indus. Eng. Chem.* **42**, 1164 (1950).

PROBLEMS

1. Calculate the diffusivity of ammonia (NH_3) in air at 76°F at 1 atmosphere.
2. Calculate the rate of diffusion of water from a bottle 3-in. in diameter and 7 in. long. Assume that the air above the bottle is dry and the air in the bottle is stagnant. The temperature is 100°F and the pressure is 1 atmosphere. Assume $D = 1.06$ ft^2/hr.
3. Ammonia gas is diffusing through a stagnant layer of air 0.01 ft thick, through a semipermeable membrane, and then through a stagnant layer of hydrogen 0.01 ft thick. The partial pressure of ammonia at the outer boundary of the air layer is

0.0075 atmosphere. The partial pressure of ammonia at the outer layer of the hydrogen layer is 0.00001 atmosphere. The temperature is 32°F and the total pressure is 1 atmosphere. The resistance of the membrane to diffusion is negligible. Diffusivity of ammonia through hydrogen is 0.294 ft²/hr. Diffusivity of ammonia through air is 0.83 ft²/hr. (a) Calculate the partial pressure of ammonia at the membrane. (b) Calculate the rate of diffusion of ammonia in lb_m/hr ft².

4. The diffusivity of benzene (C_6H_6) in air is determined experimentally by allowing liquid benzene to vaporize isothermally into air from a partially filled tube 0.01 ft² in area. The partial pressure of benzene at the mouth of the tube can be assumed negligible. The liquid level in the tube is kept constant by means of a constant head device. At 79°F, the data obtained are as follows:

Distance of liquid level from open end of tube	0.2 ft
Time of run	300 hr
Amount of benzene evaporated	0.153 lb_m
Total pressure	1 atm
Vapor pressure of benzene at 79°F	1.94 psia

Calculate the diffusivity of benzene in air at 79°F.

5. Calculate the evaporation rate of water as it flows in a thin film down the outside of a vertical circular cylinder. Dry air at 100°F and 14.7 psia flows transversely across the cylinder at the rate of 40 ft/sec. The cylinder is 2 ft long and 6 in. in diameter. The water temperature is also 100°F. Assume that the Schmidt number for water is 0.6 and $D_c = 22.5 \times 10^{-5}$ ft²/sec.

6. Calulate the diffusivity of an ethane (C_2H_6)-methane (CH_4) system at 100°F and 1 atmosphere.

CHAPTER 17

SPECIAL TOPICS

17-1 LIQUID METALS

Liquid metals are discussed as a separate class of fluids because the heat transfer equations that have been derived in previous chapters do not seem to adequately describe their behavior. A glance at the table of properties of liquid metals (Table 17-1) shows that these fluids have relatively high thermal conductivities and relatively low Prandtl numbers compared to other fluids.

The high thermal conductivity contributes to molecular heat conduction rates in moving fluids that are often comparable in magnitude to the rates of energy transport by the motion or convection of the fluid. One consequence of this is to make questionable the usual assumption that conduction in the direction of flow is negligible.

The very small values of the Prandtl numbers of liquid metals ordinarily lead to a thermal boundary layer thickness much larger than the velocity boundary layer thickness. This requires modification of the derivations of the heat transfer equations made previously for fluids with higher Prandtl number and results in unique relationships for the Nusselt number. Generally speaking, however, the fluid flow characteristics, friction factors, and universal velocity profiles for liquid metals are no different than those for other fluids.

The interest in liquid metals as a coolant is due primarily to the demand for large rates of heat removal from nuclear reactors. The high thermal conductivities and large specific heat per unit volume, as well as the favorable nuclear properties of liquid metals, make these fluids obvious ones for use in nuclear reactors. Liquid metals have the added advantage of moderate viscosity (making them easy to pump); the liquid state exists over a wide temperature range, including high temperatures where liquid metals exert moderate vapor pressures. The possibility of metering and pumping by electromagnetic means is an additional advantage.

A typical arrangement for utilizing liquid metal as a coolant in a nuclear power cycle is shown in Fig. 17-1. The liquid metal flows through the nuclear reactor, removing the energy but remaining in a liquid state. The hot liquid metal then flows to a heat exchanger where it is used to vaporize

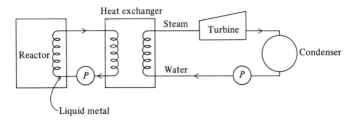

Fig. 17-1 A typical liquid metal nuclear power cycle.

water. The steam leaving the heat exchanger is used to produce work in a turbine and, after condensing, returns to the heat exchanger. In this arrangement any radioactive contamination of the fluid passing through the turbine is minimized.

Table 17-1

PROPERTIES OF LIQUID SODIUM, POTASSIUM, AND THEIR ALLOYS*

(For alloys, % K is by weight)

Materials and properties	Temperatures, °C and °F							
	100	200	300	400	500	600	700	800
	212	392	572	752	932	1112	1292	1472
Density, gm/cm³:								
Na (mp: 97.8°C)	0.927	0.905	0.882	0.858	0.834	0.809	0.783	0.757
NaK (43.4% K)	0.887	0.862	0.838	0.814	0.789	0.765	0.740	
NaK (78.6% K)	0.847	0.823	0.799	0.775	0.751	0.727	0.703	
K (mp: 63.7°C)	0.819	0.795	0.771	0.747	0.723	0.701	0.676	
Viscosity, cp:								
Na	0.705	0.450	0.345	0.284	0.243	0.210	0.186	0.165
NaK (43.3% K)	0.540	0.379	0.299	0.245	0.207	0.178	0.165	
NaK (66.9% K)	0.529	0.354	0.276	0.229	0.195	0.168	0.146	
K	0.436	0.299	0.227	0.194	0.169	0.153	0.140	
Thermal conductivity, watts/cm °C:								
Na	—	0.815	0.757	0.712	0.668	0.627	0.590	0.547
NaK (56.5% K)	—	0.249	0.262	0.269	0.271			
NaK (77.7% K)	—	0.247	0.259	0.262	0.262	0.259	0.255	
K	—	0.449	0.428	0.400	0.376	0.354		
Specific heat, cal/gm °C:								
Na	0.331	0.320	0.312	0.306	0.302	0.300	0.300	0.303
NaK (44.8% K)	0.269	0.261	0.255	0.251	0.248	0.248	0.250	
NaK (78.3% K)	0.225	0.217	0.212	0.210	0.209	0.209	0.211	
K	0.194	0.189	0.185	0.183	0.182	0.183	0.185	0.188

* From *Liquid-metals Handbook*, AEC and U.S. Dept. of the Navy, June, 1952; and "Sodium-NaK Supplement," July 1, 1955.

Table 17-1 (*cont.*)

PROPERTIES OF MISCELLANEOUS LIQUID METALS†

Element	ρ, gm/cm³	°C	c_p, cal/gm °C	°C	μ, cp	°C	k, cal/sec cm °C	°C
Aluminum (mp: 660.2°C)	2.380 2.261	660 1100	0.259	660–1000	2.9 1.4	700 800	0.247 0.290	700 790
Bismuth (mp: 271.0°C)	10.03 9.66 9.20	300 600 962	0.0340 0.0376 0.0419	271 600 1000	1.662 0.996	304 600	0.041 0.037 0.037	300 500 700
Lead (mp: 327.4°C)	10.51 10.27 10.04 9.81	400 600 800 1000	0.039 0.037 — —	327 500 — —	2.116 1.700 1.185	441 551 844	0.038 0.036 0.036	400 600 700
Lithium (mp: 179.0°C)	0.507	200	1.0	200	0.5541	208.1	0.09	218–233
Mercury (mp: 38.87°C)	13.546 13.352 13.115	20 100 200	0.033 0.0328 0.0325	0 100 200	1.68 1.21 1.01	0 100 200	0.0196 0.0261 0.0303	0 120 220
Rubidium (mp: 30°C)	1.475	39	0.0913	39	0.6734	39	0.07	39
Tin (mp: 231.9°C)	6.83	409	0.058 0.076	250 1100	1.91 1.18	240 500	0.08 0.078	240 498
Zinc (mp: 419.5°C)	6.92 6.81 6.57	419.5 600 800	0.1199 0.1173 0.1012	419.5 600 1000	2.78 1.88	500 700	0.138 0.135	500 700
Lead-bismuth eutectic (mp: 125°C)	10.46 10.19	200 400	0.035	144–358	1.7 1.29	332 500	0.023 0.027	200 320

† From *Liquid-metals Handbook*, AEC and U.S. Dept. of the Navy, June, 1952.

Because of the low pressure of metal vapors at high temperatures there have been attempts to develop power cycles using liquid metals to improve the thermal efficiency of power systems, but the attempts have met with difficulty due to the undesirable characteristics of some of these substances. Mercury, for example, has the advantage of being molten at ordinary temperatures, but both the liquid and vapor forms are highly toxic. Sodium and potassium ignite at high temperature in air, and several of the metals react violently with water or chlorinated hydrocarbons. Most of the metals solidify at room temperature and they are all relatively expensive. Complete and accurate property data do not seem to be available for all metals over a wide range of temperatures and pressures.

In spite of these disadvantages, a great deal of research continues on liquid metal heat transfer, and there is hope that liquid metals will soon find wide use, not only in conventional power cycles but also in more recently developed cycles, such as those which produce electricity directly by magneto-hydrodynamic (MHD) methods and cycles for use in space vehicle power plants. A recent survey of liquid metal heat transfer is given by Stein [2].

Sparrow and Gregg [3] discussed the effect of the low Prandtl number on the relative thicknesses of the velocity and thermal boundary layers on flat plates. Since the velocity boundary layer is so thin in such cases it might be a reasonable assumption to consider the fluid velocity constant and equal to the free stream value through the thermal boundary layer. Grosh and Cess [4] made this assumption for the flat plate and several other shapes with laminar flow of fluids with low Prandtl numbers. For the flat plate laminar flow case, the energy equation simplifies to

$$V_\infty \frac{\partial T}{\partial x} = \alpha \frac{\partial^2 T}{\partial y^2}. \quad (17\text{-}1)$$

For the case of constant fluid velocity, $u = V_\infty$, the stream function is simply $\psi = V_\infty y$. Letting the dimensionless stream function, ψ^*, equal $2y^*$, where y^* has the same meaning as in Chapter 8, $y^* = y/\sqrt{(\nu x/V_\infty)}$, Eq. (17-1) can be solved to give

$$\mathrm{Nu}_x = 0.565\sqrt{\mathrm{Re}_x\,\mathrm{Pr}}. \quad (17\text{-}2)$$

Heat transfer coefficients obtained by Eq. (8-62) for $0.005 < \mathrm{Pr} < 0.025$, are less than those obtained by Eq. (17-2) by about 7 to 12 percent. Grosh and Cess also showed that axial conduction can be neglected in this problem for $(\mathrm{Re}_x\,\mathrm{Pr}) > 50$ except in the neighborhood of the leading edge.

In Eq. (17-2) the Nusselt number is a function of the product of the Reynolds and Prandtl numbers. This product, called the *Péclet number*, occurs in almost all convective heat transfer equations for liquid metals. The product can be shown to be proportional to the ratio of the energy transported by convection to that transported by conduction.

In laminar flow through round tubes, the mean Nusselt number for liquid metals [5] appears to be lower than those predicted by the equations for ordinary fluids, such as Eq. (8-27).

For turbulent liquid metal flow through tubes the Dittus-Boelter, Seider-Tate, and Colburn equations given in Chapter 9 are, likewise, not suitable. Extensive use of analogies have been made to obtain equations for

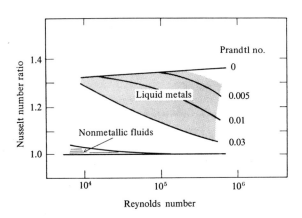

Fig. 17–2 Ratio of Nusselt number (Nu) for uniform heat flux at wall to Nu for constant wall temperature. (From C. A. Sleicher and M. Tribus, *Trans. ASME*, **79**, 789, 1957.)

this case [6–9]. For turbulent tube flow with constant wall heat flux, the equation of Lyon [7] is widely accepted:

$$\overline{Nu}_D = 7 + 0.025(Re_D\ Pr)^{0.8} \tag{17-3}$$

for

$(Re_D\ Pr) > 100$ and $\dfrac{L}{D} > 60$; properties evaluated at $T_{B\text{avg}}$.

A simplified equation which seems to fit some experimental data better than Eq. (17–3) is that of Lubarsky and Kaufman [10], valid for the same conditions as Eq. (17–3):

$$\overline{Nu}_D = 0.625(Re_D\ Pr)^{0.4}. \tag{17-4}$$

For constant wall temperature, the equation of Seban and Shimazaki [8] is useful:

$$Nu_D = 5 + 0.025(Re_D\ Pr)^{0.8} \tag{17-5}$$

for

$(Re_D\ Pr) > 100$ and $\dfrac{L}{D} > 60$; properties evaluated at $T_{B\text{avg}}$.

Liquid metals in heat transfer are extremely sensitive to the boundary conditions of the tube, in contrast to nonmetallic fluids. For example, the solutions for constant wall heat flux are quite different from those for constant wall temperature, as can be seen in Eqs. (17–3) and (17–5) and in Fig. 17–2.

It is interesting to note that the values obtained for the average Nusselt number (assuming that all of the fluid flows through the pipe with a uniform velocity—sometimes called slug flow—and that the apparent conductivity of the liquid metal is constant) are

$$\overline{Nu}_D = 8 \quad \text{for} \quad \left(\frac{q}{A}\right)_{\text{wall}} = \text{constant}, \tag{17-6}$$

$$\overline{Nu}_D = 5.8 \quad \text{for} \quad T_w = \text{constant}. \tag{17-7}$$

In this special case, the problem is identical to one of conduction to a moving solid cylinder. The somewhat similar values obtained by Eqs. (17-3) and (17-6) and by Eqs. (17-5) and (17-7) show that for low velocities (low Re_x) the energy transfer is primarily due to molecular conduction. Care should be taken in using any equation for liquid metals, since no one equation appears capable of correlating all available experimental data. Data scatter is particularly bad for liquid metal boiling and condensation. Arguments persist as to why the experimental data and theoretical equations do not agree in many cases.

17-2 NON-NEWTONIAN FLUIDS

In the other sections of this book we have been concerned with Newtonian fluids, which are fluids for which the shear stress τ is proportional to velocity gradient du/dy:

$$\tau = \mu \frac{du}{dy}. \tag{17-8}$$

This is the simplest relationship that can exist between shear stress and velocity gradient (rate of shear) in a real fluid. Fortunately many of the fluids of engineering interests (i.e. water, air, and oil) are Newtonian fluids. However, there are a significant number of fluids which do not behave according to Eq. (17-8), and these are known as non-Newtonian fluids.

The study of non-Newtonian fluids is a subdivision of the science of rheology. *Rheology* is a study of deformation and flow of gases, liquids, plastics, and elastic solids. Non-Newtonian fluids may be subdivided into three main categories: (1) time-independent fluids, (2) time-dependent fluids, and (3) viscoelastic fluids.

Time-Independent Fluids

For many fluids the shear rate depends upon the imposed shear stress but is independent of the duration of the stress; that is, their viscous characteristics do not change with time after the start of shearing, all other factors being constant. These fluids are time-independent and may usually be classified into one of the three categories described below.

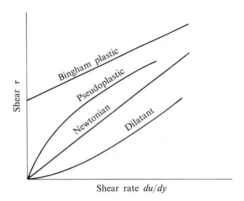

Fig. 17-3 Shear behavior of time-independent non-Newtonian fluids.

Bingham plastic fluids, are characterized by a straight line on the shear stress-shear rate plot having an intercept of τ_p on the shear stress axis (see Fig. 17-3):

$$\tau = \tau_p + \mu_p \frac{du}{dy}. \tag{17-9}$$

At shear stresses less than τ_p the fluid behaves like a solid. Many real fluids closely approximate Bingham plastic fluids, including sewage sludges, drilling muds, toothpaste, oil paints, and suspensions of solids.

Pseudoplastic fluids are characterized by the following equation:

$$\tau = K\left(\frac{du}{dy}\right)^n, \quad n < 1, \tag{17-10}$$

where K and n are constants for a given fluid. For some fluids n may not be constant over a wide range of shear rate, and in many fluids n may approach unity at very high or very low rates of shear. Examples of fluids which show pseudoplastic behavior include solutions of polymers, such as paper pulp, rubber, mayonnaise, and paint.

Dilatent fluids exhibit an increasing apparent viscosity as the shear rate increases. The shear rate stress-shear equation is similar to Eq. (17-10) except the index n is greater than unity. This type of behavior was found originally in suspensions of solids of high concentration. Osborne Reynolds suggested that at low shear rates the liquid fills the voids between the particles and lubricates them. At high shear rates the packing of the particles becomes less dense, causing a loss of lubrication and an expansion of the fluid. Suspensions of potassium silicate, sand, starch, and paint pigments are examples of dilatent fluids. Pseudoplastic and dilatent fluids are referred to as *power law* fluids since they follow a behavior described by Eq. (17-10).

Time-Dependent Fluids

In time-dependent fluids the viscous behavior depends upon the duration of the stress as well as other factors and therefore changes with time. They may be classified as either thixotropic or rheopectic.

Thixotropic fluids are characterized by a decrease in shear stress or apparent viscosity (as the duration of time under stress increases) for a fixed shear rate. Apparently the long molecules line up with the flow as time increases, thus decreasing the shear stress. Examples of thixotropic fluids include honey, mustard, ketchup, and shaving cream. It is believed that all plastics are at least partially thixotropic.

Rheopectic fluids are characterized by an increase in the shear stress or apparent viscosity (as the duration of time under stress increases) for a fixed shear rate. Their behavior is similar to that of dilatent fluids in that the particles tend to become less densely packed, causing a loss of lubrication as duration of stress increases. Examples of rheopectic fluids include suspensions of ammonium oleate and gypsum. Rheopectic fluids are not as common as thixotropic fluids.

Viscoelastic Fluids

Viscoelastic fluids possess both elastic and plastic properties. The elastic properties are dependent upon the normal stresses which may take on different values for different directions. For example, if the normal stress of a viscoelastic fluid flowing in a tube is τ_{zz} in the flow direction and τ_{xx} perpendicular to the flow, then $\tau_{zz} < \tau_{xx}$. If the fluid should leave the tube and become a jet, then the stresses will relieve themselves by increasing the diameter of the jet [12]. Techniques and models for treating viscoelastic fluids are not as well developed as other non-Newtonian fluids. Fluids which exhibit viscoelastic properties are gels such as Jell-O and puddings.

Heat Transfer

The problem of heat transfer to non-Newtonian fluids has been discussed by Metzner [13]. Generally speaking, accurate predictions can usually be made in most cases, although the computations are more complex than for the Newtonian cases. The effects of natural convection and variation of fluid properties with temperature can often be as significant as the non-Newtonian effects and should not be overlooked in making such computations.

17–3 LUBRICATION

The operation of mechanical devices depends upon lubrication for high performance, minimum wear, and long life. Fluid films confined in slots and grooves have generated pressures up to 20,000 psi in keeping metal parts from coming into contact.

The essential features of hydrodynamic lubrication can be illustrated by the converging wedge slider bearing (Fig. 17–4). For simplicity assume that the block is very long in the transverse direction so that the flow is two-dimensional. In addition, the length L is assumed very much larger than the clearance h. The lubricating fluid will be assumed to be incompressible with a constant viscosity. The coordinate system is assumed fixed with respect to the slide block, and the guide surface moves to the right with velocity V.

The hydrodynamic theory of lubrication assumes that the viscous terms of the Navier-Stokes equations—Eqs. (4–44a-c), are much greater than the inertia terms, so that

$$\frac{\rho u \, \partial u/\partial x}{\mu \, \partial^2 u/\partial y^2} \sim \frac{\rho V^2/L}{\mu V/h^2} = \frac{\rho V L}{\mu}\left(\frac{h}{L}\right)^2 \ll 1. \qquad (17\text{–}11)$$

The y-direction equation of the Navier-Stokes equations can be neglected because the velocity component v is proportional to $(h/L)u$ and hence is very small compared with u. In the x-direction equation, the term $\partial^2 u/\partial x^2$ is proportional to V/L^2 and is therefore negligible compared with $\partial^2 u/\partial y^2$, which is proportional to u/h^2. The x-direction Navier-Stokes equation then reduces to

$$\frac{dp}{dx} = \mu \frac{\partial^2 u}{\partial y^2}. \qquad (17\text{–}12)$$

Integrating twice with respect to y, the velocity distribution is

$$u = \frac{1}{2\mu}\frac{dp}{dx}y^2 + C_1 y + C_2. \qquad (17\text{–}13)$$

In the above integration dp/dx is constant in the y direction but can vary in the x direction. The boundary conditions are:

$$\begin{aligned} y &= 0; \; u = V & x &= 0; \; p = p_0 \\ y &= h; \; u = 0 & x &= L; \; p = p_0. \end{aligned} \qquad (17\text{–}14)$$

By substituting the boundary conditions into Eq. (17–13) and evaluating the constants, the velocity distribution becomes

$$u = \frac{1}{2\mu}\frac{dp}{dx}(y^2 - yh) + V\left(1 - \frac{y}{h}\right). \qquad (17\text{–}15)$$

The volume flow per unit width past any cross section is constant and is

$$\dot{Q} = \int_0^{h(x)} u \, dy. \qquad (17\text{–}16)$$

By substituting Eq. (17–15) into Eq. (17–16), the volume flow becomes

$$\dot{Q} = \frac{Vh}{2} - \frac{h^3}{12\mu}\frac{dp}{dx}. \qquad (17\text{–}17)$$

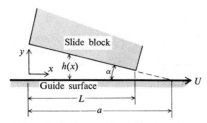

Fig. 17-4 Tapered wedge bearing.

Integrating with respect to x gives

$$p(x) = p_0 + 12\mu \int_0^L \left[\frac{V}{2h^2} - \frac{\dot{Q}}{h^3} \right] dx. \qquad (17\text{-}18)$$

Since V and $\dot{Q}$ are constants, $\dot{Q}$ can be solved from Eq. (17-18):

$$\dot{Q} = \tfrac{1}{2} V \frac{\int_0^L \dfrac{dx}{h^2}}{\int_0^L \dfrac{dx}{h^3}}. \qquad (17\text{-}19)$$

Assume the angle α in Fig. 17-4 is small, therefore, $\sin \alpha \approx \alpha$ and

$$h = \alpha(a - x). \qquad (17\text{-}20)$$

Substituting Eq. (17-20) into Eq. (17-19),

$$\dot{Q} = V\alpha a \frac{a - L}{2a - L}. \qquad (17\text{-}21)$$

By substituting $\dot{Q}$ from Eq. (17-21) and h from (17-20), the pressure can be obtained from Eq. (17-18):

$$p(x) = p_0 + 6\mu V \left(\frac{x}{h^2} \right) \left(\frac{L - x}{2a - L} \right). \qquad (17\text{-}22)$$

Typical pressure and velocity distribution are shown in Fig. 17-5 for a tapered wedge bearing. Notice that the maximum pressure occurs nearest

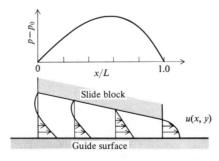

Fig. 17-5 Pressure and velocity distribution in a tapered wedge bearing.

the end with the smallest gap. Also notice that there is some backflow near the wider end. This backflow decreases as L/a increases.

A variation of the tapered wedge bearing is the journal bearing shown in Fig. 17–6(a), used to support rotating shafts. If the rotational speed of the shaft is too low to maintain a complete hydrodynamic oil film, then *hydrostatic* lubrication can be used to force lubricant under pressure into the clearance space. An example of hydrostatic lubrication is shown in Fig. 17–6(b) for the step bearing. Hydrostatic lubrication keeps the bearing surfaces separated by the externally induced lubrication pressures. Hydrostatic lubricated bearings can be designed for extremely large loads.

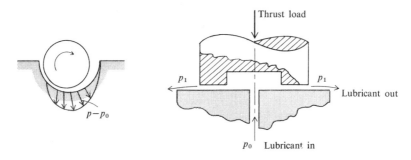

Fig. 17–6 Two types of bearings. (a) Journal bearing. (b) Step bearing.

It might be pointed out that lubricants are not necessarily limited to the conventional oils and greases. Lubricants such as gases, water, alcohol, gasoline, molten metals, and air have also been used in bearings.

17–4 FLUID-SOLID FLOWS

The transport of solid particles by flowing fluids is an example of two-phase, two-component flow which occurs in engineering practice. Pneumatic conveying of coal particles and grains over short distances has been utilized for many years. The pumping of sand and water mixtures in dredging operations and ore-water mixtures in mining has been commonplace.

More recently the transport of coal-water mixtures by pipeline has become economically competitive with other forms of transportation. For example, a 108-mile, 10-inch pipeline operated for several years to transport a coal-water mixture from Cadiz, Ohio, to the Cleveland Electric Company's Eastlake Station [17]. The pipeline handled a coal slurry of up to 60 percent concentration. An important discovery has been the fact that a coal-water slurry having a water content of up to 40 percent could be burned directly in a furnace with little loss in efficiency. This, coupled with the discovery that pumping losses for coal-water mixtures are surprisingly low, has been an important factor in the increased interest in this means of coal transport.

Forces on a Particle

In studying the dynamics of the particle suspended in a fluid stream in steady flow, it is immaterial whether the particle moves through a stationary fluid or whether the fluid flows past the stationary particle. (The turbulence level or intensity of the fluid is an important parameter, however.)

The forces acting on the particle are: (1) the force due to gravity (weight), (2) the buoyancy force due to liquid displacement, and (3) the drag force due to the relative velocity between the fluid and the particle.

The motion of the particle is determined by the resultant of these three forces. Whenever the resultant of the three forces is nonzero, there will be an acceleration of the particle. The particle will continue to accelerate until there is a balance of the forces acting on the particle. With the forces on the particle balanced, the particle will have a constant velocity relative to the fluid. For example, consider a small sphere dropped into a deep pool of water. Initially the gravity force (downward) will exceed the sum of the buoyancy force and the drag force (upward), and the particle velocity will increase. As the particle velocity increases, the drag force will increase. Eventually the drag force will become large enough that the sum of the drag and buoyancy forces equals the gravity force. The particle will then fall at a steady velocity. This steady velocity is called the *terminal velocity of fall*. An expression for the terminal velocity can easily be derived. For the case of a spherical particle the force balance is

$$\text{gravity force} + \text{buoyancy force} + \text{drag force} = 0,$$

$$-\frac{\pi D^3}{6} g \rho_P + \frac{\pi D^3}{6} g \rho_F + \frac{C_D \rho_F (V_F - V_P)^2}{2}\left(\frac{\pi D^2}{4}\right) = 0, \quad (17\text{--}23)$$

where

ρ_P = particle density,
ρ_F = fluid density,
C_D = drag coefficient = drag/$(\tfrac{1}{2}\rho V_\infty^2 A)$,
V_P = particle velocity,
V_F = fluid velocity, and
V_∞ = relative velocity = $(V_F - V_p)$.

Equation (17–23) can be simplified and solved for the velocity of the particle V_P. For the case being described, the fluid velocity $V_F = 0$, and V_P, the particle velocity, is the terminal velocity of fall:

$$V_T = \left[\frac{4gD(\rho_P - \rho_F)}{3\rho_F C_D}\right]^{1/2}. \quad (17\text{--}24)$$

The drag coefficient for spheres was given in Fig. 9–15 as a function of the Reynolds number, $\text{Re}_D = V_\infty D/\nu$, where V_∞ is the relative velocity of the particle and the fluid. Terminal velocity calculations require a trial and error procedure since C_D in Eq. (17–24) is dependent upon the value of the relative velocity.

If a particle is to be transported upward by a fluid, the relative velocity between the fluid and the particle must exceed the terminal velocity. For horizontal transport, the diffusion of the particles must be sufficient to keep them all from settling to the bottom of the duct, and turbulent flow is necessary. A special exception to this requirement occurs in the case of non-Newtonian flow.

In an actual flow of a fluid in a solid suspension, several factors must be considered:

1. When many particles are present in a flowing fluid, they interact with each other, affecting the value of the drag coefficient.

2. Particles are usually not spherical in shape, and drag coefficients are difficult to estimate with accuracy.

3. Most suspensions are made up of particles with a variety of sizes and shapes, which complicates the expressions for the forces on the particles.

4. The fluid velocity is not uniform across the flow cross section, resulting in particle spin and also complicating the description of the relative velocities.

5. Turbulence is ordinarily present in two-phase flows, and small particles are buffeted about randomly by the turbulent eddies.

6. The particles, which are often more dense than the fluid, may follow a different flow path than the fluid whenever there is a change in the direction of flow. This can sometimes lead to erosion of the duct at bends and contractions.

In spite of the difficulties enumerated above, theoretical equations, such as Eq. (17–24), and drag coefficient curves, such as Fig. 9–15, have been used extensively as the basis of semiempirical equations for describing such flows.

Description of Nonspherical Particles

An example of one means of overcoming the complexities encountered in attempting to predict the characteristics of real particle flows is the way in which we describe deviations in the shape of nonspherical particles. The amount of deviation in the shape of a particle from that of a sphere is called its *sphericity*, defined as the ratio of the surface area of a sphere having the volume of the particle to the actual surface area of the particle. Values of sphericity for certain types of particles are shown in Table 17–2. Also shown in the table is the ratio of equivalent diameter (D_E) to *average diameter* (D_{avg}) of the particle, where the average diameter has been determined by a screen analysis. The *equivalent diameter* is the diameter of a sphere having the same volume as the particle.

A set of curves showing values of C_D to use in Eq. (17–22) for nonspherical particles is shown in Fig. 17–7. The Reynolds number used as the abscissa of Fig. 17–7 is based on the equivalent diameter.

Table 17–2
SPHERICITY AND THE VALUE OF D_E RELATED TO SCREEN SIZE*

Shape	Sphericity	D_E/D_{avg}
Sphere	1.00	1.00
Octahedron	0.817	0.965
Cube	0.806	1.24
Prisms:		
$a \times a \times 2a$	0.767	1.564
$a \times 2a \times 2a$	0.761	0.985
$a \times 2a \times 3a$	0.725	1.127
Cylinders:		
$h = 2r$	0.874	1.135
$h = 3r$	0.860	1.31
$h = 10r$	0.691	1.96
$h = 20r$	0.580	2.592
Disks:		
$h = 1.33r$	0.858	1.00
$h = r$	0.827	0.909
$h = r/3$	0.594	0.630
$h = r/10$	0.323	0.422
$h = r/15$	0.254	0.368

* From *Unit Operations*, Brown and associates, copyright 1950, John Wiley & Sons, used by permission.

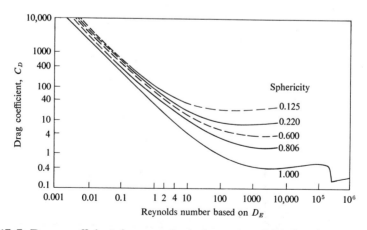

Fig. 17–7 Drag coefficient for nonspherical particles. (From *Unit Operations*, Brown and Associates, Copyright 1950. Used by permission of John Wiley & Sons, Inc.)

The Durand Equation

An empirical equation which is useful for the predicting of pressure loss due to friction in a water-solid suspension flow is the Durand equation [17]:

$$\frac{p_M - p_W}{C p_W} = 81 \left[\frac{gD\left(\frac{\rho_p}{62.4} - 1\right)}{V_{avg}^2} \frac{1}{C_D} \right]^{1.5}, \quad (17\text{--}25)$$

where

p_M = pressure drop of mixture (frictional),
p_W = pressure drop of clear water (frictional),
C = concentration, by volume (fraction),
D = pipe diameter, and
V_{avg} = average flow velocity.

The drag coefficient C_D in Eq. (17–25) is estimated from the terminal velocity of fall by

$$\frac{1}{\sqrt{C_D}} = \frac{V_T}{\sqrt{\frac{4}{3} g D_E \left(\frac{\rho_p}{62.4} - 1\right)}}, \quad (17\text{--}26)$$

where D_E is the particle equivalent diameter.

Figure 17–8 shows the pressure loss (in millimeters of clear water) as a function of average velocity (in meters per second). Sedimentation occurs to

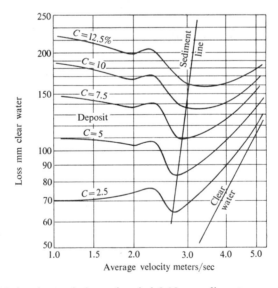

Fig. 17–8 Friction in sand slurry (graded 2.08-mm diameter; up to 12.5 percent concentration) in a 6-in. pipe. [17] (Used with permission of the American Society of Mechanical Engineers.)

the left of the sediment line. The Durand equation, Eq. (17–23), estimates the pressure loss at the minimum value. Higher pressure losses (higher velocities) are necessary to insure against sedimentation.

The use of the Durand equation for air-solid flows is discussed in Stepenoff [20]. A more theoretical treatment and complete survey of the field of gas-solid flows is given by Soo [21].

Heat Transfer to Particles

When there is a temperature difference between a spherical particle and an infinite surrounding fluid, heat is transferred by convection. Drake suggested that the Nusselt number can be predicted by

$$\overline{\mathrm{Nu}} = 2.0 + 0.459(\mathrm{Re})^{0.55}(\mathrm{Pr})^{0.33}, \tag{17-27}$$

for $1 < \mathrm{Re} < 70{,}000$, and $0.6 < \mathrm{Pr} < 400$.

For very small values of Reynolds number ($\mathrm{Re} < 0.1$), the Nusselt number is seen to be

$$\overline{\mathrm{Nu}} = \frac{hD}{k} = 2.0, \tag{17-28}$$

which is the value for pure conduction to a quiescent fluid.

Analogous to the heat transfer equation of Eq. (17–27), Froessling (Ref. [10] in Chap. 16) found for small drops of liquids and solid spheres the following mass transfer equation for the Sherwood number:

$$\mathrm{Sh} = \frac{h_D D}{D_c} = 2.0 + 0.552(\mathrm{Re})^{0.5}(\mathrm{Sc})^{0.33}, \tag{17-29}$$

where h_D is the mass transfer coefficient, D_c the diffusivity, and Sc the Schmidt number (ν/D_c).

17–5 FLOW THROUGH POROUS MEDIA

The question of flow behavior of gases and liquids through porous media has always attracted a considerable amount of attention among engineers. The most intensive interest with petroleum engineers has probably been in connection with the removal of crude oil from underground porous sedimentary formations such as sandstone, limestone, and dolomite. Chemical engineers have an interest in packed beds of granular particles used in chemical processes and in flows through filter cakes. The civil engineer is interested in the flow of water through sand filters in sanitary installations or through soil formations underlying dams. More recently, aerospace and mechanical engineers have become interested in the flow of gases through porous propellants and the transpiration cooling of porous-wall reentry vehicles and rocket nozzles.

In general there are two types of porous media: consolidated and unconsolidated. A *consolidated porous media* is simply a solid with holes in it. Examples of consolidated porous media are sandstone, limestone, and a loaf of bread. *Unconsolidated porous media* are those made up of separate solid particles. Examples are beach sand, tobacco in a cigarette, and a tube filled with loose gravel.

One of the most important characteristics of a porous medium is the porosity. The *total porosity* is defined as the ratio of void volume to total volume. It can be calculated if one knows the apparent density of the porous medium, ρ_a, and the density of the pore-free solid, ρ_s:

$$\text{total porosity} = 1 - \frac{\rho_a}{\rho_s}. \tag{17-30}$$

However, in many consolidated porous media the voids may not necessarily be interconnected. Thus, some void spaces are not effective in allowing fluid to flow through the porous medium. In these cases it is more convenient to deal with an effective porosity. The *effective porosity*, ϵ, is the ratio of the interconnected void volume to the total volume. Some materials like lava have a high total porosity but a small effective porosity. For unconsolidated porous media the total porosity and effective porosity are equal.

Another characteristic of porous media is the mean pore diameter, δ. This is in reality an idealized characteristic because no single dimension can describe the size or geometric shape of the void spaces. The *mean pore diameter* is the mean diameter the voids would have if they were transformed to short circular tubes having the same effective cross-sectional area as the actual porous medium.

Mathematical models have been used to predict the flow behavior in porous media. The simplest model that has been used to depict the porous medium is the bundle of capillary tubes. See, for example, Scheidegger [23]. While the capillary model is of some value for low-speed flow through porous media, it falls short of adequately predicting high-speed flow through porous media. High-speed flow through porous media means, in general, flows for which the Reynolds number, based on local velocity in the pores and on mean pore diameter, is greater than unity.

The capillary-orifice model of Blick [24] is an attempt to extend the usefulness of the capillary model to the high-speed regime. The capillary-orifice model assumes that the porous media is a bundle of capillary tubes with orifice plates spaced along the tube, one mean pore diameter apart.

For the flow of a fluid through a porous medium, the conservation equations for one-dimensional flow [24, 25] can be written as follows.

Conservation of mass:

$$\epsilon \frac{\partial \rho}{\partial t} + \frac{\partial \rho v}{\partial x} = 0 \tag{17-31}$$

Conservation of momentum:

$$-\frac{\partial p}{\partial x} = \frac{\mu}{K}v + \rho\beta v^2 + \frac{\rho v}{\epsilon^2}\frac{\partial v}{\partial x} - \rho g \cos\theta \quad (17\text{-}32)$$

Conservation of energy:

fluid: $\rho c_p \left(\frac{\partial T}{\partial t} + \frac{v}{\epsilon}\frac{\partial T}{\partial x}\right) = k\frac{\partial^2 T}{\partial x^2} + \frac{\partial p}{\partial t}$

$$+ \frac{v}{\epsilon}\frac{\partial p}{\partial x} + \mu\Phi + \frac{hS}{\epsilon}(T_s - T) \quad (17\text{-}33)$$

solid: $\quad hS(T - T_s) + k_s(1 - \epsilon)\frac{\partial^2 T_s}{\partial x^2} = \rho_s c_s(1 - \epsilon)\frac{\partial T_s}{\partial t}, \quad (17\text{-}34)$

where
    T = fluid temperature,
    T_s = solid temperature,
    k = fluid thermal conductivity,
    k_s = effective solid thermal conductivity,
    S = specific internal area (per unit volume),
    ρ = fluid density,
    ρ_s = solid density,
    c_s = solid specific heat,
    h = heat transfer coefficient,
    K = permeability of the porous medium,
    β = inertial resistance coefficient,
    v = superficial or filter velocity (volume flow per unit cross-sectional area of the solid plus fluid),
    θ = angle between x direction and direction of gravity, g, and
    $\mu\Phi$ = energy dissipated per unit volume = $(v/\epsilon)[(\mu/K)v + \beta\rho v^2]$.

The *permeability*, K, of a porous medium is one of its most useful properties. In the low velocity region it is a direct measure of the ease with which a fluid will flow through the medium. The capillary model predicts K by the following equation:

$$K = \frac{\delta^2 \epsilon}{32}. \quad (17\text{-}35)$$

Thus, large mean pore diameters and large porosities are associated with large permeabilities. Equation (17-35) for K is somewhat inaccurate, but it is useful for making estimates of K. For accurate values of the permeability one must resort to experiments. It can be seen from Eq. (17-35) that the permeability has the units of (length)^2. Quite often petroleum and civil engineers will use the Darcy unit in describing the permeability of a porous medium. The *Darcy* is defined as the permeability of a porous medium which will allow a single-phase fluid of one centipoise viscosity to flow through it under conditions of viscous flow at the rate of 1 cubic centimeter per second per square centimeter of cross-sectional area under a pressure gradient of 1 atmosphere per centimeter.

The inertial resistance coefficient β can be estimated from the following equation derived from the capillary-orifice model [24]:

$$\beta = \frac{\rho C_D}{2\delta\epsilon^2}, \qquad (17\text{--}36)$$

where C_D is the drag coefficient of the orifice plate. The drag coefficient can be estimated by assuming it is equivalent to the drag coefficient of a circular disk at the same Reynolds number based on the mean pore diameter, δ, and mean pore velocity, v/ϵ (see Fig. 9–15).

For many cases of engineering interest the velocity is low enough in the porous medium so that the equation for the conservation of momentum for steady flow, Eq. (17–32), can be simplified to

$$v = -(K/\mu)\left(\frac{dp}{dx} - \rho g \cos\theta\right). \qquad (17\text{--}37)$$

This equation is called the *modified Darcy equation*. (The *Darcy equation* is simply Eq. (17–37) with the gravity term omitted.) The Darcy and modified Darcy equations have been widely used for predicting velocities and flow rates through porous media.

This section has presented only the bare fundamentals of flow through porous media. Many of the more advanced topics, such as two-phase flow and rarified flow through a porous medium, will not be considered in this volume.

17-6 HEAT TRANSFER AT CRYOGENIC TEMPERATURES

The word *cryogenic* comes from the Greek word meaning "icy cold." It is used today to describe the production of very low temperatures and the effects created by low temperatures. More exactly, *cryogenic temperature* is usually taken to be any temperature below $-150°C$ ($-238°F$). Substances used to obtain or maintain these temperatures are referred to as *cryogens*. These are normally liquified gases. Common cryogens are liquified nitrogen, oxygen, hydrogen, and helium. The properties of these and other cryogens are given in Table 17–3. Their normal boiling points are shown in Fig. 17–9.

Cryogens have become particularly important in recent years due to their use in space propulsion systems, in medicine and surgery, and in studies of basic physical phenomena at low temperatures. The storage of liquified gases, particularly natural gas, has become commercially important, primarily due

Table 17-3
PROPERTIES OF CRYOGENS‡

	Normal boiling point (at 1 atm) °F	Normal melting point (at 1 atm) °F	Liquid density (at nbp)† lb/ft³	Vapor density (at nbp)† lb/ft³	Vapor density (at 70°F, 1 atm) lb/ft³	Enthalpy of vaporization (at nbp)† Btu/lb	Critical temperature °F	Critical pressure psia
Helium 4	−452.1	−453.5*	7.80	1.06	0.01035	9.9	−450.2	33.2
Hydrogen	−423.0	−434.5	4.43	0.0803	0.0522	194.2	−399.8	188
Deuterium	−417.3	−426.0	10.7	0.16	0.01041	134	−390.8	239
Neon	−410.7	−415.4	74.9	0.593	0.0522	39.4	−379.7	395
Nitrogen	−320.5	−345.9	50.4	0.2756	0.0725	85.9	−232.8	492
Carbon monoxide	−312.6	−337.1	50.6	0.312	0.0725	92.8	−218.2	509
Fluorine	−306.2	−363.2	94.5	0.537	0.0989	77.7	−200	808
Argon	−302.6	−308.7	87.4	0.37	0.1033	70.2	−187.6	706
Oxygen	−297.3	−361.8	71.2	0.296	0.0828	93.5	−181.8	730
Methane	−258.6	−296.4	26.3	0.11	0.0416	219	−116.5	673
Krypton	−244.0	−250.9	135.0	0.52	0.2149	46.8	−81.4	720
Nitric oxide	−241.0	−262.6	79.3	0.204	0.0777	199	−137.2	945

* At 1514 psia.   † nbp = normal boiling point.
‡ Courtesy *Oil and Gas Equipment*. [26] Used with permission.

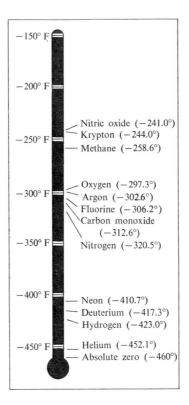

Fig. 17-9 Comparison of normal boiling points of cryogens. (Courtesy *Oil and Gas Equipment*. [26] Used with permission.)

to the fact that the liquid phase occupies much less volume per pound than the gaseous phase. Natural gas can now be stored easily in liquid form near large population centers to serve as reserve capacity in peak demand period. Natural gas in liquid form is being transported in increasing quantities and is having a significant effect on some commercial markets.

Many cryogens can be treated as ordinary liquids and many of the equations which have been derived in this text are applicable. In this section the problems that are more or less unique to heat transfer at cryogenic temperatures will be discussed.

One problem of dealing with solid substances at cryogenic temperatures is that of large property variations. Figure 17-10 shows the variation of the thermal conductivity with temperature for several metals at cryogenic temperatures [27]. This large variation complicates the analytical solution of heat transfer problems and forces the engineer to utilize numerical methods in his computations.

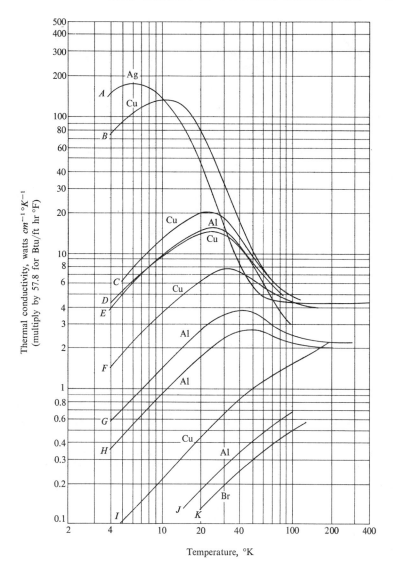

Fig. 17-10 Low temperature thermal conductivity of metals. *A*—silver 99.999 percent pure; *B*—high-purity copper; *C*—coalesced copper; *D*—copper, electrolytic tough pitch; *E*—aluminum single crystal; *F*—free-matching tellurium copper; *G*—aluminum, 1100°F; *H*—aluminum, 6063-T5; *I*—copper, phosphorus deoxidized; *J*—aluminum, 2024-T4; *K*—free-machining leaded brass. (From Scott's *Cryogenic Engineering*. Copyright 1959, D. Van Nostrand Company, Inc., Princeton, N.J.)

In convective heat transfer problems with most cryogens, the major difficulty appears to occur due to the fact that the thermodynamic critical point of the fluid may be approached or exceeded in many cases. This is a region of large property variations and for any fluid this makes the computations for flow and heat transfer difficult.

The cryogens are always in a hostile environment, since by definition they are much colder than their surroundings. Boiling occurs very readily due to this large temperature difference. Film boiling is a common occurrence particularly in cooldown of equipment. Two-phase flow and the accompanying unsteadiness can be expected in many situations. The boiling correlations for noncryogens appear to be marginally suitable for initial design work with cryogens; however, for close predictions the experiments with cryogens should be utilized.

A cryogen near its boiling point can be cooled by bubbling a noncondensable foreign gas up through it. This is called *injection cooling*. The cryogen evaporates into the gas bubble, absorbing energy from the surrounding fluid and thus lowering its temperature. This is an effective means of subcooling a saturated liquid to suppress unwanted boiling [28].

Frost formation and ice buildup are common occurrences with these substances and should not be overlooked. Many gases, particularly water vapor, condense and often freeze when brought in contact with the container or pipe which has a cryogen inside.

The only cryogen which behaves in an unusual manner, as regards its fluid flow and heat transfer characteristics, is a form of normal helium, helium II. Below a temperature of about $2.2°K$, referred to as the *lambda point*, the fluid has very unusual characteristics, such as superconductivity and superfluidity, that is, extremely large value of thermal conductivity and extremely low value of viscosity. This behavior, not fully understood, presents the physicist and engineer with a challenging area of work.

The storage of liquified gases at low temperatures has resulted in a demand for insulations with much lower thermal conductances than ordinary insulations. This has led to the development of new insulation materials and techniques. The most significant result of this effort has been the development of *superinsulations* with apparent thermal conductivities as low as 2×10^{-4} Btu/hr ft °F, two orders of magnitude below most insulations [37]. The superinsulations consist of many layers, as many as 50 typically, of highly reflective material separated by low conductivity spacers of some type and evacuated to low pressures (less than 10^{-4} mm of Hg). With the superinsulations cryogens such as hydrogen can be shipped by railway or truck tanker over long distances with small losses. Other important insulations for cryogenic service include expanded foams, powders such as perlite, and evacuated spaces. For purposes of comparison, Figure 17–11 shows the apparent thermal conductivities of typical cryogenic insulation materials.

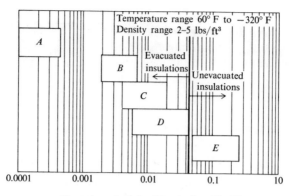

Fig. 17-11 Apparent thermal conductivities of typical cryogenic insulation material. *A*—multilayer insulations; *B*—opacified powders; *C*—glass fibers; *D*—powders; *E*—foams, powders, and fibers. [37] (Used with permission of the American Society of Mechanical Engineers.)

A number of texts on cryogenic engineering have been published and can be consulted for more specific details [27, 29, 30, 31].

17-7 MAGNETOHYDRODYNAMICS

Magnetohydrodynamics is the study of the interactions of electromagnetic fields with the velocity fields of electrically conducting fluids, such as an ionized gas or mercury. Interest in this field has been spurred by developments in nuclear energy and reentry aerodynamics. Other names given to this field of study include *electrodynamics, plasma physics, plasma dynamics, magnetogasdynamics, magnetoaerodynamics,* and *magnetofluidmechanics.* Even though the last name seems more logical, the term magnetohydrodynamics seems to be more common. It has been shortened to MHD.

Consider the flow of an electrically conducting fluid, shown in Fig. 17-12. A magnetic field, **B**, is applied in the positive *y* direction perpendicular to the flow field **V**, which is in the *x* direction. The side walls are electrically

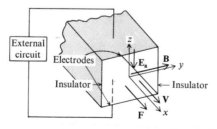

Fig. 17-12 Flow of a conducting fluid in a magnetic field.

insulated but permeable to the magnetic field, while the top and bottom walls act as electrodes. It is assumed that the external circuit applies an electric field of strength $\mathbf{E}_a$ in the negative z direction.

The interaction between the magnetic field $\mathbf{B}$ and the velocity field $\mathbf{V}$ of the conducting fluid induces an electric field $\mathbf{E}_i$, equal to the cross product of $\mathbf{V}$ and $\mathbf{B}$:

$$\mathbf{E}_i = \mathbf{V} \times \mathbf{B}. \quad (17\text{-}38)$$

Using the "right-hand screw rule" for cross products, the induced electric field is directed in the positive z direction. Therefore, the net electric field $\mathbf{E}$ is the sum of the applied electric field $-\mathbf{E}_a$ and the induced field $\mathbf{E}_i$:

$$\mathbf{E} = \mathbf{E}_i - \mathbf{E}_a = \mathbf{V} \times \mathbf{B} - \mathbf{E}_a. \quad (17\text{-}39)$$

Under the influence of a strong magnetic field the electrical conductivity becomes a tensor since the gas becomes anisotropic, having properties which depend on direction. For simplicity, assuming that the conducting fluid remains isotropic under the influence of the magnetic field, the electrical conductivity of the fluid can be labeled by a scalar quantity, σ. The current $\mathbf{j}$ through the fluid is given by Ohm's law:

$$\mathbf{j} = \sigma \mathbf{E} = \sigma(\mathbf{V} \times \mathbf{B} - \mathbf{E}_a). \quad (17\text{-}40)$$

In solid conductors a body force, or *Lorentz force* $\mathbf{F}$, is induced by the current cutting the lines of the magnetic field. The same thing happens in a flowing conducting fluid where the force is given by

$$\mathbf{F} = \mathbf{j} \times \mathbf{B} = \sigma(\mathbf{V} \times \mathbf{B} - \mathbf{E}_a) \times \mathbf{B}. \quad (17\text{-}41)$$

For the example shown in Fig. 17–12, the force will be in the positive or negative x direction depending on the relative magnitudes of $\mathbf{E}_a$ and $\mathbf{V} \times \mathbf{B}$.

If the external circuit were a simple electrical load, e.g., a resistance, then the applied field $\mathbf{E}_a$ would be in the negative z direction and smaller in magnitude than $\mathbf{V} \times \mathbf{B}$. Then, from Eq. (17–41), the Lorentz force would be in the negative x direction and would tend to decelerate the fluid. In addition, from Eq. (17–40), the net current flow is in the positive z direction. This fluid deceleration can therefore be used for power generation. Advantages of such a system over the conventional power system are the elimination of the moving parts associated with a rotating armature and the high theoretical thermal efficiencies associated with the high-temperature MHD generators.

If, however, the external circuit provided an electric field, $\mathbf{E}_a$, greater than $\mathbf{V} \times \mathbf{B}$, then the Lorentz force would be in the positive x direction, which would accelerate the fluid. Such a device is known as an accelerator or an electromagnetic propulsion engine.

Other applications of MHD devices include the pinching of hot plasma away from the walls of nuclear fusion devices, as drag-increasing and aerodynamic cooling devices during the reentry of hypersonic vehicles into the atmosphere and as flow meters.

The conservation equations for MHD include the effects of the Lorentz force and ohmic or joule heating. For the device shown in Fig. 17-12, the one-dimensional conservation equations for steady inviscid flow may be written as

Mass:
$$\rho A V = \text{constant.} \qquad (17\text{-}42)$$

Momentum:
$$\rho V \frac{dV}{dx} = -\frac{dp}{dx} + (\mathbf{j} \times \mathbf{B})_x \qquad (17\text{-}43a)$$

or
$$\rho V \frac{dV}{dx} = -\frac{dp}{dx} + \sigma(E_a - VB)B. \qquad (17\text{-}43b)$$

Energy:
$$\rho V \left[c_p \frac{dT}{dx} + V \frac{dV}{dx} \right] = \mathbf{j} \cdot \mathbf{E}_a \qquad (17\text{-}44a)$$

or
$$\rho V \left[c_p \frac{dT}{dx} + V \frac{dV}{dx} \right] = \sigma(E_a - VB)E_a. \qquad (17\text{-}44b)$$

From Eqs. (17-43) and (17-44) it is seen that the sign of the term $(E_a - VB)$ determines whether the flow is accelerated or decelerated and whether the stagnation enthalpy increases or decreases.

No attempt will be made here to solve the above set of MHD equations due to the number of assumptions and qualifications that would have to be made in order to obtain a tractable solution. Rather than present a complete treatise on the subject, the intent in this section has been to show some of the basic concepts and conservation equations of MHD.

17-8 FLOW IN OPEN CHANNELS

Man has long been interested in the problems involved with flow through ditches, canals, sewer lines, and rivers. These are all examples of the general area of open-channel flows. Open-channel flows can be classified in different ways, such as laminar or turbulent, steady or unsteady, uniform or varied (Fig. 17-13), or subcritical or supercritical.

Most open-channel flows of practical interest are turbulent flows. Exceptions are the laminar flows found in the drainage of streets, airport runways, etc. Quite often open-channel flows are unsteady, but the time variations of the flow variables are so slow that the flow can be treated as steady.

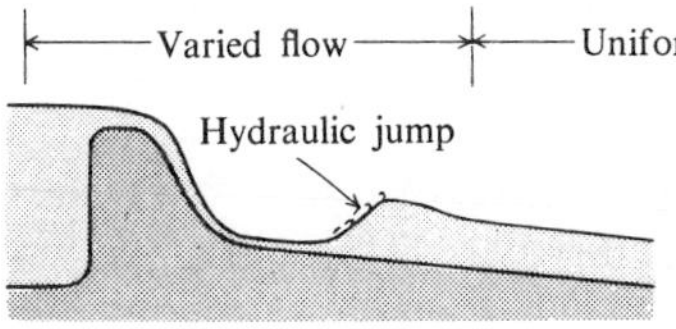

Fig. 17-13 Types of open-channel flow.

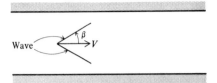

Fig. 17-14 Surface wave in supercritical flow.

The terms subcritical and supercritical are somewhat analogous to subsonic and supersonic flow of compressible fluids. A *subcritical flow* is one in which small surface waves can propagate upstream. A *supercritical* flow is one in which such waves will not propagate upstream but will be swept downstream. Referring to Fig. 17-14, the angle between the wave and flow direction (β) is found by the following equation:

$$\beta = \sin^{-1}\left(\frac{\sqrt{gy}}{V}\right), \qquad (17\text{-}45)$$

where y = depth of fluid, g = local acceleration of gravity, and V = flow velocity. The wave angle β is analogous to the angle of the Mach cone, Eq. (11-10), for a compressible fluid. The term $\sqrt{gy}$ is the velocity of propagation of a small surface wave, hence the term $V/\sqrt{gy}$ in Eq. (17-45) is analogous to the Mach number for a compressible fluid. This analogy has some usefulness in studies of compressible flow by use of a *water table*, a device in which a shallow open channel flow is created in the laboratory.

A flow is subcritical or supercritical if the depth of the fluid in the channel is greater than or less than the *critical depth*, respectively. The critical depth is given by the following equation:

$$y_c = (\dot{Q}^2/g)^{1/3}, \qquad (17\text{-}46)$$

where $\dot{Q}$ = volume flow rate per unit channel width, usually given in ft³/sec ft.

A *hydraulic jump* is a rapid increase in depth of flow (see Fig. 17-13). In all such instances the flow upstream is supercritical and the flow downstream is subcritical. The hydraulic jump is analogous to the normal shock wave in a compressible fluid flow.

Uniform and varied flow are illustrated in Fig. 17-13. *Uniform flow* is a flow in which there is no fluid acceleration, and *varied flow* is a flow in which there are fluid accelerations. Consider the uniform flow of a liquid in an open channel (Fig. 17-15). For steady uniform flow the velocity will not change in the flow direction, hence, from conservation of mass, the cross-sectional flow area will not vary. This means that the free surface will be parallel to the channel floor in uniform flow.

17-8

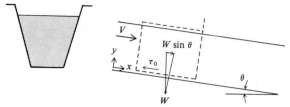

Fig. 17-15 Uniform flow in an open channel.

It is desirable to predict the shear stress-velocity relationships in this situation so that flow velocities might be estimated for various types of open channels. Applying the momentum theorem to the control volume in Fig. 17-15, with P the wetted perimeter, the following is obtained:

$$W \sin \theta - \tau_0 P \, dx = 0. \tag{17-47}$$

The velocities and pressures have been neglected in Eq. (17-47) because they do not change in the flow direction. The slope S is equal to $\tan \theta$, and for the small slopes encountered in practice, $\tan \theta = \sin \theta$. The weight of the liquid in the control volume $W = \rho g A \, dx$. Solving for τ_0,

$$\tau_0 = \rho g \left(\frac{A}{P}\right) S = \rho g R_H S, \tag{17-48}$$

where $R_H = A/P$ is the *hydraulic radius*. Just as in pipe flow a friction factor can be defined by

$$\tau_0 = \frac{f \rho V^2}{8}. \tag{17-49}$$

Solving for the velocity from Eqs. (17-48) and (17-49),

$$V = C\sqrt{R_H S},$$

where

$$C = \sqrt{\frac{8g}{f}}. \tag{17-50}$$

C is called the *Chézy coefficient* and Eq. (17-50) is known as the *Chézy equation* after the French engineer, Antoine Chézy, who established this equation in 1775.

One of the simplest relations that can be used to determine the Chézy coefficient is the *Manning formula* [32]:

$$C = \frac{1.486 R_H^{1/6}}{n},$$

where n is Manning's roughness factor and some typical values are listed in Table 17-4 [33].

Table 17-4
VALUES OF MANNING'S ROUGHNESS FACTOR n*

Type of channel	n
Well-planed timber	0.009
Neat cement plaster	0.010
Cement plaster, smooth iron pipe	0.011
Unplaned timber, ordinary iron pipes	0.012
Best brickwork, new sewer pipe	0.013
Average brickwork, foul iron pipe, foul planks	0.015
Unfinished concrete, poor brickwork	0.017
Canals with firm gravel	0.020
Ordinary earth canals free from large stones and heavy weeds	0.025
Canals and rivers with many stones and weeds	0.03–0.04

* From R. Manning, "On the Flow of Water in Open Channels and Pipes," *Trans. Inst. Civil Engrs. Ireland*, 1891.

17-9 RAREFIED GAS DYNAMICS

All of the equations considered so far in this text have been based upon the assumption that the material under consideration could be treated as a continuum. This assumption leads to reasonable results when the molecules are spaced at relatively close distances. When enough molecules are in the vicinity of a point in space, then collisions and energy exchange between the molecules will occur frequently and a condition of local equilibrium will exist at that point. This permits the assignment of local property values (such as pressure and temperature) to each point. The assumption of no fluid slip at a solid-liquid interface is based on the concept of continuum behavior, as is the assumption that the fluid next to a solid surface has the same temperature as the surface.

In those cases where molecules are relatively far apart, the molecules may travel large distances between collisions and thus may not be in equilibrium with the other molecules in the vicinity. The *molecular mean free path* is the mean distance that a group of molecules moves between collisions. The number of molecules in the group must be sufficiently large to give a good statistical average. The symbol λ will be used for the molecular mean free path, thus

$$\lambda = \frac{\text{total distance traveled by } N \text{ molecules between collisions}}{N}. \quad (17\text{-}51)$$

The mean free path λ of the gas particles has been determined by Jeans [38]:

$$\lambda = \frac{1}{\sqrt{2\pi n d^2}}, \quad (17\text{-}52)$$

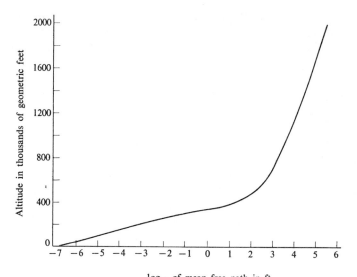

Fig. 17-16 Mean free path vs. altitude. (From R. A. Minzner et al., "The ARDC Model Atmosphere," Air Force Geophysics Survey 115, 1959.)

where $n = \rho/m$ = number of gas particles per unit volume (ρ = volume mass density and m = mean mass of gas particles) and d = mean diameter of gas particles. Thus, the mean free path varies inversely with particle density and the characteristic dimension of the particle.

The mean free path can also be expressed [39] in terms of the more measurable quantities, speed of sound a, kinematic viscosity ν, and the ratio of specific heats λ:

$$\lambda = 1.255\sqrt{\gamma}(\nu/a). \tag{17-53}$$

Figure 17-16 contains values of mean free path of air. Thus, for low altitudes, the mean free path is truly negligible in comparison with the body dimensions, and therefore the formulas of gas dynamics are sufficient for aerodynamic calculations. However, at extremely high altitudes, where the pressure may be only one-millionth of the pressure at the surface of the earth, the mean free path will be comparable to the body dimensions.

At first it might appear that when the mean free path becomes large, the number of molecules present is negligible. However, when the mean free path is 10 feet, there are still approximately 10^{13} molecules in a cubic inch. Aerodynamic forces will still exist, but intermolecular collisions will play little part in the underlying momentum and energy transfer processes.

The ratio of the molecular mean free path of a characteristic length of the object or flow field is a dimensionless quantity which is given the name *Knudsen number*.

$$\text{Kn} = \frac{\lambda}{l}. \tag{17-54}$$

The Knudsen number will be shown to be an important parameter in the study of rarefied flows.

In any collection of gas molecules there are molecules with different velocities present. As these molecules collide they change speed and direction. If a large number of molecules are in equilibrium, their velocities are distributed in a Maxwellian distribution, named after James Clerk Maxwell, who derived the relationship in 1859.

Gas molecules with a specified distribution of velocities may be studied by *kinetic theory*. In a simple theory, gas molecules are considered to have elastic impacts and to travel in straight lines between impacts. The temperature of the gas at a point is fixed by the average kinetic energy of the molecules around that point. The pressure is determined by the impact (and resulting momentum change) of molecules striking a surface.

Two important results from kinetic energy will now be utilized. The first of these is a prediction of the *mean molecular velocity*, $\bar{v}_M$. According to simple kinetic theory,

$$\bar{v}_M = \sqrt{\frac{8R_g T}{\pi}}. \tag{17-55}$$

The second important relationship from kinetic theory is a prediction of the viscosity of a gas. According to the theory,

$$\mu = C\rho\lambda\bar{v}_M, \tag{17-56}$$

where C is a constant. It can also be shown that the product $(\rho\lambda)$ is independent of pressure. Using Eqs. (17-55) and (11-8), the Mach number can be shown to be

$$M = \frac{V}{a} = \frac{V}{\bar{v}_M}\sqrt{\frac{8}{\gamma\pi}}. \tag{17-57}$$

The Reynolds number can be combined with Eq. (17-56) to give

$$Re = \frac{V\rho l}{\mu} = \frac{Vl}{C\bar{v}_M \lambda}. \tag{17-58}$$

The ratio of the Mach number to the Reynolds number is, therefore, given by

$$\frac{M}{Re} = C\sqrt{\frac{8}{\gamma\pi}}\frac{\lambda}{l} = C\sqrt{\frac{8}{\gamma\pi}}\, Kn. \tag{17-59}$$

Thus, for a given ratio of specific heats, the Knudsen number can be considered to be a constant times the ratio of the Mach number to the Reynolds number. The constant C is approximately equal to 0.5. Equation (17-59) can be rewritten as

$$Kn = \sqrt{\frac{\gamma\pi}{2}}\frac{M}{Re}. \tag{17-60}$$

In flows at moderately high Reynolds number, boundary layer assumptions are valid and the boundary layer thickness δ may be selected as the characteristic length of importance. The Knudsen number for such cases is then defined by

$$\text{Kn} = \frac{\lambda}{\delta} \quad \text{(boundary layer flow)}. \tag{17-61}$$

It has been shown that the boundary layer thickness at the rear of a flat plate is proportional to the square root of the Reynolds number, Re_L:

$$\frac{\delta}{L} \sim \frac{1}{\sqrt{\text{Re}_L}}. \tag{17-62}$$

Thus,

$$\text{Kn} = \frac{\lambda}{\delta} \sim \frac{\lambda \sqrt{\text{Re}_L}}{L} \sim \frac{1}{C}\sqrt{\frac{\gamma\pi}{8}} \frac{M}{\text{Re}_L}\sqrt{\text{Re}_L} \sim \frac{M}{\sqrt{\text{Re}_L}}. \tag{17-63}$$

Therefore, in flow at high Reynolds numbers (boundary layer flow) the Knudsen number is proportional to the Mach number divided by the square root of the Reynolds number based on length. This relationship is assumed to be valid for objects of arbitrary shape. It has been found in such cases that continuum conditions exist when

$$\frac{M}{\sqrt{\text{Re}_L}} < 0.01. \tag{17-64}$$

Under continuum conditions a molecule in the boundary layer will most likely undergo numerous collisions with other molecules before striking the wall.

At relatively small values of Reynolds number, boundary layer flow does not exist and the characteristic length of importance in the Knudsen number is the length related to the size of the solid object. Thus,

$$\text{Kn} = \frac{\lambda}{L} \sim \frac{M}{\text{Re}_L}.$$

It has been found in such cases that a *free molecular flow* condition exists when

$$\frac{M}{\text{Re}_L} > 10. \tag{17-65}$$

In free molecular flow, molecular collisions most likely occur at large distances from the solid surface. A molecule in the vicinity of the surface and traveling toward the surface is more likely to strike the surface than to collide with another molecule.

In between the ranges of continuum flow and free molecular flow are two regions referred to as the slip flow regime and the transition regime.

Slip flow is the name given to the flow condition near a wall for which the mean free path λ is small but not negligible compared to the body dimension L or the thickness δ of the boundary layer. If, on the average, a gas molecule does not acquire the momentum of the wall at one collision, "slip" is said to exist. Therefore, in the case of slip, the molecules near the boundary will have a mean velocity different from the velocity of the wall.

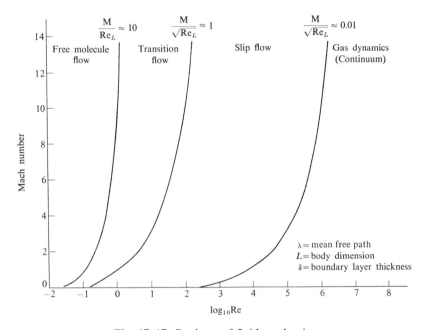

Fig. 17-17 Regimes of fluid mechanics.

The interval $1/100 < (M/\sqrt{\mathrm{Re}_L}) < 1$ is considered to be the range for the slip flow. This realm of fluid mechanics occupies the region shown in Fig. 17-17. The region to the right of this slip flow region belongs to the realm of conventional gas dynamics, in which the usual boundary conditions exist at the wall.

The region between the free molecule flow and the slip flow is called *transition flow*. This is the realm of fluid mechanics where the collision between molecules and the collision of molecules with the wall are of equal importance. The related problems are extremely complicated, and very little theoretical or experimental data are available in this regime. However, it is certain that as far as the fluid itself is concerned, the characteristic parameters are still the Mach number M and the Reynolds number Re as used in conventional gas dynamics.

Techniques based on the kinetic theory of gases have been developed for calculating pressure forces, skin friction forces, and heat transfer in the free molecule regime. The techniques require some knowledge of the types of molecular reflection from the surface (specular or diffuse) and of the amount of energy the molecules transfer when they collide with the surface.

REFERENCES

1. *Liquid Metals Handbook*, Atomic Energy Commission and U.S. Navy (1952 and later revisions).
2. Ralph Stein, "Liquid Metal Heat Transfer," *Advances in Heat Transfer*, Vol. III, Academic Press (1966).
3. E. M. Sparrow and J. L. Gregg, *J. Aeronautical Sci.* **24**, 852 (1957).
4. R. J. Grosh and R. D. Cess, *Trans. ASME* **80**, 667 (1958).
5. H. A. Johnson et al., *Trans. ASME* **75**, 1191 (1953).
6. R. C. Martinelli, *Trans. ASME* **69**, 947 (1947).
7. R. N. Lyon, *Chem. Eng. Progress* **47**, 75 (1951).
8. R. A. Seban and T. T. Shimazaki, *Trans. ASME* **73**, 803 (1951).
9. R. G. Deissler, *Trans. ASME* **73**, 101 (1951); NACA Rept. 1210 (1955); NACA TN 3145 (1954).
10. B. Lubarsky and S. J. Kaufman, NACA Rept. 1270 (1956).
11. C. A. Sleicher, Jr. and M. Tribus, *Trans. ASME* **79**, 789 (1957).
11a. O. E. Dwyer, "Recent Developments in Liquid Metal Heat Transfer," *Atomic Energy Rev.* **4**, 3 (1966).
12. P. A. Longwell, *Mechanics of Fluid Flow*, McGraw-Hill, New York (1966).
13. A. B. Metzner, "Heat Transfer in Non-Newtonian Flow," *Advances in Heat Transfer*, Vol. II, Academic Press (1965).
14. D. D. Fuller, "Theory and Practice of Lubrication for Engineers," Wiley, New York (1956).
15. M. C. Shaw and E. F. Macks, *Analysis and Lubrication of Bearings*, McGraw-Hill, New York (1949).
16. J. I. Clower, *Lubricants and Lubrication*, McGraw-Hill, New York (1939).
17. A. J. Stephanoff, "Pumping Solid-Liquid Mixtures," *Mech. Eng.* (Sept. 1964), p. 29.
18. M. R. Tek, *Handbook of Fluid Dynamics*, ed. by V. L. Streeter, McGraw-Hill, New York (1961), pp. 17–31.

19. G. G. Brown et al., *Unit Operations*, Wiley, New York (1950).
20. A. J. Stepanoff, *Pumps and Blowers—Two-Phase Flow*, Wiley, New York (1964).
21. S. L. Soo, *Fluid Dynamics of Multiphase Systems*, Blaisdell Publishing Co., Waltham, Mass. (1967).
22. W. E. Ranz and W. R. Marshall, Jr., *Chem. Eng. Progress*, **48**, 141 (1952).
23. A. E. Scheidegger, *Physics of Flow Through Porous Media*, University of Toronto Press, Toronto (1960).
24. E. F. Blick, "Capillary-Orifice Model for High-Speed Flow through Porous Media," *I & EC Process Design and Development*, **5**, 90 (1966).
25. E. F. Blick, "High Speed Flow Through Porous Media," PhD Dissertation University of Oklahoma (1963).
26. J. D. Parker, "The Basics of Cryogenics," *Oil and Gas Equipment Technical Manual*, Petroleum Publishing Co., Tulsa (1967).
27. R. B. Scott, *Cryogenic Engineering*, Van Nostrand, New York (1959).
28. P. S. Larsen et al., "Cooling of Cryogenic Liquids by Gas Injection," *Advances in Cryogenic Engineering*, Vol. 8 (1963).
29. R. W. Vance, *Cryogenic Technology*, Wiley, New York (1963).
30. K. D. Timmerhaus, ed., *Advances in Cryogenic Engineering*, Plenum Press, New York (annual).
31. R. Barron, *Cryogenic Systems*, McGraw-Hill, New York (1967).
32. R. Manning, "On the Flow of Water in Open Channels and Pipes," *Trans. Inst. Civil Engrs. Ireland* **20**, 161 (1891).
33. A. T. Ippen, "Mechanics of Liquids," *Mark's Mechanical Engineers' Handbook*, 6th ed., McGraw-Hill, New York (1958), pp. 3–81.
34. A. B. Cambel, "Plasma Physics and Magnetofluid Mechanics," McGraw-Hill, New York (1963).
35. W. F. Hughes and F. S. Young, "Electromagnetodynamics of Fluids," Wiley, New York (1966).
36. W. R. Sears, ed., "Magneto Fluid Dynamics," AIAA Selected Reprints, Vol. 2, American Institute of Aeronautics and Astronautics, New York (1967).
37. P. E. Glaser, I. A. Black, and P. Doherty, "Multilayer Insulation," *Mech. Eng.*, 23 (Aug. 1965).
38. J. Jeans, *The Dynamical Theory of Gases*, 4th ed., Dover, New York (1954).
39. R. A. Minzner et al., "The ARDC Model Atmosphere," Air Force Geophysics Survey 115, (1959).
40. Hsue-Shen Tsien, "Superaerodynamics, Mechanics of Rarefied Gases," *J. Aeronautical Sci.* **13**, 653 (1946).
41. G. N. Patterson, *Molecular Flow of Gases*, Wiley, New York (1956).

42. E. F. Blick, "Forces on Bodies of Revolution in Free Molecular Flow by the Newtonian-Diffuse Method," McDonnell Aircraft Corp., Report 6670 (1959).
43. R. W. Truitt, *Fundamentals of Aerodynamic Heating*, Ronald Press, New York (1960).
44. J. R. Stalder and V. J. Zurick, "Theoretical Aerodynamic Characteristics of Bodies in a Free-Molecular-Flow Field," NACA TN 2423 (1951).
45. W. A. Gustafson, "The Newtonian Diffuse Method for Computing Aerodynamic Forces," Lockheed Aircraft Corp. Missile Systems Div., Tech. Memo. 5132 (1958).
46. A. K. Oppenheim, *J. Aeronautical Sci.* **20,** 49 (1953).

APPENDIX 1

LIST OF SYMBOLS AND ABBREVIATIONS

Composite Symbols

A_p	Projected area or profile area
Bo	Biot number, hL/k_s, where k_s is the thermal conductivity of the solid
Btu	British thermal unit
Ca	Cauchy number, V^2/a^2
C_D	Coefficient of drag, $\text{drag}/\frac{1}{2}\rho V_\infty^2 A$
Cv	Cavitation number
C_f	Coefficient of friction, $\tau_w/\frac{1}{2}\rho V_\infty^2$
°C	Degrees centigrade
D_c	Diffusion coefficient
D_E	Equivalent diameter
D_H	Hydraulic diameter
Ek	Eckert number, $V_\infty^2/c_p \Delta T$
Eu	Euler number, $\Delta p/\rho V^2$
Fo	Fourier number, $\alpha t/L^2$
Fr	Froude number, V^2/gL
°F	Degrees Fahrenheit
F_t	Correction factor for LMTD
Gr	Grashof number, $(g\beta \Delta T L^3)/\nu^2$
Gz	Graetz number, $(\text{RePr}) \div L/D$
i.d.	Inside diameter
IPTS	International practical temperature scale
KE	Kinetic energy per unit mass
K_L	Empirical loss coefficient
Kn	Knudsen number λ/L
°K	Degrees Kelvin
Le	Lewis number, α/D_c
LMTD	Logarithmic mean temperature difference
Nu	Nusselt number, hL/k
NTU	Number of transfer units, $AU/C_{\min}$, for constant U
Δp_E	Pressure drop in entrance region
Pr	Prandtl number, $\mu c_p/k$
Pé	Péclet number, RePr

PE	Potential energy per unit mass
$\dot{Q}$	Volume flow rate
R_g	Specific gas constant
Re	Reynolds number, $\rho VL/\mu$
R_H	Hydraulic radius
°R	Degrees Rankine
$\dot{R}$	$(dr/dt)_{r=R}$
$\ddot{R}$	$(d^2r/dt^2)_{r=R}$
Sh	Sherwood number, $h_D L/D_c$
St	Stanton number, $h/V\rho c_p$
Sk	Stokes number, $\Delta pL/\mu V$
Sc	Schmidt number, v/D_c
SI	International System of Units
Sl	Strouhal number, L/tV
$\dot{W}$	Work rate
We	Weber number, $\rho V^2 L/\sigma$
WP	Wetted perimeter
X_F	Flowing mass quality
X_S	Static mass quality
c_p	Specific heat at constant pressure
c_v	Specific heat at constant volume
c.s.	Control surface
cgs	Centimeter-gram-second
c.v.	Control volume
div	Divergence
f_{TB}	Friction factor for tube banks
ft	Feet
g_c	Dimensional constant
g_0	Standard gravitational acceleration
h_D	Mass transfer coefficient
in.	Inch
h_{fg}	Enthalpy of vaporization
lb_f	Pounds force
lb_m	Pounds mass
mks	Meter kilogram second
$\dot{m}$	Mass flow rate
mp	Melting point
m_a	Mass transfer rate of component a per unit area
$\dot{q}$	Heat flow rate
$\dot{q}'$	Heat flow rate per unit length, $\dot{q}/L$
$\dot{q}''$	Heat flux, heat flow rate per unit area, $\dot{q}/A$
$\dot{q}'''$	Energy generation rate per unit volume

LIST OF SYMBOLS AND ABBREVIATIONS

$\dot{\mathbf{q}}_c$	Heat conduction flux vector
$\dot{\mathbf{q}}_r$	Heat radiation flux vector
y^*	A dimensionless distance
ϵ_H	Eddy diffusivity of heat
ϵ_M	Eddy diffusivity of momentum
ψ^*	A dimensionless stream function
Vol	Volume

English Letters

A	Area
B	Magnetic field strength
B	Bulk modulus
B	An extensive property
C	Flow stream capacity rate, $\dot{m}c$
C	Volume concentration
C	Chézy coefficient
C	Capacitance
D	Diameter
E	Young's modulus
E	Total emissive power
E	Electric field strength
E	Electrical potential
F	Force
F	The dimension force
F	Radiation configuration factor
G	Mass velocity, mass per unit area
G	Total irradiation
I	Electrical current
I	Radiant intensity
J	Conversion factor, 778.28 ft lb$_f$/Btu
J	Radiosity
K	Mass absorption coefficient
K	Compressibility
K	Constant for non-Newtonian behavior
K	Permeability
L	The length of an object or a characteristic length
L	The dimension length
M	A parameter in conduction $= (\Delta x)^2/\alpha \, \Delta T$
M	Mach number, V/a
M	The dimension mass
N	Frequency
N	A parameter in fin equations, $\sqrt{hP/kA}$
N	Molal diffusion rate

P	A point in space
P	Heat exchanger parameter, an effectiveness
P	Electrical power
Q	Volume flow
Q	The dimension charge
R	Reaction
R	Heat exchanger parameter, ratio of fluid capacity rates
R	Radius
R	Resistance
R	Rth component of body force per unit volume
S	Shape factor
T	Temperature
T	The dimension temperature
U	Overall coefficient of heat transfer
V	Velocity
V	Electrical voltage
W	Work
W	Molecular weight
X	x component of body force per unit volume
Y	y component of body force per unit volume
Z	z component of body force per unit volume
a	Acceleration
a	Speed of sound
b	A specific property
c	Specific heat
e	Specific internal energy
e	Fin efficiency
f	Force
f	Moody friction factor, or dimensionless pressure gradient, $\dfrac{-\left(\dfrac{dp}{dz}\right)\dfrac{2}{D}}{\rho V_{\text{avg}}^2}$
f	A function
g	Local gravitational acceleration
h	Head of fluid
h	Height or distance
h	Heat transfer coefficient
h	Planck's constant
i	Number of fundamental dimensions
j	Specific enthalpy
j	Electric current
k	Thermal conductivity
k	The reduction factor in dimensional analysis
m	Mass
m	Number of variables

m	Poisson's ratio
m	Manning roughness factor
n	Direction normal to a line or surface
n	Number of π groups, $n = m - k$
n	Exponent for pseudoplastic fluids
p	Pressure
q	Heat flow
$\dot{r}$	Rate of production per unit volume
r	Radial distance from axis or origin
r	Recovery factor
s	Distance
s	Specific entropy
t	Time
t	The dimension time
u	x coordinate component of velocity
v	y coordinate component of velocity
w	z coordinate component of velocity
w	Width of fin
x,y,z	Coordinate distances

Greek Letters

α	Linear coefficient of thermal expansion
α	Thermal diffusivity, $k/\rho c_p$
α	Void fraction
α	Absorptance
α	Kinetic energy correction factor
β	Inertial resistance coefficient
β	Thermal expansion coefficient
γ	Ratio of specific heats, c_p/c_v
δ	Boundary layer thickness
δ	Condensate thickness
δ	Deflection angle
ϵ	Heat exchanger effectiveness
ϵ	Effective porosity
ϵ	Emittance
θ	Angle
θ	Coordinate of body force per unit volume in θ direction
λ	Mean free path
λ	Wavelength
μ	Absolute viscosity or viscosity coefficient
μ	Micron, 10^{-6} meters
μ	Micro, or one millionth
ν	Kinematic viscosity, μ/ρ

ρ	Density
ρ	Reflectance
σ	Surface tension
σ	Stefan-Boltzmann constant
τ	Stress
τ	Optical distance
τ	Transmittance
τ	An arbitrary variable
ω	Mass fraction of species
φ	Angle
φ	Component of body force per unit volume in φ direction
Φ	The potential function
Φ	The viscous dissipation function
ψ	The stream function

Common Subscripts

atm	Atmospheric
avg	Average
AW	Adiabatic wall
max	Maximum
min	Minimum
sat	Saturated
sub	Subcooled
sys	System
B	Bulk
B	Blackbody
D	Dynamic
G	Gas
G	Gray body
M	Mean
T	Isothermal
T	Thermal
T	Total
S	Isentropic
W	Wall
c	Cold fluid
e	Electrical
f	Film
f	Friction
g	Gas
h	Hot fluid
i	Inside
i	Inlet

l	Liquid
m	Model
n	Normal
o	Outside
o	Outlet
p	Prototype
r	Radiant
r	Relative
s	Solid
s	Surface
s	Sunlight
t	Thermal
t	Tangent
v	Vapor
w	Wall
x	Based on distance x
x	In the x direction
y	In the y direction
z	In the z direction
∞	Free undisturbed stream value
λ	Monochromatic
0	Initial

Special Symbols

$*$	Dimensionless quantity
$'$	Fluctuating quantity
—	Time average value
$\cdot$	Per unit time or rate
∇	Del operator
∇^2	Laplacian operator
$\triangleq$	Has the dimension of or has the units of
$\dfrac{D}{Dt}$	Substantial derivative
$\star$	Critical condition, where $M = 1$
$\star R$	Critical condition, Rayleigh flow
$\star F$	Critical condition, Fanno flow
$\mathscr{R}$	Universal gas constant

APPENDIX 2

$g_c = 32.2 \dfrac{lb_m \cdot ft}{lb_f \cdot s^2}$ *(handwritten)*

USEFUL CONVERSION FACTORS

Length

1 foot = 30.48 cm
1 inch = 2.54 cm
1 meter = 39.37 inches
1 micron = 10^{-6} meters = 3.281×10^{-6} feet
1 mile = 5280 feet

Volume

1 U.S. gallon (liquid) = 231 cubic inches = 0.1337 cubic feet
1 barrel (oil) = 42 gallons
1 barrel (U.S. liquid) = 31.5 gallons
1 liter = 1000.028 cubic centimeters = 0.26417762 gallons (U.S.)
1 British gallon = 1.20094 U.S. gallons

Mass

1 kilogram = 2.20462 pounds mass
1 pound mass = 7000 grains = 453.5924 grams
1 slug = 32.17 pounds mass

Force

1 dyne = 2.248×10^{-6} pounds force
1 poundal = 0.031081 pounds force

Energy

1 Btu = 778.28 foot-pounds force
1 kilocalorie = 10^3 calories = 3.968 Btu
1 joule = 9.47×10^{-4} Btu = 10^7 ergs = 0.73756 foot-pounds force
1 erg = 1 dyne centimeter
1 kilowatt-hour = 3413 Btu = 2.6552×10^6 foot-pounds force = 3.6709×10^5 kilogram-meters
1 horsepower-hour = 2545 Btu = 0.7457 kilowatt-hours = 1.98×10^6 foot-pounds force

Pressure

1 atmosphere = 14.6959 pounds per square inch absolute = 2116 pounds force per square foot = 1.01325×10^6 dynes per square centimeter

$1 \, Pa = 1.451 \times 10^{-4} \dfrac{lb_f}{in^2}$ *(handwritten)*

USEFUL CONVERSION FACTORS 579

Temperature

1°R difference = 1°F difference = $\frac{5}{9}$°C difference = $\frac{5}{9}$°K difference
degrees Fahrenheit = $\frac{9}{5}$ (degrees Centigrade) + 32

Thermal Conductivity

$$1\,\frac{\text{Btu}}{\text{hr ft °F}} = 0.004134\,\frac{\text{cal}}{\text{sec cm °C}} = 0.01731\,\frac{\text{watts}}{\text{cm °C}}$$

$$1\,\frac{\text{cal}}{\text{sec cm °C}} = 241.9\,\frac{\text{Btu}}{\text{hr ft °F}} = 360\,\frac{\text{kg cal}}{\text{hr meter °C}}$$

Viscosity, absolute

1 poise = 100 centipoises = 1 gram per second centimeter

$$1\,\frac{\text{lb mass}}{\text{sec ft}} = 1490\text{ centipoises} = 0.0311\,\frac{\text{lb force sec}}{\text{ft}^2}$$

$$= 5350\,\frac{\text{kg}}{\text{hr meter}}$$

$$1\,\frac{\text{lb force sec}}{\text{ft}^2} = 47{,}800\text{ centipoises} = 172{,}000\,\frac{\text{kg}}{\text{hr meter}}$$

Specific Heat

$$1\,\frac{\text{cal}}{\text{gm °C}} = 1\,\frac{\text{Btu}}{\text{lb mass °F}}$$

$$1\text{ hp} = 33{,}000\,\frac{\text{ft lb}}{\text{min}} = 550\,\frac{\text{ft lb}}{\text{s}} = 746\text{ watts}$$

APPENDIX 3

THERMOPHYSICAL PROPERTIES

The following pages contain thermophysical properties of various substances important in engineering practice. Although great care was used in assembling and calculating these data, they are intended primarily to furnish approximate values for use by students and professors in the classroom. The data are reliable enough to be used for making actual engineering approximations, but where very precise values are desired or where great reliability must be placed on the values of the properties used, original sources of property data should be utilized.

The authors have deliberately presented the property data in a variety of ways and different sets of units have been utilized. For example, both curves and tables are used to present data. Viscosity is presented in mass units in one appendix and in force units in another, and both seconds and hours are utilized. It is hoped that this will give the student useful experience in obtaining property values.

The user is warned to be careful in location of decimal points. For example, where the coefficient of thermal expansion is given by $\beta \times 10^4$, the value read from the table must be multiplied by 10^{-4}.

Properties of gases are given at one atmosphere. These satisfactorily approximate property values at moderate pressures, with the exception of values of density and kinematic viscosity.

Properties of liquids not at saturation conditions can be approximated by the saturation values at the desired temperature providing the pressure is not too much greater than the saturation pressure. This small variation of properties with pressure in a liquid can be seen in the water data in Table A3–2, Properties of Steam and Water.

Table A3-1

STANDARD ATMOSPHERE TABLE
Based on ICAO and NACA Standards

	h, ft	T, °F	p, lb$_f$/ft^2	$\rho \times 10^3$, slugs/ft^3	a, ft/sec	$\mu \times 10^7$, slug/ft sec
Troposphere	0	59.0	2116	2.377	1117	3.74
	1,000	55.4	2041	2.308	1113	3.72
	2,000	51.9	1968	2.241	1109	3.70
	3,000	48.3	1897	2.175	1105	3.68
	4,000	44.7	1828	2.111	1101	3.66
	5,000	41.2	1761	2.048	1098	3.64
	6,000	37.6	1696	1.987	1094	3.62
	7,000	34.0	1633	1.927	1090	3.60
	8,000	30.5	1572	1.868	1086	3.58
	9,000	26.9	1513	1.811	1082	3.56
	10,000	23.3	1455	1.755	1078	3.54
	15,000	5.5	1194	1.496	1058	3.43
	20,000	−12.3	972	1.266	1037	3.32
	25,000	−30.1	785	1.065	1016	3.21
	30,000	−48.0	628	0.889	995	3.10
	35,000	−65.8	498	0.737	973	2.99
	36,089	−69.7	473	0.706	968	2.96
Stratosphere	40,000	−69.7	392	0.585	968	2.96
	45,000	−69.7	308	0.460	968	2.96
	50,000	−69.7	242	0.362	968	2.96
	60,000	−69.7	150	0.224	968	2.96
	70,000	−69.7	93.7	0.140	968	2.96
	80,000	−69.7	58.1	0.087	968	2.96

APPENDIX 3

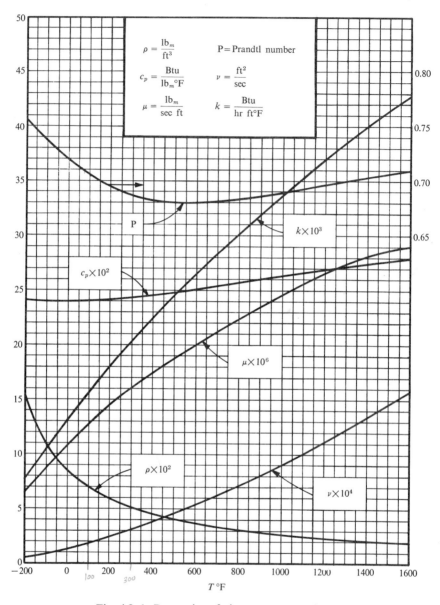

Fig. A3–1 Properties of air at one atmosphere.

THERMOPHYSICAL PROPERTIES 583

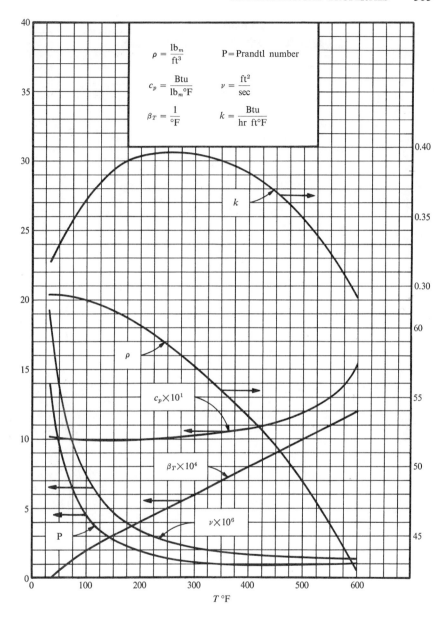

Fig. A3–2 Properties of saturated water.

Table A3-2
PROPERTIES OF STEAM AND WATER

Temperature °F		Pressure, psia					
		1	10	100	1000	5000	7500
32	c_p	1.007	1.007	1.007	0.999	0.969	0.953
	μ	366.1	366.1	366.1	365.3	361.9	359.8
	k	328.6	328.6	328.9	331.9	344.1	350.8
	Pr	13.0	13.0	12.97	12.73	11.80	11.31
	ρ	62.42	62.42	62.42	62.62	63.3	65.4
100 Liquid	c_p	0.998	0.998	0.998	0.994	0.979	0.972
	μ	142.0	142.0	142.0	142.1	142.3	142.5
	k	363.3	363.3	363.6	366.6	378.3	384.1
	Pr	4.52	4.52	4.51	4.46	4.27	4.18
	ρ	62.00	62.00	62.00	62.19	62.9	64.5
Vapor 200	c_p	0.453	0.475	1.005	1.002	0.989	0.982
	μ	2.5	2.5	62.6	62.9	64.0	64.7
	k	13.8	14.0	391.8	394.4	404.9	410.6
	Pr	0.94	0.97	1.86	1.85	1.81	1.79
	ρ	0.00255	0.0257	60.13	60.13	61.0	62.9
400	c_p	0.464	0.470	0.436	1.072	1.040	1.025
	μ	3.4	3.4	3.4	27.6	28.9	29.8
	k	19.4	19.4	20.0	384.9	398.6	406.4
	Pr	0.95	0.96	1.05	0.89	0.87	0.87
	ρ	0.00195	0.0196	0.203	53.91	55.2	57.5
600	c_p	0.479	0.481	0.499	0.888	1.253	1.170
	μ	4.4	4.4	4.4	4.4	19.1	20.0
	k	26.0	26.1	26.4	34.1	333.7	349.3
	Pr	0.93	0.93	0.95	1.32	0.83	0.78
	ρ	0.00158	0.0159	0.161	1.95	45.7	50.5
800	c_p	0.497	0.497	0.505	0.605	2.419	1.794
	μ	5.3	5.3	5.3	5.5	8.4	
	k	33.6	33.6	33.9	37.9	129.6	223.2
	Pr	0.91	0.91	0.92	1.01	2.68	1.29
	ρ		0.0133	0.134	1.45	16.9	31.4

Table A3-2 (continued)

Temperature °F		Pressure, psia					
		1	10	100	1000	5000	7500
1000	c_p	0.515	0.515	0.519	0.566	0.969	1.314
	μ	6.3	6.3	6.3	6.4	7.4	8.5
	k	41.7	41.8	42.1	45.5	70.7	102.9
	Pr	0.89	0.89	0.90	0.92	1.16	1.28
	ρ		0.0115	0.116	1.21	7.62	13.6
1500	c_p	0.559	0.559	0.561	0.580	0.668	0.726
	μ	8.6	8.6	8.6	8.7	9.2	9.6
	k	63.7	63.7	64.0	67.1	82.0	92.2
	Pr	0.88	0.88	0.88	0.87	0.87	0.88
	ρ		0.00857	0.858	0.867	4.54	6.98
Saturated liquid	c_p	0.998	1.004	1.039	1.286		
	μ	139.4	65.3	34.3	19.7		
	k	364.0	390.4	394.7	327.6		
	Pr	4.4	1.9	1.0	0.9		
	ρ	61.97	60.27	56.37	46.32		
Saturated vapor	c_p	0.450	0.475	0.582	1.191		
	μ	2.0	2.4	3.0	4.0		
	k	11.6	13.8	18.4	36.5		
	Pr	0.9	1.0	1.1	1.5		
	ρ	0.00300	0.0260	0.226	2.24		
Saturation temperature °F		101.74	193.21	327.82	544.58		

c_p Btu/lb$_m$ °F
μ (lb$_f$ sec/ft^2) × 10^7
k (Btu/hr ft °F) × 10^3
Pr dimensionless
ρ lb$_m$/ft^3

Compiled from *The 1967 ASME Steam Tables*, used with the permission of The American Society of Mechanical Engineers.

Table A3–3 SATURATION PRESSURE AND ENTHALPY OF VAPORIZATION OF WATER

Temperature, °F	p_{sat}, psia	h_{fg}, Btu/lb$_m$
32.018	0.08865	1075.5
50.0	0.17796	1065.3
100	0.94924	1037.1
200	11.526	977.9
212	14.696	970.3
300	67.005	910.0
400	247.259	825.9
500	680.86	714.3
600	1543.2	550.6
700	3094.3	172.7
705.47	3208.2	0

Table A3–4 PROPERTIES OF SODIUM

Saturated Liquid

Temperature °F	ρ $\dfrac{lb_m}{ft^3}$	c_p $\dfrac{Btu}{lb_m \, °F}$	μ $\dfrac{lb_m}{ft \, hr}$	k $\dfrac{Btu}{hr \, ft \, °F}$	α $\dfrac{ft^2}{hr}$	Pr	σ $\dfrac{lb_f}{ft}$	$\beta \times 10^4$ $\dfrac{1}{°F}$
210	57.8	0.331	1.67	50.5	2.64	0.0109	0.0135	1.39
300	57.3	0.325	1.33	48.9	2.63	0.00886	0.0131	1.42
600	54.8	0.310	0.804	43.8	2.58	0.00570	0.0120	1.50
900	52.3	0.302	0.588	39.1	2.48	0.00454	0.0109	1.59
1200	49.8	0.300	0.472	34.8	2.33	0.00407	0.00972	1.69
1500	47.2	0.304	0.400	30.8	2.15	0.00393	0.00858	1.77
1800	44.6	0.314	0.350	27.3	1.95	0.00403	0.00743	1.88
2100	42.2	0.330	0.314	24.1	1.73	0.00430	0.00629	1.98
2400	39.7	0.352	0.286	21.3	1.52	0.00474	0.00515	2.09

Saturated Vapor

Temperature °F	h_{fg} $\dfrac{Btu}{lb_m}$	ρ $\dfrac{lb_m}{ft^3}$	c_p $\dfrac{Btu}{lb_m \, °F}$	μ $\dfrac{lb_m}{ft \, hr}$	k $\dfrac{Btu}{hr \, ft \, °F}$	α $\dfrac{ft^2}{hr}$	Pr
500	1901	0.117×10^{-6}	0.349	0.0382	0.0158	0.387×10^6	0.846
800	1849	0.242×10^{-4}	0.522	0.0407	0.0240	0.188×10^5	0.888
1100	1785	0.613×10^{-3}	0.623	0.0433	0.0304	0.794×10^2	0.888
1400	1717	0.529×10^{-2}	0.638	0.0457	0.0350	0.103×10^2	0.833
1700	1648	0.242×10^{-1}	0.613	0.0482	0.0380	0.225×10	0.779
2000	1578	0.746×10^{-1}	0.587	0.0507	0.0391	0.892	0.760
2300	1498	0.178	0.571	0.0531	0.0385	0.377	0.786

Table A3–5
APPROXIMATE VALUE OF THERMOPHYSICAL PROPERTIES OF SOME NONMETALLIC SOLIDS

	Temperature °F	k $\dfrac{\text{Btu}}{\text{hr ft °F}}$	ρ $\dfrac{\text{lb}_m}{\text{ft}^3}$	c_p $\dfrac{\text{Btu}}{\text{lb}_m \text{°F}}$	α $\dfrac{\text{ft}^2}{\text{hr}}$
Acrylic (transparent Plexiglas)	70	0.112	74	0.365	0.00415
Asbestos (36 lb/ft³)	70	0.092	36	0.195	0.0013
Brick (masonry)	70	0.30	112	0.20	0.013
Bakelite	70	0.134	80	0.38	0.0044
Clay	70	0.74	90	0.21	0.039
Coal (anthracite)	70	0.15	85	0.30	0.0059
Concrete	70	0.54	140	0.21	0.021
Cork board	70	0.025	10	0.4	0.0063
Diatomaceous earth	70	0.040	14	0.21	0.014
Felt, hair	70	0.025	8.2		
Fiber glass laminates					
Silicone	200	0.085	115	0.24	0.0031
Polyester	200	0.080	110	0.27	0.0027
Phenolic	200	0.070	98	0.25	0.0029
Glass					
Silica (fused quartz)	70	0.88	137	0.174	0.037
Borosilicate crown	70	0.72	136	0.182	0.029
Soda-lime	70	0.54	154	0.189	0.019
Pyrex	70	0.68	138	0.18	0.027
Ice	−150	2.04	57	0.33	0.11
	32	1.28	57	0.49	0.045
Magnesia (85%)	70	0.033	17		
Marble	70	1.6	160	0.193	0.057
Nylon	70	0.14	71	0.4	0.0049
Rock wool	70	0.023	8		
Rubber					
Hard	70	0.106	70	0.4	0.0038
Natural	70	0.085	60	0.5	0.0028
Neoprene	70	0.121	76.8	0.4	0.0039
Sandstone (dry)	70	1.1	140	0.17	0.046
Santocel	70	0.013	5.3	0.2	0.012
Teflon	70	0.14	135	0.4	0.0026
Wood (typical)	70	0.10	35	0.6	0.0048

Note: Properties change markedly with moisture, compaction, manufacturer, and in some cases with direction.

Table A3-6
REPRESENTATIVE THERMOPHYSICAL PROPERTIES OF SEVERAL LIQUIDS AT NEAR SATURATION PRESSURES

Liquid	Temperature °F	ρ $\frac{lb_m}{ft^3}$	c_p $\frac{Btu}{lb_m \cdot °F}$	μ $\frac{lb_m}{hr \cdot ft}$	k $\frac{Btu}{hr \cdot ft \cdot °F}$	$\nu \times 10^3$ $\frac{ft^2}{hr}$	Pr	$\beta \times 10^4$ $\frac{1}{°F}$	h_{fg} $\frac{Btu}{lb_m}$
Ammonia	−50	43.5	1.00	0.714	0.316	16.4	2.26	9.6	604
	0	41.3	1.08	0.612	0.315	14.8	2.10	10.8	569
	50	39.0	1.11	0.559	0.306	14.3	2.03	12.6	527
	100	36.4	1.16	0.484	0.285	13.3	1.97	15.4	478
Bismuth	600	625	0.0345	3.92	9.5	6.26	0.0142	0.65	
	1000	608	0.0369	2.67	9.0	4.39	0.0110	0.70	
	1400	591	0.0393	1.91	9.0	3.23	0.0083		
Carbon Dioxide	−50	71	0.50	0.327	0.054	4.61	3.03	18.8	140
	0	64	0.53	0.284	0.066	4.43	2.28	26.8	117
	50	54	0.79	0.212	0.056	3.93	2.99	49.1	84
Ethyl alcohol	68	49.4	0.680	2.91	0.105	58.9	18.8		
Ethylene glycol 100%	20	70.7		203		2870			
	100	68.8	0.58	25.1	0.159	364	91.6		
Aqueous solution 50% by weight	20	67.2		26.7		297			
	100	65.8	0.81	7.65	0.242	116	25.6		
	200	63.4	0.86	0.58	0.237	9.15	2.10		
Freon 12 CCl_2F_2	−20	93.0	0.214	0.900	0.040	9.68	4.82	1.03	
	20	89.2	0.220	0.756	0.042	8.48	3.96	1.34	
	60	83.0	0.231	0.648	0.042	7.81	3.56	2.1	
	100	78.5	0.240	0.576	0.040	7.33	3.46	2.5	
Gasoline Mil-F-5572	−50	47.3		2.78		58.8			
	0	45.8	0.465	1.74	0.0815	38.0	9.92		
	100	42.7	0.526	0.93	0.0789	21.8	6.20		156
Glycerin	50	79.3	0.554	9216	0.165	116000	31000	2.71	
	100	78.1	0.600	677	0.163	8660	2490	2.77	

THERMOPHYSICAL PROPERTIES

Fluid	T (°F)	ρ	cp	C	D	E	F	G	H
Hydraulic fluid Mil-H-5606	0	55.0	0.410			3660			
	100	52.5	0.467			443		4.0	
	200	50.0	0.523			221			
Hydrogen (para-hydrogen)	−430	4.61	1.86	20.1		9.65	1.30		155.7
	−410	3.71	3.76	2.32		6.25	1.11		193.6
Isopropyl alcohol	100	48	0.66	0.0445	0.0681	113	40.6		152
Jet fluid JP-4	0	49.9	0.441	0.0232	0.0591	78.2	21.1		
	100	47.2	0.502	5.44	0.0636	33.2	9.99		
Lithium	400	31.5	1.04	3.90	0.0785	45.7	0.0576		
	800	30.5	1.00	1.57	0.0885	40.0	0.0550		
Mercury	0	852	0.033	1.44	0.0815	5.17	0.0364		
	300	823	0.033	1.22	0.0789	3.10	0.0124	1.0	
	600	800	0.032	4.41	26.0	2.54	0.0079		
Methane (CH_4)	−300	28.5	0.80	2.55	22.2	16.6	4.53		186
	−200	23.2	0.87	2.03	4.0	8.36	1.94		
	−120	14.3	—	0.532	6.8	4.04	—		
Methyl alcohol	68	50.5	0.601	0.194	8.2	28.5	7.15		
Nitrogen	−340	53.5	0.478	0.058	0.094	11.1	3.16		90.3
	−300	46.6	0.505	1.44	0.087	5.49	1.86		78.2
	−260	38.3	0.617	0.594	0.051	4.73	2.41		57.2
	−240	30.7	0.717	0.256	0.121	5.54	3.77		34.0
Oil (light)	100	56.0	0.46	0.181	0.0898	982	3.33		
	200	54.0	0.51	0.170	0.0694	167	61.2	4.0	
	300	51.8	0.54	55.0	0.0463	58	22.2		
Oxygen	−360	81.3	0.401	9.0	0.0324	22.9	6.84		106.0
	−300	71.4	0.405	3.0	0.076	6.61	2.21		92.2
	−240	59.4	0.438	1.86	0.074	4.60	1.96		73.2
	−200	47.6	0.556	0.472	0.073	5.02	3.18		46.3
Potassium	200	50.6	0.192	0.273	0.109	21.3	0.00781		
	600	47.8	0.187	0.239	0.0866	11.4	0.00428		
	1000	44.0	0.179	1.079	0.0612	8.43	0.00313		
				0.545	0.0418				
				0.371	26.5				
					23.8				
					21.2				

589

Table A3-7
REPRESENTATIVE VALUES OF THERMOPHYSICAL PROPERTIES OF PURE METALS*

	Temperature °F	k $\dfrac{\text{Btu}}{\text{hr ft °F}}$	ρ $\dfrac{\text{lb}_m}{\text{ft}^3}$	c_p $\dfrac{\text{Btu}}{\text{lb}_m \text{ °F}}$	α $\dfrac{\text{ft}^2}{\text{hr}}$	Melting point °F
Aluminum	−200	144		0.212		1220
	100	137	169	0.225	3.60	
Beryllium	100	95	117	0.46	1.77	2340
Bismuth	100	4.5	612	0.03	0.245	520
Cadmium	100	53	540	0.056	1.75	610
Chromium	100	52	448	0.12	0.967	3435
Cobalt	100	75	550	0.10	1.36	2715
Copper	100	221	558	0.094	4.21	1981
	600	208		0.098		
Gold	100	171	1206	0.031	4.57	1945
Iron	0	47	491	0.10	0.957	2802
	200	39		0.11		
	1000	24		0.16		
Lead	100	20	705	0.031	0.915	621
Lithium	100	46	33	0.85	1.64	354
Magnesium	100	78	109	0.24	2.98	1200
Molybdenum	100	87	638	0.065	2.10	4760
Nickel	100	40	556	0.10	0.719	2651
Platinum	100	41	1339	0.031	0.987	3224
Silver	100	235	653	0.06	6.00	1761
	600	208				
Sodium	200	50	58	0.33	2.61	208
Tin	200	37	456	0.055	1.48	449
Titanium	100	11	290	0.13	0.438	3300
Tungsten	100	96	1204	0.032	2.49	6170
	600	76		0.034		
	2400	63		0.039		
Zinc	100	65	446	0.088	1.66	787

* Values of thermophysical properties change with purity, method of forming, heat treatment, and temperature.

Table A3-8
REPRESENTATIVE THERMOPHYSICAL PROPERTIES OF SOME METAL ALLOYS

	Temperature °F	k $\dfrac{\text{Btu}}{\text{hr ft °F}}$	ρ $\dfrac{\text{lb}_m}{\text{ft}^3}$	c_p $\dfrac{\text{Btu}}{\text{lb}_m \text{ °F}}$	α $\dfrac{\text{ft}^2}{\text{hr}}$
Aluminum alloy (as received)					
7075-T6	−200	51	177	0.152	1.89
	+200	79	174	0.215	2.11
2024-T4	−200	51	175	0.151	1.93
	+200	78	173	0.217	2.08
Brass and bronze					
Commercial bronze	68	109	549	0.082	2.42
Red brass	68	92	546	0.093	1.81
Cartridge brass	68	70	532	0.093	1.41
Free-cutting brass	68	67	530	0.093	1.36
Admiralty metal	68	64	532		
Aluminum bronze	68	48	510	0.098	0.96
Phosphor bronze	68	47	552		
Bearing bronze	68	100	550		
Constantan	68	12.5	557	0.098	0.228
Inconel-X	68	6.9	515	0.103	0.13
Iron (cast, 4% carbon)	68	30	454	0.10	0.660
Steel					
SAE 1095	100	34	490	0.111	0.625
SAE 1010	68	37.1	491	0.102	0.74
Stainless type 301	68	8.6	494	0.109	0.159
Stainless type 347	−250	7.1	497	0.080	0.178
	200	9.0	492	0.116	0.157
	1600	14.9	467	0.175	0.182

Table A3–9
STEEL-PIPE DIMENSIONS*

Nominal pipe size, in.	Outside diam, in.	Schedule No.	Wall thickness, in.	Inside diam, in.	Cross-sectional area metal, sq in.	Inside cross-sectional area, sq ft
$\frac{1}{4}$	0.540	40	0.088	0.364	0.125	0.00072
		80	0.119	0.302	0.157	0.00050
$\frac{3}{8}$	0.675	40	0.091	0.493	0.167	0.00133
		80	0.126	0.423	0.217	0.00098
$\frac{1}{2}$	0.840	40	0.109	0.622	0.250	0.00211
		80	0.147	0.546	0.320	0.00163
$\frac{3}{4}$	1.050	40	0.113	0.824	0.333	0.00371
		80	0.154	0.742	0.433	0.00300
1	1.315	40	0.133	1.049	0.494	0.00600
		80	0.179	0.957	0.639	0.00499
$1\frac{1}{2}$	1.900	40	0.145	1.610	0.799	0.01414
		80	0.200	1.500	1.068	0.01225
2	2.375	40	0.154	2.067	1.075	0.02330
		80	0.218	1.939	1.477	0.02050
$2\frac{1}{2}$	2.875	40	0.203	2.469	1.704	0.03322
		80	0.276	2.323	2.254	0.02942
3	3.500	40	0.216	3.068	2.228	0.05130
		80	0.300	2.900	3.016	0.04587
4	4.500	40	0.237	4.026	3.173	0.08840
		80	0.337	3.826	4.407	0.07986
5	5.563	40	0.258	5.047	4.304	0.1390
		80	0.375	4.813	6.112	0.1263
6	6.625	40	0.280	6.065	5.584	0.2006
		80	0.432	5.761	8.405	0.1810
8	8.625	40	0.322	7.981	8.396	0.3474
		80	0.500	7.625	12.76	0.3171
10	10.75	40	0.365	10.020	11.90	0.5475
		60	0.500	9.750	16.10	0.5185

* Based on A.S.A. Standards B36.10.

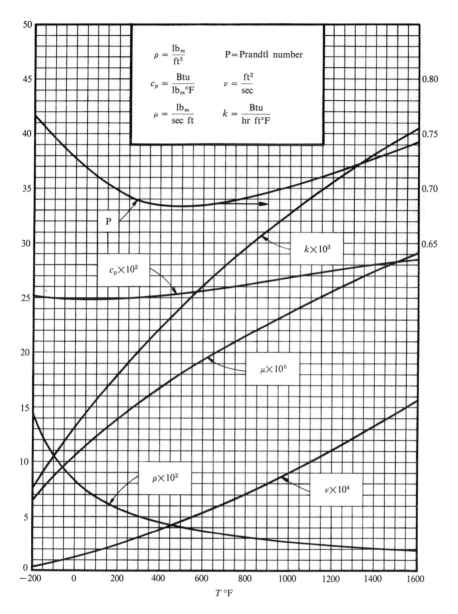

Fig. A3-3 Properties of nitrogen at one atmosphere.

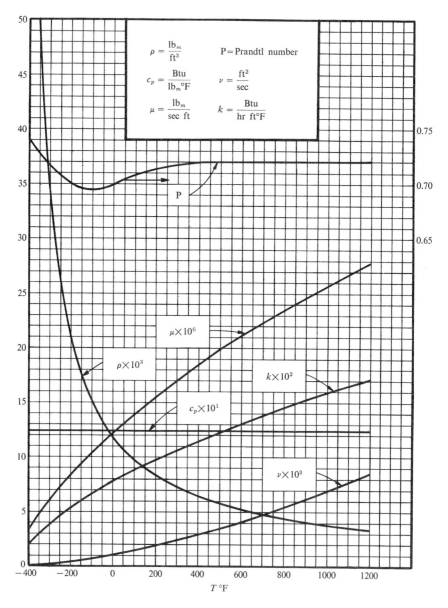

Fig. A3–4 Properties of helium at one atmosphere.

THERMOPHYSICAL PROPERTIES 595

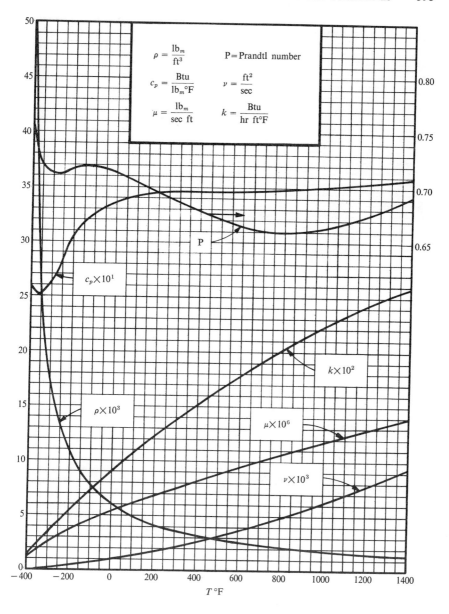

Fig. A3–5 Properties of hydrogen at one atmosphere.

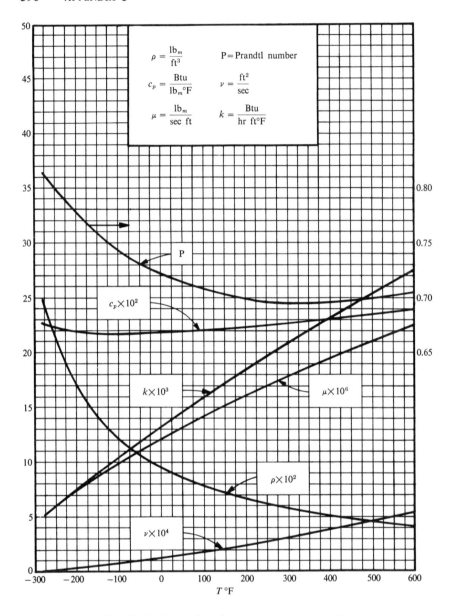

Fig. A3-6 Properties of oxygen at one atmosphere.

THERMOPHYSICAL PROPERTIES 597

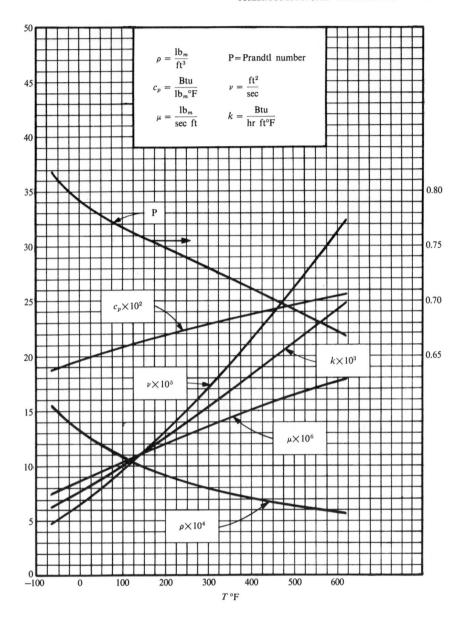

Fig. A3-7 Properties of carbon dioxide at one atmosphere.

INDEX

INDEX

Abbreviations, 571
Absolute dimensional systems, 5
Absolute temperature scale, 10
Absorptance, 484
Absorption, 483
Absorptivity, 484
Acceleration, 50
  convective, 52
  local, 52
  normal, 78
  tangential, 78
Adiabatic bulk modulus, 29
Adiabatic surface, 17
Adiabatic wall temperature, 327
Air, properties at one atmosphere, 582
Alloys, properties, 591
Analogies, 64, 212
  between electrical and thermal conduction systems, 383, 457
  between electrical and thermal radiation systems, 500
  Colburn, 228
  Reynolds, 212, 227
  von Kármán, 228
Analog computer solution to free convection problem, 293
Analytical approach to problems, 64
Angle factor, 492
Annular flow, 355
Apparent stresses, 224
Apparent thermal conductivity, 15, 289
Appendix, 571
Approximate method for boundary flow, 208, 279
Archimedes principle, 36
Arithmetic temperature difference, 420

Atmosphere, standard table, 581
Average heat transfer coefficient, 207, 208

Baker's flow regime plot, 357
Barometer, 35
Beer's law, 510
Benard cells, 288
Bernoulli equation, 147
  comparison with energy equation, 151
  for a streamline, 148
Bingham plastic, 540
Biot number, 137, 456
Blackbody, 485
Blackbody radiation functions, 488
Blasius solution, 200
Body force, 22
Boiling, 345
  film, 348, 349
  subcooled or local, 348
Boundary layer, 166
  condensation, 372
  displacement thickness, 170
  equations, 170
  free convection, 272, 274
  separation, 251
  thermal, 172
  thickness, 168, 199, 201
  transition to turbulence, 171
Bourdon-tube pressure gage, 23
British Gravitational System of Units, 6
British thermal unit, 12
Bromley's film boiling correlation, 354
Bubble agitation mechanism, 350
Bubble and droplet behavior, 337

Bubble dynamics, 337, 345
Buckingham pi theorem, 127
Buffer layer, 242
Bulk coefficient of viscosity, 58
Bulk kinetic energy, 95
Bulk modulus of elasticity, 27
   secant, 27
   tangent, 27
   water, table of values, 28
Bulk motion and diffusion, 523
Bulk property, 95
Bulk temperature, 96
Buoyancy, 36
   Archimedes principle, 36
   center of, 39
   forces, 271
Burnout, 348, 364, 365

Capillary flow, 186
Carbon dioxide, property curve, 597
Cauchy number, 137
Cavitation, 366
Cavitation number, 367
Celsius temperature scale, 10
Centigrade temperature scale, 10
Chézy coefficient, 561
Chézy equation, 561
Clausius-Clapeyron relationship, 342
Coefficient of cubical expansion, 29
Coefficient of thermal expansion, 29
Coefficient of viscosity, 58
Coefficient of volume expansion, 29
Colburn analogy, 244
Colburn equation, 245
Colburn j factor, 228, 531
Combined conduction, convection, and radiation, 506
Compact heat exchangers, 415
Compressibility, 26
   effect on drag coefficient, 304
   isentropic, 27
   isothermal, 27
Condensation, 368
   boundary layer analysis, 372
   film, 368, 369
   dropwise, 368, 376

   laminar, on horizontal tubes, 375
   mixed, 369
   Nusselt analysis, 369
Condensate Reynolds number, 372
Conduction, steady, 425
   composite slabs, 383
   convective boundaries, 444
   cylindrical shapes, 385
   extended surfaces, 444
   internal conversion present, 440
   numerical solutions, 432
   shape factor, 428
   solids, 425
   spheres, 390
Conduction, transient, 451
   cylinders, 462
   equivalent electrical circuit, 458
   Heisler charts, 458
   infinite plate, 460
   numerical methods, 466
   semi-infinite solid, 454
   spheres, 464
   time constant, 457
   two- and three-dimensional bodies, 459
Configuration factor, 492
Conformal transformation, 160
Conservation of energy, 12
   overall energy balance, 92
Conservation equations, 68
Conservation of mass, equation of, 75
   finite system, 70
   in differential form, 73
   in dimensionless form, 116
   law of, 70
   one-dimensional form, 307
Conservation of momentum, 81
Continuity equation, 75
   in cylindrical coordinates, 76
   in dimensionless form, 116
   in rectangular coordinates, 75
   in spherical coordinates, 77
Continuum, 2, 565
Control surface, 21, 68
Control volume, 68
Convected energy, 101

Convective heat transfer coefficient, 130
Convective terms, 101
Converging-diverging nozzle flow, 312
Conversion of electrical to thermal energy, 440
Conversion factors, 4, 578
Coordinate systems, 46
Correlation between fluctuating components, 223
Critical conditions in compressible flow, 310
Critical depth, 560
Critical heat flux, 348, 353
Critical pressure ratio for converging-diverging nozzles, 311
Cross flow heat exchanger, 392, 400, 405, 406
Cryogenic, 552
Cryogenic temperature, 552
Cryogens, 552
  normal boiling points, 554
  properties of, 553
Cylinder, average Nusselt number, 257, 258, 269
  drag, in crossflow, 115
  flow net, ideal flow, 154
  heat transfer by convection, 254
  Heisler chart for transient conduction, 462, 463
  local Nusselt number, 255
  pressure distribution, 156
  steady conduction, 385

Dalton's law of partial pressures, 520
Darcy, definition, 551
Darcy equation, 552
Darcy friction factor, 181
Density, 8
  stagnation or total value in isentropic flow, 305
Departure from nucleate boiling, 348, 363, 365
Deviatoric stresses, 55
Diabatic flows and effects of friction, 320

Differential optical depth, 511
Diffuse surface, 490, 567
Diffuser, definition, 307
  variation of velocity and area, 308, 309
Diffusion coefficient, 520
Diffusion, mass, 519
Diffusivity, mass, 520, 522
Dilatent fluids, 540
Dimension, definition, 3
Dimensional analysis, 64, 110
Dimensional and Unit Systems, 6
Dimensional constant, g_c, 7
Dimensionless groups, 110, 137
  interpretation, 136, 137
  local value, 125
Dimensional Homogeneity, Principle of, 5, 7, 110, 128
Dimensions of common variables, table, 112, 113
Dimensions of steel pipe, 592
Dirt factor, 387
Disk, drag coefficient curve, 252
Displacement thickness, 170
Dittus-Boelter equation, 245
Divergence, 57, 75
DNB, 348, 363, 365
Drag, 116
  form, 117
  induced, 117
  pressure, 117
  profile, 117
  skin friction, 116
  wake, 117
Drag coefficient, 119
  circular cylinders, 121
  nonspherical particles, 547
  spheres and disks, 253
Droplet behavior, 337
Dropwise condensation, 368, 376
Dryout, 363, 365
Durand equation, 548
Dynamic pressure, 152
Dynamic similarity, 133
Dynamic temperature, 153
Dynamic viscosity, 58

604 INDEX

Eckert number, 125, 137, 327
Eddy diffusivity of heat, 226
Eddy diffusivity of momentum, 225
Eddy viscosity, 225
Effectiveness, heat exchanger, 401
   comparison of flow arrangement, 407
   effect of number of shell-passes, 407
   equation for counterflow, 404
   equations for various types of exchangers, 406
     plots for various types of exchangers, 405
Effective porosity, 550
Electric field strength, 558
Electrical coils, heating, 442
Electromagnetic spectrum, 482
Elevation head, 147
Elevation pressure, 32
Emission of radiation, 483, 484
   monochromatic, 485
Emission mean absorption coefficient, 512
Emissivity, 489
Emittance, 489
Empirical equations, 64
Enclosed spaces, free convection in, 288
Energy equation, cylindrical coordinates, 102
   differential form, 97, 101
   dimensionless form, 123
   ideal gas, 109
   spherical coordinates, 102
   steady flow, 96
Engineering English System of Units 7,
Enthalpy, 94
Enthalpy of vaporization of water, 586
Entrance region, 174
   Langhaar's equation, 178
   prediction of pressure drop in laminar flow, 177
   thermal, laminar flow, 193
   transition to turbulence, 175
Equation of state, ideal gas, 26
   Van der Waals, 41
Equilibrium radius of a bubble, 342
Equimolal counterdiffusion, 522

Equivalent diameter, 546
Error function, 453, 455
Euler equations, 142, 146, 308
Euler number, 137
Evaporation, 345
   theory of boiling, 351
   in two-phase flow, 362
Extended surfaces, 444
Extensive property, 68

Fahrenheit temperature scale, 11
Fanning friction factor, 181
Fanno flow, 323
Fick's law, 519, 520, 528
Film boiling, 348, 349, 354
Film coefficient, 130
Film condensation, 368, 369
Film temperature, 194, 207
Finite difference method, 432
Finned radiator surface, 506
First law of thermodynamics, 12
   for infinitesimal system, 97
Flat plate, boundary layer flow, 166
   boundary layer thickness, 201
   drag, 203
   film temperature, 207
   integral method, 208
   temperature profiles, 206
   turbulent flow, 227
   velocity profiles, 201
Flow around bodies, 248
Flow energy, 93
Flow net, 154
Flow patterns, 355
Flow regimes, defined by Mach number, 302
Flow regimes in two-phase flow, 355
Flow separation, 251
Flow work, 93
Fluid, definition, 1
Fluid capacity rate, 395
Fluid stress, 53
   deviatoric, 55
   sign convention, 53
   stress matrix, 55
Fluid-solid flows, 544

Flux algebra, 498
Flux plot, 426
Fog flow, 356
Forced convection, definition, 271
Forces on a particle, 545
Form drag, 117
Fouling factors, 387
Fourier Law, 13
Fourier number, 137, 453
Fourier-Biot conduction equation, 14
Free convection, 271
  boundary layer, 272
  boundary layer equations, 274
  boundary layer thickness, 276
  constant wall heat flux, 279
  enclosed spaces, 288
  horizontal cylinders, 285
  inclined enclosures, 289
  integral energy equation, 280
  integral momentum equation, 280
  Nusselt number for vertical plate, 279, 280
  Nusselt numbers for various situations, 292
  Pohlhausen solution, 274
  vertical cylinders, 282
  vertical layers, 289
  vertical surfaces, 272
  wires and small cylinders, 283, 287
Free molecular flow, 565
Free vortex, 158
Friction factor, 179
  Darcy, 181
  Fanning, 181
  laminar tube flow, 182
  length required for constant value, 230
  Moody, 181, 231
  Nikuradse equation, 230
Friction heat, 99
Friction velocity, 241
Frictionless fluid flow, 142
Froude number, 137
Fins, 444
  adiabatic tip, 446
  efficiency, 448
  fictitious length, 447
  non-adiabatic tip, 447
  profile area, 449
  usefulness, 448

Gas, definition, 1, 2
Gas bubble, 338
Gas radiation, 509
Gaussian error function, 453, 455
Geometrical factor, 492
Graetz number, 137, 192
Grashof number, 124, 137
  for flow classification, 271
  modified, 279
Gravity, local, 8
Gravity, standard acceleration, 7
Gray body, 489
Gray gas approximation, 513

Hagen-Poiseuille flow, 186
Half body, flow past, 159
Harmonic function, 426
Head, definition, 33
Heat, definition, 1, 12, 13
Heat conduction flux vector, 100
Heat exchangers, 382, 391
  compact, 415
  complex flow patterns, 397
  correction factor, 399
  counterflow, 392, 393
  effectiveness, 401
  fluid capacity rate, 395
  log mean temperature difference, 395
  mean temperature difference, 391
  NTU approach, 401
  parallel flow, 393
  periodic flow, 410
  pressure drop, 410
  shell and tube, 397
  storage type, 409
Heat flux, 15
Heat transfer, 1
  conduction, 13
  convection, 13
  radiation, 13

INDEX

Heat transfer coefficient, 130, 190
   overall, 382, 384
Heat transfer to particles, 549
Heat transfer in turbulent flow through conduits, 244
Helium property curve, 594
High speed flow and heat transfer, 299
Hydraulic diameter, 234, 246
Hydraulic jump, 560
Hydraulic radius, 561
Hydrogen property curve, 595
Hydrometer, 38
Hydrostatic lubrication, 544
Hyperbolic cosine, 446
Hyperbolic tangent, 447
Hypersonic flow regime, 302

Ideal fluid flow, 142
Ideal gas, 26
Incompressible flow regime, 302
Induced drag, 117
Inertial resistance coefficient, 552
Infrared radiation, 483
Injection cooling, 556
Integration by parts, 209
Integral form of boundary layer equations, 210
Integral methods, in forced convection, 208
   in free convection, 280
Intensity, 490
   change in gases, 510
Intensity of turbulence, 63
   effect on transition, 172
Interception factor, 492
International practical temperature scale, 11
International system of units, 6
Irradiation, 483
Irrotational flow, 144
Isentropic flow, 305
   flow variables, 307
   parameters for air, 306
Isothermal bulk modulus, 28
Iteration, to solve numerical equations, 434

j factor, 228, 531

Kelvin temperature scale, 10
Kinematic similarity, 133
Kinematic viscosity, 113
Kinetic energy correction factor, 95
Kinetic theory, 564
Kirchhoff's laws, 490
Knudsen number, 137, 563

Lambda point, 556
Laminar flow, 61
Laminar flow and heat transfer entrance region, 191
   fully developed tube flow, 184
Laminar sublayer, 242
Langhaar, equation for entrance length, 178
   prediction of pressure drop in entrance region, laminar flow, 177, 178
Laplace equation, 143, 426
Laplacian operator, 143
Leidenfrost phenomenon, 354
Lewis number, 137, 529
Lift, 116
Lift coefficient, 119
Liquid, definition, 1
Liquid metals, 534
   properties of, 535, 536
Liquid-gas flow, 355
Liquids, properties, 588
Local acceleration of gravity, 8, 9
Local boiling, 348
Local property, 10
Lockhart and Martinelli method for two-phase pressure drop, 359
Log mean temperature difference, 395
   correction for complex flow, 399
   equation for 1-2 exchanger, 399
Lorentz force, 558
Loss coefficients, 237
Lost work, 237
Lubrication, 541
Lumps, finite difference, 433

Mach cone, 303
Mach number, 137, 302, 327
   effect on drag, 120

Mach wave, 303
Magnetic field strength, 557
Magnetohydrodynamics, 557
Manning formula, 561
Manometer, 33
Marker and Cell method, 344
Mass, 5
Mass, standard kilogram, 6
Mass transfer, 519
Mass transfer coefficient, 526
Mass transfer cooling, 528
Material derivative, 51
MHD, 557
Maximum heat flow through insulation, 390
Maxwellian distribution, 564
Mean free path, 562
   vs. altitude, 563
Mean molecular velocity, 564
Mean pore diameter, 550
Mean temperature difference, 391
   true, 409
   with variable U, 397
Mechanical equivalent of heat, 12
Metacenter, 40
Metacentric height, 40
Metals, properties, alloys, 591
   liquid, 534
   pure, 590
Meter, orifice, 163
   turbine, 73
   venturi, 163
Microlayer theory, 351
Micron, 482
Mist flow, 356
Mixed condensation, 369
Mixed convection, 271
Models, 135
Modified Darcy equation, 552
Molecular mean free path, 562
Momentum, 81
   conservation of, 81
   correction factor, 85
   cylindrical coordinates, 90
   differential form, 87
   spherical coordinates, 91
Monochromatic emissive power, 485

Monochromatic emittance, 489
Moody friction factor, 181
Multiphase behavior, 337, 355

Natural convection, 271
Navier-Stokes equations, 89
   in dimensionless form, 116, 123, 126
Newton, unit of force, 6
Newtonian fluids, 59
Newton's law of cooling, 190
Newton's law of viscosity, 58, 59
Newton's second law, 5
Nikuradse, equation for friction factor, 230
   velocity profile for turbulent flow, 243
Nitrogen property curve, 593
Node, finite difference, 432
Nonblack body, 489
Non-metallic solids, properties, 587
Non-Newtonian fluids, 529
Nonspherical particles, 546
Non-viscous flow, 142
Nozzle, definition, 307
   flexible plate, 312
   flow at various exit pressures, 312
   variation of velocity and area, 308, 309
NTU, 401, 403
Nucleate boiling, 347
   correlation, 351
   departure from, 348
   mechanism, 350
Nucleation, 347
Numerical solutions in steady conduction, 432
Numerical solutions in transient conduction, 466
   explicit method, 466, 471
   implicit method, 471
Nusselt number, 126, 138, 141
Nusselt's analysis of laminar film condensation, 369

Opaque surface, 484
Open channel flow, 559
Optical depth, 511
Optically thick gases, 512
Optically thin gases, 512

Orifice meter, 163
Ostrach solutions to free convection on vertical flat plate, 278
Overall heat transfer coefficient, 382, 384
Oxygen property curve, 596

Particle derivative, 51
Pascal's law, 22
Passive temperature control, 505
Pathline, 77
Peak heat flux in boiling, 348, 353
Péclet number, 138, 537
Permeability, 551
Phase velocity ratio, 358
Phenomenological laws, 65
Pi groups, 110, 127
Pi theorem, 127
Piezometric head, 33
  pressure, 33
Pipe dimensions, steel, 592
Pitot tube, 162, 163
Pitot-static tube, 163
  in compressible flow, 335, 336
Plate, transient conduction, 460
Planck mean absorption coefficient, 512
Planck's equation for blackbody emission, 485
Pohlhausen solution to free convection equations, 274
Pohlhausen solution for laminar flow along a flat plate, 204
Point of separation, 252
Poiseuille flow, 186
Pool boiling, 345
Porosity, 550
Porous media, 549
Potential flow, 143
Potential pressure, 152
Pound force, 7
Pound mass, 7
Poundal, 6
Power law fluids, 540
Prandtl number, 125, 138
  effect on recovery factor, 329
  effect on velocity profiles, 206, 211, 329, 330
  turbulent, 227

Prandtl's boundary layer concept, 167
Prandtl's boundary layer equations, 170
Prandtl's velocity defect law, 244
Pressure, 21
  absolute, 23, 24
  atmospheric, 23
  dynamic, 152
  elevation, 32
  gage, 23, 24
  head, 33
  piezometric, 33
  potential, 152
  stagnation, 149
  stagnation or total value in isentropic flow, 305
  static, 152
  terminology, 24
  total, 152
  transducers, 24, 25
  vapor, 23
Pressure distribution around a circular cylinder, 156
Pressure distribution around a streamlined body, 156
Pressure drag, 117
Pressure drop in components, 234, 237
Pressure drop in two-phase flow, 359
Pressure field in a static fluid, 31
Pressure gradient, effect on velocity profile, 248
Principle of Archimedes, 36
Product solutions, 459
Properties, air at one atmosphere, 582
  alloys, 591
  carbon dioxide curve, 597
  enthalpy of water, 586
  helium curve, 594
  hydrogen curve, 595
  liquids, 588
  nitrogen curve, 593
  non-metallic solids, 587
  oxygen curve, 596
  pure metals, 590
  saturated water, 583
  saturation pressure and enthalpy of water, 586
  sodium, 586
  steam and water, 584

Property, concept of, 10
  curves and tables, 580
  extensive, 68
  local, 10
Pseudoplastic fluids, 540
Pure metals, properties, 590

Quality, two-phase flow, 358

Radiance, 490
Radiant exchange, 492
  between non-black surfaces, 499
Radiation functions, 488
Radiation heat flux vector, 100, 511
Radiation heat transfer coefficient, 354
Radiosity, 485
Rankine temperature scale, 11
Rarefied gas dynamics, 562
Rayleigh flow, 320
Rayleigh method of dimensional
    analysis, 140
Rayleigh number, 274
Reciprocity relationship, 493
Recovery factor, 328
  for laminar boundary layers, 329
Recovery temperature, 327
Reduction factor, 127
Reference temperature, 332
Reflectance, 484
Reflection, 484
Reflectivity, 484
Regions of turbulent boundary layer
    flow, 242
Relative intensity of turbulence, 63
Relative roughness of pipes, 232, 233
Relaxation, 434
  residual, 436
Repeating variables, 128
Residual, relaxation, 436
Resistance of valves and fittings, 235
Reynolds analogy, 212, 227
Reynolds number, 116
  condensate, 372
  criterion for transition in conduit
      flow, 186
  criterion for transition in flat plate
      flow, 171
  effect on velocity profile, 243

Reynolds stresses, 224
Rheology, 539
Rheopectic fluids, 541
Rod flow, 194
Rohsenow's correlation for nucleate
    boiling, 352
Rosseland mean absorption coefficient,
    512
Rotational flow, 144
Roughness
  effect on drag coefficient, 253
Roughness factor, Manning, 561

Saturated boiling, 365
Saturation pressure of water, 586
Scale buildup in tubes, 389
Schmidt number, 138, 529
Second Law of Thermodynamics, 10
Second viscosity coefficient, 58
Semi-infinite solid, conduction, 454
Separation, 157, 250
Shadowgraph, 304
Shaft work, 92
Shape factor, 428, 429
Shear stress in turbulent flow, 224
Shear work, 92
Shell and tube heat exchanger, 397, 411
Sherwood number, 138, 530
Shock waves, 303
  attached and detached, 319
  normal and oblique, 313
  normal shock relations, 315
  oblique shocks, 316
  weak solution, 320
Sieder-Tate equation, 245, 246
Similarity, complete, 135
  dynamic, 134
  kinematic, 133
  thermal, 133
Similarity methods, 197
  in free convection, 274, 278
Similitude, 133
Sink flow, 158
Skin friction coefficient, 180, 203
Skin friction drag, 116
Slip, 143
Slip flow, 566
Slip, two-phase flow, 358

Slug, unit of mass, 6
Slug flow, 194
Sodium, properties, 586
Source flow, 157
Special topics, 534
Species conservation equation, 526
Specific gravity, 9
Specific volume, 8
Specific weight, 9
Specular, 490, 567
Speed of sound, 299, 301
Sphere, drag coefficient, 252, 304
  heat transfer, 260, 549
  mass transfer, 549
  transient conduction, 464
Sphericity, 546
Splashing drop, 344
Stability, of floating and submerged objects, 39
Stagnation pressure, 149, 152
Stagnation temperature, 153
  in isentropic flow, 305
Standard acceleration of gravity, 7, 8
Standard atmosphere table, 581
Stanton number, 138, 213, 245
Static pressure, 152
Static temperature, 153
Statics, fluid, 21
Steady flow energy equation, 96, 151
Stokes' law of friction, 56
  in cylindrical coordinates, 57
  in spherical coordinates, 57
Stokes number, 138
Streakline, 77
Stream function, 79
  in cylindrical coordinates, 104
  in rectangular coordinates, 79
Streamline flow, 61
Streamlines, 77
  equation of, 79
Streamtube, 81
Stress matrix, 55
Strouhal number, 138
Subcooled boiling, 348
  effect of velocity and inlet subcooling, 365
Subcritical open channel flow, 560

Subscripts, 576
Subsonic flow regime, 302
Substantial derivative, 50, 51
Supercavitation, 368
Superconductivity, 556
Supercritical open channel flow, 560
Superficial velocity, 358
Superfluidity, 556
Superinsulations, 556
Superposition, 158, 428
Surface coefficient of convective heat transfer, 130
Surface force, 22
Surface tension, 338
  values for saturated water, 340
  values for various liquids, 341
Surface wave propagation, 560
Steel pipe dimensions, 592
Stefan-Boltzmann constant, 486
Stefan-Boltzmann equation, 487
Steam and water, properties, 584
Streamlining, 252
Symbols, list, 571
System, 68

Taylor series, 432
Technical English system of units, 6
Temperature, 10
  adiabatic wall, 327
  boundary layer, 173
  bulk, 96
  cryogenic, 552
  dynamic, 153, 327
  film, 194, 207
  gradient, 14
  high speed flow, 332
  mixing cup, 96
  recovery, 327
  reference, 332
  stagnation, 153, 305
  static, 153
  total, 153
Terminal velocity of fall, 545
Thermal analyzer, 458
Thermal boundary layer, 172
Thermal circuit, 383

INDEX   611

Thermal conductivity, 14
　apparent, 12, 289, 507
　metals at low temperatures, 555
　superinsulations, 556
　typical cryogenic materials, 557
　values for common substances, 16, 580
Thermal control of spacecraft, 503
Thermal diffusivity, 125, 426
Thermal expansion coefficient, 28
Thermal radiation, 482
　in gases, 509
Thermal resistance, 383
Thermal similarity, 133
Thermometer error, 508
Thixotropic fluids, 541
Throat, converging-diverging nozzle, 310
Time smoothing, 221
Time-dependent fluids, 539, 541
Time-independent fluids, 539
Torricelli equation, 150
Total emissive power, 484
Total head, 147
Total pressure, 152
　isentropic flow, 305
　local value, 152
Total temperature, 153
　isentropic flow, 305
Transition flow, 566
Transition flow heat transfer, 246
Transition to turbulence
　along flat plate, 171
　cylinders, 252
　effect of intensity of turbulence, 172
　free convection, 274
Transmission, 484
Transmittance, 484
Transonic flow regime, 302
Tube banks, 260
Tube bundles, 260
Turbine flowmeter, 73
Turbulence level, 63
Turbulent core, 242
Turbulent flow, 61
　in conduits, 230
Turbulent flow and heat transfer, 221

Turbulent Prandtl number, 227
Turning angle of an oblique shock, 316
Two-phase flow, 355
　flow regimes, 355
　with heat addition, 363
　pressure drop, 359

Uniform open channel flow, 560
Unit, 3
Units and dimensions, 3
Unit thermal convective conductance, 130
Universal gas constant, 302
Universal velocity profiles, 240, 242

Vacuum, 24
Vapor bubble, 337
Vapor pressure
　water, 586
Varied open channel flow, 560
Velocity, 48
　around a cylinder, 155
　fluctuating component, 62
　instantaneous component, 62
　mean component, 62
　RMS value, 63
Velocity defect law, 244
Velocity head, 147
Velocity potential, 143
Velocity pressure, 152
Velocity of propagation of surface waves, 560
Velocity of sound, 299, 301
Venturi meter, 163
View factor, 492
Virtual stresses, 224
Viscoelastic fluids, 541
Viscosity, 58
　absolute, 58
　dynamic, 58
　kinematic, 113
　Newton's law of, 58
Viscosity, eddy, 225
Viscous dissipation function, 99
　cylindrical coordinates, 102, 185
　spherical coordinates, 102
　two dimensions, 122

Viscous flow, 61
Void fraction, 358
von Kármán analogy, 228
von Kármán equations for turbulent boundary layer, 242
von Kármán momentum integral equation, 210, 220
Vortex, free, 158
Vorticity vector, 144

Wake, 117
Wake drag, 117
Wall shear stress, 180

Water, properties of saturated, 583
Water table, 560
Weber number, 138
Weight, 8
Wien's displacement law, 487
Wien's distribution law, 487
Work, 12
  flow, 93
  shaft, 92
  shear, 92

Zeroth law of thermodynamics, 10